Climate Change and Environmental Sustainability-Volume 3

Climate Change and Environmental Sustainability-Volume 3

Editors

Bao-Jie He
Ayyoob Sharifi
Chi Feng
Jun Yang

MDPI • Basel • Beijing • Wuhan • Barcelona • Belgrade • Manchester • Tokyo • Cluj • Tianjin

Editors
Bao-Jie He
Chongqing University
China

Ayyoob Sharifi
Hiroshima University
Japan

Chi Feng
Chongqing University
China

Jun Yang
Northeastern University
China

Editorial Office
MDPI
St. Alban-Anlage 66
4052 Basel, Switzerland

This is a reprint of articles from the Topics published online in the open access journal *Atmosphere* (ISSN 2073-4433), *Buildings* (ISSN 2075-5309), *Land* (ISSN 2073-445X), *Remote Sensing* (ISSN 2072-4292), *Sustainability* (ISSN 2071-1050) (available at: https://www.mdpi.com/topics/Climate_Environmental).

For citation purposes, cite each article independently as indicated on the article page online and as indicated below:

LastName, A.A.; LastName, B.B.; LastName, C.C. Article Title. *Journal Name* **Year**, *Volume Number*, Page Range.

ISBN 978-3-0365-3008-6 (Hbk)
ISBN 978-3-0365-3009-3 (PDF)

201 k words

© 2022 by the authors. Articles in this book are Open Access and distributed under the Creative Commons Attribution (CC BY) license, which allows users to download, copy and build upon published articles, as long as the author and publisher are properly credited, which ensures maximum dissemination and a wider impact of our publications.

The book as a whole is distributed by MDPI under the terms and conditions of the Creative Commons license CC BY-NC-ND.

Contents

About the Editors vii

Preface to "Climate Change and Environmental Sustainability-Volume 3" ix

Lingxue Yu, Jiuchun Yang, Kun Bu, Tingxiang Liu, Yue Jiao, Guangshuai Li, Luoman Pu and Shuwen Zhang
Impacts of Saline-Alkali Land Improvement on Regional Climate: Process, Mechanisms, and Implications
Reprinted from: *Remote Sens.* **2021**, *13*, 3407, doi:10.3390/rs13173407 1

Dou Zhang, Xiaolei Geng, Wanxu Chen, Lei Fang, Rui Yao, Xiangrong Wang and Xiao Zhou
Inconsistency of Global Vegetation Dynamics Driven by Climate Change: Evidences from Spatial Regression
Reprinted from: *Remote Sens.* **2021**, *13*, 3442, doi:10.3390/rs13173442 17

Jianhu Wang, Juan Li, Jiyuan Yin, Wei Tan and Yuchen Liu
Sea Level Seasonal, Interannual and Decadal Variability in the Tropical Pacific Ocean
Reprinted from: *Remote Sens.* **2021**, *13*, 3809, doi:10.3390/rs13193809 37

Li Liu, Ran Huang, Jiefeng Cheng, Weiwei Liu, Yan Chen, Qi Shao, Dingding Duan, Pengliang Wei, Yuanyuan Chen and Jingfeng Huang
Monitoring Meteorological Drought in Southern China Using Remote Sensing Data
Reprinted from: *Remote Sens.* **2021**, *13*, 3858, doi:10.3390/ rs13193858 59

Baoan Hu, Zhijie Zhang, Hairong Han, Zuzheng Li, Xiaoqin Cheng, Fengfeng Kang and Huifeng Wu
The Grain for Green Program Intensifies Trade-Offs between Ecosystem Services in Midwestern Shanxi, China
Reprinted from: *Remote Sens.* **2021**, *13*, 3966, doi:10.3390/rs13193966 81

Peijie Wei, Shengyun Chen, Minghui Wu, Yinglan Jia, Haojie Xu and Deming Liu
Increased Ecosystem Carbon Storage between 2001 and 2019 in the Northeastern Margin of the Qinghai-Tibet Plateau
Reprinted from: *Remote Sens.* **2021**, *13*, 3986, doi:10.3390/rs13193986 97

Ukkyo Jeong and Hyunkee Hong
Comparison of Total Column and Surface Mixing Ratio of Carbon Monoxide Derived from the TROPOMI/Sentinel-5 Precursor with In-Situ Measurements from Extensive Ground-Based Network over South Korea
Reprinted from: *Remote Sens.* **2021**, *13*, 3987, doi:10.3390/rs13193987 115

Sheng Yan, Jianyu Liu, Xihui Gu and Dongdong Kong
Global Runoff Signatures Changes and Their Response to Atmospheric Environment, GRACE Water Storage, and Dams
Reprinted from: *Remote Sens.* **2021**, *13*, 4084, doi:10.3390/rs13204084 131

Siqi Gao, Guotao Dong, Xiaohui Jiang, Tong Nie, Huijuan Yin and Xinwei Guo
Quantification of Natural and Anthropogenic Driving Forces of Vegetation Changes in the Three-River Headwater Region during 1982–2015 Based on Geographical Detector Model
Reprinted from: *Remote Sens.* **2021**, *13*, 4175, doi:10.3390/rs13204175 149

Dexuan Sha, Younghyun Koo, Xin Miao, Anusha Srirenganathan, Hai Lan, Shorojit Biswas, Qian Liu, Alberto M. Mestas-Nuñez, Hongjie Xie and Chaowei Yang
Spatiotemporal Analysis of Sea Ice Leads in the Arctic Ocean Retrieved from IceBridge Laxon Line Data 2012–2018
Reprinted from: *Remote Sens.* **2021**, *13*, 4177, doi:10.3390/rs13204177 . . . . **173**

Ying Zhang, Zhaohui Chi, Fengming Hui, Teng Li, Xuying Liu, Baogang Zhang, Xiao Cheng and Zhuoqi Chen
Accuracy Evaluation on Geolocation of the Chinese First Polar Microsatellite (Ice Pathfinder) Imagery
Reprinted from: *Remote Sens.* **2021**, *13*, 4278, doi:10.3390/rs13214278 . . . . **191**

Ziqi Zhao, Ayyoob Sharifi, Xin Dong, Lidu Shen and Bao-Jie He
Spatial Variability and Temporal Heterogeneity of Surface Urban Heat Island Patterns and the Suitability of Local Climate Zones for Land Surface Temperature Characterization
Reprinted from: *Remote Sens.* **2021**, *13*, 4338, doi:10.3390/rs13214338 . . . . **211**

Andrew K. Marondedze and Brigitta Schütt
Predicting the Impact of Future Land Use and Climate Change on Potential Soil Erosion Risk in an Urban District of the Harare Metropolitan Province, Zimbabwe
Reprinted from: *Remote Sens.* **2021**, *13*, 4360, doi:10.3390/rs13214360 . . . . **241**

Yi Zhang, Yilin Liu, Xinyuan Zhang, Haijun Huang, Keyu Qin, Zechao Bai and Xinghua Zhou
Correlation Analysis between Land-Use/Cover Change and Coastal Subsidence in the Yellow River Delta, China: Reviewing the Past and Prospecting the Future
Reprinted from: *Remote Sens.* **2021**, *13*, 4563, doi:10.3390/rs13224563 . . . . **271**

Yue Jiao, Kun Bu, Jiuchun Yang, Guangshuai Li, Lidu Shen, Tingxiang Liu, Lingxue Yu, Shuwen Zhang and Hengqing Zhang
Biophysical Effects of Temperate Forests in Regulating Regional Temperature and Precipitation Pattern across Northeast China
Reprinted from: *Remote Sens.* **2021**, *13*, 4767, doi:10.3390/rs13234767 . . . . **285**

About the Editors

Bao-Jie He is a Research Professor of Urban Climate and Built Environment at the School of Architecture and Urban Planning, Chongqing University, China. Prior to joining Chongqing University, Bao-Jie He was a PhD researcher at the Faculty of Built Environment, University of New South Wales, Australia. Bao-Jie is working on the Cool Cities and Communities and Net Zero Carbon Built Environment project. Bao-Jie has strong academic capability and has published around 80 peer-reviewed papers in high-ranking journals and given oral presentations at reputable conferences. Bao-Jie acts as the Topic Editor-in-Chief, Leading Guest Editor, Associate Editor, Editorial Board Member, Conference Chair, Sessional Chair, and Scientific Committee of a variety of international journals and conferences. Dr. He received the Green Talents Award (Germany) in 2021 and National Scholarship for Outstanding Self-Funded Foreign Students (China) in 2019. Dr. He was ranked as one of the 100,000 global scientists (both single-year and career top 2%) by the Mendeley, 2021.

Ayyoob Sharifi works at the Graduate School of Humanities and Social Sciences, Hiroshima University. He also has a cross-appointment at the Graduate School of Advanced Science and Engineering. Ayyoob's research is mainly at the interface of urbanism and climate change mitigation and adaptation. He actively contributes to global climate change research programs such as the Future Earth and is currently serving as a lead author for the Sixth Assessment Report (AR6) of the Intergovernmental Panel on Climate Change (IPCC). Before joining Hiroshima University, he was the Executive Director of the Global Carbon Project (GCP)—a Future Earth core project—leading the urban flagship activity of the project, which is focused on conducting cutting-edge research to support climate change mitigation and adaptation in cities.

Chi Feng received his joint PhD training in South China University of Technology (China) and KU Leuven (Belgium). He is now a research professor in the School of Architecture and Urban Planning, Chongqing University (China) and is leading a research group of more than 10 members. His research topics cover coupled heat and moisture transfer in porous building materials, as well as the hygrothermal performance of building envelopes and built environments. He has led eight international and national research projects, including the China–Europe round robin campaign on material property determination (nine countries participated), the National Natural Science Foundation of China, and the National Key R&D Program of China. He has published more than 60 peer-reviewed journal/conference papers at home and abroad. He is drafting two Chinese standards and participating in another nine international/national ones.

Jun Yang is working at the Urban Climate and Human Settlements Lab, Northeastern University (Shenyang China). His research expertise involves urban climate zones, urban ecology, urban human settlements, and sustainability. As PI or Co-PI, he has been involved in 50 research projects, receiving a total of 15 million RMB from EGOV.CN (e.g., NFC, MOST, and MOE) since 2002. He has authored and co-authored more than 160 papers and book chapters and published more than 50 English papers and more than 110 Chinese papers in academic journals. He is now the Associate Editor of the *Social Sciences* section of the *International Journal of Environmental Science and Technology*, sits on the Editorial Board of *PLOS One*, *PLOS Climate*, and *Frontiers in Built Environment*, is a Lead Guest Editor of *Complexity*, and a Guest Editor of *IEEE Journal of Selected Topics in Applied Earth Observations and Remote Sensing*.

Preface to "Climate Change and Environmental Sustainability-Volume 3"

The Earth's climate is changing; the global average temperature is estimated to already be about 1.1 °C above pre-industrial levels. Indeed, we are now living in conditions of a climate emergency. Climate change leads to many adverse events, such as extreme heat, flooding, bushfire, drought, and many other associated economic and social consequences. Further warming is projected to occur in the coming decades, and climate-induced impacts may exceed the capacity of society to cope and adaptive in a 1.5 °C or 2 °C world. Therefore, urgent actions should be taken to address climate change and avoid irreversible environmental damages.

Climate change is interrelated with many other challenges such as urbanisation, population increase and economic growth. For instance, cities are now the main settlements of human being and are major sources of greenhouse gas emissions that are key contributors to climate change. Moreover, rapid and unregulated urbanisation in some contexts further causes urban problems such as environmental pollution, traffic congestion, urban flooding and heat island intensification. In the absence of well-designed measures, increasing urbanisation trends in the next two–three decades are likely to further aggravate such problems. Overall, climate change and many other challenges have deteriorated the sustainable development of the world.

The United Nations proposed the Sustainable Development Goals in 2015. Goal 13, Climate Action, emphasises the need for urgent action to combat climate change and its impacts in order to enhance sustainability. To achieve this, there is a need to develop a holistic framework that considers mitigation—the decarbonisation of society—to address the challenge of climate change from the root, and adaptation—an immediate action—to increase the resilience of and protect society from climate-induced hazards. The framework prioritises the transformation of the traditional methods of environmental modifications in various fields, including transportation, industry, building, energy generation, agriculture, land use and forestry, towards sustainable ones to limit greenhouse gas emissions. The framework also highlights the significance of sustainable environmental planning and design for adaptation in order to reduce climate-induced threats and risks. Moreover, it encourages the involvement and participation of all stakeholders to accelerate climate change mitigation and adaptation progress by developing sound climate-related governance systems.

The framework also calls for the support and engagement of all societal stakeholders. To support the achievement and implementation of the framework, this book focuses on climate change and environmental sustainability by covering four key aspects, including climate change mitigation and adaptation, sustainable urban–rural planning and design, decarbonisation of the built environment in addition to climate-related governance and challenges. Climate change mitigation and adaptation covers topics of greenhouse gas emissions and measurement, climate-related disasters and reduction, risk and vulnerability assessment and visualisation, impacts of climate change on health and well-being, ecosystem services and carbon sequestration, sustainable transport and climate change mitigation and adaptation, sustainable building and construction, industry decarbonisation and economic growth, renewable and clean energy potential and implementation in addition to environmental, economic and social benefits of climate change mitigation.

Sustainable urban–rural planning and design deals with questions of climate change and regional economic development, territorial spatial planning and carbon neutrality, urban overheating mitigation and adaptation, water-sensitive urban design, smart development for urban habitats, sustainable land use and planning, low-carbon cities and communities, wind-sensitive urban planning and design, nature-based solutions, urban morphology and environmental performance in addition to innovative technologies, models, methods and tools for spatial planning. Decarbonisation of the built environment addresses issues of climate-related impacts on the built environment, the health and well-being of occupants, demands on energy, materials and water, assessment methods, systems and tools, sustainable energy, materials and water systems, energy-efficient design technologies and appliances, smart technology and sustainable operation, the uptake and integration of clean energy, innovative materials for carbon reduction and environmental regulation, building demolition and material recycling and reusing in addition to sustainable building retrofitting and assessment. Climate-related governance and challenges concerns problems of targets, pathways and roadmaps towards carbon neutrality, pathways for climate resilience and future sustainability, challenges, opportunities and solutions for climate resilience, the development and challenges climate change governance coalitions (networks), co-benefits and synergies between adaptation and mitigation measures, conflicts and trade-offs between adaptation and mitigation measures, mapping, accounting and trading carbon emissions, governance models, policies, regulations and programs, financing urban climate change mitigation, education, policy and advocacy of climate change mitigation and adaptation in addition to the impacts and lessons of COVID-19 and similar crises.

Overall, this book aims to introduce innovative systems, ideas, pathways, solutions, strategies, technologies, pilot cases and exemplars that are relevant to measuring and assessing the impact of climate change, mitigation and adaptation strategies and techniques in addition to public participation and governance. The outcomes of this book are expected to support decision makers and stakeholders to address climate change and promote environmental sustainability. Lastly, this book aims to provide support for the implementation of the United Nations Sustainable Development Goals and carbon neutrality in efforts aimed at achieving a more resilient, liveable and sustainable future.

To cope with the challenges of climate change, it is essential to assess climate-induced impacts and explore possible solutions. Remote sensing techniques are capable of monitoring, collecting, interpreting, and mapping the physical characteristics of Earth's surface and its associated spatiotemporal variations. These techniques outperform many data acquisition techniques in overcoming spatial and geographic constraints. The adoption of remote sensing techniques strengthens the capacity for climate change mitigation and adaptation and facilitates evidence-based climate governance. This book presents the use of different kinds of remote sensing techniques to obtain original data across global, regional, city, or local scales for exploring climate-related issues such as sea level variation, sea ice dynamics, drought, extreme heat and precipitation, ecosystem services and carbon sequestration, forest and vegetation cover, coastal subsidence, atmospheric carbon monoxide, soil erosion and runoff, and urban heat islands. This book is important to demonstrate the use of remote sensing techniques for revealing climate-related risks and vulnerabilities. Meanwhile, results reported in this book provide a good understanding of the climate emergency situations, drivers, and solutions. We expect the book to benefit decision makers, practitioners, and researchers in different fields such as climate modeling and prediction, forest ecosystem, land management, urban planning and design, urban governance, and institutional operations.

Prof. Bao-Jie He acknowledges that Project NO. 2021CDJQY-004 is supported by the Fundamental Research Funds for the Central Universities and that Project NO. 2022ZA01 is supported by the State Key Laboratory of Subtropical Building Science, South China University of Technology, China. We appreciate the assistance from Mr. Lifeng Xiong, Mr. Wei Wang, Ms. Xueke Chen, and Ms. Anxian Chen at the School of Architecture and Urban Planning, Chongqing University, China.

Bao-Jie He, Ayyoob Sharifi , Chi Feng, Jun Yang
Editors

Article

Impacts of Saline-Alkali Land Improvement on Regional Climate: Process, Mechanisms, and Implications

Lingxue Yu [1,2], Jiuchun Yang [1,*], Kun Bu [1], Tingxiang Liu [3], Yue Jiao [1,4], Guangshuai Li [1,3], Luoman Pu [5] and Shuwen Zhang [1]

1. Remote Sensing and Geographic Information Research Center, Northeast Institute of Geography and Agroecology, Chinese Academy of Sciences, Changchun 130102, China; yulingxue@iga.ac.cn (L.Y.); bukun@iga.ac.cn (K.B.); jiaoyue@iga.ac.cn (Y.J.); liguangshuai@iga.ac.cn (G.L.); zhangshuwen@iga.ac.cn (S.Z.)
2. Remote Sensing and Geographic Information Research Center, Changchun Jingyuetan Remote Sensing Observation Station, Chinese Academy of Sciences, Changchun 130102, China
3. College of Geography Science, Changchun Normal University, Changchun 130031, China; liutingxiang@ccsfu.edu.cn
4. School of Life Science, Liaoning Normal University, Dalian 116029, China
5. School of Public Administration, Hainan University, Haikou 570228, China; 994424@hainanu.edu.cn
* Correspondence: yangjiuchun@iga.ac.cn

Abstract: Studying land use change and its associated climate effects is important to understand the role of human activities in the regulation of climate systems. By coupling remote sensing measurements with a high-resolution regional climate model, this study evaluated the land surface changes and corresponding climate impact caused by planting rice on saline-alkali land in western Jilin (China). Our results showed that paddy field expansion became the dominant land use change in western Jilin from 2015 to 2019, 25% of which was converted from saline-alkali land; this percentage is expected to increase in the near future. We found that saline-alkali land reclamation to paddy fields significantly increased the leaf area index (LAI), particularly in July and August, whereas it decreased albedo, mainly in May and June. Our simulation results showed that planting rice on saline-alkali land can help decrease the air temperature and increase the relative humidity. The temperature and humidity effects showed different magnitudes during the growing season and were most significant in July and August, followed by September and June. The nonradiative process, rather than the radiative process, played a dominant role in regulating the regional climate in this case, and the biophysical competition between evapotranspiration (ET) and albedo determined the temperature and relative humidity response differences during the growing season.

Keywords: land use and land cover changes; regional climate; regional climate model; remote sensing

1. Introduction

Both land use changes and CO_2 emissions have been documented as dominant driving factors influencing the climate system at different scales from global to regional [1–5]. However, at the regional scale, some studies emphasize that climate change induced by land use change is even greater than climate change induced by greenhouse gases [6–8]. Understanding the predominant regional land use change as well as the mechanisms by which it affects climate through altering energy, momentum, and water exchange processes is crucial to fully clarify how humans modify and regulate climate [9–11]. In addition, studying land use changes and estimating their climate impact is also a major requirement for the sustainable development of agriculture [12–15].

Global land use trajectories show the transition from pre-settlement natural ecosystems to intensive complex composite ecosystems, in which urbanization and intensive agriculture plays an increasingly important role in supporting the unprecedented population and its associated crop requirements [16–19]. Based on satellite big data, Kuang et al.

detected significant global urbanization since the beginning of the 21st century [19]. The human-induced cropland expansion has been examined as the major driver for a series of land-use changes, including deforestation, grassland, and wetland reclamation across both tropical and temperature regions [18,20,21]. Through using land-use management, China and India lead in the global widespread greening since 2000, among which, fertilization use and irrigation drive the greening and food production increase in croplands [22]. Although some studies have investigated the regional climate impacts caused by urban expansion and agricultural intensification, mainly the land use conversion from forest, grassland, and wetland to farmland, agricultural practice, and agricultural irrigation [1,7,23–30], attention has rarely been given to exploring the connection between the vegetation restoration caused by saline-alkali land improvement and regional climate responses and the role in seasonal transitions in temperature and energy balance.

As a widely distributed area of the black soil region, the Northeast China has experienced unprecedented agricultural intensification since the 1950s and has become the major grain-producing area in China [31–33]. However, the ecological environment has become increasingly vulnerable across the transitional climate and ecological zone of northern China due to high-intensity land development [34]. As a result, some grassland or farmland in subarid, ecologically fragile areas such as western Jilin has degraded into saline-alkali land and resulted in severe environmental and ecological problems [35]. Recently, the development of saline-alkali soil improvement technology has made planting rice in saline-alkali land possible, and this will become the dominant land use in the near future with sufficient policy support [36]. Saline-alkali land improvement substantially changes the surface biophysical and biochemical properties and influences the interactions between the land surface and atmosphere [37]. However, comprehensive evaluations of how saline-alkali land improvements influence surface plant physiological and optical parameters and further affect the local climate are still lacking.

Satellite observations provide detailed Earth surface information and have become the most commonly used approach to study environmental change [38–40]. Some studies have used remote sensing measurements to investigate the response of surface temperature to land use changes such as afforestation and urbanization [41,42]. However, it is difficult to identify the mechanisms corresponding to climate impact. High-resolution regional climate modelling involving a land surface model can accurately represent the energy and moisture exchanges at the surface/atmosphere interface and has become an efficient way to simulate climate effects based on historic and future land use changes [43–48]. Precise land surface properties are essential in simulating the interactions between the surface and the atmosphere and have been widely documented [49,50]. Coupling spatially continuous satellite observations with regional climate models has become the state-of-the-art approach to study climate impacts due to land surface changes [51].

Therefore, in this study, we quantify the air temperature and relative humidity impacts related to changes from regional typical and novel land use changes—from saline-alkali land to paddy fields. First, the historic and future projected land use changes in western Jilin were analysed. We then evaluated the influence of saline-alkali land improvement on two crucial surface parameters, including albedo and leaf area index (LAI). Finally, the climate responses, mechanisms and implications for saline-alkali land improvement were simulated and further analysed by coupling the land surface model into the regional climate model. Through this study, our results can provide suggestions for regional agricultural development.

2. Materials and Methods
2.1. Study Area

Western Jilin is located in the western part of Jilin Province in Northeast China, extending from 43°59′27″N to 46°18′5″N latitude and 121°37′31″E to 126°10′43″E longitude (Figure 1). With a total area of 46,900 km², western Jilin contains ten county-level cities, including Zhenlai, Taobei, Taonan, Tongyu, Da'an, Qian'an, Changling, Qianguo, Ningjiang and Fuyu. The landform of western Jilin is an alluvial and proluvial plain with an average altitude of 160 m. The climate is dominated by a temperate continental climate with distinct seasonal variation. The average annual temperature is 4–5 °C, and the annual precipitation is 350–500 mm. The water resources are rich, the main rivers of which are the Taoer River, Nenjiang River and Songhua River. Affected by landform and climate, the soil is mainly light chernozem and meadow soil. From the perspective of ecological zoning, western Jilin is located in the ecotone between agriculture and animal husbandry.

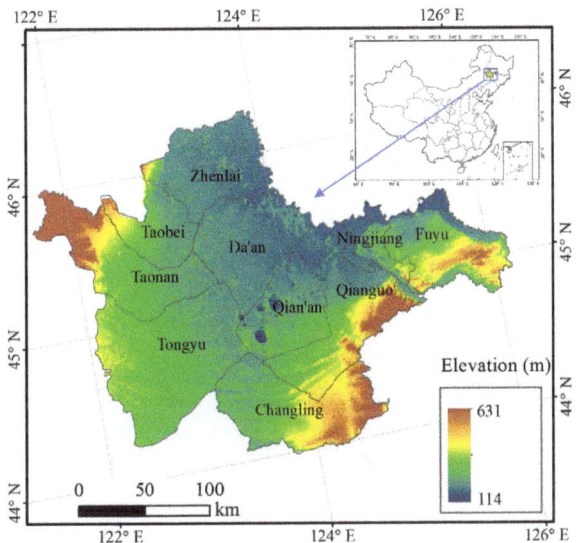

Figure 1. Geographic location of west Jilin.

The long-term agricultural intensification that began in the 1950s contributed to severe land degradation; as a result, increasing amounts of land were converted to saline-alkali land, and the ecological environment has became increasingly vulnerable. Since the 2010s, saline-alkali soil improvement technology has been promoted to cultivate rice on unused saline-alkali land and has become a new regional land use change characteristic. As a result, western Jilin has become the ideal area to fully understand how human activities regulate or modify climate at the regional scale.

2.2. Data Processing

2.2.1. Land Use and Land Cover (LULC) Data

In this study, we used time series land use datasets produced by the Chinese Academy of Sciences (CAS) (downloaded from http://www.resdc.cn/ (accessed on 25 June 2021)) to describe the land use pattern in western Jilin. Two periods, 1975 to 2015 and 2015 to 2019, were used to analyse land use change over the past 45 years. Based on the land use maps from 2015 and 2019, we extracted the unchanged pure grids (where the dominant type is the only land use type in that grid) at a 1 km × 1 km spatial resolution for paddy fields and saline-alkali land. There were 1621 and 1478 pure grids for saline-alkali land and paddy fields in our study area, respectively (Figure S1). These pure pixels were used to extract the interannual cycle of surface properties, including LAI and albedo.

In this study, we used the CAS LULC data from 2015 to represent the land use pattern in China. The European Space Agency (ESA) Climate Change Initiate (CCI) land use and land cover dataset from 2015 was used to fill in the land use data outside Northeast China. Both the CAS and CCI LULC data were converted to USGS 24-category land use categories at a resolution of 1 km. The fraction of each LULC type, the dominant LULC type and the land mask layer at each grid were then obtained through spatial statistical analysis using ArcGIS and Python processing.

2.2.2. Land Surface Parameters Datasets

LAI and albedo are two dominant surface biogeophysical parameters influencing the energy budget and water cycling. The LAI determines the vegetation transpiration and CO_2 exchange of the vegetation canopy, whereas the surface albedo can affect the shortwave radiation absorbed by the surface. Temporally and spatially continuous LAI and albedo data can help represent the interactions between the land surface and atmosphere more accurately. In this study, we used MODIS products, including the MOD15A2 8-day composite LAI dataset and the MCD43B3 daily albedo dataset, to represent the spatial heterogeneity of LAI and albedo. To match the temporal resolution and projection of Weather Research and Forecasting (WRF) preprocessing, the 8-day composite or daily datasets were first aggregated monthly, and the projection was transformed into a Lambert equal area projection. To avoid the influence of climate variation, we used the 2015–2019 five-year average monthly LAI and albedo to update the original corresponding dataset in the WRF preprocessing.

2.2.3. Climate Forcing Dataset

ERA5 reanalysis datasets were used to force the WRF model at both the surface and pressure levels. ERA5 is the fifth generation of ECMWF atmospheric reanalysis of the global climate, which began with the FGGE reanalysis produced in the 1980s, followed by ERA-15, ERA-40 and most recently ERA-Interim. These data have a high spatial resolution of 0.25 × 0.25 degrees and a high temporal resolution, which can reach three hours. The long-term (1950 to the most current) ERA5 datasets have been widely applied to historic and future climate change research. At the surface level, 19 surface variables including the 10 m u component of wind, 10 m v component of wind, 2 m dewpoint temperature, 2 m temperature, land sea mask, mean sea level pressure, sea ice cover, sea surface temperature, skin temperature, snow depth, soil temperature at four soil layers, surface pressure, volumetric soil water at four soil layers were used, and at the pressure level, six variables including geopotential, relative humidity, specific humidity, temperature, u component of wind, and v component of wind at 37 vertical levels were used for the meteorological forcing.

2.2.4. Meteorological Observation Dataset

The monthly dataset of surface climate data in China from the China Meteorological Data Service Center (CMDSC) (http://cdc.cma.gov.cn/ (accessed on 25 June 2021)) was used to validate the efficiency of our simulation. This monthly dataset spans a period from 1951 to the present and includes 23 meteorological variables. Two climate variables, including air temperature (at 2 m) and relative humidity (at 2 m) at six meteorological stations covering western Jilin were selected to compare the observed results with the model-simulated results. The six meteorological stations included Baicheng, Fuyu, Qianguo, Tongyu, Qian'an and Changling.

2.3. Regional Climate Sumulation

The Weather Research and Forecasting (WRF) model has been used in a broad range of applications, including regional climate research and forecast research across scales ranging from metres to thousands of kilometres [52]. Based on different dynamic solvers, the WRF system contains an advanced research WRF (ARW) core and a nonhydrostatic mesoscale model (NMM) core. In this study, we used the flexible, efficient and state-of-the-art atmospheric ARW simulation system version 3.6 to perform the numerical simulation.

We designed two domains in our experiments (Figure 2), which had horizontal resolutions of 30 km and 10 km, respectively. The first domain included all of Northeast China, whereas the second domain included our study area: western Jilin and its surrounding areas. The two scenarios were designed to simulate the impact of saline-alkali land improvements on the regional climate. In the control scenario, the land use and land cover (LULC) data from 2015 were used to calculate the LULC-related land surface variables, including the dominant LULC type, the fraction of each LULC type, and the land mask at each domain resolution. The 2015–2019 5-year average monthly LAI and albedo were used to replace the corresponding initial model geostatistical datasets. In the sensitivity experiments, all the saline-alkali land in 2015 was converted to paddy fields, referring to herbaceous wetlands from the 24-category USGS land use categories. The seasonally varied LAI and albedo on the converted saline-alkali land were updated based on the statistical characteristics of pure paddy fields (Table 1). The initial boundary conditions and the physical parameterization schemes were held constant in both the control and sensitivity experiments. Therefore, the regional climate impacts due to saline-alkali land improvement can be identified.

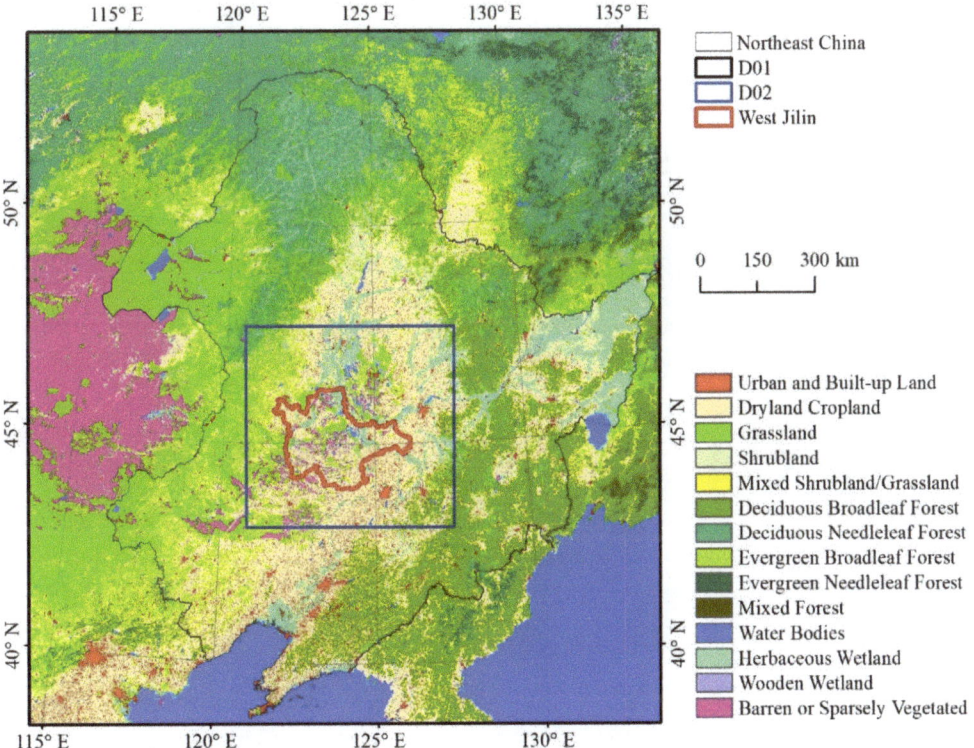

Figure 2. The land cover pattern for domain 1 (D01) and domain 2 (D02) in our experiments.

Table 1. The regional monthly mean albedo and LAI for paddy fields. (When the saline-alkali land was converted to the paddy field, the albedo and LAI of the saline-alkali land were replaced by the corresponding values shown in this table).

Paddy Field	May	June	July	August	September	October
Albedo	0.08	0.09	0.18	0.19	0.20	0.21
LAI	0.17	1.23	4.20	3.20	1.15	0.29

The main schemes used in the simulation were as follows: the microphysics was a WRF Single-Moment 3-class scheme, both the longwave radiation and shortwave radiation were CAM schemes, the surface layer was an MM5 similarity scheme, the land surface was represented by the Noah Land Surface Model, the planetary boundary layer was a Yonsei University scheme, and the cumulus parameterization was a Kain-Fritsch scheme. Given that the growing season in western Jilin extends from late May to late September, the two experiments were initialized on 1 May to 1 October of each year from 2015 to 2019. The results from the first month were used to spin up the model, and the results from following months (June, July, August and September) were used for analysis in this paper.

3. Results

3.1. Land Use Changes in Western Jilin: Historic, Current and Future

Dry farmland, grassland and saline-alkali land are three dominant land use types in western Jilin, accounting for approximately 75~78% of the total area. Remarkable land use conversions were identified from 1975 to 2015 in western Jilin due to both human activities and natural environmental changes (Figure 3). Extensive grassland degradation to saline-alkali land was observed mainly in Da'an, Tongyu, Changling and Zhenlai. In Qian'an and Qianguo, a large number of grasslands have been reclaimed to dry farmland. Notable paddy field expansion from dry farmland or wetland was observed in the main rice planting areas such as Zhenlai, Taobei and Qianguo. In addition, notable built-up land expansion was also detected from 1975 to 2015.

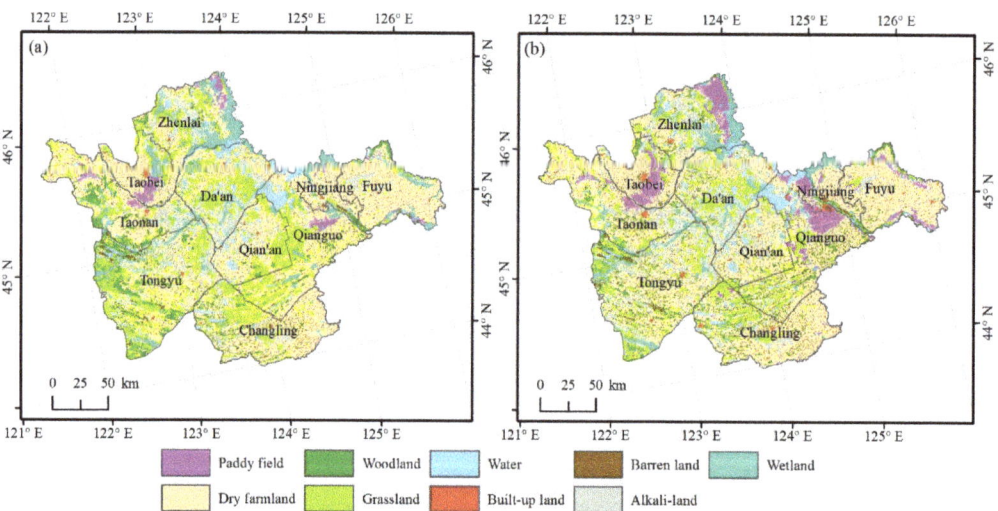

Figure 3. Land use maps of west Jilin from 1975 (**a**) and 2015 (**b**).

A land use transition matrix was created to determine the specific land use conversions in western Jilin (Table 2). From 1975 to 2015, grassland decreased by 3228 km^2, 43.9% of which was reclaimed to dry farmland, and 39.4% of which degraded to saline-alkali land. Paddy fields and dry farmland increased by 1738 and 620 km^2, respectively, indicating that agricultural development was the dominant driving factor of the land use changes in western Jilin during our study period. Expanded paddy fields were primarily converted from dry farmland (52.8%) and wetlands (33%). It should be noted that approximately 6% of the paddy field expansion was converted from saline-alkali land due to novel techniques of planting rice on saline-alkali land. As a result, degraded saline-alkali land has become an important reserved cultivated land resource in these regions. In addition, wetland and water decreased by 636 and 200 km^2, respectively, whereas built-up land, woodland and barren land increased by 437, 317 and 32 km^2, respectively.

Recent remote sensing measurement records from 2015 to 2019 showed that paddy field expansion continued at an increasing rate. In those four years, paddy fields increased by 2031 km^2, compared with 1738 km^2 in the past 40 years from 1975 to 2015 (Figure 4).

A total of 37.1% of the increase in paddy fields was converted from dry farmland, most of which were located in areas adjacent to already existing paddy fields. Saline-alkali land reclamation has become the second most important approach to paddy field expansion, with a 520.9 km^2 increase in western Jilin. An increased plant area is a direct way to guarantee or increase food production and is beneficial due to the low land cost of saline-alkali land and provincial government support. Paddy field development on saline-alkali land in western Jilin has become and will continue to be the main trend in future land use management.

Table 2. Transition matrix of land use categories from 1975 to 2015 in western Jilin (km^2).

1975	2015									
	Woodland	Grassland	Water	Built-up Land	Barren Land	Saline-Alkali Land	Wetland	Paddy Field	Dry Farmland	Total
Woodland	1309.8	256.5	0.5	8.3	6.1	39.1	15.2	6.9	819.0	2461.3
Grassland	892.6	3801.1	31.6	32.3	56.2	1273.5	152.8	174.1	1415.7	7830.1
Water	1.4	29.1	1801.2	13.1	1.2	89.9	176.2	15.4	51.0	2178.4
Built-up land	1.7	1.1	0.3	1316.4	0.2	1.7	0.3	4.0	65.2	1390.8
Barren land	4.8	7.8	0.1	0.3	181.7	0.3	11.5	0.0	9.7	216.2
Saline-alkali land	3.8	240.3	95.5	55.3	1.2	5798.5	117.3	102.5	91.1	6505.5
Wetland	12.7	155.7	31.7	8.6	0.3	150.0	1982.2	573.6	198.0	3112.6
Paddy field	0.2	1.4	5.3	13.5	0.0	0.1	0.0	763.7	36.3	820.3
Dry farmland	551.0	109.6	12.6	379.6	0.8	73.8	20.7	917.8	20,316.5	22,382.3
Total	2777.9	4602.4	1978.6	1827.3	247.7	7426.8	2476.2	2557.9	23,002.5	46,897.3

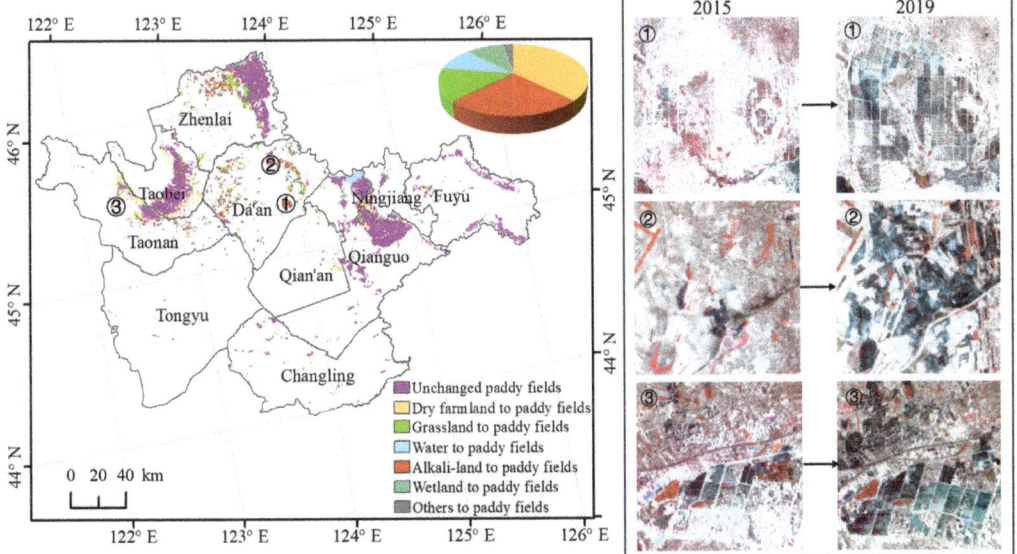

Figure 4. The paddy fields increase in western Jilin from 2015 to 2019.

3.2. Impact of Saline-Alkali Land Development to Paddy Fields on Land Surface Geophysical Parameters

The distribution of both albedo and LAI showed notable spatial heterogeneity across western Jilin (Figure 5). Previous studies have suggested that the land use distribution, vegetation coverage and background geophysical conditions may explain why this region shows this spatial variation. By comparing the land use pattern and the LAI/albedo pattern, we found that the saline-alkali land-dominated regions such as Da'an, Qian'an and Tongyu showed higher albedo and lower LAI than those of other regions. For the main region of paddy fields such as the northern part of Qianguo, Taobei and the eastern part of Zhenlai, the albedo was lower and the LAI was higher. As our study focused on the climate impact due to the conversion from saline-alkali land to paddy fields, the albedo and LAI for the pure saline-alkali land and paddy fields were separated. The albedo for the saline-alkali land during the growing season was 0.21 ± 0.02, whereas it was 0.15 ± 0.007 for the paddy field. The LAI for the saline-alkali land during the growing season was 0.57 ± 0.19, whereas it was 1.99 ± 0.24 for the paddy field.

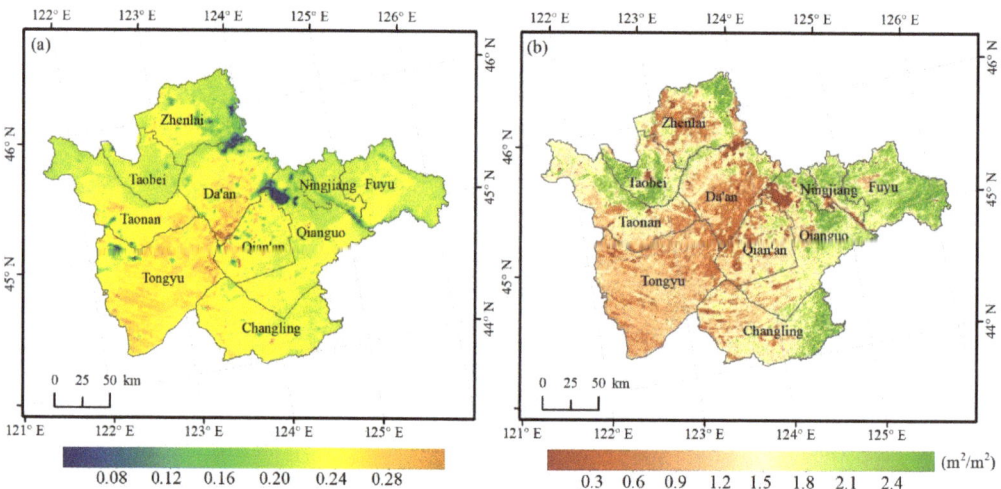

Figure 5. Spatial distribution of albedo (**a**) and LAI (**b**) for the growing season (June to September) across western Jilin.

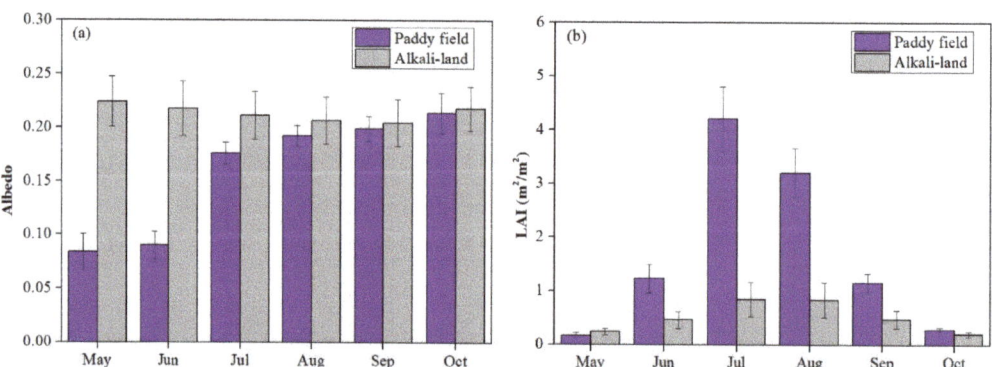

Figure 6. The seasonal variations in albedo (**a**) and LAI (**b**) for paddy fields and saline-alkali land in western Jilin.

The monthly mean albedo and LAI based on zonal statistics (Figure 6a,b) indicated that the seasonal variation in albedo/LAI for the paddy field was much greater than that for the saline-alkali land. At the beginning of the growing season, including May and June, the albedo in the paddy field was less than 0.10; however, it was greater than 0.2 for the saline-alkali land. With the tillering, heading and fruiting of rice and vegetation growth in saline-alkali land, the albedo differences from July to October decreased. The LAI variations paralleled the air temperature changes, which increased from May on and reached their peak value in July and then decreased gradually to October. However, the differences in LAI between the paddy fields and saline-alkali land were not synchronized with albedo. Instead of occurring in May and June, the distinct differences in LAI mainly occurred in July and August (3.35 m^2/m^2 and 2.37 m^2/m^2, respectively). These asynchronous differences in surface parameters between paddy fields and saline-alkali land are likely to bring distinct seasonal climate responses to saline-alkali land improvements.

3.3. Impact of Saline-Alkali Land Development to Paddy Fields on Air Temperature and Relative Humidity

Prior to the analysis of climate impacts caused by land use changes, the model-simulated results were first validated by the observed air temperature (T-2 m) and relative humidity (Rh-2 m). Our results showed a cooling bias of 0.21 °C for T-2 m and a drying bias of 2.02% for Rh-2 m in western Jilin, indicating that our model captured the patterns of T-2 m and Rh-2 m well in our experiments.

The differences in T2-m and Rh-2 m between the control and sensitivity experiments showed that local T-2 m and Rh-2 m responded to saline-alkali land improvement, i.e., T-2 m and Rh-2 m mainly changed where saline-alkali land development occurred. At the pixel scale, the relationship between the improved saline-alkali land fraction and changes in T-2 m and Rh-2 m showed a nonlinear relationship (Figure 7a,b).

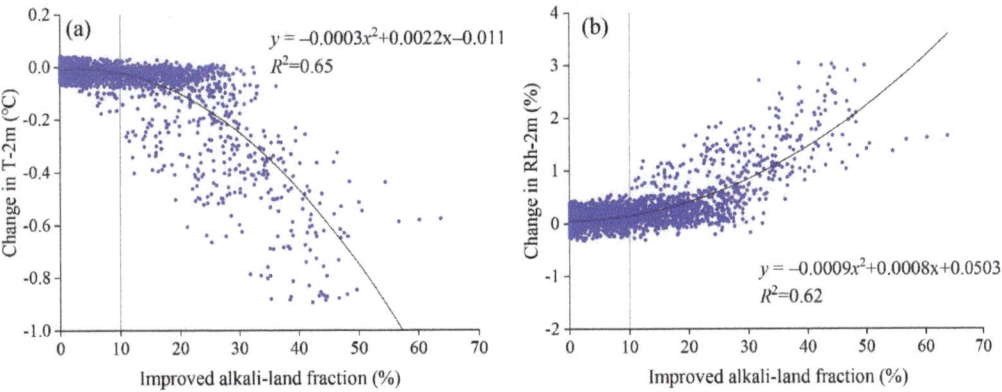

Figure 7. Scatter plots of cell-based changes (SEN minus CTL) in the improved saline-alkali land fraction and changes in simulated (**a**) surface air temperature (T-2 m) and (**b**) relative humidity (Rh-2 m) for JJAS (June-July August-September) over western Jilin.

When the improved saline-alkali land fraction was less than 10%, the changes in T-2 m and Rh-2 m were relatively small. With the increase in the improved saline-alkali land fraction, the temperature-humidity effect became increasingly significant. As the improved saline-alkali land fraction increased by 50%, the T-2 m decreased by 0.65 °C during the growing season. In contrast to T-2 m, a 50% increase in improved saline-alkali land could lead to a 2.16% increase in Rh-2 m.

The land use change from saline-alkali land to paddy fields contributed to varied land surface changes among different months during the growing season, which was shown in Section 3.2. We used the pixels with the dominant type converted from saline-alkali land to paddy fields in our experiments to further investigate the seasonal variations in temperature responses and their biogeophysical mechanisms. Our results showed that the saline-alkali land improvements brought consistent temperature cooling and relative humidity increases during the growing season from June to September (Figure 8). The most significant cooling was observed in July and August with the T-2 m decreasing by 0.66 °C (mean value) and 0.67 °C (mean value), respectively. The temperature cooling effect was also detected in September and June and declined by 0.47 °C and 0.27 °C, respectively. The Rh-2 m increased by 2.35% and 2.11% in July and August, respectively, whereas it increased by 0.94% and 0.93% in June and September, respectively.

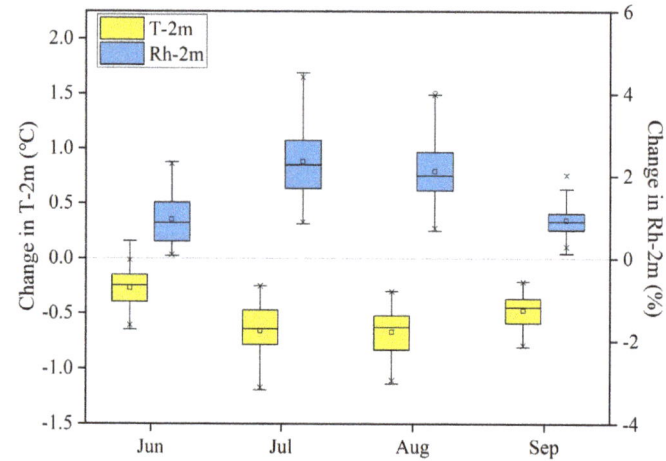

Figure 8. Differences (SEN-CTL) in simulated regional mean air temperature and relative humidity for the grids with dominant land use types converted from saline-alkali land to wetland for June, July, August, and September in western Jilin.

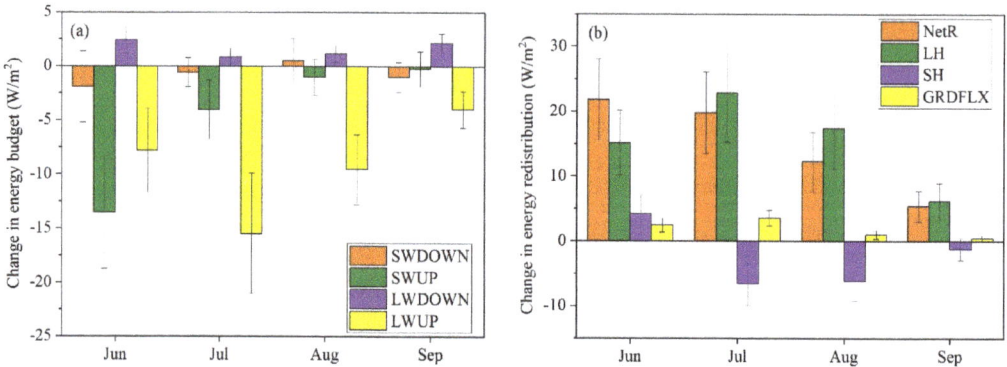

Figure 9. Differences (SEN-CTL) in simulated regional means of (**a**) incoming shortwave radiation (SWDOWN), outcoming shortwave radiation (SWUP), incoming longwave radiation (LWDOWN), outcoming longwave radiation (LWUP), (**b**) net radiation (NetR), latent heat flux (LH), sensible heat flux (SH), and ground heat flux (GRDFLX) for the grids with dominant land use type converted from saline-alkali land to wetland for June, July, August, September of western Jilin. The error bar is the standard deviation of changes in each variable.

From an energy balance perspective (Figure 9), the saline-alkali land improvement decreased the reflected solar radiation by decreasing the albedo, particularly in June and July. The upward shortwave radiation decreased by -13.56 ± 5.18 W/m^2 and -4.03 ± 2.72 W/m^2 in June and July, respectively, indicating warming effects through the absorption more solar radiation. From an energy redistribution perspective, the increase in LAI enhanced vegetation transpiration, resulting in the latent heat flux increasing by 15.17 ± 5.05 W/m^2, 22.82 ± 7.55 W/m^2, 17.41 ± 6.35 W/m^2, and 6.11 ± 2.84 W/m^2

from June to September, which contributed to the cooling effect. The interaction between these two processes explained the variations in the T-2 m and Rh-2 m changes. In June, a large part of the cooling effect caused by evapotranspiration enhancement was counteracted by the warming effect caused by the significant decline in albedo, resulting in a lower magnitude change in T-2 m relative to that in other months. Because both the evapotranspiration (ET) increases and solar radiation increases were larger in July than in August, the offset between the two processes led to similar T-2 m changes. However, with the relatively small decrease in LH in September, the decline in the magnitude of T-2 m is even larger than that in June, which can be attributed to the similar albedo (Figure 6).

4. Discussion

4.1. Saline-Alkali Land Development and Its Impact on Surface Parameters

By using a time series of land use datasets and a spatial overlay analysis, our study evaluated the land use changes from 1975 to 2015 and 2015 to 2019. The results showed that cropland expansion was the dominant land use change in western Jilin from 1975 to 2015, which is consistent with previous studies [18,35,53]. The application of the Three North Shelterbelt Project and Natural Forest Protection Project explained the woodland increase in western Jilin [54]. High-intensity agricultural development also brought increases in grassland degradation, wetland loss and unused saline-alkali land. Driven by economic interest and combined with better irrigation conditions, part of the rain-fed farmland has turned into paddy fields, and this is characteristic not only for western Jilin but also for all of Northeast China [55–58]. It should be noted that the total increase in paddy field area from 2015 to 2019 was even larger than that from 1975 to 2015, 25% of which was converted from saline-alkali land. As an important reserved cultivated land, unused saline-alkali land has more potential for development in the future. Previous studies have shown that saline-alkali paddy fields produce 1500 kg rice in the first reclamation year and could reach 8000 kg/ha after 5 years [59], implying that the rice yield could increase from 11,140 t to 59,414 t when the saline-alkali land in western Jilin is completely improved to paddy fields after 5 years.

In addition, we found seasonal LAI and albedo variation for both paddy fields and saline-alkali land, which is consistent with previous studies [23,60]. We also detected variable albedo decreases from May to September when saline-alkali land was converted into paddy fields; the largest decrease occurred in May and June. In comparison with albedo, the LAI increase was mainly concentrated in July and August. A lower albedo helps the surface absorb more solar radiation and has a warming effect, whereas a higher LAI enhances evapotranspiration and tends to cool the surface [3,41,61–64]. The offset between the albedo warming effect and the ET cooling effect determines the final temperature impact due to saline-alkali land improvements, indicating that the albedo and ET changes caused by the conversion from saline-alkali land to paddy fields were similar to those of afforestation [65–67].

4.2. Impact of Saline-Alkali Land Development on Regional Climate and Corresponding Mechanisms

Recent studies have reported that vegetation greening or vegetation growth brings significant surface cooling in China [41,48,60,67–69]. Cao et al. found that forest restoration attributed to the Grain to Green (GTG) programme lowered the 2-m air temperature of the Loess Plateau in summer [69]. Zhang et al. found that the cropland greenness increases in spring contributed to cooling and wetting effects, whereas the crop greenness decline in summer led to warming and drying effects on the North China Plain [48]. However, Yu et al. found that crop greening in the Northeast China Plain in the first two decades of the 21st century cooled the surface temperature in summer [68]. Our results showed that saline-alkali land improvement could accelerate surface greening and help decrease the air temperature and increase the relative humidity, which is consistent with previous studies.

In ecologically vulnerable regions, numerous studies have focused on land degradation as well as its climate implications [34,70]. They have revealed that land degradation or desertification led to notable warming over northern China. As the opposite trend to land degradation, saline-alkali land improvement brought significant cooling, which can help slow climate warming in this area. This result suggests that the energy redistribution to latent heat and sensible heat dominated the temperature impacts. In addition, our results also found that the cooling effects varied among months during the growing season. Air temperature cooling was most obvious in July and August, followed by September and June, implying that interactions between energy balance and energy partitioning vary during the growing season. Liu et al. found similar surface temperature differences between saline-alkali land and paddy fields using remote sensing observations [37]. It should be noted that their change in magnitude of the surface temperature is greater than our results, which can be explained by the following two observations. One is that the surface temperature response to land use change is usually more sensitive than the air temperature response [68]. The other is that Liu et al. [37] used pure pixels at 1 km to calculate the surface temperature differences between saline-alkali land and paddy fields, whereas our model simulations using a 10 km horizontal resolution involved mixed pixels. For this reason, we used the regression model to reconstruct the relationship between the change in T-2 m/Rh-2 m and the saline-alkali land improvement fraction (Figure 7). Zhang et al. found that the change in T2 was directly correlated with the change in green vegetation fraction (GVF) in cropland [48]. However, our results cannot be fitted by a linear equation. With the increase in the saline-alkali land fraction, T-2 m decreased at an increasing rate, particularly for pixels with a saline-alkali land improvement fraction (SALIF) greater than 10%. When the SALIF reached a threshold, the SALIF changed the dominant land use type to paddy fields and influenced all the surface parameters, including LAI and albedo.

4.3. Uncertainties and Future Works

There are a few points that should be addressed in the future. First, this study used the 24 USGS land use and land cover categories to represent the land use pattern. The saline-alkali land and paddy fields were not included in the list of the 24 types and were characterized by the surface parameters of bare/sparse vegetation and herbaceous wetlands, respectively. Although saline-alkali land is a type of bare/sparse vegetation and paddy fields are a type of herbaceous wetland, some specific properties, including albedo, LAI, soil moisture, etc. for saline-alkali land and paddy fields cannot be fully expressed by this categorization. For this reason, we used the remote sensing observed albedo and LAI to update the original values. Second, in addition to LAI and albedo, the surface parameters, including surface roughness, root depth, green vegetation fraction etc., can influence land/atmosphere interactions [48,71], which should be localized in future studies to help decrease the model uncertainties. Thus, in future works, the land use and land cover categories should be expanded in the land surface model to better represent the land surface processes and their climate effects.

Finally, our work evaluated the climate impact due to saline-alkali land improvement to paddy fields, which helped to slow climate warming and increase the relative humidity. Note that in addition to regulating climate, saline-alkali land improvement can also bring other environmental or ecological benefits by producing more rice, regulating hydrology, improving soil, increasing aesthetics and promoting tourism, all of which should be comprehensively evaluated in our next works.

5. Conclusions

Using satellite-based land use maps for 1975, 2015 and 2019, this study analysed the spatial-temporal changes in land use in western Jilin. From 1975 to 2015, grass degradation (3228 km^2) and paddy field expansion (1738 km^2) dominated the land use changes, whereas from 2015 to 2019, paddy fields increased by 2031 km^2 and became the main land use change characteristic in this region. Interestingly, our results showed that 25% of the paddy field increase in the latter period was reclaimed from saline-alkali land, and that percentage is expected to increase in the future. The saline-alkali land and paddy fields have distinct differences in regard to LAI and albedo during the growing season. We found that saline-alkali land reclamation to paddy fields significantly increased LAI, particularly in July and August, whereas it decreased albedo, mainly in May and June.

By coupling remote sensing measurements and regional climate simulations, we also evaluated the climate impact caused by the potential saline-alkali land improvement in western Jilin. We found that saline-alkali land improvement to paddy fields can help decrease the air temperature and increase the relative humidity. The temperature and humidity effects showed different magnitudes during the growing season, which were the most significant in July and August, followed by September and June. Evapotranspiration (ET), rather than albedo, played a dominant role in regulating the regional climate, and the interaction between ET and albedo determined the temperature and relative humidity response variations during the growing season.

Supplementary Materials: The following are available online at https://www.mdpi.com/article/10.3390/rs13173407/s1, Figure S1: Pure pixels for saline-alkali land (a) and paddy field (b) at a 1 km × 1 km spatial resolution.

Author Contributions: Conceptualization, L.Y.; Investigation, J.Y.; methodology, T.L.; software, K.B.; Visualization, K.B.; validation, Y.J. and G.L.; writing—original draft preparation, L.Y.; writing—review and editing, J.Y., T.L. and L.P.; Resources, S.Z.; project administration, L.Y. All authors have read and agreed to the published version of the manuscript.

Funding: This study was supported by the National Natural Science Foundation of China (42071025), the Science and Technology Basic Resources Investigation Program of China (2017FY101301), a Strategic Priority Research Program (A) of the Chinese Academy of Sciences (XDA2003020301), Natural Science Foundation of Hainan Province (321QN187) and 14th Five-year Network Security and Informatization Plan of Chinese Academy of Sciences (WX145XQ06-07).

Institutional Review Board Statement: The study did not involve humans or animals.

Informed Consent Statement: The study did not involve humans.

Data Availability Statement: The Landsat OLI 8 images are available from GloVis website (https://glovis.usgs.gov/ (accessed on 25 June 2021)). The CCI land cover data in 2015 are available from ESA (http://www.esa-landcover-cci.org/ (accessed on 25 June 2021)). Monthly LAI and FVC data were obtained from product GLASS, which are available from University of Maryland (http://www.glass.umd.edu/ (accessed on 25 June 2021)). The ERA5 and ERA5-Land reanalysis data are obtained from ECMWF (https://cds.climate.copernicus.eu/cdsapp#!/search?type = dataset&text = ERA5 (accessed on 25 June 2021)). The meteorological observation data are obtained from the China Meteorological Data Service Center (CMDC; http://data.cma.cn/en (accessed on 25 June 2021)) and National Climate Data Center of United States (https://gis.ncdc.noaa.gov/ (accessed on 25 June 2021)).

Acknowledgments: In this section, you can acknowledge any support given which is not covered by the author contribution or funding sections. This may include administrative and technical support, or donations in kind (e.g., materials used for experiments).

Conflicts of Interest: The authors declare no conflict of interest.

References

1. Alter, R.E.; Douglas, H.C.; Winter, J.M.; Eltahir, E.A.B. Twentieth Century Regional Climate Change During the Summer in the Central United States Attributed to Agricultural Intensification. *Geophys. Res. Lett.* **2018**, *45*, 1586–1594. [CrossRef]
2. Friedlingstein, P.; Jones, M.W.; O'Sullivan, M.; Andrew, R.M.; Hauck, J.; Peters, G.P.; Peters, W.; Pongratz, J.; Sitch, S.; Le Quere, C.; et al. Global Carbon Budget 2019. *Earth Syst. Sci. Data* **2019**, *11*, 1783–1838. [CrossRef]
3. Liang, S.L.; Kustas, W.; Schaepman-Strub, G.; Li, X.W. Impacts of Climate Change and Land Use Changes on Land Surface Radiation and Energy Budgets. *IEEE J-STARS* **2010**, *3*, 219–224. [CrossRef]
4. Feddema, J.J.; Oleson, K.W.; Bonan, G.B.; Mearns, L.O.; Buja, L.E.; Meehl, G.A.; Washington, W.M. The importance of land-cover change in simulating future climates. *Science* **2005**, *310*, 1674–1678. [CrossRef]

5. Mahmood, R.; Quintanar, A.I.; Conner, G.; Leeper, R.; Dobler, S.; Pielke, R.A.; Beltran-Przekurat, A.; Hubbard, K.G.; Niyogi, D.; Bonan, G.; et al. Impacts of Land Use/Land Cover Change on Climate and Future Research Priorities. *Bull. Am. Meteorol. Soc.* **2010**, *91*, 37–46. [CrossRef]
6. Euskirchen, E.S.; McGuire, A.D.; Kicklighter, D.W.; Zhuang, Q.; Clein, J.S.; Dargaville, R.J.; Dye, D.G.; Kimball, J.S.; McDonald, K.C.; Melillo, J.M.; et al. Importance of recent shifts in soil thermal dynamics on growing season length, productivity, and carbon sequestration in terrestrial high-latitude ecosystems. *Glob. Chang. Biol.* **2006**, *12*, 731–750. [CrossRef]
7. Kalnay, E.; Cai, M. Impact of urbanization and land-use change on climate. *Nature* **2003**, *423*, 528–531. [CrossRef]
8. Pielke, R.A., Sr.; Marland, G.; Betts, R.A.; Chase, T.N.; Eastman, J.L.; Niles, J.O.; Niyogi, D.D.; Running, S.W. The influence of land-use change and landscape dynamics on the climate system: Relevance to climate-change policy beyond the radiative effect of greenhouse gases. *Philos. Trans. R. Soc. London. Ser. A Math. Phys. Eng. Sci.* **2002**, *360*, 1705–1719. [CrossRef]
9. Seneviratne, S.I.; Luthi, D.; Litschi, M.; Schar, C. Land-atmosphere coupling and climate change in Europe. *Nature* **2006**, *443*, 205–209. [CrossRef]
10. Baldocchi, D. Managing land and climate. *Nat. Clim. Chang.* **2014**, *4*, 330–331. [CrossRef]
11. Bright, R.M.; Davin, E.; O'Halloran, T.; Pongratz, J.; Zhao, K.G.; Cescatti, A. Local temperature response to land cover and management change driven by non-radiative processes. *Nat. Clim. Chang.* **2017**, *7*, 296–302. [CrossRef]
12. Li, F.; Zhou, M.J.; Shao, J.Q.; Chen, Z.H.; Wei, X.L.; Yang, J.C. Maize, wheat and rice production potential changes in China under the background of climate change. *Agric. Syst.* **2020**, *182*, 102853. [CrossRef]
13. Morton, J.F. The impact of climate change on smallholder and subsistence agriculture. *Proc. Natl. Acad. Sci. USA* **2007**, *104*, 19680–19685. [CrossRef]
14. Piao, S.L.; Ciais, P.; Huang, Y.; Shen, Z.H.; Peng, S.S.; Li, J.S.; Zhou, L.P.; Liu, H.Y.; Ma, Y.C.; Ding, Y.H.; et al. The impacts of climate change on water resources and agriculture in China. *Nature* **2010**, *467*, 43–51. [CrossRef] [PubMed]
15. Webb, N.P.; Marshall, N.A.; Stringer, L.C.; Reed, M.S.; Chappell, A.; Herrick, J.E. Land degradation and climate change: Building climate resilience in agriculture. *Front. Ecol. Environ.* **2017**, *15*, 450–459. [CrossRef]
16. Foley, J.A.; DeFries, R.; Asner, G.P.; Barford, C.; Bonan, G.; Carpenter, S.R.; Chapin, F.S.; Coe, M.T.; Daily, G.C.; Gibbs, H.K.; et al. Global consequences of land use. *Science* **2005**, *309*, 570–574. [CrossRef] [PubMed]
17. Turner, B.L.; Lambin, E.F.; Reenberg, A. The emergence of land change science for global environmental change and sustainability. *Proc. Natl. Acad. Sci. USA* **2007**, *104*, 20666–20671. [CrossRef]
18. Liu, J.Y.; Zhang, Z.X.; Xu, X.L.; Kuang, W.H.; Zhou, W.C.; Zhang, S.W.; Li, R.D.; Yan, C.Z.; Yu, D.S.; Wu, S.X.; et al. Spatial patterns and driving forces of land use change in China during the early 21st century. *J. Geogr. Sci.* **2010**, *20*, 483–494. [CrossRef]
19. Kuang, W.; Du, G.; Lu, D.; Dou, Y.; Li, X.; Zhang, S.; Chi, W.; Dong, J.; Chen, G.; Yin, Z.; et al. Global observation of urban expansion and land-cover dynamics using satellite big-data. *Sci. Bull.* **2021**, *66*, 297–300. [CrossRef]
20. Macedo, M.N.; DeFries, R.S.; Morton, D.C.; Stickler, C.M.; Galford, G.L.; Shimabukuro, Y.E. Decoupling of deforestation and soy production in the southern Amazon during the late 2000s. *Proc. Natl. Acad. Sci. USA* **2012**, *109*, 1341–1346. [CrossRef] [PubMed]
21. Morton, D.C.; DeFries, R.S.; Shimabukuro, Y.E.; Anderson, L.O.; Arai, E.; Espirito-Santo, F.D.; Freitas, R.; Morisette, J. Cropland expansion changes deforestation dynamics in the southern Brazilian Amazon. *Proc. Natl. Acad. Sci. USA* **2006**, *103*, 14637–14641. [CrossRef]
22. Chen, C.; Park, T.; Wang, X.; Piao, S.; Xu, B.; Chaturvedi, R.K.; Fuchs, R.; Brovkin, V.; Ciais, P.; Fensholt, R.; et al. China and India lead in greening of the world through land-use management. *Nat. Sustain.* **2019**, *2*, 122–129. [CrossRef] [PubMed]
23. Cooley, H.S.; Riley, W.J.; Torn, M.S.; He, Y. Impact of agricultural practice on regional climate in a coupled land surface mesoscale model. *J. Geophys. Res.-Atmos* **2005**, *110*. [CrossRef]
24. Sakai, R.K.; Fitzjarrald, D.R.; Moraes, O.L.L.; Staebler, R.M.; Acevedo, O.C.; Czikowsky, M.J.; Da Silva, R.; Brait, E.; Miranda, V. Land-use change effects on local energy, water, and carbon balances in an Amazonian agricultural field. *Glob. Chang. Biol.* **2004**, *10*, 895–907. [CrossRef]
25. Bonan, G.B. Effects of land use on the climate of the United States. *Clim. Chang.* **1997**, *37*, 449–486. [CrossRef]
26. Diffenbaugh, N.S. Influence of modern land cover on the climate of the United States. *Clim. Dyn.* **2009**, *33*, 945–958. [CrossRef]
27. Kueppers, L.M.; Snyder, M.A.; Sloan, L.C. Irrigation cooling effect: Regional climate forcing by land-use change. *Geophys. Res. Lett.* **2007**, *34*. [CrossRef]
28. Thiery, W.; Visser, A.J.; Fischer, E.M.; Hauser, M.; Hirsch, A.L.; Lawrence, D.M.; Lejeune, Q.; Davin, E.L.; Seneviratne, S.I. Warming of hot extremes alleviated by expanding irrigation. *Nat. Commun.* **2020**, *11*, 1–7. [CrossRef] [PubMed]
29. Mishra, V.; Ambika, A.K.; Asoka, A.; Aadhar, S.; Buzan, J.; Kumar, R.; Huber, M. Moist heat stress extremes in India enhanced by irrigation. *Nat. Geosci.* **2020**, *13*, 722–728. [CrossRef]
30. Lamptey, B.L.; Barron, E.J.; Pollard, D. Impacts of agriculture and urbanization on the climate of the Northeastern United States. *Glob. Planet Chang.* **2005**, *49*, 203–221. [CrossRef]
31. Liu, T.; Zhang, S.; Yu, L.; Bu, K.; Yang, J.; Chang, L. Simulation of regional temperature change effect of land cover change in agroforestry ecotone of Nenjiang River Basin in China. *Theor. Appl. Climatol.* **2017**, *128*, 971–981. [CrossRef]
32. Wang, Z.; Liu, Z.; Song, K.; Zhang, B.; Zhang, S.; Liu, D.; Ren, C.; Yang, F. Land use changes in Northeast China driven by human activities and climatic variation. *Chin. Geogr. Sci.* **2009**, *19*, 225–230. [CrossRef]
33. Zhang, S.W.; Zhang, Y.Z.; Li, Y.; Chang, L.P. *Temporal and Spatial Characteristics of Land Use/Cover in Northeast China*; Science Press: Beijing, China, 2006.

34. Zhang, J.; Dong, W.; Fu, C. Impact of land surface degradation in northern China and southern Mongolia on regional climate. *Chin. Sci. Bull.* **2005**, *50*, 75–81. [CrossRef]
35. Li, F.; Shuwen, Z.; Jiuchun, Y.; Kun, B.; Qing, W.; Junmei, T.; Liping, C. The effects of population density changes on ecosystem services value: A case study in Western Jilin, China. *Ecol. Indic.* **2016**, *61*, 328–337. [CrossRef]
36. Ma, S.; Liu, W.; Chang, G. Three-Dimensional Improvement Method for Saline-Alkali Land, Involves Planting Saline-Alkali-Tolerant Rice in Saline-Alkali Land, Crushing and Burying Harvested Straw in Saline-Alkali Land, after Harvesting Saline-Alkali-Tolerant Rice. China Patent No CN112189394-A, 8 January 2021.
37. Liu, T.; Yu, L.; Zhang, S. Land Surface Temperature Response to Irrigated Paddy Field Expansion: A Case Study of Semi-arid Western Jilin Province, China. *Sci. Rep.* **2019**, *9*, 5278. [CrossRef]
38. Liang, S. Recent algorithm developments in quantitative remote sensing of land surfaces. In Proceedings of the 2003 IEEE International Geoscience and Remote Sensing, Toulouse, France, 21–25 July 2003; pp. 558–560.
39. Yang, J.; Gong, P.; Fu, R.; Zhang, M.; Chen, J.; Liang, S.; Xu, B.; Shi, J.; Dickinson, R. The role of satellite remote sensing in climate change studies. *Nat. Clim. Chang.* **2013**, *3*, 875–883. [CrossRef]
40. Song, X.; Hansen, M.C.; Stehman, S.V.; Potapov, P.V.; Tyukavina, A.; Vermote, E.F.; Townshend, J.R. Global land change from 1982 to 2016. *Nature* **2018**, *560*, 639–643. [CrossRef]
41. Peng, S.; Piao, S.; Zeng, Z.; Ciais, P.; Zhou, L.; Li, L.; Myneni, R.B.; Yin, Y.; Zeng, H. Afforestation in China cools local land surface temperature. *Proc. Natl. Acad. Sci. USA* **2014**, *111*, 2915–2919. [CrossRef] [PubMed]
42. Zhou, D.; Li, D.; Sun, G.; Zhang, L.; Liu, Y.; Hao, L. Contrasting effects of urbanization and agriculture on surface temperature in eastern China. *J. Geophys. Res.-Atmos.* **2016**, *121*, 9597–9606. [CrossRef]
43. Davin, E.L.; Maisonnave, E.; Seneviratne, S.I. Is land surface processes representation a possible weak link in current Regional Climate Models? *Environ. Res. Lett.* **2016**, *11*, 074027. [CrossRef]
44. Jin, J.M.; Miller, N.L.; Schlegel, N. Sensitivity Study of Four Land Surface Schemes in the WRF Model. *Adv. Meteorol.* **2010**, *2010*, 1–11. [CrossRef]
45. Lakshmi, V.; Hong, S.; Small, E.E.; Chen, F. The influence of the land surface on hydrometeorology and ecology: New advances from modeling and satellite remote sensing. *Hydrol. Res.* **2011**, *42*, 95–112. [CrossRef]
46. Sellers, P.J.; Dickinson, R.E.; Randall, D.A.; Betts, A.K.; Hall, F.G.; Berry, J.A.; Collatz, G.J.; Denning, A.S.; Mooney, H.A.; Nobre, C.A.; et al. Modeling the exchanges of energy, water, and carbon between continents and the atmosphere. *Science* **1997**, *275*, 502–509. [CrossRef] [PubMed]
47. Zhang, X.; Xiong, Z.; Zhang, X.; Shi, Y.; Liu, J.; Shao, Q.; Yan, X. Using multi-model ensembles to improve the simulated effects of land use/cover change on temperature: A case study over northeast China. *Clim. Dyn.* **2016**, *46*, 765–778. [CrossRef]
48. Zhang, X.; Tang, Q.; Zheng, J.; Ge, Q. Warming/cooling effects of cropland greenness changes during 1982-2006 in the North China Plain. *Environ. Res. Lett.* **2013**, *8*, 024038. [CrossRef]
49. Case, J.L.; Crosson, W.L.; Kumar, S.V.; Lapenta, W.M.; Peters-Lidard, C.D. Impacts of High-Resolution Land Surface Initialization on Regional Sensible Weather Forecasts from the WRF Model. *J. Hydrometeorol.* **2008**, *9*, 1249–1266. [CrossRef]
50. Miller, J.; Barlage, M.; Zeng, X.B.; Wei, H.L.; Mitchell, K.; Tarpley, D. Sensitivity of the NCEP/Noah land surface model to the MODIS green vegetation fraction data set. *Geophys. Res. Lett.* **2006**, *33*. [CrossRef]
51. Yu, L.; Xue, Y.; Diallo, I. Vegetation greening in China and its effect on summer regional climate. *Sci. Bull.* **2021**, *66*, 13–17. [CrossRef]
52. Skamarock, W.C.; Klemp, J.B.; Dudhia, J.; Gill, D.; Barker, D.M.; Duda, M.G.; Huang, X.; Wang, W.; Powers, J.G. *A Description of the Advanced Research WRF Version 3*. NCAR Technical Note, NCAR/TN-475+STR; National Center for Atmospheric Research: Boulder, CO, USA, 2008.
53. Liu, J.; Kuang, W.; Zhang, Z.; Xu, X.; Qin, Y.; Ning, J.; Zhou, W.; Zhang, S.; Li, R.; Yan, C.; et al. Spatiotemporal characteristics, patterns, and causes of land-use changes in China since the late 1980s. *J. Geogr. Sci.* **2014**, *24*, 195–210. [CrossRef]
54. Cai, D.; Ge, Q.; Wang, X.; Liu, B.; Goudie, A.; Hu, S. Contributions of ecological programs to vegetation restoration in arid and semiarid China. *Environ. Res. Lett.* **2020**, *15*, 114046. [CrossRef]
55. Yu, L.; Liu, T. The Impact of Artificial Wetland Expansion on Local Temperature in the Growing Season—The Case Study of the Sanjiang Plain, China. *Remote Sens.* **2019**, *11*, 2915. [CrossRef]
56. Liu, T.; Yu, L.; Zhang, S. Impacts of Wetland Reclamation and Paddy Field Expansion on Observed Local Temperature Trends in the Sanjiang Plain of China. *J. Geophys. Res.-Earth* **2019**, *124*, 414–426. [CrossRef]
57. Ning, J.; Liu, J.; Kuang, W.; Xu, X.; Zhang, S.; Yan, C.; Li, R.; Wu, S.; Hu, Y.; Du, G.; et al. Spatiotemporal patterns and characteristics of land-use change in China during 2010–2015. *J. Geogr. Sci.* **2018**, *28*, 547–562. [CrossRef]
58. Yan, F.; Yu, L.; Yang, C.; Zhang, S. Paddy Field Expansion and Aggregation since the Mid-1950s in a Cold Region and Its Possible Causes. *Remote Sens.* **2016**, *10*, 384. [CrossRef]
59. Zhang, M. Research on the modified mechanism and key technology by Plangting paddy in soda saline alkali soil in Western Jilin Province. Ph.D. Thesis, Jilin Agricultural University, Changchun, China, 2015. (In Chinese).
60. Yu, L.; Liu, T.; Zhang, S. Temporal and Spatial Changes in Snow Cover and the Corresponding Radiative Forcing Analysis in Siberia from the 1970s to the 2010s. *Adv. Meteorol.* **2017**, *2017*, 1–11. [CrossRef]
61. Shen, M.; Piao, S.; Jeong, S.; Zhou, L.; Zeng, Z.; Ciais, P.; Chen, D.; Huang, M.; Jin, C.; Li, L.; et al. Evaporative cooling over the Tibetan Plateau induced by vegetation growth. *Proc. Natl. Acad. Sci. USA* **2015**, *112*, 9299–9304. [CrossRef] [PubMed]

62. Zeng, Z.; Piao, S.; Li, L.; Zhou, L.; Ciais, P.; Wang, T.; Li, Y.; Lian, X.; Wood, E.F.; Friedlingstein, P.; et al. Climate mitigation from vegetation biophysical feedbacks during the past three decades. *Nat. Clim. Chang.* **2017**, *7*, 432–436. [CrossRef]
63. Barnes, C.A.; Roy, D.P. Radiative forcing over the conterminous United States due to contemporary land cover land use albedo change. *Geophys. Res. Lett.* **2008**, *35*. [CrossRef]
64. Bright, R.M.; Zhao, K.; Jackson, R.B.; Cherubini, F. Quantifying surface albedo and other direct biogeophysical climate forcings of forestry activities. *Glob. Chang. Biol.* **2015**, *21*, 3246–3266. [CrossRef]
65. Huang, L.; Zhai, J.; Liu, J.; Sun, C. The moderating or amplifying biophysical effects of afforestation on CO_2-induced cooling depend on the local background climate regimes in China. *Agric. For. Meteorol.* **2018**, *260*, 193–203. [CrossRef]
66. Betts, R.A. Climate Science Afforestation cools more or less. *Nat. Geosci.* **2011**, *4*, 504–505. [CrossRef]
67. Li, Y.; Piao, S.L.; Chen, A.P.; Wang, X.H.; Ciais, P.; Li, L.Z.X. Local and teleconnected temperature effects of afforestation and vegetation greening in China. *Natl. Sci. Rev.* **2020**, *7*, 897–912. [CrossRef]
68. Yu, L.; Liu, Y.; Liu, T.; Yan, F. Impact of recent vegetation greening on temperature and precipitation over China. *Agric. For. Meteorol.* **2020**, *295*, 108197. [CrossRef]
69. Cao, Q.; Wu, J.; Yu, D.; Wang, W. The biophysical effects of the vegetation restoration program on regional climate metrics in the Loess Plateau, China. *Agric. For. Meteorol.* **2019**, *268*, 169–180. [CrossRef]
70. Xue, Y. The impact of desertification in the Mongolian and the Inner Mongolian grassland on the regional climate. *J. Clim.* **1996**, *9*, 2173–2189. [CrossRef]
71. Liu, Y.; Guo, W.; Huang, H.; Ge, J.; Qiu, B. Estimating global aerodynamic parameters in 1982–2017 using remote-sensing data and a turbulent transfer model. *Remote Sens. Environ.* **2021**, *260*, 112428. [CrossRef]

Article

Inconsistency of Global Vegetation Dynamics Driven by Climate Change: Evidences from Spatial Regression

Dou Zhang [1], Xiaolei Geng [1], Wanxu Chen [2], Lei Fang [1], Rui Yao [3], Xiangrong Wang [1,*] and Xiao Zhou [4]

[1] Department of Environmental Science and Engineering, Fudan University, Shanghai 200438, China; 19110740011@fudan.edu.cn (D.Z.); 19110740020@fudan.edu.cn (X.G.); fanglei@fudan.edu.cn (L.F.)
[2] Department of Geography, School of Geography and Information Engineering, China University of Geosciences, Wuhan 430074, China; cugcwx@cug.edu.cn
[3] Hubei Key Laboratory of Critical Zone Evolution, School of Geography and Information Engineering, China University of Geosciences, Wuhan 430074, China; yaorui123@cug.edu.cn
[4] School of Public Administration, China University of Geosciences, Wuhan 430074, China; zhouxiao@cug.edu.cn
* Correspondence: xrxrwang@fudan.edu.cn

Citation: Zhang, D.; Geng, X.; Chen, W.; Fang, L.; Yao, R.; Wang, X.; Zhou, X. Inconsistency of Global Vegetation Dynamics Driven by Climate Change: Evidences from Spatial Regression. *Remote Sens.* 2021, 13, 3442. https://doi.org/10.3390/rs13173442

Academic Editors: Baojie He, Ayyoob Sharifi, Chi Feng and Jun Yang

Received: 26 July 2021
Accepted: 26 August 2021
Published: 30 August 2021

Publisher's Note: MDPI stays neutral with regard to jurisdictional claims in published maps and institutional affiliations.

Copyright: © 2021 by the authors. Licensee MDPI, Basel, Switzerland. This article is an open access article distributed under the terms and conditions of the Creative Commons Attribution (CC BY) license (https://creativecommons.org/licenses/by/4.0/).

Abstract: Global greening over the past 30 years since 1980s has been confirmed by numerous studies. However, a single-dimensional indicator and non-spatial modelling approaches might exacerbate uncertainties in our understanding of global change. Thus, comprehensive monitoring for vegetation's various properties and spatially explicit models are required. In this study, we used the newest enhanced vegetation index (EVI) products of Moderate Resolution Imaging Spectroradiometer (MODIS) Collection 6 to detect the inconsistency trend of annual peak and average global vegetation growth using the Mann–Kendall test method. We explored the climatic factors that affect vegetation growth change from 2001 to 2018 using the spatial lag model (SLM), spatial error model (SEM) and geographically weighted regression model (GWR). The results showed that EVI_{max} and EVI_{mean} in global vegetated areas consistently showed linear increasing trends during 2001–2018, with the global averaged trend of 0.0022 yr^{-1} ($p < 0.05$) and 0.0030 yr^{-1} ($p < 0.05$). Greening mainly occurred in the croplands and forests of China, India, North America and Europe, while browning was almost in the grasslands of Brazil and Africa (18.16% vs. 3.08% and 40.73% vs. 2.45%). In addition, 32.47% of the global vegetated area experienced inconsistent trends in EVI_{max} and EVI_{mean}. Overall, precipitation and mean temperature had positive impacts on vegetation variation, while potential evapotranspiration and vapour pressure had negative impacts. The GWR revealed that the responses of EVI to climate change were inconsistent in an arid or humid area, in cropland or grassland. Climate change could affect vegetation characteristics by changing plant phenology, consequently rendering the inconsistency between peak and mean greening. In addition, anthropogenic activities, including land cover change and land use management, also could lead to the differences between annual peak and mean vegetation variations.

Keywords: global vegetation growth; climate change; inconsistent greening trend; spatial autocorrelation and heterogeneity; spatial regression models

1. Introduction

Terrestrial vegetation controls the cycle of carbon, water and energy between land soil and atmosphere through biochemical processes such as photosynthesis and evapotranspiration and is strongly influenced by hydrothermal conditions and climate change and can affect the climate system in return [1–3]. In addition to directly providing ecosystem services such as food, raw materials, and landscape aesthetics, vegetation also has the potential of climate regulation, carbon sequestration and air purification, which are of great importance to maintaining the stability of a terrestrial ecosystem under global change [4,5]. Thus, systematic monitoring, detecting and quantifying vegetation dynamics and their response and feedback to global change have elicited a wide range of attention in multiple subjects.

It is currently unlikely to obtain vegetation properties and variations at the global scale using ground-based observations [5]. Vegetation indices (VIs) products, such as normalized difference vegetation index (NDVI) and enhanced vegetation index (EVI) derived from satellite observations, provide a possibility for long-term vegetation monitoring on a large scale [6,7]. VIs products from the Advanced Very-High-Resolution Radiometer (AVHRR) sensor, Système Pour l'Observation de la Terre VEGETATION (SPOT-VGT) and Moderate Resolution Imaging Spectroradiometer (MODIS) sensors on board different satellites have been widely used in vegetation growth monitoring and crop mapping [8–10]. Although providing the longest record of NDVI data (1981–present), AVHRR has the problems of orbit drifts and data inconsistency between various sensors. SPOT-VGT data discontinuity was also found in some areas due to sensor differences in spectral response functions [11]. MODIS data is considered to have a higher temporal consistency than AVHRR and SPOT-VGT data due to no orbit drift and sensor shifts problems and are usually used as reference data [12–14]. However, the negative influences of sensor degradation have been captured in MODIS-Terra Collection 5 VIs products [15]. Consequently, MODIS Collection 6 VIs products with the improved algorithm were released to address the sensor degradation problem, and the effects have been supported by some studies [16,17]. Compared to NDVI, EVI minimizes canopy-soil and atmospheric influences and improves sensitivity over dense vegetation conditions, and its reliability has been recognized by comparison analysis of multiple VIs products [10].

The overall greening trend of global vegetation has been widely reported and discussed since the early 1980s, supported by comparisons of multiple satellite observations [18], forest inventories [19] and process-based model simulations [20,21]. Pan et al. [22] explored the increasing global browning trend hidden in overall greening during 1982–2013 by using the ensemble empirical mode decomposition method and piecewise linear regression models; Gao et al. [23] verified the significant trend of global cultivated land greening during 1982–2015 by using two long-term satellite LAI datasets; Zhang et al. [24] found that the previous browning trend monitored by MODIS Terra-C5 needed to be reconsidered due to sensor degradation, while the trend from MODIS Terra-C6 was confirmed by AVHRR and enhanced land carbon sink data [25]. Although a lot of studies have re-examined global greening trend, whether there is a consistent trend between annual peak and average growth of vegetation remains unknown. The annual peak vegetation growth closely related to environmental change is critical in characterizing the capacity of terrestrial ecosystem productivity [20]. Vegetation growth is an ever-changing dynamic process. The annual average vegetation growth, which was selected in the most previous studies, can only reflect the overall state of a certain year but obscure details of vegetation growth. To comprehensively understand the changing ecosystems, it is necessary to track vegetation growth change from multi-dimensions.

Climate change characterized by global warming has been a dominant driver of greening over 28% of the global vegetation regions [21]. However, the global vegetation–climate relationship is complex and has firm spatial heterogeneity [26]. Warming has noticeable positive effects on vegetation growth in the temperate and arctic regions [27], while rising temperatures could limit vegetation growth in tropical regions [28] where the ambient temperature is the proximity of the physiological optimum [29]. In arid and semi-arid regions, vegetation is more sensitive to precipitation change than temperature due to water limitation. In contrast, in humid regions, the impacts of precipitation on vegetation variation are weaker than temperature [30]. In sum, the climate–vegetation relationship geographically varies with the climatic environment [5,23]. In addition, although temperature and precipitation have been the hot spots for a long time, the impacts of climate change on vegetation are more complex than that. For example, the impact of other climatic factors, such as vapour pressure deficit on vegetation growth, have been highlighted, which could offset the effect of rising CO_2 concentration, suggested to be considered in evaluating vegetation responses to climate change [31].

At present, most studies used the non-spatial regression methods to depict the relationship between vegetation variation and climate change, such as multiple linear regression [32] and partial correlation analysis [33]. However, the theoretical assumption of these methods is that the spatial data is statistically independent and uniformly distributed [34,35], which ignores influences of spatial autocorrelation. According to the first law of geography, spatial autocorrelation is ubiquitous, and the correlation between adjacent locations is usually stronger than that between distant locations [36,37]. Influenced by the spatial interaction and spatial diffusion of endogenous or exogenous factors, geographic data may no longer be independent of each other, but related to each other [38]. For example, the flow characteristics of the atmosphere and water supporting vegetation growth can suggest that vegetation growth are absolutely not spatially independent [37,39,40]. Previous empirical research within ecology have revealed that regression coefficients may radically shift between non-spatial and spatial (taking autocorrelation into account) regression modelling, resulting in erroneous conclusions [39,41–44]. Given that spatial autocorrelation exists extensively in spatial data, it is important to use spatially explicit models to explore the climate-driving mechanism of vegetation growth. Therefore, the aims of this study are: (1) to examine inconsistency trends of global vegetation in peak and average growth during 2001–2018; (2) to reveal the overall relationship between climate change and vegetation growth using spatial regression models (at the global level); (3) to find out the spatial heterogeneity characteristic of climate driving using the geographically weighted regression model (at the local level).

2. Materials and Methods

2.1. Global Datasets and Pre-Processing

EVI, land cover and climatic datasets used in this study are displayed in Table 1. The raw MODIS data were mosaicked, re-projected and converted to Geo TIFF from HDF format using the MODIS Reprojection Tool. We averaged monthly EVI as the annual EVI_{mean} and extracted the maximum monthly EVI as the annual EVI_{max}. In the spatial regression analysis, we resampled EVI data into 0.5° × 0.5° spatial resolution by using the bilinear algorithm to match the climatic gridded data as well as the basic analytical unit. Global terrestrial land cover data are from the International Geosphere-Biosphere Programme (IGBP) classification layer of MCD12Q1 (Figure 1). To eliminate the direct effects of transformation between vegetated and non-vegetated land during 2001–2018, we masked out non-vegetated areas including permanent wetlands, urban and built-up lands, permanent snow and ice, barren and water bodies by overlaying the IGBP land cover data during the period of 2001–2018. Global annual climatic gridded data were generated by extracting the maximum, minimum, mean or sum value from the monthly climatic gridded data. For example, annual precipitation data were generated by summating monthly precipitation, while annual maximum temperature data were obtained by calculating the maximum value of the monthly maximum temperature. All data pre-processing was accomplished in R x64 4.0.2 and RStudio (http://www.r-project.org/, accessed on 18 September 2020).

Table 1. Global datasets and sources.

Dataset	Spatial Resolution	Temporal Resolution	Time Span	Source
MODIS-Terra Collection 6 EVI (MOD13A3)	1 km	Monthly	2001–2018	The Level-1 and Atmosphere Archive and Distribution System Distributed Active Archive Center (LAADS DAAC) (https://ladsweb.modaps.eosdis.nasa.gov/search, accessed on 6 August 2020).
Land Cover (MCD12Q1)	500 m	Yearly	2001–2018	
Precipitation	0.5°	Monthly	2001–2018	The Climatic Research Unit Time-Series version 4.03 (CRU TS4.03) datasets (https://data.ceda.ac.uk/badc/cru/data/cru_ts, accessed on 11 February 2021)
Maximum temperature	0.5°	Monthly	2001–2018	
Mean temperature	0.5°	Monthly	2001–2018	
Minimum temperature	0.5°	Monthly	2001–2018	
Potential evapotranspiration	0.5°	Monthly	2001–2018	
Vapour pressure	0.5°	Monthly	2001–2018	
Wet day frequency	0.5°	Monthly	2001–2018	
Diurnal temperature range	0.5°	Monthly	2001–2018	
Frost day frequency	0.5°	Monthly	2001–2018	

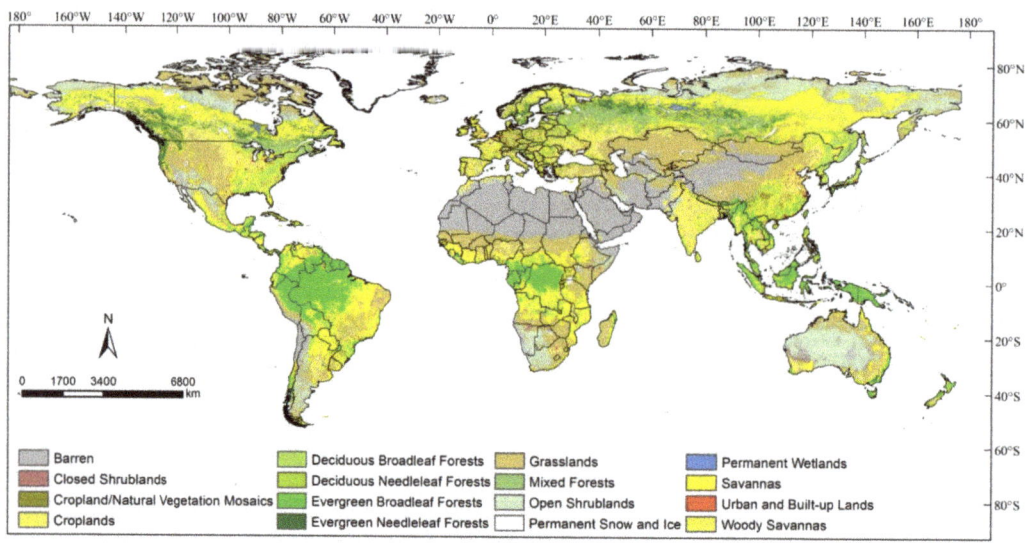

Figure 1. Global land cover of IGBP for 2018.

2.2. Methods

A methodological flowchart of this study is shown in Figure 2. Firstly, to find out the greening and browning areas and the area with inconsistent changes, we detected the trend of global vegetation growth variation during 2001–2018 from two dimensions of annual peak and mean using the Mann–Kendall test method recommended by the World Meteorological Organization in Section 2.2.1. Secondly, to prove our theoretical hypothesis that spatial autocorrelation exists in vegetation growth change, the global univariate Moran's I was first used to diagnose the spatial autocorrelation in the response variable in Section 2.2.2. The response variables were the changes of EVI_{max} and EVI_{mean}

from 2001 to 2018, and the explanatory variables are the changes of climatic variables from 2001 to 2018.

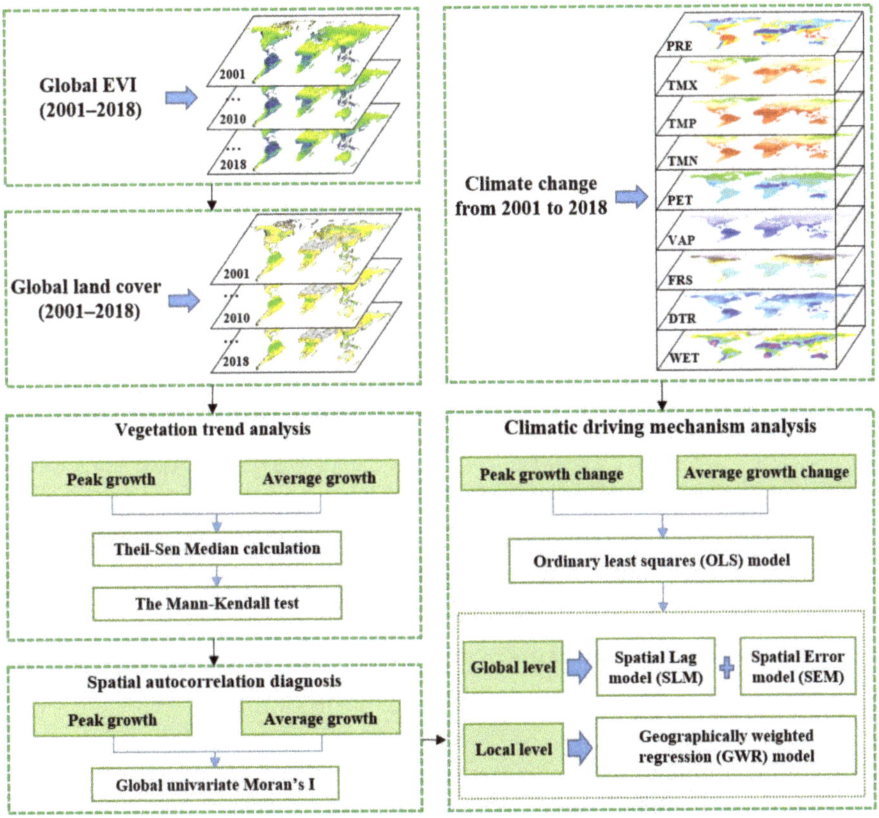

Figure 2. A methodological flowchart of this research.

Thirdly, to incorporate spatial autocorrelation in the response of vegetation growth to climate change, we used multiple spatial regression models to analyse the spatial relationship between climate change and vegetation variation at the global level. The first step is to perform the ordinary least squares (OLS) model, which is under the assumption that there is no autocorrelation [45] because the spatial lag model (SLM) and the spatial error model (SEM) are developed from the OLS by incorporating spatial autocorrelation into the regression by means of a spatial weight matrix [34]. SLM was used when spatial autocorrelation exists in dependent variables. SEM was used when spatial autocorrelation exists in the residual. The second step is to select SLM or SEM based on the statistical significance of the Lagrange multiplier (LM)-lag and LM-error, or robust LM-lag and robust LM-error. The third step is to determine the better performance model, with a larger maximum likelihood logarithm (LIK) and a smaller Akaike information criterion (AIC) and Schwarz criterion (SC).

Finally, the relationship between variables varies with the change of geographical location due to the differences in the natural environment and human disturbance in different regions. This changing relationship also needs to be considered in the spatial analysis [46]. To reveal the differentiated local characteristics hidden in the overall correlation, we used the geographically weighted regression (GWR) model to measure the spatial heterogeneity of the relationship between vegetation growth and climate change by obtaining local regression results at each spatial unit in Section 2.2.4. The spatial autocorrelation analysis and the OLS, SLM and SEM were conducted in GeoDa software (http://geodacenter.github.io/, accessed on 8 February 2021). The GWR model was conducted in GWR4 software (https://gwr4.software.informer.com/, accessed on 11 March 2021).

2.2.1. Trend Detection of Vegetation Growth

The linear trend of global vegetation growth during 2001–2018 at each grid cell were estimated using Sen's slope method, also known as the Theil–Sen median method [47]. The method is a robust non-parametric statistical trend calculation method that has high computational efficiency and is insensitive to measurement error and outlier data [48]. Sen' slope was calculated by Equation (1):

$$Slope = \text{Median}\left(\frac{x_j - x_i}{j - i}\right), \forall j > i \quad (1)$$

where the median refers to the mean value of all the slopes, and x_i and x_j represent the EVI values of years i and j.

Then the Mann–Kendall test method was used to test the significance of the Sen' slope [49,50]. The significant confidence level with $p < 0.05$ corresponds to the absolute value of the Z statistic >1.96. The Sen' slope and Mann–Kendall test for EVI data at each grid cell were accomplished in R x64 4.0.2.

2.2.2. Spatial Autocorrelation Analysis

Moran's I statistic is arguably the most commonly used indicator of global spatial autocorrelation. It was initially suggested by Moran [51], and popularized through the classic work on spatial autocorrelation by Cliff and Ord [35]. For an observation at location i, this is expressed as $z_i = x_i - \bar{x}$, where $\bar{x}$ is the mean of variable x. Moran's I statistic is then:

$$I = \frac{\sum_i \sum_j w_{ij} z_i z_j / \sum_i \sum_j w_{ij}}{\sum_i z_i^2 / n} \quad (2)$$

where n is the number of observations, w_{ij} is the elements of the spatial weights matrix, x_i and x_j are the observed value of the location i and its surrounding location j, $\bar{x}$ is the mean of variable x.

At a given significance level, when Moran's I > 0, it indicates a positive correlation between the observed values, and similar attributes cluster together. That is, the high value is adjacent to the high value, and the low value is adjacent to the low value; when I < 0, it indicates a negative correlation between the observed values, and the observations are dispersed; when I = 0, the observed value is randomly distributed.

2.2.3. Spatial Regression at the Global Level

Anselin [34] put forward the general form of spatial regression. When $\rho = 0$, $\beta \neq 0$, $\alpha = 0$, the model is the ordinary least squares (OLS) model; When $\rho \neq 0$, $\beta \neq 0$, $\alpha = 0$, the model is a spatial lag model (SLM), that is, the dependent variable of a location is not only related to the independent variable of the location, but also related to the dependent variable of the neighbourhood. When $\rho = 0$, $\beta \neq 0$, $\alpha \neq 0$, the model is a spatial error model (SEM), that is, the dependent variable of a location is not only related to the independent variable of the location, but also related to other variables not considered at adjacent regions.

$$y = \rho W_1 y + \beta x + \varepsilon \varepsilon = \alpha W_2 \varepsilon + \mu \mu \sim N\left(0, \sigma^2 I_n\right) \quad (3)$$

where y is the dependent variable, x is the independent variable, W_1 is the spatial weight matrix of the dependent variable, ρ is the coefficient of the spatial lag variable $W_1 y$, β is the coefficient of x, ε is the residual, W_2 is the spatial weight matrix of ε, α is the coefficient of ε, μ is the random error of normal distribution, σ is the variance of μ.

2.2.4. Spatial Regression at the Local Level

The GWR model is essentially the locally weighted least squares regression model, which is an extension of the OLS model [52]. In GWR, the weight of the observations is no longer constant during the regression process but is weighted by the degree of adjacency to location i [53]. The model structure is as follows:

$$y_i = \beta_0(\mu_i, v_i) + \sum_{j=1}^{k} \beta_j(\mu_i, v_i) x_{ij} + \varepsilon_i \quad (4)$$

where, (μ_i, v_i) is the co-ordinate of i, $\beta_j(\mu_i, v_i)$ is the jth regression parameter of i.

3. Results

3.1. Temporal Trend of Global Vegetation Growth

EVI_{max} can reflect the potential productivity of terrestrial vegetation, and EVI_{mean} represents the average vegetation growth state in a year, which depicts two aspects of vegetation change. Globally, peak (EVI_{max}) and average (EVI_{mean}) growth in global vegetation consistently showed linear increasing trends during 2001–2018, with the global averaged trend of 0.0030 yr^{-1} ($p < 0.05$) and 0.0022 yr^{-1} ($p < 0.05$). In terms of EVI_{max}, 18.16% of the global vegetated areas showed a statistically significant (Mann–Kendall test, $p < 0.05$) greening during 2001–2018, and 3.08% of the global vegetated areas showed a statistically significant (Mann–Kendall test, $p < 0.05$) browning during 2001–2018 (Figure 3a). By overlaying land cover, we found that the most dramatic greening occurred mainly in those areas with cropland agricultural activities, such as Northern China, India, Central-North America and Southeast Europe. The fastest degradation areas were mainly the grassland of Africa and South America, such as Tanzania, Nigeria and Brazil.

In terms of the EVI_{mean}, 40.73% of the global vegetated areas showed a statistically significant (Mann–Kendall test, $p < 0.05$) greening during 2001–2018, and 2.45% of the global vegetated areas showed a statistically significant (Mann–Kendall test, $p < 0.05$) degradation trend (Figure 3b). The most obvious greening areas mainly occurred in China, India, Canada and Europe, covering a variety of land types including cropland, shrubland, forests and savannas. Vegetation degradation areas were mainly grassland and savannas in areas of South America and southern Africa.

3.2. Inconsistent Global Vegetation Growth in Terms of EVI_{max} and EVI_{mean}

It is worth noting that EVI_{max} and EVI_{mean} experienced consistent changes in 15.97% of the global vegetated areas (14.99% for greening and 0.98% for browning) (Figure 4). However, 32.47% of the global vegetated areas experienced inconsistent changes for EVI_{max} and EVI_{mean}. Specifically, 25.74% of the global vegetated areas that experienced significant greening in EVI_{mean}, with no increase in EVI_{max} simultaneously, occurred mainly in of Europe, Russia, Central Africa, North America and China. On the contrary, 3.17% of the global vegetated areas that experienced significant browning in EVI_{max}, with no increase in EVI_{mean} simultaneously, were mainly distributed in Northern Canada, Eastern Russia, Southern Australia, and were scattered in South America and Africa. There was also 2.10% of the global vegetated area showing a significant browning in EVI_{max}, with no decrease in EVI_{mean} simultaneously, scattered in Africa, South America and Canada, especially in Argentina and Madagascar. While 1.46% of the global vegetated area showed a decreased trend for EVI_{mean}, with no decrease in EVI_{max} simultaneously, such as Central Africa, Central Russia, Eastern Canada and Eastern Brazil. In addition, we found that EVI_{max} and EVI_{mean} had opposite trends in some areas; however, these areas together accounted for only 0.56% of the inconsistent area. EVI_{mean} increased but EVI_{max} decreased in 0.52%

of the inconsistent area, while EVI_{max} increased but EVI_{mean} decreased in 0.04% of the inconsistent area.

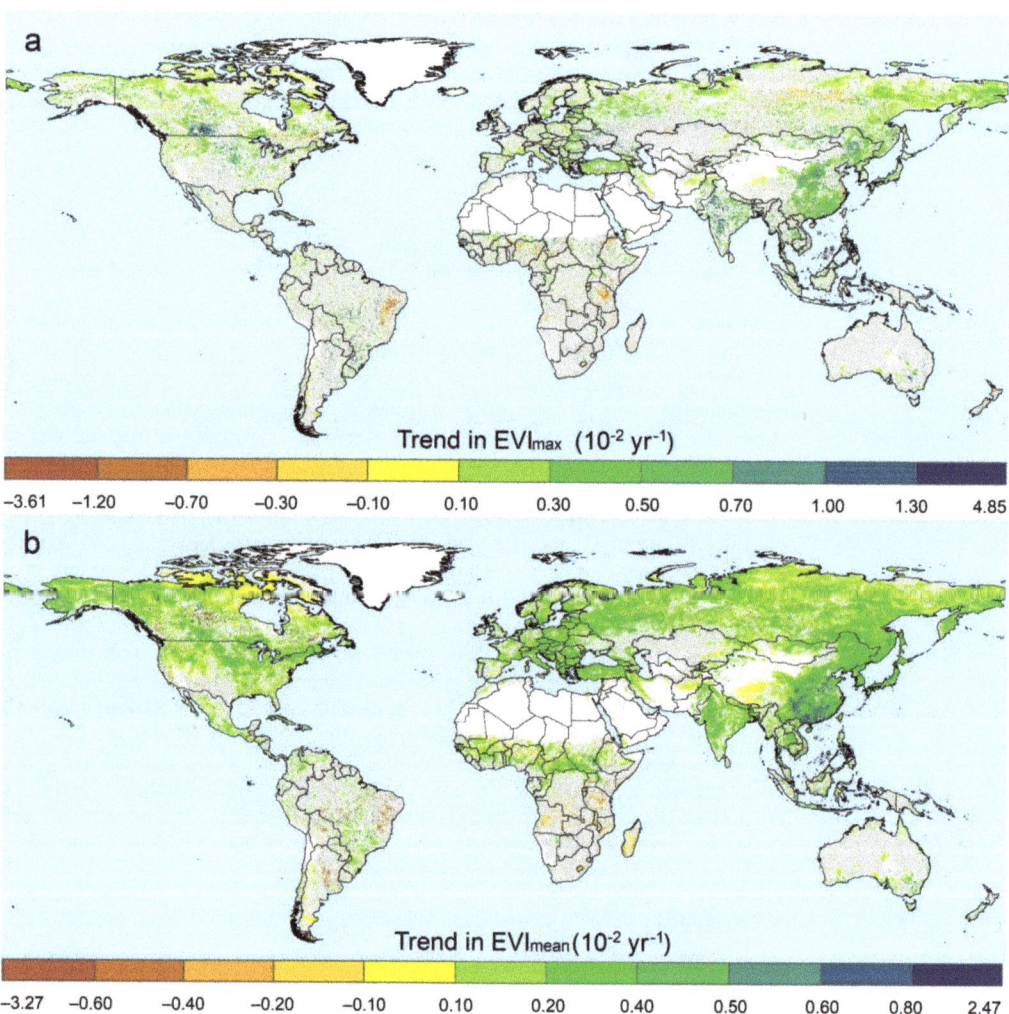

Figure 3. Trends of global vegetation growth during 2001–2018. (**a**) Trend in EVI_{max} at each grid. (**b**) Trend in EVI_{mean} at each grid. White indicates non-vegetated areas including barren, permanent snow and ice, permanent wetlands, and urban and built-up lands. Vegetated areas with statistically insignificant (Mann-Kendall test, $p < 0.05$) are colored grey.

3.3. Relationship between Climate Change and Vegetation Growth

In this study, the Moran's I values for the EVI_{max} and EVI_{mean} changes during 2001–2018 were 0.273 and 0.549 ($p < 0.05$), showing that a significant positive spatial autocorrelation existed in EVI changes in the past 18 years. The multicollinearity condition number was 4.74 (<30) from the OLS model estimation results, indicating no multicollinearity problems in the explanatory variables. The test statistics of LM-lag, LM-error, robust LM-lag and robust LM-error that form the OLS model were all significant ($p < 0.0001$); thus, SLM and SEM were both built to analyse spatial global correlation between vegetation and climate factors. By comparing the index of LIK, AIC and SC, SLM has the largest LIK value and the smallest AIC and SC value, showing that SLM is superior to SEM and OLS in this study (Table 2). R-squared represents the degree to which nine climatic factors explained the global vegetation change in the past 18 years. R^2 in SLM were 0.5370 and 0.2499 for EVI_{mean} and EVI_{max}, respectively, indicating climatic drivers and the spatial lags of EVI could explain more than half and the 24.99% changes of EVI_{mean} and EVI_{max}, respectively.

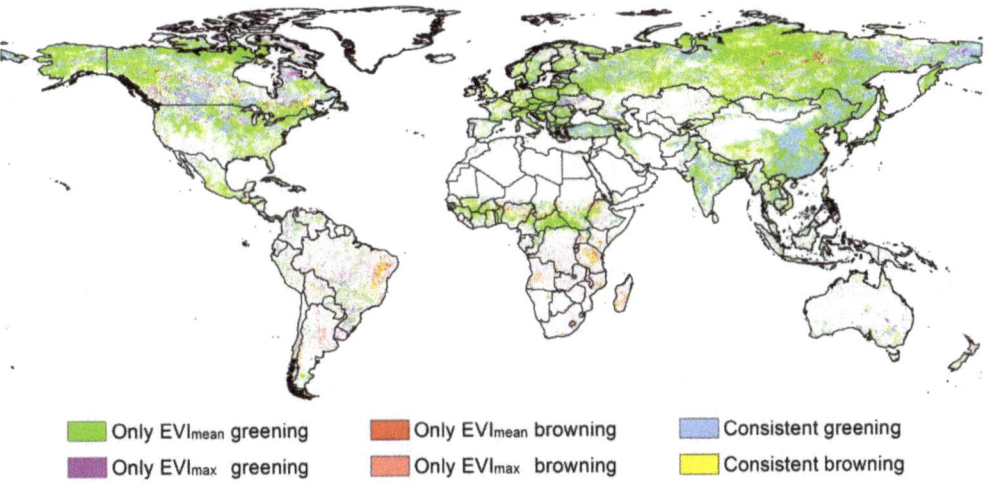

Figure 4. Inconsistency between EVI_{max} and EVI_{mean} variations during 2001–2018.

Table 2. Comparison of spatial regression models.

Dependent Variables	Model	R^2	LIK	AIC	SC
EVI_{max}	OLS	0.0229	96,424.5	−192,829	−192,739
	SLM	0.2499	103,235.0	−206,448	−206,349
	SEM	0.2498	103,225.8	−206,432	−206,342
EVI_{mean}	OLS	0.0817	84,046.7	−168,073	−167,983
	SLM	0.5370	102,864.0	−205,706	−205,606
	SEM	0.5372	102,851.1	−205,682	−205,592

Notes: R-squared (R^2), maximum likelihood logarithm (LIK), Akaike information criterion (AIC), Schwarz criterion (SC), Ordinary least square (OLS), Spatial lag model (SLM), Spatial error model (SEM).

The regression results of the best fit model (SLM) in this study for EVI_{max} and EVI_{mean} were shown in Table 3. The spatial lag of EVI_{max} in SLM passed the statistical significance test ($p < 0.05$) during 2001–2018, proving that changes in EVI were correlated not only to these climatic factors but also to EVI variation in its adjacent areas. The coefficient indicates that the vegetation in a certain location might change by 0.7472 units for every 1-unit change of vegetation in its adjacent areas. Precipitation and mean temperature had a statistically significant ($p < 0.05$) positive correlation with EVI_{max} during 2001–2018, while potential evapotranspiration and vapour pressure had a statistically significant negative correlation with EVI_{max}. Compared to EVI_{max}, EVI_{mean} had a higher R-squared than EVI_{max}, 0.5370 vs 0.2499 (Table 2), indicating a higher climatic driving explanation of EVI_{mean}. Except for precipitation and mean temperature, minimum temperature also had a statistically significant ($p < 0.05$) positive correlation with EVI_{mean} change. The coefficient of spatial lag for EVI_{mean} was 0.8721, suggesting EVI_{mean} had a higher correlation with its adjacent EVI_{mean} and, therefore, a stronger spatial dependence than EVI_{max}.

Table 3. Regression results of spatial lag model (SLM).

Dependent Variables	Independent Variables	Coefficient	Standard Error	Z Statistic	Probability
EVI_{max}	Lag term	0.7472	0.0059	126.9690	0.00000
	Constant	-	-	-	>0.05
	PRE	0.0105	0.0038	2.7355	0.00623
	TMX	-	-	-	>0.05
	TMP	0.0060	0.0029	2.0398	0.04137
	TMN	-	-	-	>0.05
	PET	−0.0142	0.0027	−5.2688	0.00000
	WET	-	-	-	>0.05
	VAP	−0.0110	0.0027	−4.1029	0.00004
	FRS	-	-	-	>0.05
	DTR	-	-	-	>0.05
	$EVI_{max} = 0.0105 PRE + 0.0060 TMP - 0.0142 PET - 0.0110 VAP + 0.7472 \text{Lag term} + \varepsilon$				
EVI_{mean}	Lag term	0.8721	0.0038	227.8330	0.00000
	Constant	-	-	-	>0.05
	PRE	0.0165	0.0038	4.3219	0.00002
	TMX	-	-	-	>0.05
	TMP	0.0111	0.0029	3.7669	0.00017
	TMN	0.0079	0.0015	5.0860	0.00000
	PET	−0.0137	0.0027	−5.0794	0.00000
	WET	-	-	-	>0.05
	VAP	−0.0138	0.0027	−5.1319	0.00000
	FRS	-	-	-	>0.05
	DTR	-	-	-	>0.05
	$EVI_{mean} = 0.0165 PRE + 0.0111 TMP + 0.0079 TMN - 0.0137 PET - 0.0138 VAP + 0.8721 \text{Lag term} + \varepsilon$				

Notes: Lag term here is the spatial lag variable of the dependent variable obtained by the spatial weight matrix in 2.2.3; PRE = precipitation, TMX = maximum temperature, TMP = mean temperature, TMN = minimum temperature, PET = potential evapotranspiration, WET = wet day frequency, VAP = vapour pressure, FRS = frost day frequency, DTR = diurnal temperature range.

3.4. Spatial Heterogeneity of the Climatic Driving

After examining the relationship at the global level by SLM, we then focused on the representative areas with significant greening or browning identified in the previous trend analysis. We further explored how climate change affects vegetation in these areas with the GWR model results. Five highly representative regions were selected for further analysis, located in China, India, North America, Brazil and Africa. The areas in China, India and North America were characterized by vegetation greening, while the areas in Africa and Brazil were characterized by vegetation browning. Local coefficients of climatic drivers for EVI_{max} and EVI_{mean} were mapped in Figures 5 and 6, respectively.

As shown in Figure 5, in Northern China, EVI_{max} was positively affected by precipitation, potential evapotranspiration, minimum temperature and humid days, and was limited by maximum temperature and vapour pressure. However, in Southern China, maximum temperature and vapour pressure had positive effects, as well as potential evapotranspiration and humid days. The reason was that there are semi-arid areas in the north where precipitation is low and potential evapotranspiration is far greater than precipitation and the high temperature would lead to insufficient rainwater irrigation for crops and grass. It indicated that, due to the difference in climatic conditions (an arid or humid area) and vegetation types (cropland or grassland), the responses of EVI_{max} to climate change in South and North China are inconsistent. In India, except minimum temperature, all other factors had positive impacts on the EVI_{max}. In North America, EVI_{max} was positively influenced by potential evapotranspiration and humid day but negative by precipitation and vapour pressure. In Brazil, the degradation of the EVI_{max} might be due to the decrease in maximum temperature, potential evapotranspiration and humid days and the increase in mean temperature and diurnal temperature range. In Africa, the degradation of

EVI$_{max}$ might be due to the joint decrease in mean temperature, humid days and diurnal temperature range and the increase in evapotranspiration and maximum temperature.

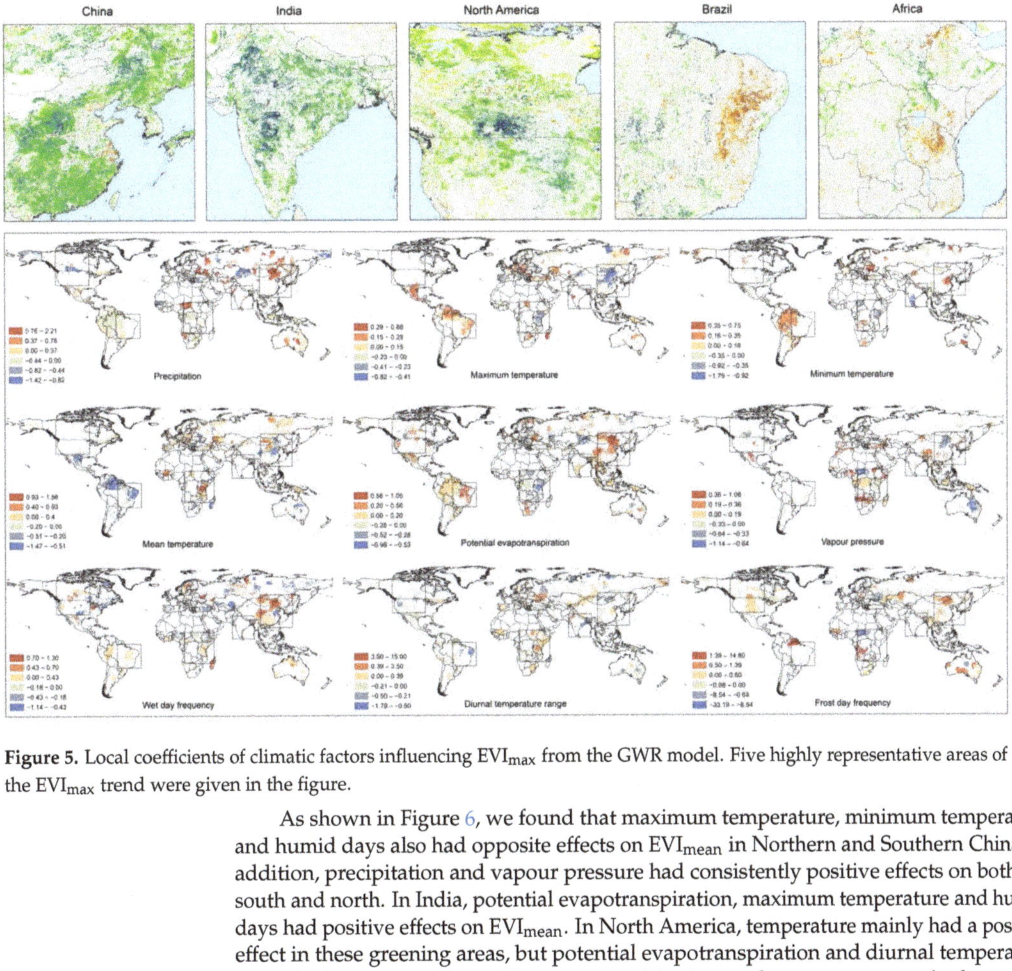

Figure 5. Local coefficients of climatic factors influencing EVI$_{max}$ from the GWR model. Five highly representative areas of the EVI$_{max}$ trend were given in the figure.

As shown in Figure 6, we found that maximum temperature, minimum temperature and humid days also had opposite effects on EVI$_{mean}$ in Northern and Southern China. In addition, precipitation and vapour pressure had consistently positive effects on both the south and north. In India, potential evapotranspiration, maximum temperature and humid days had positive effects on EVI$_{mean}$. In North America, temperature mainly had a positive effect in these greening areas, but potential evapotranspiration and diurnal temperature range had negative effects. However, precipitation and vapour pressure had opposite impacts on Canada and the United States, with negative and positive, respectively. In Brazil, all browning areas of EVI$_{mean}$ occurred in grassland, which might be affected by the decrease in precipitation, mean temperature, potential evapotranspiration and wet days. In Africa, degradation of vegetation also occurred in grassland and was caused by the decrease in precipitation, mean temperature, vapour pressure and wet days.

We defined the climate factor having the largest absolute value of the local coefficient with a statistical significance ($p < 0.05$) as the dominant climatic factor influencing vegetation change (Figure 7a,b). In the meantime, the local coefficients of the dominant climatic factors of EVI$_{max}$ and EVI$_{mean}$ were mapped in Figure 7c,d. We found that the EVI$_{max}$ change was strongly influenced by precipitation in 14.02% of the global vegetated areas during 2001–2018, followed by vapour pressure (12.56%), minimum temperature (9.87%), mean temperature (9.10%) and potential evapotranspiration (6.96%) (Figure 7a). 21.06% of the global vegetated areas where peak vegetation growth had no significant correlation ($p > 0.05$) with any climatic factor. In terms of the EVI$_{mean}$, the global vegetated areas were strongly affected by mean temperature (17.36%) and precipitation (16.97%) (Figure 7b). There were 11.86% of the global vegetated areas where EVI_{mean} change had no significant correlation ($p > 0.05$) with any climatic factor.

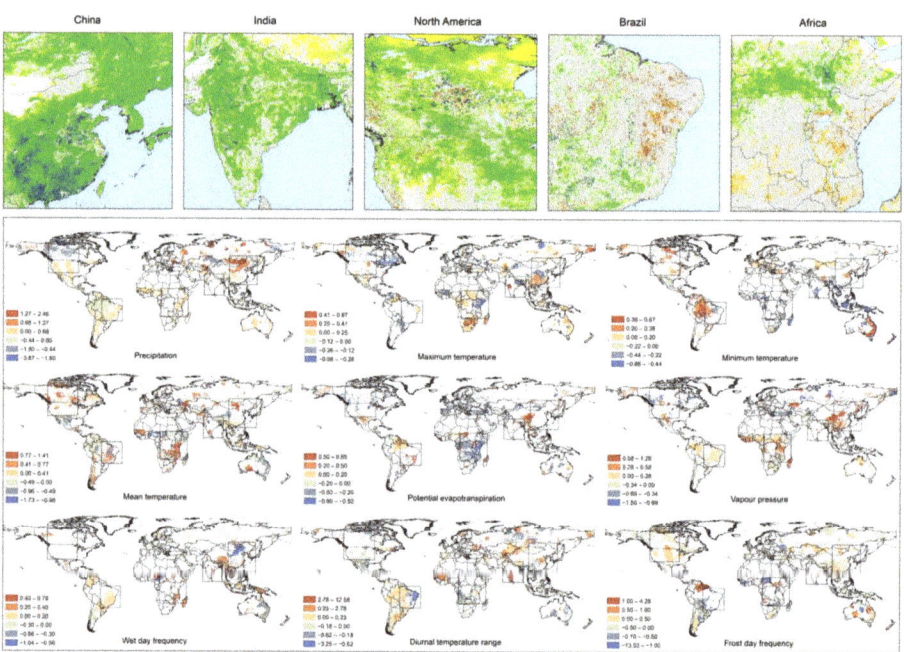

Figure 6. Local coefficients of all climatic factors influencing EVI_{mean} from the GWR model. Five highly representative areas of the EVI_{mean} trend were given in the figure.

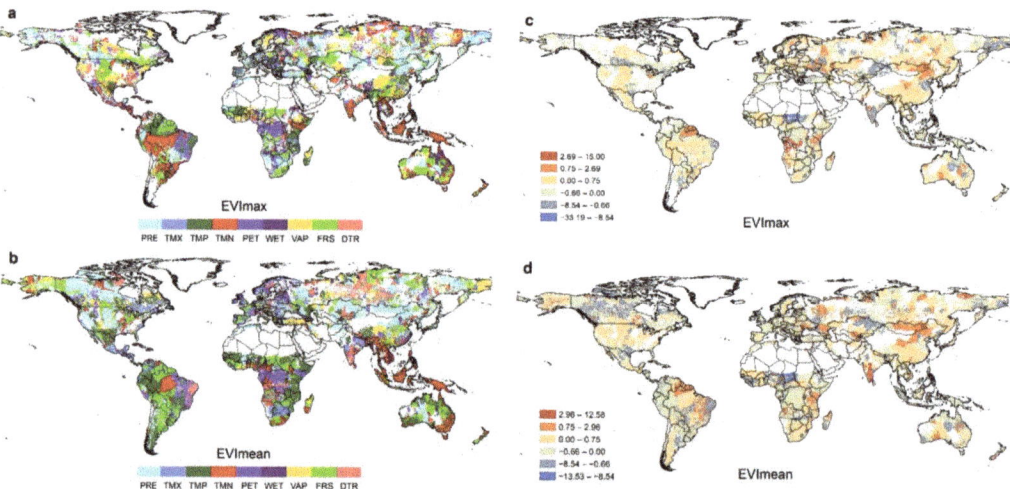

Figure 7. Dominant climatic drivers influencing (**a**) EVI_{max} and (**b**) EVI_{mean}; local coefficients of the dominant climatic factors for (**c**) EVI_{max} and (**d**) EVI_{mean} at each grid.

4. Discussion

4.1. Comparison of Global Vegetation Trend Results and Uncertainties

The trends of EVI_{mean} and EVI_{max} during 2001–2018 were detected in this study and were compared with previous studies (Table 4). Zhang et al. [24] found that the global vegetation has an increasing trend of 0.0028 yr^{-1} and 0.0023 yr^{-1} at a significant level ($p < 0.0001$) in EVI_{mean} and EVI_{max} during 2001–2015, respectively. The results were consistent with our results of 0.0022 yr^{-1} and 0.0030 yr^{-1} in EVI_{mean} and EVI_{max}, respectively. Moreover, the results from NDVI also showed the greening trend, with 0.0022 yr^{-1} and 0.0015 yr^{-1} in $NDVI_{mean}$ and $NDVI_{max}$ during 2001–2015, respectively [24], 0.0013 yr^{-1} in $NDVI_{max}$ during 1982–2011 [20], and 0.0012 yr^{-1} in the growing season NDVI during 1982–2013 [54]. Overall, the global greening trends were found in both our and previous studies.

However, there were some differences in global vegetation greening or browning areas due to different satellite products, vegetation greenness indicators and time range. For example, Ding [57] found that 18.9% of the global vegetated area had a greening trend for EVI_{mean} during 2000–2015, while ~3% of browning (vs. 40.73% and 2.45% in this study); Chen [56] found that 34.1% of the global vegetated area showed a greening and 4.85% of browning for LAI from MODIS during 2000–2017, while 22.42% of greening and 13.54% of browning for LAI from AVHRR during 2000–2016. Furthermore, although the same satellite-derived data (MODIS Terra-C6 EVI) were used, there were still differences in area ratios of greening and browning between our study and the relevant studies, which might be due to the following reasons: (1) different vegetation monitoring time range; (2) different spatial resolution of EVI data; (3) different land cover data and starting reference year; (4) different trend analysis methods.

Table 4. Comparison of global vegetation change trend results.

Time Range	Index	Datasets	Spatial Resolution	Averaged Trend	Greening Area Ratio	Browning Area Ratio	References
1982–2011	$NDVI_{max}$	$GIMMS_{3g}$	1/12°	0.0013 yr^{-1} **	-	-	[20]
1982–2013	$NDVI_{gs}$	$GIMMS_{3g}$	1/12°	0.0012 yr^{-1} ***	48% *	8% *	[54]
1982–2014	LAI_{gs}	$GIMMS_{3g}$	1/12°	0.032 $m^2 m^{-2} yr^{-1}$ **	35% **	4% **	[21]
1982–2015	NDVI	$GIMMS_{3g}$	1/12°	-	50% **	8% **	[55]
	LAI	$GIMMS_{3g}$	1/12°	-	23% **	15% **	
1982–2016	LAI	AVHRR	1/12°	-	40.91% *	10.59% *	
2000–2017	LAI	MODIS C6	500 m	-	34.1% *	4.85% *	[56]
2000–2016	LAI	AVHRR	1/12°	-	22.42% *	13.54% *	
2001–2015	NDVI	MODIS Terra-C6	0.05°	0.0022 yr^{-1} ****	23.1% **	10.5% **	[24]
	$NDVI_{max}$	MODIS Terra-C6	0.05°	0.0015 yr^{-1} ****	-	-	
	EVI	MODIS Terra-C6	0.05°	0.0028 yr^{-1} ****	22.8% **	3.3% **	
	EVI_{max}	MODIS Terra-C6	0.05°	0.0023 yr^{-1} ****	-	-	
2001–2013	NDVI	MODIS Aqua-C6	0.05°	-	12.1% **	-	
	EVI	MODIS Aqua-C6	0.05°	-	14.3% **	-	
	NDVI	$GIMMS_{3g}$	1/12°	-	13.8% **	-	
2000–2015	NDVI	MODIS Terra-C6	0.05°	-	~16% **	~5% **	[57]
	EVI	MODIS Terra-C6	0.05°	-	18.9% **	~3% **	
	LAI	MODIS Terra-C6	0.05°	-	~17% **	~3% **	
2000–2015	NDVI	MODIS Terra-C6	0.05°	0.0023 yr^{-1} ****	28.6% **	5.4% **	[58]
2001–2018	EVI	MODIS Terra-C6	1 km	0.0022 yr^{-1} **	40.73% **	2.45% **	This study
	EVI_{max}	MODIS Terra-C6	1 km	0.0030 yr^{-1} **	18.16% **	3.08% **	

Notes: $NDVI_{max}$ and EVI_{max} refer to the annual maximum NDVI and EVI, respectively, and unsubscripted ones represent the annual mean values; $NDVI_{gs}$ and LAI_{gs} refer to NDVI and LAI for growing season, respectively; **** $p < 0.0001$, *** $p < 0.01$, ** $p < 0.05$, * $p < 0.1$.

4.2. Potential Causes Inducing Inconsistencies in Vegetation Change

Our study found that only 15.97% of the global vegetated area experienced a consistent change (enhancement or degradation) in peak and mean growth of vegetation, and 32.47% of the global vegetated area experienced inconsistent changes in peak and mean vegetation growth. The first potential cause could be changes in plant phenology or growing season induced by climate change in these areas. In some areas, the annual EVI_{mean} showed an increasing trend while the annual EVI_{max} remained unchanged and even decreased, indicating that with time going on, the monthly EVI, which are larger than the original annual EVI_{mean}, increased during 2001–2018, and in this case the growth circle of vegetation might be prolonged. In contrast, for the areas with increased EVI_{max} but no significant change or decrease in EVI_{mean}, the growth circle of vegetation might be shortened. This speculation can be supported by the consistent evidence for plant phenological changes from in situ and satellite observations [59]. The earlier beginning of the growing season and autumn postponement has been widely reported in Europe, North America and China since the 1980s [60–63]. Moreover, with the stagnation of warming, spring green-up advancement's trend might have slowed down or even reversed since the 2000s [64,65]. However, this speculation still is of great uncertainty due to the unclear vegetation phenological information and its variations in recent 20 years, which needs to be investigated in detail in further studies.

Plant phenology changes are determined by various biological and environmental factors such as nutrient and water availability, temperature and photoperiod. Temperature is generally regarded as one of the most critical controls of plant phenology through multiple processes, such as inducing the plant endodormancy by cold temperature [66] and breaking the ecodormancy by the accumulated warming [59]. For vegetation in pasture regions, precipitation variation had a significant limiting effect on its productivity [67], this might result in the inconsistency of peak and mean greenness due to variations of annual maximum and minimum rainfall. What is more, the impact of climatic conditions on vegetation growth is more complex due to the interactions with other environmental and climatic factors [68].

In addition, anthropogenic activities, including land cover change and land use management, also could lead to the differences between annual peak and mean vegetation variations. On the one hand, the conversion between different vegetation types might directly result in the observed inconsistency between the peak and mean greenness, due to the different responses to environmental changes determined by different vegetation properties, such as thermal adaptability and photosynthetic efficiency [69]. For example, crops have higher photosynthetic efficiency than other non-crops [20]. On the other hand, land-use intensity and management could also explain the inconsistency largely. For example, anthropic seasonal irrigation and fertilization could improve the peak greenness and productivity of croplands [20,56,70], but the annual mean greenness might not enhance simultaneously.

4.3. Spatial Heterogeneity of Vegetation Growth Driven by Climate Change

Consistent with previous findings, we found that rainfall and temperature significantly impact vegetation both in peak and mean growth [20,21]. However, except for precipitation and temperature, vegetation growth was also found to be significantly correlated with other climatic factors such as potential evapotranspiration and vapour pressure in our study (Table 2). Furthermore, previous studies had neglected the spatial heterogeneity of this response. Using the GWR model, we revealed the spatial pattern of each climatic factor influencing vegetation variation. We found that not all regions had the strongest correlation between vegetation change and precipitation and temperature, which were generally considered to be the main climatic factors. For example, potential evapotranspiration contributed dominantly to the trend of EVI_{max} and EVI_{mean} over 6.96% and 7.54% of the global vegetated area, respectively. Significant positive effects of potential evapotranspiration occurred mainly in Brazil, Northern Europe and Northeast China, while adverse effects were mainly found in Africa (Figure 7). Vapour pressure had dominant contribution to the trend of EVI_{max} and EVI_{mean} in 12.56% and 8.95% of the global vegetation. The greening induced by climate change in the Tibetan Plateau was mainly attributed to increasing vapour pressure and temperature in this study (Figure 7), whereas that was rising in a previous study [21].

4.4. Limitations and Further Directions

Data continuity and accuracy of the VIs products are critical to accurately detect subtle changes in vegetation, which is important to the assessment of vegetation dynamics [12]. VIs are a kind of spectral vegetation index derived from satellite remote sensing, which is calculated by spectral band reflectance [14]. Thus, unlike LAI, VIs cannot be directly validated and calibrated continuously in time series by in situ measurements. A number of comparisons have been made between different satellite-based VIs datasets [12,14,16,71,72] and MODIS VIs data with improved technology, and no shifts of sensor is considered to be superior to other products in terms of data temporal consistency and has been widely used for reference data [10]. The latest MODIS Collection 6 (C6) VIs data providing several algorithmic improvements and calibration adjustments were released for correcting the negative influence of sensor degradation found in MODIS Collection 5 (C5) VIs data [17]. After release of MODIS C6 products, some studies began to evaluate differences between C5 and C6 VIs both on a local and global scale to verify that C6 products had eliminated effects of sensor degradation, and highlighted the need of re-analysing some previous results based on MODIS C5 VIs products [17,24,73]. Therefore, we selected MODIS C6 data to examine global vegetation growth trend in this study. However, the differences between MODIS EVI and NDVI were not sufficiently considered to reduce the uncertainty in detecting trend analysis in this study. Although EVI can improve reflectance sensitivity in dense vegetation areas, NDVI and EVI are generally regarded as two complementary datasets for providing more effective assessment of global vegetation dynamics [24]. Thus, it is necessary to evaluate the differences between the two datasets for monitoring vegetation dynamics using both EVI and NDVI data in the future. Moreover, research on the climatic driving mechanism of vegetation dynamic should be combined with VIs and other vegetation indicators, such as LAI and net primary productivity, which can be simulated in the process-based models, because either non-spatial or spatial regression analysis cannot explain the climate-driving mechanism from the ecological processes of vegetation growth but can only provide a hint of correlation.

5. Conclusions

This study detected the trend of global peak and average vegetation growth during 2001–2018 and mapped the inconsistency in vegetation growth, and the climatic factors that affected the inconsistency of vegetation growth were explored. The results showed that in terms of EVI_{max}, 18.16% of the global vegetated areas are greening and 3.08% are browning, and in terms of EVI_{mean}, there are 40.73% and 2.45%, respectively. The most dramatic greening of EVI_{max} occurred mainly in those areas with cropland agricultural activities, and the fastest degradation areas of EVI_{max} were mainly grassland and savannas of Africa and South America. Through mapping the consistency of global vegetation growth, it was found that from 2001 to 2018, 32.47% of the global vegetated area experienced inconsistent trends in EVI_{max} and EVI_{mean}.

The SLM was proved to be more suitable than SEM and OLS in this study for spatial regression at the global level. We found that precipitation and mean temperature had a statistically significant ($p < 0.05$) positive correlation with EVI_{max} and EVI_{mean} during 2001–2018, while potential evapotranspiration and vapour pressure had a statistically significant negative correlation. The results of SLM indicated that there was spatial autocorrelation in both EVI_{max} and EVI_{mean} change, which means the changes in EVI were correlated not only to these climatic factors but also to EVI variation in its adjacent areas.

The SLM results only indicated a correlation between EVI changes and climatic drivers on the whole but failed to reveal the spatial heterogeneity of climatic drivers.

The GWR model was used to explore the spatial heterogeneity of climatic drivers influencing vegetation change by obtaining regression results for each spatial unit. The results showed that the EVI_{max} change was strongly influenced by precipitation in 14.02% of the global vegetated areas during 2001–2018, followed by vapour pressure (12.56%), minimum temperature (9.87%), mean temperature (9.10%) and potential evapotranspiration (6.96%). In terms of the EVI_{mean}, the global vegetated areas were strongly affected by mean temperature (17.36%) and precipitation (16.97%). In China, maximum temperature and vapour pressure had opposite effects on EVI_{max} in the north and south, and maximum temperature and humid days also had opposite effects on EVI_{mean}. Due to the difference in climatic conditions (arid or humid area) and vegetation types (cropland or grassland), the responses of EVI to climate change were inconsistent. Climate change could affect vegetation characteristics by changing plant phenology, consequently rendering the inconsistency between peak and mean greening. In addition, anthropogenic activities, including land cover change and land use management, also could lead to the differences between annual peak and mean vegetation variations.

Author Contributions: Conceptualization, D.Z. and X.W.; data curation, W.C.; formal analysis, D.Z.; funding acquisition, X.W.; investigation, D.Z., X.G., W.C., L.F. and X.W.; methodology, D.Z. and W.C.; project administration, X.W.; resources, R.Y.; software, X.G.; supervision, X.W.; validation, D.Z., X.G. and X.W.; visualization, L.F.; writing—original draft, D.Z.; writing—review and editing, X.G., W.C., R.Y. and X.Z. All authors have read and agreed to the published version of the manuscript.

Funding: This research was funded by the National Key Research and Development Program of China, grant number 2016YFC0502700.

Institutional Review Board Statement: Not applicable.

Informed Consent Statement: Not applicable.

Data Availability Statement: The data that support the findings of this study are available from the corresponding author upon reasonable request.

Acknowledgments: The authors are grateful to the Level-1 and Atmosphere Archive and Distribution System Distributed Active Archive Center (LAADS DAAC) (https://ladsweb.modaps.eosdis.nasa.gov/search) and the Climatic Research Unit Time-Series version 4.03 (CRU TS4.03) datasets (https://data.ceda.ac.uk/badc/cru/data/cru_ts) for the data access.

Conflicts of Interest: The authors declare no conflict of interest.

References

1. Law, B.E.; Falge, E.; Gu, L.; Baldocchi, D.D.; Bakwin, P.; Berbigier, P.; Davis, K.; Dolman, A.J.; Falk, M.; Fuentes, J.D.; et al. Environmental controls over carbon dioxide and water vapor exchange of terrestrial vegetation. *Agric. For. Meteorol.* **2002**, *113*, 97–120. [CrossRef]
2. Nemani, R.R.; Keeling, C.D.; Hashimoto, H.; Jolly, W.M.; Piper, S.C.; Tucker, C.J.; Myneni, R.B.; Running, S.W. Climate-driven increases in global terrestrial net primary production from 1982 to 1999. *Science* **2003**, *300*, 1560–1563. [CrossRef]
3. Friedlingstein, P.; Cox, P.; Betts, R.; Bopp, L.; Von Bloh, W.; Brovkin, V.; Cadule, P.; Doney, S.; Eby, M.; Fung, I.; et al. Climate-carbon cycle feedback analysis: Results from the (CMIP)-M-4 model intercomparison. *J. Clim.* **2006**, *19*, 3337–3353. [CrossRef]
4. Lee, X.; Goulden, M.L.; Hollinger, D.Y.; Barr, A.; Black, T.A.; Bohrer, G.; Bracho, R.; Drake, B.; Goldstein, A.; Gu, L.; et al. Observed increase in local cooling effect of deforestation at higher latitudes. *Nature* **2011**, *479*, 384–387. [CrossRef] [PubMed]
5. Piao, S.; Wang, X.; Park, T.; Chen, C.; Lian, X.; He, Y.; Myneni, R.B. Characteristics, drivers and feedbacks of global greening. *Nat. Rev. Earth Environ.* **2019**, *1*, 14–27. [CrossRef]
6. Li, Y.; Zhao, M.; Motesharrei, S.; Mu, Q.; Kalnay, E.; Li, S. Local cooling and warming effects of forests based on satellite observations. *Nat. Commun.* **2015**, *6*, 6603. [CrossRef] [PubMed]
7. Forzieri, G.; Alkama, R.; Miralles, D.G.; Cescatti, A. Satellites reveal contrasting responses of regional climate to the widespread greening of Earth. *Science* **2017**, *356*, 1180–1184. [CrossRef]
8. Tucker, C.J.; Pinzon, J.E.; Brown, M.E.; Slayback, D.A.; Pak, E.W.; Mahoney, R.; Vermote, E.F.; El Saleous, N. An extended AVHRR 8-km NDVI dataset compatible with MODIS and SPOT vegetation NDVI data. *Int. J. Remote Sens.* **2005**, *26*, 4485–4498. [CrossRef]

9. Maisongrande, P.; Duchemin, B.; Dedieu, G. VEGETATION/SPOT: An operational mission for the Earth monitoring; presentation of new standard products. *Int. J. Remote Sens.* **2004**, *25*, 9–14. [CrossRef]
10. Huete, A.; Didan, K.; Miura, T.; Rodriguez, E.P.; Gao, X.; Ferreira, L.G. Overview of the radiometric and biophysical performance of the MODIS vegetation indices. *Remote Sens. Environ.* **2002**, *83*, 195–213. [CrossRef]
11. Fensholt, R.; Rasmussen, K.; Nielsen, T.T.; Mbow, C. Evaluation of earth observation based long term vegetation trends—Intercomparing NDVI time series trend analysis consistency of Sahel from AVHRR GIMMS, Terra MODIS and SPOT VGT data. *Remote Sens. Environ.* **2009**, *113*, 1886–1898. [CrossRef]
12. Tian, F.; Fensholt, R.; Verbesselt, J.; Grogan, K.; Horion, S.; Wang, Y. Evaluating temporal consistency of long-term global NDVI datasets for trend analysis. *Remote Sens. Environ.* **2015**, *163*, 326–340. [CrossRef]
13. Beck, H.E.; McVicar, T.R.; van Dijk, A.I.J.M.; Schellekens, J.; de Jeu, R.A.M.; Bruijnzeel, L.A. Global evaluation of four AVHRR–NDVI data sets: Intercomparison and assessment against Landsat imagery. *Remote Sens. Environ.* **2011**, *115*, 2547–2563. [CrossRef]
14. Tarnavsky, E.; Garrigues, S.; Brown, M.E. Multiscale geostatistical analysis of AVHRR, SPOT-VGT, and MODIS global NDVI products. *Remote Sens. Environ.* **2008**, *112*, 535–549. [CrossRef]
15. Wang, D.; Morton, D.; Masek, J.; Wu, A.; Nagol, J.; Xiong, X.; Levy, R.; Vermote, E.; Wolfe, R. Impact of sensor degradation on the MODIS NDVI time series. *Remote Sens. Environ.* **2012**, *119*, 55–61. [CrossRef]
16. Kern, A.; Marjanović, H.; Barcza, Z. Evaluation of the Quality of NDVI3g Dataset against Collection 6 MODIS NDVI in Central Europe between 2000 and 2013. *Remote Sens.* **2016**, *8*, 955. [CrossRef]
17. Lyapustin, A.; Wang, Y.; Xiong, X.; Meister, G.; Platnick, S.; Levy, R.; Franz, B.; Korkin, S.; Hilker, T.; Tucker, J.; et al. Scientific impact of MODIS C5 calibration degradation and C6+ improvements. *Atmos. Meas. Tech.* **2014**, *7*, 4353–4365. [CrossRef]
18. Zhu, Z.; Bi, J.; Pan, Y.; Ganguly, S.; Anav, A.; Xu, L.; Samanta, A.; Piao, S.; Nemani, R.R.; Myneni, R.B. Global Data Sets of Vegetation Leaf Area Index (LAI)3g and Fraction of Photosynthetically Active Radiation (FPAR)3g Derived from Global Inventory Modeling and Mapping Studies (GIMMS) Normalized Difference Vegetation Index (NDVI3g) for the Period 1981 to 2011. *Remote Sens.* **2013**, *5*, 927–948.
19. Pan, Y.; Birdsey, R.A.; Fang, J.; Houghton, R.; Kauppi, P.E.; Kurz, W.A.; Phillips, O.L.; Shvidenko, A.; Lewis, S.L.; Canadell, J.G.; et al. A Large and Persistent Carbon Sink in the World's Forests. *Science* **2011**, *333*, 988–993. [CrossRef]
20. Huang, K.; Xia, J.; Wang, Y.; Ahlstrom, A.; Chen, J.; Cook, R.B.; Cui, E.; Fang, Y.; Fisher, J.B.; Huntzinger, D.N.; et al. Enhanced peak growth of global vegetation and its key mechanisms. *Nat. Ecol. Evol.* **2018**, *2*, 1897–1905. [CrossRef] [PubMed]
21. Zhu, Z.; Piao, S.; Myneni, R.B.; Huang, M.; Zeng, Z.; Canadell, J.G.; Ciais, P.; Sitch, S.; Friedlingstein, P.; Arneth, A.; et al. Greening of the Earth and its drivers. *Nat. Clim. Chang.* **2016**, *6*, 791–795. [CrossRef]
22. Pan, N.; Feng, X.; Fu, B.; Wang, S.; Ji, F.; Pan, S. Increasing global vegetation browning hidden in overall vegetation greening: Insights from time-varying trends. *Remote Sens. Environ.* **2018**, *214*, 59–72. [CrossRef]
23. Gao, M.D.; Piao, S.L.; Chen, A.P.; Yang, H.; Liu, Q.; Fu, Y.S.H.; Janssens, I.A. Divergent changes in the elevational gradient of vegetation activities over the last 30 years. *Nat. Commun.* **2019**, *10*, 2970. [CrossRef]
24. Zhang, Y.; Song, C.; Band, L.E.; Sun, G.; Li, J. Reanalysis of global terrestrial vegetation trends from MODIS products: Browning or greening? *Remote Sens. Environ.* **2017**, *191*, 145–155. [CrossRef]
25. Le Quere, C.; Moriarty, R.; Andrew, R.M.; Canadell, J.G.; Sitch, S.; Korsbakken, J.I.; Friedlingstein, P.; Peters, G.P.; Andres, R.J.; Boden, T.A.; et al. Global Carbon Budget 2015. *Earth Syst. Sci. Data* **2015**, *7*, 349–396. [CrossRef]
26. Myers-Smith, I.H.; Kerby, J.T.; Phoenix, G.K.; Bjerke, J.W.; Epstein, H.E.; Assmann, J.J.; John, C.; Andreu-Hayles, L.; Angers-Blondin, S.; Beck, P.S.A.; et al. Complexity revealed in the greening of the Arctic. *Nat. Clim. Chang.* **2020**, *10*, 106–117. [CrossRef]
27. Keenan, T.F.; Riley, W.J. Greening of the land surface in the world's cold regions consistent with recent warming. *Nat. Clim. Chang.* **2018**, *8*, 825–828. [CrossRef] [PubMed]
28. Corlett, R.T. Impacts of warming on tropical lowland rainforests. *Trends Ecol. Evol.* **2011**, *26*, 606–613. [CrossRef]
29. Huang, M.; Piao, S.; Ciais, P.; Penuelas, J.; Wang, X.; Keenan, T.F.; Peng, S.; Berry, J.A.; Wang, K.; Mao, J.; et al. Air temperature optima of vegetation productivity across global biomes. *Nat. Ecol. Evol.* **2019**, *3*, 772–779. [CrossRef] [PubMed]
30. Shen, M.; Piao, S.; Cong, N.; Zhang, G.; Jassens, I.A. Precipitation impacts on vegetation spring phenology on the Tibetan Plateau. *Glob. Chang. Biol.* **2015**, *21*, 3647–3656. [CrossRef]
31. Yuan, W.; Zheng, Y.; Piao, S.; Ciais, P.; Lombardozzi, D.; Wang, Y.; Ryu, Y.; Chen, G.; Dong, W.; Hu, Z.; et al. Increased atmospheric vapor pressure deficit reduces global vegetation growth. *Sci. Adv.* **2019**, *5*, eaax1396. [CrossRef]
32. Hua, W.; Chen, H.; Zhou, L.; Xie, Z.; Qin, M.; Li, X.; Ma, H.; Huang, Q.; Sun, S. Observational Quantification of Climatic and Human Influences on Vegetation Greening in China. *Remote Sens.* **2017**, *9*, 425. [CrossRef]
33. Wei, H.; Zhao, X.; Liang, S.L.; Zhou, T.; Wu, D.H.; Tang, B.J. Effects of Warming Hiatuses on Vegetation Growth in the Northern Hemisphere. *Remote Sens.* **2018**, *10*, 683. [CrossRef]
34. Anselin, L. *Spatial Econometrics: Methods and Models*; Kluwer Academic Publishers: Dordrecht, The Netherlands, 1988.
35. Cliff, A.D.; Ord, J.K. *Spatial Processes: Models and Applications*; Pion: London, UK, 1981.
36. Kissling, W.D.; Carl, G. Spatial autocorrelation and the selection of simultaneous autoregressive models. *Glob. Ecol. Biogeogr.* **2007**, *17*, 59–71. [CrossRef]
37. Tobler, W. On the First Law of Geography: A Reply. *Ann. Assoc. Am. Geogr.* **2004**, *94*, 304–310. [CrossRef]
38. Anselin, L.; Li, X. Tobler's Law in a Multivariate World. *Geogr. Anal.* **2020**, *52*, 494–510. [CrossRef]

39. Beale, C.M.; Lennon, J.J.; Yearsley, J.M.; Brewer, M.J.; Elston, D.A. Regression analysis of spatial data. *Ecol. Lett.* **2010**, *13*, 246–264. [CrossRef]
40. Zeng, Z.; Piao, S.; Li, L.Z.X.; Zhou, L.; Ciais, P.; Wang, T.; Li, Y.; Lian, X.; Wood, E.F.; Friedlingstein, P.; et al. Climate mitigation from vegetation biophysical feedbacks during the past three decades. *Nat. Clim. Chang.* **2017**, *7*, 432–436. [CrossRef]
41. Begueria, S.; Pueyo, Y. A comparison of simultaneous autoregressive and generalized least squares models for dealing with spatial autocorrelation. *Glob. Ecol. Biogeogr.* **2009**, *18*, 273–279. [CrossRef]
42. Dormann, C.F. Effects of incorporating spatial autocorrelation into the analysis of species distribution data. *Glob. Ecol. Biogeogr.* **2007**, *16*, 129–138. [CrossRef]
43. Mauricio Bini, L.; Diniz-Filho, J.A.F.; Rangel, T.F.L.V.B.; Akre, T.S.B.; Albaladejo, R.G.; Albuquerque, F.S.; Aparicio, A.; Araújo, M.B.; Baselga, A.; Beck, J.; et al. Coefficient shifts in geographical ecology: An empirical evaluation of spatial and non-spatial regression. *Ecography* **2009**, *32*, 193–204. [CrossRef]
44. Ver Hoef, J.M.; Peterson, E.E.; Hooten, M.B.; Hanks, E.M.; Fortin, M.J. Spatial autoregressive models for statistical inference from ecological data. *Ecol. Monogr.* **2018**, *88*, 36–59. [CrossRef]
45. Chen, W.; Chi, G.; Li, J. The spatial association of ecosystem services with land use and land cover change at the county level in China, 1995–2015. *Sci. Total Environ.* **2019**, *669*, 459–470. [CrossRef]
46. Griffith, D.A. Spatial-Filtering-Based Contributions to a Critique of Geographically Weighted Regression (GWR). *Environ. Plan. A Econ. Space* **2008**, *40*, 2751–2769. [CrossRef]
47. Sen, P.K. Estimates of the Regression Coefficient Based on Kendall's Tau. *J. Am. Stat. Assoc.* **1968**, *63*, 1379–1389. [CrossRef]
48. Jiang, W.; Yuan, L.; Wang, W.; Cao, R.; Zhang, Y.; Shen, W. Spatio-temporal analysis of vegetation variation in the Yellow River Basin. *Ecol. Indic.* **2015**, *51*, 117–126. [CrossRef]
49. Mann, H.B. Nonparametric Tests against Trend. *Econometrica* **1945**, *13*, 245–259. [CrossRef]
50. Kendall, M.G. *Rank Correlation Methods*; Charles Griffin: London, UK, 1975.
51. Moran, P.A.P. The Interpretation of Statistical Maps. *J. R. Stat. Soc. Ser. B-Stat. Methodol.* **1948**, *10*, 243–251. [CrossRef]
52. Brunsdon, C.; Fotheringham, S.; Charlton, M. Geographically weighted regression-modelling spatial non-stationarity. *J. R. Stat. Soc. Ser. Statistician.* **1998**, *47*, 431–443. [CrossRef]
53. Foody, G.M. Geographical weighting as a further refinement to regression modelling: An example focused on the NDVI–rainfall relationship. *Remote. Sens. Environment.* **2003**, *88*, 283–293. [CrossRef]
54. Zhao, L.; Dai, A.; Dong, B. Changes in global vegetation activity and its driving factors during 1982–2013. *Agric. For. Meteorol.* **2018**, *249*, 198–209. [CrossRef]
55. Yuan, X.L.; Hamdi, R.; Ochege, F.U.; Kurban, A.; De Maeyer, P. The sensitivity of global surface air temperature to vegetation greenness. *Int. J. Climatol.* **2021**, *41*, 483–496. [CrossRef]
56. Chen, C.; Park, T.; Wang, X.; Piao, S.; Xu, B.; Chaturvedi, R.K.; Fuchs, R.; Brovkin, V.; Ciais, P.; Fensholt, R.; et al. China and India lead in greening of the world through land-use management. *Nat. Sustain.* **2019**, *2*, 122–129. [CrossRef]
57. Ding, Z.; Peng, J.; Qiu, S.; Zhao, Y. Nearly Half of Global Vegetated Area Experienced Inconsistent Vegetation Growth in Terms of Greenness, Cover, and Productivity. *Earths Future* **2020**, *8*, e2020EF001618. [CrossRef]
58. Zhang, Y.; Song, C.; Band, L.E.; Sun, G. No Proportional Increase of Terrestrial Gross Carbon Sequestration from the Greening Earth. *J. Geophys. Res. Biogeosci.* **2019**, *124*, 2540–2553. [CrossRef]
59. Piao, S.; Liu, Q.; Chen, A.; Janssens, I.A.; Fu, Y.; Dai, J.; Liu, L.; Lian, X.; Shen, M.; Zhu, X. Plant phenology and global climate change: Current progresses and challenges. *Glob. Chang. Biol.* **2019**, *25*, 1922–1940. [CrossRef]
60. Fu, Y.H.; Piao, S.; Op de Beeck, M.; Cong, N.; Zhao, H.; Zhang, Y.; Menzel, A.; Janssens, I.A. Recent spring phenology shifts in western Central Europe based on multiscale observations. *Glob. Ecol. Biogeogr.* **2014**, *23*, 1255–1263. [CrossRef]
61. Gallinat, A.S.; Primack, R.B.; Wagner, D.L. Autumn, the neglected season in climate change research. *Trends Ecol. Evol.* **2015**, *30*, 169–176. [CrossRef]
62. Ge, Q.; Wang, H.; Rutishauser, T.; Dai, J. Phenological response to climate change in China: A meta-analysis. *Glob. Chang. Biol.* **2015**, *21*, 265–274. [CrossRef]
63. Gill, A.L.; Gallinat, A.S.; Sanders-DeMott, R.; Rigden, A.J.; Short Gianotti, D.J.; Mantooth, J.A.; Templer, P.H. Changes in autumn senescence in northern hemisphere deciduous trees: A meta-analysis of autumn phenology studies. *Ann. Bot.* **2015**, *116*, 875–888. [CrossRef]
64. Wang, X.; Piao, S.; Xu, X.; Ciais, P.; MacBean, N.; Myneni, R.B.; Li, L. Has the advancing onset of spring vegetation green-up slowed down or changed abruptly over the last three decades? *Glob. Ecol. Biogeogr.* **2015**, *24*, 621–631. [CrossRef]
65. Park, H.; Jeong, S.-J.; Ho, C.-H.; Park, C.-E.; Kim, J. Slowdown of spring green-up advancements in boreal forests. *Remote Sens. Environ.* **2018**, *217*, 191–202. [CrossRef]
66. Delpierre, N.; Vitasse, Y.; Chuine, I.; Guillemot, J.; Bazot, S.; Rutishauser, T.; Rathgeber, C.B.K. Temperate and boreal forest tree phenology: From organ-scale processes to terrestrial ecosystem models. *Ann. For. Sci.* **2016**, *73*, 5–25. [CrossRef]
67. Sloat, L.L.; Gerber, J.S.; Samberg, L.H.; Smith, W.K.; Herrero, M.; Ferreira, L.G.; Godde, C.M.; West, P.C. Increasing importance of precipitation variability on global livestock grazing lands. *Nat. Clim. Chang.* **2018**, *8*, 214–218. [CrossRef]
68. Flynn, D.F.B.; Wolkovich, E.M. Temperature and photoperiod drive spring phenology across all species in a temperate forest community. *New Phytol.* **2018**, *219*, 1353–1362. [CrossRef] [PubMed]

69. Walther, S.; Guanter, L.; Heim, B.; Jung, M.; Duveiller, G.; Wolanin, A.; Sachs, T. Assessing the dynamics of vegetation productivity in circumpolar regions with different satellite indicators of greenness and photosynthesis. *Biogeosciences* **2018**, *15*, 6221–6256. [CrossRef]
70. Gao, X.; Liang, S.; He, B. Detected global agricultural greening from satellite data. *Agric. For. Meteorol.* **2019**, *276–277*, 107652. [CrossRef]
71. Bai, Y.; Yang, Y.; Jiang, H. Intercomparison of AVHRR GIMMS3g, Terra MODIS, and SPOT-VGT NDVI Products over the Mongolian Plateau. *Remote Sens.* **2019**, *11*, 2030. [CrossRef]
72. Fensholt, R.; Proud, S.R. Evaluation of Earth Observation based global long term vegetation trends—Comparing GIMMS and MODIS global NDVI time series. *Remote Sens. Environ.* **2012**, *119*, 131–147. [CrossRef]
73. Heck, E.; de Beurs, K.M.; Owsley, B.C.; Henebry, G.M. Evaluation of the MODIS collections 5 and 6 for change analysis of vegetation and land surface temperature dynamics in North and South America. *ISPRS J. Photogramm. Remote Sens.* **2019**, *156*, 121–134. [CrossRef]

Article

Sea Level Seasonal, Interannual and Decadal Variability in the Tropical Pacific Ocean

Jianhu Wang [1], Juan Li [1,*], Jiyuan Yin [1], Wei Tan [1,2] and Yuchen Liu [1]

1. College of Ocean Science and Engineering, Shandong University of Science and Technology, Qingdao 266590, China; 201801021217@sdust.edu.cn (J.W.); 201801021226@sdust.edu.cn (J.Y.); tanwei01@sdust.edu.cn (W.T.); 201901210119@sdust.edu.cn (Y.L.)
2. Laboratory for Regional Oceanography and Numerical Modeling, Pilot National Laboratory for Marine Science and Technology, Qingdao 266590, China
* Correspondence: juanli0419@sdust.edu.cn; Tel.: +86-532-80698656

Citation: Wang, J.; Li, J.; Yin, J.; Tan, W.; Liu, Y. Sea Level Seasonal, Interannual and Decadal Variability in the Tropical Pacific Ocean. *Remote Sens.* **2021**, *13*, 3809. https://doi.org/10.3390/rs13193809

Academic Editor: Vladimir N. Kudryavtsev

Received: 2 September 2021
Accepted: 18 September 2021
Published: 23 September 2021

Publisher's Note: MDPI stays neutral with regard to jurisdictional claims in published maps and institutional affiliations.

Copyright: © 2021 by the authors. Licensee MDPI, Basel, Switzerland. This article is an open access article distributed under the terms and conditions of the Creative Commons Attribution (CC BY) license (https://creativecommons.org/licenses/by/4.0/).

Abstract: The satellite altimeter data, temperature and salinity data, and 1.5-layer reduced gravity model are used to quantitatively evaluate the contributions of the steric effect and the dynamic process to sea level variations in the Tropical Pacific Ocean (TPO) on different time scales. Concurrently, it also analyses the influence of wind forcing over the different regions of the Pacific Ocean on the sea level variations in the TPO. Seasonal sea level variations in the TPO were the most important in the middle and eastern regions of the 5°–15°N latitude zone, explaining 40–60% of the monthly mean sea level variations. Both the steric effect and dynamic process jointly affected the seasonal sea level variations. Among them, the steric effect was dominant, contributing over 70% in most regions of the TPO, while the dynamic process primarily acted near the equator and southwest regions, contributing approximately 55–85%. At the same time, the seasonal dynamic sea level variations were caused by the combined actions of primarily local wind forcing, alongside subtropical north Pacific wind forcing. On the interannual to decadal time scale, the sea level interannual variations were significant in the northwestern, southwestern, and middle eastern regions of the TPO and explained 45–60% of the monthly mean sea level variations. The decadal sea level variations were the most intense in the eastern Philippine Sea, contributing 25–45% to the monthly mean sea level variations. The steric effect and the dynamic process can explain 100% of the interannual to decadal sea level variations. The contribution of the steric effect was generally high, accounting for more than 85% in the regions near the equator. The impact of the dynamic process was mainly concentrated in the northwest, northeast, and southern regions of the TPO, contributing approximately 55–80%. Local wind forcing is the leading role of interannual to decadal sea level variations. The combined actions of El Niño–Southern Oscillation (ENSO) and the Pacific Decadal Oscillation (PDO) can explain 90% of the interannual to decadal sea level variations in the northwestern and eastern of the TPO.

Keywords: tropical pacific ocean; sea level variations; steric effect; dynamic process

1. Introduction

In the 21st century, due to rising sea levels and frequent extreme sea-level events, the storm surges, coastal erosion, flood risks, and economic losses faced by coastal cities worldwide will continue to increase in frequency and severity [1,2]. During 1993–2009, the global mean sea level rise rate was 3.2 ± 0.4 mm/yr, mainly caused by thermal expansion and variation in the quality of seawater transported from the land to the ocean [3]. However, regional sea level variations are primarily related to large-scale climate variations on the monthly decadal time scales. For example, the sea level anomalies in the Tropical Pacific Ocean (TPO) exceed 30 cm due to the interannual variations of El Niño–Southern Oscillation (ENSO) [4], and its anomalies exceed the 21 cm rising in global mean sea level during 1880–2009 [3]. Due to the impact of climate variations, there are significant regional differences in the rate of sea level rise on decadal time scales [5]. For example, in the past few decades, the western TPO has had the highest rate of sea level rise in the world, while the eastern TPO is the region with the fastest rate of sea level decline [3,6]. During 1993–2009, the rate of sea level rise in the western TPO was three times the global mean sea level rise rate [7–9]. Many studies have shown that in the Pacific Ocean, the steric effect has a greater contribution to the sea level variations [10–12]. However, the contribution of the dynamic process to the sea level variations should not be ignored [13].

Sea level variations reveal significant multi-time-scale variations in the TPO [14–19]. On seasonal time scales, the sea level variations in the northeastern TPO are the most dramatic, and its seasonal signals account for more than 60% of the total sea level variation signals [20]. Among them, the seasonal variation amplitude of the steric sea level and the total sea level is relatively consistent, indicating that the steric effect has a greater influence on the seasonal sea level variations in this area [21,22]. Secondly, the dynamic process is mainly driven by buoyancy flux and local wind stress, influencing the seasonal sea level variations in the TPO [23,24]. In addition, the sea surface wind stress is the driving force for the seasonal and interannual sea level variations in the TPO [25,26]. In some low-latitude regions of the Pacific Ocean (such as the western TPO), the seasonal-interannual sea level variations are mainly caused by the steric effect created by the baroclinic Rossby waves driven by wind stress. These wind-driven baroclinic Rossby waves are closely related to the tropical Pacific Ocean's climate mode (PDO/IPO) [27]. The sea level anomaly signals propagate westward in the form of baroclinic Rossby waves during the season in the northern regions of the TPO. After arriving in the Philippines, they are transformed into coastal waves (CTWs, Kelvin waves) and then enter the eastern regions of the South China Sea along with the Philippine Islands [28].

On the interannual time scale, variations in the sea level are most significant for the core region of the western Pacific Warm Pool, the central and eastern Pacific Ocean, and especially in the eastern regions of the Philippines and New Guinea. The significant periods of the sea-level low-frequency sequences are concentrated in 30 months and 52 months [29,30]. The sea level interannual variations are mainly affected by ENSO in the TPO [31–36]. At the same time, the wind field, the heat flux at the air-sea interface, the vertical ocean thermal structure, and the circulation variations all influence the interannual sea level variations in the TPO [29]. During El Niño events, the mass redistributed by the relaxed trade winds over the TPO eventually results in a significant decline in sea levels across the western TPO and a considerable rise in the eastern TPO [5,37,38]. Convergence and divergence anomalies of the wind field can explain the interannual sea level variations in the TPO during different types of El Niño. The continuous rise of sea levels is caused by the constant weakening of the divergent wind field in the eastern TPO, while anomalous westerly winds in the western Pacific Ocean cause sea levels to rise in the center of the eastern TPO [30]. In addition, the effect of heat flux on the interannual sea level variations cannot be ignored in the eastern TPO. The north–south movement of the bifurcation of the north Equatorial Current and variations in the intensity of the equatorial current system will also induce interannual sea level variations in the TPO. Among them, the heat flux primarily contributes to the northeastern and western regions of the TPO, reaching up to 60% and 50%, respectively [20].

The dominant factors influencing the interannual sea level variations are regionally distinct in the TPO. In the northern TPO, namely the regions around the Hawaiian Islands, Rossby waves propagating westward and the abnormal cooling of surface seawater caused by trade wind anomalies can pass through the density of the mixed layer to affect the interannual sea level variations [39]. The local response of surface heating and the eastern boundary forcing is significant in explaining the interannual sea level variations in the northeastern TPO. In the southeastern TPO, eastern boundary forcing primarily contributes to the interannual sea level variations [40]. Ocean dynamic buoyancy-driven processes play a vital role in the interannual sea level variations [41]. The contribution of local Ekman pumping to the interannual sea level variations in the southeastern TPO is relatively small, while the contribution to the southwest regions of the sea cannot be ignored [21]. The first baroclinic Rossby waves, caused by wind stress, significantly impact the interannual sea level variations in the western Pacific Ocean [42]. In addition, the long-distance adjustment to wind stress forcing of the oceans strongly influences interannual sea level variations in the western TPO. Therefore, the contribution of local responses to ocean surface warming and wind forcing cannot be ignored [21].

The impact of ENSO on the climate of the northwest Pacific Ocean is not static, but is modulated by decadal processes [43–46]. During the warm phase of the Pacific Decadal Oscillation (PDO), the relationship between ENSO and the East Asian winter monsoon is weak. In contrast, during the cold phase of PDO, ENSO strongly influences the East Asian winter monsoon [47]. In addition, PDO and north Pacific Circulation Oscillation (NPGO) can also affect the interannual sea level variations by modulating subsurface sea temperatures and salinity in the TPO [46–49].

The interannual climate system variations, the atmosphere–ocean coupling, and the ENSO phenomenon are the most significant in the TPO; it is also the region with the most dramatic decadal and long-term sea level variations. Since the early 1990s, the decadal sea level variations in the northwestern TPO have increased significantly. From 1991 to 2005, the standard deviation of sea level variations was 2.84 cm, decreasing to 1.12 cm between 1963 and 1976 and increasing to 1.31 cm between 1977 and 1990 [8,50]. During the El Niño period, the sea level in the eastern TPO is abnormally high (tens of centimeters), while the sea level in the western TPO is unusually low. More than 50% of the abnormal variations are explained by the decadal modulation of ENSO [51,52]. In the western TPO, the decadal sea level variations are greatly affected by PDO [5,9,16], contributing 53% to the first mode of decadal sea level variations in the TPO. The correlation coefficient between the time series of the first mode and the PDO index reaches 0.59 [19].

The decadal signals are mainly driven by the wind stress curl of the TPO. Concurrently, the atmospheric circulation related to the variations of the ENSO-PDO phase relationship also enhances the decadal oscillation of sea level [32,53,54]. According to the fifth phase of the Coupled Model Intercomparison Project (CMIP5), abnormal wind stress and wind stress curl are the leading causes of interannual and decadal sea level variations in the TPO. Although remote forcing may also induce abnormal sea level variations [55]. In addition, the significant decadal sea level variations in the TPO during 1993–2015 were mainly related to variations in the heat content of the upper ocean [56]. In the western TPO, the intensified decadal sea level variations result from the "out-of-phase" relationship between the Indian Ocean and the central and eastern Tropical Pacific since 1985, which has an "in-phase" effect on sea level variations in the western TPO [57]. Secondly, the Interdecadal Pacific Oscillation (IPO) also contributes to the decadal sea level variations in the western TPO [50]. The region with the most extensive decadal sea level variations in the eastern TPO does not coincide with the region with drastic interannual variations. The eastern TPO has the largest interannual sea level variations, but the decadal variations are minimal, mainly due to the ENSO.

The steric effect significantly contributes to sea level variations in the TPO [10–12]. At the same time, the contribution of the dynamic process to sea level variations is also important [13]. However, recent studies have not comprehensively investigated the geographic and temporal role of the steric effect and the dynamic process in influencing sea level variations in the TPO. This paper analyzes the contribution and influence mechanisms of the steric effect and the dynamic process to the seasonal, interannual, and decadal sea level variations across the Tropical Pacific.

2. Data and Methods
2.1. Datasets

The monthly sea level anomaly (SLA) is obtained from the merged (TOPEX, Jason-1/2, ERS-1/2, Envisat, GFO and CryoSat-2) Ssalto/Duacs altimeter products, which are remapped and distributed by the Archiving, Validation and Interpretation of Satellite Oceanographic (https://www.aviso.altimetry.fr/en/data/products/sea-surface-height-products/global.html, accessed on 22 September 2021). This dataset has a 0.25° × 0.25° horizontal resolution, and is available since 1993. The monthly subsurface temperature and salinity from the EN4.2.1 are used to calculate the steric sea level (SSL). EN4.2.1 is released by the UK Met Office Hadley Centre (https://www.metoffice.gov.uk/hadobs/en4/download-en4-2-1.html, accessed on 22 September 2021) with 1.0° × 1.0° horizontal resolution and 42 levels in vertical.

To clarify the role of ENSO-related and PDO-related processes in the sea-level variations, the Multivariate ENSO Index (MEI, https://psl.noaa.gov/enso/mei/, accessed on 22 September 2021) and PDO index (http://ds.data.jma.go.jp/tcc/tcc/products/elnino/decadal/pdo.html, accessed on 22 September 2021) are used. The MEI is a relatively complicated ENSO index. It is defined as the leading principle component of combined EOF based on five different variables (sea level pressure, sea surface temperature, zonal and meridional surface wind, and outgoing longwave radiation) over the tropical Pacific (30°S–30°N and 100°E–70°W). In comparison with some single-variable indices, such as Nino3.4 index, Southern Oscillation Index (SOI), etc., the MEI can portray a more realistic coupled ocean–atmosphere processes. The PDO index also depends on the EOF analysis, which is defined as the leading principal component of sea surface temperature in the north Pacific (north of 20°N). For the overlapping period, all the aforementioned data are adopted from 1993 to 2019 in this paper. Before analysis, the linear trend is removed for all the variables.

In addition, the monthly surface wind stress from the European Centre for Medium-Range Weather Forecasts (ECMWF) Reanalysis V5 (ERA-5) (https://www.ecmwf.int/en/forecasts/datasets/reanalysis-datasets/era5, accessed on 22 September 2021) to drive the 1.5 layer nonlinear reduced-gravity model. ERA-5 is available from 1979 to present, and has several sets of horizontal resolution. To match the model resolution, 0.25° × 0.25° is selected in this paper.

2.2. Methods

2.2.1. Calculation of Explained Variances Percentage

To quantify the contribution of one variable (h_x) to another variable (h_y), the explained variances percentage (skill) is defined as:

$$S = \left(1 - \frac{\langle (h_y - h_x) \rangle^2}{\langle h_y^2 \rangle}\right) \times 100\% \qquad (1)$$

where S is the skill, h_x is the independent variable, h_y is the dependent variable, and $\langle \cdots \rangle$ represents the average over time. A large (small) skill suggests a significant (poor) contribution of h_x to h_y.

2.2.2. Multiple Variable Linear Regression Method

To isolate the contribution of interannual and decadal processes to the SCS sea-level change, a multiple variable linear regression method is performed in this study, and expressed as [5]:

$$h = a_0 + a_1 ICI + a_2 DCI + noises \qquad (2)$$

where h is the SLA considered as a dependent variable, *ICI* (interannual climate index) and *DCI* (decadal climate index) are two independent variables, a_1 and a_2 are regression coefficients, *noises* is the error. The *DCI* indicates a low-pass filtered PDO index, which has experienced two consecutive (25- and 37-month) running averages. The *ICI* is a high-pass filtered MEI, which is calculated as the difference between the original MEI and low-pass filtered MEI [5].

2.2.3. Steric Sea Level

The SSL (η_s) includes thermosteric (η_t) and halosteric height (η_h), which are calculated as [58]:

$$\eta_t = \int_{-H}^{0} -\frac{1}{\rho}\frac{\partial \rho}{\partial T} \cdot \Delta T dz \tag{3}$$

$$\eta_h = \int_{-H}^{0} -\frac{1}{\rho}\frac{\partial \rho}{\partial S} \cdot \Delta S dz \tag{4}$$

$$\eta_s = \eta_t + \eta_h \tag{5}$$

in which T is the temperature, S is the salinity, H = 800 m is the lower limit of the vertical integration, dz is the thickness of each layer, ΔT and ΔS are the temperature and salinity anomaly relative to the mean values from 1993 to 2019.

2.2.4. 1.5-Layer Reduced-Gravity Model

The 1.5-layer nonlinear reduced-gravity model is performed in this paper to investigate the role of the dynamic processes [41]. The momentum and continuity equations are described as [41]:

$$\frac{\partial u}{\partial t} + \zeta k \times u = -\nabla E + A_h \nabla^2 u + \frac{\tau}{\rho_0 h} + \frac{\epsilon}{H} u \tag{6}$$

$$\frac{\partial h}{\partial t} + \nabla \cdot (h u) = 0 \tag{7}$$

in which $g' = g\Delta\rho/\rho$ is the reduced gravity acceleration, h is the time-varying upper layer thickness. The detailed interpretation of the other variables and model configuration can be found in Li et al. (2020). Derived from the 1.5-layer reduced-gravity model, the dynamic sea level (DSL) is defined as $DSL = g'h/g$. Four experiments are conducted in this paper, including one control experiment and three sensitivity experiments (Table 1). In the control experiment, the whole model domain (40°S–65°N, 100°E–70°W) is driven by the real wind stress from ERA-Interim, which is referred to as Exp 0. Meanwhile, in the two sensitivity experiments, the real wind stress only covers the tropical Pacific Ocean (Exp 1: 20°S–20°N, 100°E–70°W) and north Pacific (Exp 2: 20°N–65°N, 100°E–70°W), and the other areas are forced by monthly climatological wind stress which is calculated from 1981 to 2010. All the four experiments are integrated from 1979 to 2019, but only the outputs during 1993–2019. Table 1 presents the numerical experiments of the 1.5-layer reduced gravity model, analyzed in accordance with the observed data as described in Section 2.1.

Table 1. Numerical experiments of the 1.5-layer reduced gravity model.

Model Experiments	Descriptions
Exp 0	Control experiment, model domain (30°S–65°N, 100°E–70°W, NPO) closed lateral boundaries, originally forced by monthly ERA-5 wind stress.
Exp 1	The model domain as Exp 0. Originally forced by monthly wind stress over the north Pacific (north of 20°N, NNP); monthly climatological wind over other regions
Exp 2	The model domain as Exp 0. Originally forced by monthly wind stress over the tropical Pacific Ocean (20°S–20°N, 100°E–70°W, TPO); monthly climatological wind over other regions

3. Seasonal Sea Level Variations in the TPO

To extract the seasonal variation, a 6–18-month band-pass filter is performed in this section. The sea level variations exhibit significant spatial differences in the TPO (Figure 1a). The region with the most dramatic seasonal sea level variations is located near Clipperton Island (10.33°N, 109.22°W) in the northeastern TPO, with variances up to 50 cm^2. There is also a narrow and long eastward latitude zone from 170°W between 5° and 15°N, with variances of 30–45 cm². Additionally, seasonal sea level variations in the East Philippines (5°–15°N, 120°–128°E) and the Coral Sea (18°–13°S, 145°–165°E) are significant, with variances greater than 20 cm². Overall, the seasonal variation amplitude of other regions of the Tropical Pacific is small, with variances of less than 10 cm². Using Equation (1), we quantitatively evaluate the contributions of the seasonal sea level variations to the monthly mean sea level variations (Figure 1). Seasonal sea level variations across the long and narrow latitude zone between 5° and 15°N in the northern TPO explains 40–60% of the monthly mean sea level variations (Figure 1b). In most northwest and southwestern TPO regions, seasonal signals contribute 35% to 50% of the monthly mean sea level. While in the southeastern TPO, the proportion of seasonal sea level variations is relatively smaller (<15%).

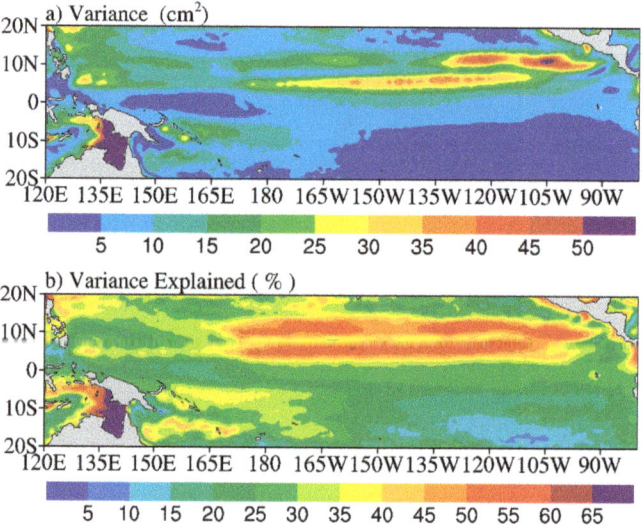

Figure 1. (a) The seasonal sea level variances and (b) explained variances in the TPO.

To explore the contribution of the steric effect and the dynamic process to the seasonal sea level variations in the TPO (Figure 2), we use Equation (1) to calculate the influence of the steric effect and the dynamic process on sea level variations (Figure 2). The steric sea level is calculated using Equation (5), and the dynamic sea level is simulated by a 1.5-layer reduced gravity model.

Seasonal sea level variations in the TPO mainly result from the steric effect [21], especially in the central and eastern TPO, where the steric effect contributes to more than 90% of the seasonal sea level variations (Figure 2a). For the northernmost and southeastern Tropical Pacific, the contribution of the steric effect is small, even indicating a negative contribution. The dynamic process mainly acts on two long and narrow latitude zones in the northernmost central Tropical Pacific, contributing 55–85% (Figure 2b). While in most other regions of the TPO, the contribution of the dynamic process is less than 50%, and in some regions even revealing a significant negative contribution. Compared with the steric effect, the contribution of the dynamic process in the TPO is significantly weaker; therefore, the steric effect is the dominant process influencing seasonal sea level variations in the TPO. This is consistent with the previous conclusion that the steric effect plays a leading role [25].

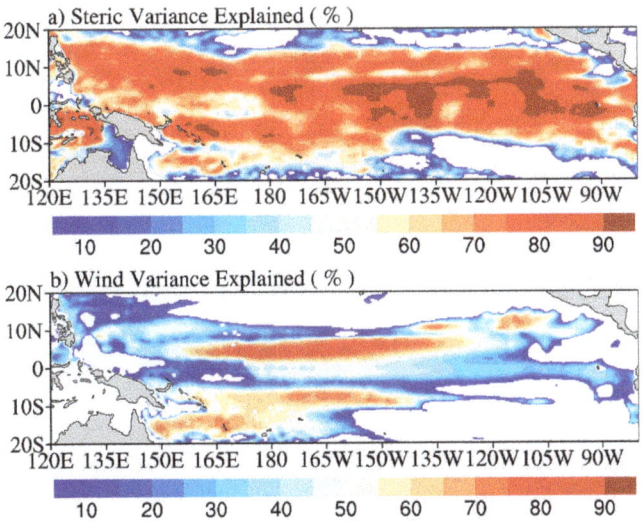

Figure 2. The seasonal sea level explained variances of (**a**) the steric effect and (**b**) the dynamic process.

Figure 3a reveals the time–longitude features of seasonal sea level anomalies averaged from 20°S–20°N in the TPO. Positive sea level anomalies generally appear from June to December, spanning the entire TPO in the meridian direction. In the western TPO, the sea level anomalies start to appear in May, with the maximum positive sea level anomalies (>3 cm) appearing in the eastern Philippines 130°–135°E. This is distinct to the eastern TPO, where the sea level anomalies are relatively small (<1.5 cm). In other regions of TPO, especially in the west, the sea level reveals negative anomalies, with the greatest occurring from January to March, even exceeding −3 cm. The TPO is bounded by June, with negative sea level anomalies occurring before June and positive sea level anomalies after. Seasonal sea level variations propagate from west to east in the central and eastern TPO (Figure 3a), mainly caused by the steric effect, which is induced by the baroclinic Rossby waves, which in turn are driven by wind stress (Figure 3b). It is worth noting that in the meridional direction of 135°–140°E, the sea level variations are abnormal, likely resulting from variances of seasonal variations between Australia and New Guinea reaching more than 50 cm^2. As this geographical location is closed, sea level variations are mainly affected by tropical cyclones, and the occurrence of tropical cyclones is closely related to the monsoon [59,60].

To explore the influence of the steric effect and the dynamic process on seasonal sea level variations in the TPO and their temporal and spatial variations, we calculated the steric sea level anomalies based on Equation (5) (Figure 3b) and the dynamic sea level anomalies simulated by a 1.5-layer reduced gravity model (Figure 3c). The spatial distribution of the steric sea level variations presents a structure similar to the altimeter sea level variations, indicating that the steric variations suitably capture the characteristics of the altimeter variations. The sea level anomalies are only minor in the middle east of the TPO (<2 cm). However, the steric sea level variations between 135°E and 140°E are completely different from the altimeter variations, confirming that the sea level anomalies in this area are indeed not caused by the steric effect but tropical cyclones. Secondly, the spatial distribution of the dynamic sea level variations is quite different from the structure of altimeter sea level variations. Dynamic sea level anomaly signals propagate westward in the form of baroclinic Rossby waves on the scale of seasonal variations. Concurrently, in the meridional direction of 120°E–135°W, the positive and negative sea level anomaly signals anomaly signals displayed an obvious arc-shaped trend. From west to east along the TPO, negative sea level anomalies occur earlier in the year (from September to January). Moreover, the seasonal sea level variations are dominated by the steric effect in most regions.

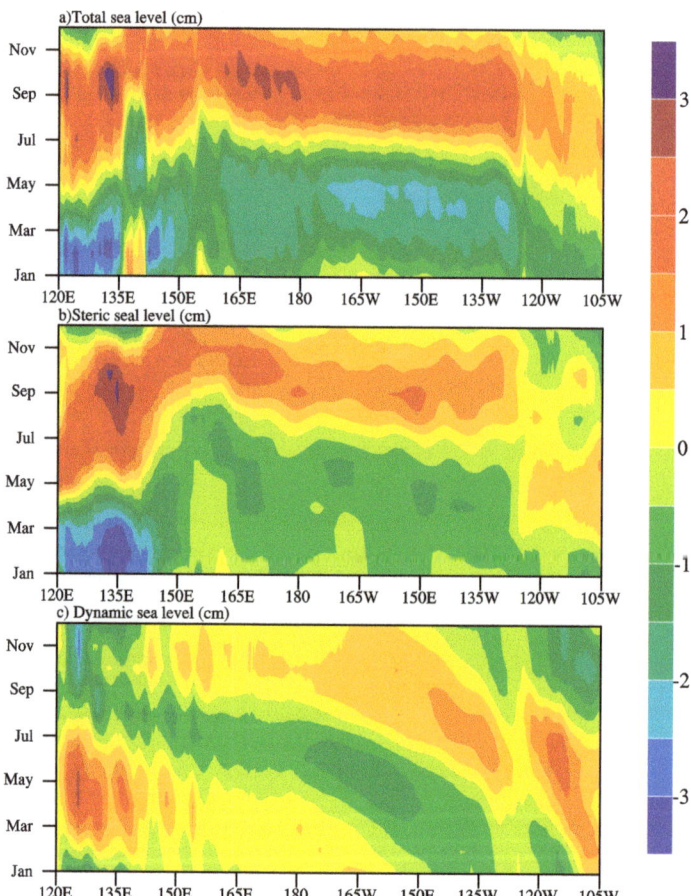

Figure 3. Time–longitude plot of the seasonal sea level anomalies (cm) averaged along the 20°S–20°N latitudinal band from (**a**) altimetric observations, (**b**) the steric effect, and (**c**) the 1.5-layer model.

Conducting sensitivity experiments in specific regions is an important method for exploring the mechanisms of sea level variations. Experiments isolating only various wind fields and ignoring other factors can highlight the role of dynamic processes in sea level variations [59]. The seasonal sea level variations in the TPO are closely related to the wind forcing in the Pacific. Therefore, it is necessary to explore the relative effects on seasonal sea level variations in the TPO between the wind forcing in the TPO and other regions. Based on the above questions, we designed three sensitivity experiments (Table 1) to evaluate the relative effects of wind forcing on the seasonal sea level variations in different regions of the TPO.

On the basis of the comparison of the simulated results (Figure 4), the dynamic sea level derived from the control experiment (Exp 0) contributes the most to the seasonal sea level variations in the central northern Tropical Pacific (Figure 4). A long and narrow belt across the meridian explains more than 80% of the variance, while a second narrow belt is formed in eastern New Guinea (explaining 55–75%). Additionally, wind forcing significantly contributes to the variance (65–75%) in the Coral Sea (18°–13°S, 145°–165°E) and around Clipperton Island (10.33°N, 109.22°W). Nevertheless, the real monthly wind forcing explains approximately 25% of the variances in other regions and even reaches negative contributions in the north and southeast (Figure 4a). The local wind forcing in the TPO plays a leading role in seasonal sea level variations, through the comparison the simulated results of EXP 0 and EXP 1 (Figure 4a,b). If the subtropical north Pacific models the real monthly wind field, and TPO is the climatic wind field (Figure 4c), the contribution of the subtropical north Pacific wind field to seasonal sea level variations is generally consistent with the NPO wind field. Still, the variances are generally minor, with a maximum of only 75%, forming a long and narrow belt. The variances of the coral region are less than 55%, and the area of the negative contribution region in the north and southeast region expanded. This indicates that compared with the contribution of TPO wind forcing to seasonal sea level variations, the contribution of subtropical north Pacific Ocean wind forcing is relatively small. Overall, on seasonal time scales, the dynamic sea level variations in the TPO are caused by the combined effects of local wind forcing and subtropical north Pacific Ocean wind forcing, of which local wind forcing plays a leading role.

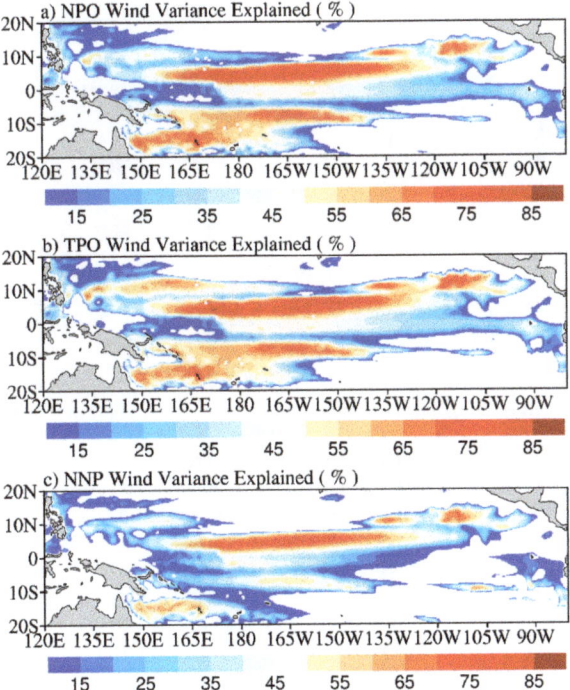

Figure 4. Contributions of different wind forcing to seasonal sea level variations: (**a**) the real monthly wind field; (**b**) the real wind field in the TPO and the climatic wind field in other regions; (**c**) the real wind field in other regions and the climatic wind field in the TPO.

4. Interannual to Decadal Sea Level Variations in the TPO

4.1. Interannual Sea Level Variations in the TPO

There are significant spatial differences in the interannual sea level variations in the TPO (Figure 5a). Significant interannual variations are presented in the northwest, southwest, and east-central parts of the TPO. These variances exceed 40 cm^2, with the maximum value in the northwest and southwest regions surpassing 45 cm^2 (Figure 5a).

In these regions with high sea level interannual variation variances, the contribution of the interannual sea level signals to the monthly mean variations is as high as 45–60% (Figure 5b), and their spatial patterns are consistent with the interannual sea level variations. However, it is worth noting that the significant interannual sea level variations in the equatorial eastern Pacific can be explained by the interannual response of Kelvin waves caused by wind anomalies (Figure 5a). Variances of the interannual sea level variations are lower in other regions, less than 20 cm^2 (Figure 5a). Except for the northeastern TPO, the contribution of the interannual signals to the monthly mean variations remains relatively minor (below 45%) (Figure 5b). In the northeastern TPO, while the interannual sea level variations are not large (Figure 5a), their contribution to monthly mean variations explains up to 60% (Figure 5b), further indicating that the sea level variations in this region are not significant.

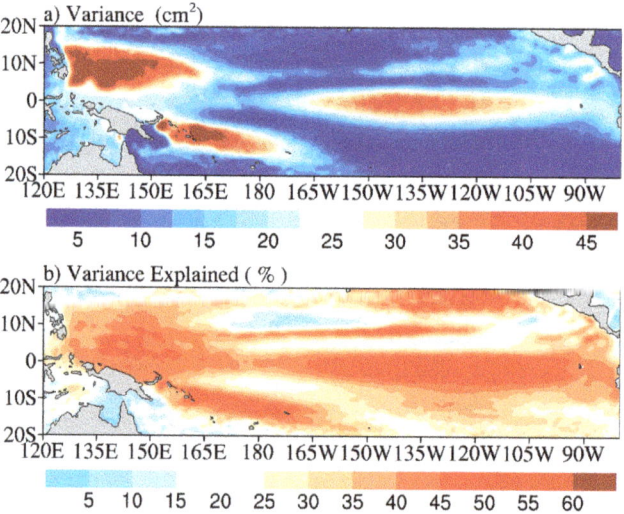

Figure 5. (a) The interannual sea level variances and (b) explained variances in the TPO.

From the perspective of atmospheric forcing, variations in the wind field and heat flux caused variations in ocean density field and circulation. These can induce variations in heat content and other factors, in turn causing sea level variations. Therefore, the root causes of the interannual variations are the variations of the wind field and heat flux [12,29]. The contribution of the steric effect to the interannual sea level variations in the TPO is generally significant (Figure 6), as high as 95% in the central TPO. However, there is a small area of less than 15% and even a negative contribution in the northernmost and southeastern regions. The dynamic process contributes significantly (55–80%) to the interannual sea level variations in the TPO, mainly influencing the northwest, northeast, and southern regions, and contributed less (<45%) in other regions. This is consistent with the results of previous studies [12,20,41]. The contributions of the steric effect and the dynamic process to the interannual sea level variations vary. For instance, while the contribution of the dynamic process is unevenly distributed spatially, its effect in most regions exceeds 85%. Therefore, in most regions of TPO, the steric effect and the dynamic process explains the interannual sea level variations. Similar conclusions have been reflected in previous studies [19,40,54].

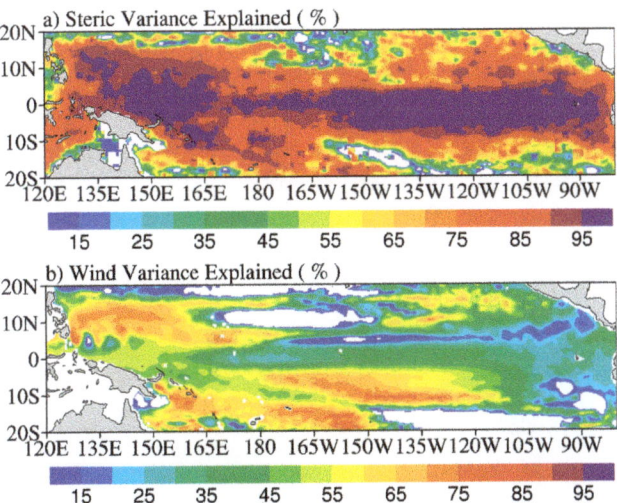

Figure 6. The interannual sea level explained variances of (**a**) the steric effects and (**b**) the dynamic process.

The interannual sea level fluctuation signals of TPO are affected by the sea surface wind forcing. The dominant contribution to interannual sea level variations in western TPO is the wind-induced first baroclinic Rossby waves [7,24,32,36]. Therefore, exploring the relative effects of wind forcing in the TPO and the wind forcing in other regions on the interannual sea level variations is necessary. The variances explained by the wind forcing in different regions from the sensitivity control experiments (Figure 7) reveal that the real monthly wind forcing contributed to the interannual sea level variations. These are larger in the northwest, south, and southwest (>65%), only reaching 65% in the northeast. However, its contribution is minor and even negative in other regions. When the TPO is modeled with the real monthly wind field, and other regions are the climatic wind fields (Figure 7b), the contribution of local wind forcing to the interannual sea level variations in the TPO is almost the same as the real wind forcing. The variances are also the same, indicating that the local wind field in the TPO plays a leading role in influencing the interannual sea level variations. However, when the subtropical north Pacific Ocean is modeled with the real monthly wind field and TPO is the climatic wind field (Figure 7c), the subtropical north Pacific Ocean wind field explains less than 15% of the variances of the interannual sea level variations. This indicates that the subtropical north Pacific wind forcing barely contributes to this region. Therefore, the local wind forcing in the TPO is the leading role of the interannual sea level variations.

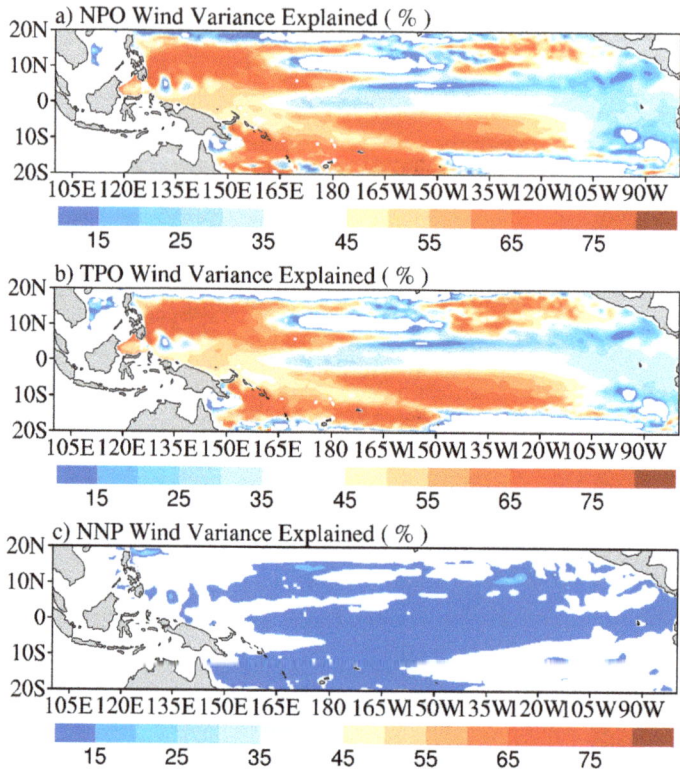

Figure 7. Contributions of different wind forcing to interannual sea level variations: (**a**) the real monthly wind field; (**b**) the real wind field in the TPO and the climatic wind field in other regions; (**c**) the real wind field in other regions and the climatic wind field in the TPO.

4.2. Decadal Sea Level Variations in the TPO

There are significant spatial differences in decadal sea level variations in the TPO. They are most dramatic in the eastern Philippines of the northwestern TPO, with a maximum variance of more than 40 cm^2. In the southwest, central, eastern, and northeastern regions, these variances range from 10 cm^2 to 20 cm^2, and are generally lower in other regions (<10 cm^2) (Figure 8a). The contribution of decadal signals to monthly mean sea level variations in the northeastern and southeastern TPO is as high as 50–70%, with some regions exceeding 70% (Figure 8b). Decadal contributions are the smallest in the north-central and southwestern regions, as low as 25% (Figure 8b). The spatial characteristics of the contribution of decadal signals to monthly mean sea level variations are quite different from those of decadal sea level variations.

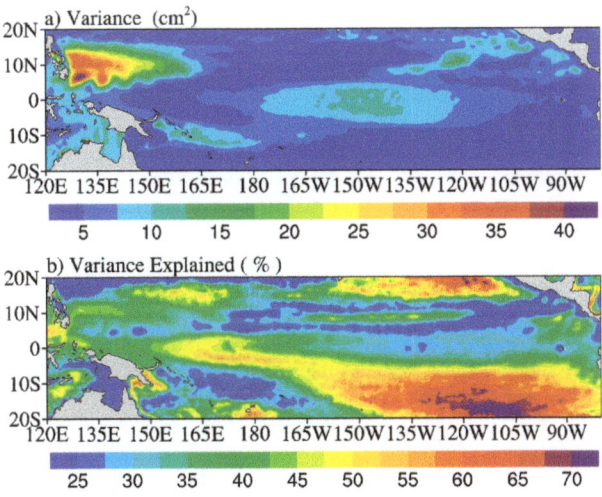

Figure 8. (a) The decadal sea level variances and (b) explained variances in the TPO.

The decadal sea level variations in the TPO are mainly controlled by large-scale ocean–atmosphere processes such as PDO and the decadal variations of ENSO [53,61]. ENSO and PDO primarily affect sea level variations through variations in trade winds in the Pacific Ocean, alongside numerous other factors, such as the effect of heat flux and so on [29]. Similar to the interannual time scales, on decadal time scales, the contribution of the hematocrit effect to the sea level variations in the TPO is dominant, and the maximum contribution exceeds 95% (Figure 9). At the same time, the dynamic process mainly acts on the northwestern and southwestern of the TPO, which is consistent with previous studies [62], with contributions of approximately 55–70%. While the contribution of the dynamic process is generally lower in the eastern TPO, it is mainly dominated by the steric effect (Figure 9). There are considerable differences between the steric effect and the intensity of the dynamic process, i.e., the dynamic process only contributed more than the steric in the central and western areas, and its greatest contribution is no more than 65%. In the central and western regions of TPO, the combined effect of the hematocrit effect (major contributor) and dynamic process explains 100% of the decadal sea level variations.

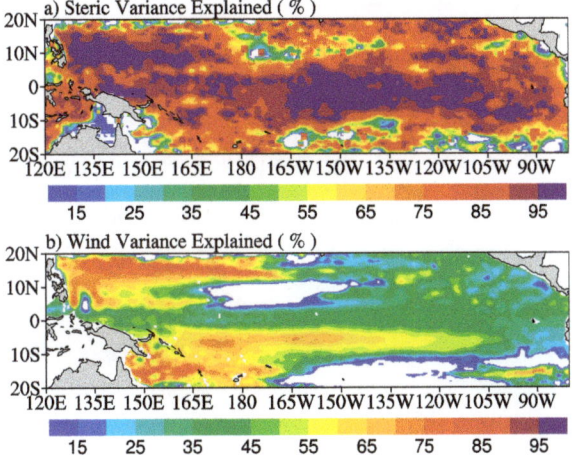

Figure 9. The decadal sea level explained variances of (a) the steric effect and (b) the dynamic process.

The decadal signals are mainly driven by wind stress in the TPO. Abnormal wind stress and wind stress curl are considered to be the leading causes of sea level variations in the TPO on interannual and decadal scales [32,53–55]. Therefore, it is necessary to explore the relative effect of wind forcing in different regions of the Pacific Ocean on the sea level variations in the TPO. The contribution of real monthly wind forcing to the decadal sea level variations in the TPO is greater in the northwest and southwest regions (>65%) (Figure 10). The contribution in other regions is minor and in some regions it is even negative. If TPO models the real monthly wind field and other regions include the climatic wind fields (Figure 10b), the contribution of local wind forcing in the TPO to the decadal sea level variations is almost the same as the real wind forcing. Furthermore, the variances of the variations were almost the same, revealing that the local wind forcing in the TPO plays a significant role in the decadal sea level variations. If the subtropical north Pacific Ocean includes the real monthly wind field and TPO is the climatic wind field (Figure 10c), then the wind field of subtropical north Pacific contributes little or even negatively to the decadal sea level variations in most regions of the TPO. Therefore, local wind forcing in the TPO is likely the leading role of the decadal sea level variations.

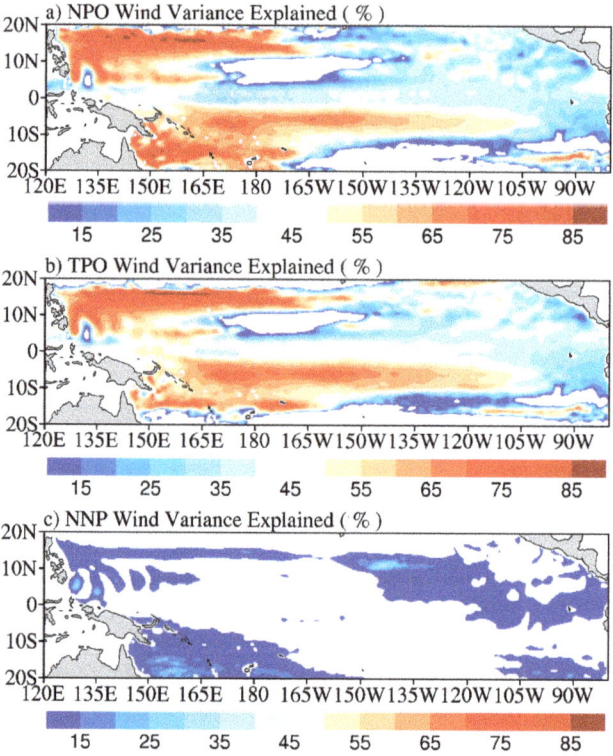

Figure 10. Contributions of different wind forcing to decadal sea level variations: (**a**) the real monthly wind field; (**b**) the real wind field in the TPO and the climatic wind field in other regions; (**c**) the real wind field in other regions and the climatic wind field in the TPO.

4.3. Interannual to Decadal Sea Level Variations in the TPO

The eastern and western TPO has recently exhibited a seesaw spatial pattern of annual mean sea level anomalies (Figure 11a). Sea level anomalies and negative sea level anomalies alternately appeared during 1993–1995, especially when ENSO occurred. Negative anomalies (blue) occurred during El Niño: 1997, 1997, 2002–2005, 2007, 2014–2016, and 2018–2019. On the contrary, positive anomalies (red) occurred during La Niña: 1996, 1998–2001, 2008–2013, and 2017 (Figure 11a). The meridional mean sea level anomalies in the TPO are significant (>3 cm or <−3 cm), and mainly concentrates in a small latitude zone, roughly between 120°E–140°E in the eastern Philippines (5°–15°N, 120°–128°E) and western New Guinea (10°S–0°, 135°–150°E). The eastward signals gradually weaken, while the negative sea level anomalies form in a long and narrow belt along the zonal direction. This phenomenon also confirms that the interannual sea level variations in the TPO and its low-frequency modulation are driven by ENSO [32,33,36,61]. It is worth noting that from the entire time–longitude diagram, the sea level anomalies in the TPO have the characteristics of an east–west reverse-phase variation in the zonal direction of west longitude 170°W. In contrast to this region, this phenomenon is consistent with the results of previous studies [42,62,63].

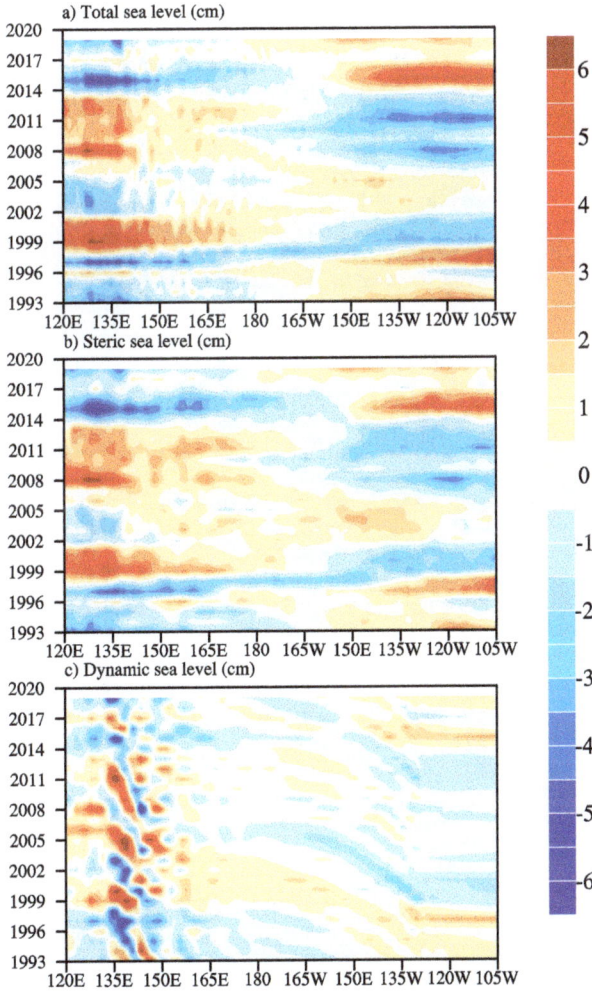

Figure 11. Time–longitude plot of the interannual to decadal sea level anomalies (cm) averaged along the 20°S–20°N latitudinal band from (**a**) altimetric observation, (**b**) steric effect, and (**c**) the 1.5-layer model.

The sea level variations induced by the steric effect strongly reflect the characteristics of the altimeter sea level variations. Its spatial distribution presents a structure similar to the altimeter sea level variations, which is driven by ENSO (Figure 11), which indicates that the steric sea level variations have a dominant effect on the altimeter sea level variations. At the same time, the spatial pattern of the low-frequency dynamic sea level variations in the latitude 165°E–135°W region is roughly the same as the altimeter sea level variations. However, the magnitude of the variations is small, indicating that the dynamic sea level variations affect the low-frequency altimeter sea level variations. The dynamic process mainly involves the interannual to decadal sea level variations in the TPO through the westward propagation of Rossby waves (Figure 11c). Therefore, similar to the sea level variations from interannual to decadal time scales, the steric effect and the dynamic process works together to influence the sea level variations in the TPO, with the steric effect being dominant.

4.4. Interannual and Decadal Sea Level Fingerprints of TPO

The contribution of the PDO to sea level variations can be inferred from the increase of the sea level climate index. Based on the PDO index and Multivariate ENSO Index (MEI), using multiple regression methods, Zhang and Church (2012) defined new interannual and decadal climate indexes to explain the low-frequency sea level variations in the Pacific Ocean since 1993. They concluded that 60% of the observed low-frequency sea level variations originated from internal climate models (ENSO and PDO) [64].

Therefore, in order to explore the impact of ENSO and PDO on sea level variations in the TPO, we used the multiple regression model (MVLR) to simulate the interannual and decadal sea level fingerprints of TPO (Figure 12a,b). The annual sea level fingerprint is expressed as the regression of sea level relative to ICI, equivalent to the regression coefficient a1 in Equation (2) and closely related to ENSO. For the interannual sea level fingerprint, the regions with the most significant regression results are western and eastern TPO. The maximum interannual sea level fingerprint reaches 80 mm/unit ICI in the eastern regions, and the largest in the western sea is −80 mm/unit ICI; this spatial feature is similar to a "seesaw". During El Niño Taimasa, the abnormal movement of the zonal wind and the corresponding wind stress curl lead to the "seesaw" pattern of the sea level of TPO, manifesting as sea level rise in the eastern TPO and a decline in the warm pool area (10°S–10°N, 60°W–150°E). In the northwest and southeastern TPO, the sea level meridional seesaw mode with 5°N as the fulcrum is also related to the air–sea coupling mode [30,65–67]. Secondly, we used Equation (1) to calculate the contribution of ENSO-related sea level variations to interannual to decadal sea level variations (Figure 12c), which corresponded to the interannual sea level fingerprint. In the eastern and western regions of the TPO, ENSO-related sea level variation explanation reaches more than 65%, while the variance in the north American coast reached 60%. The contribution of ENSO in the regions near the C-shaped east–west boundary of the "seesaw" is low, less than 10%. Therefore, the ENSO can better explain the interannual sea level variations in the western and eastern TPO and the coast of north America. The interannual sea level fingerprint and the spatial characteristics of its contribution are consistent with the impact of the ENSO dynamic on the interannual sea level variations in the TPO [31,68].

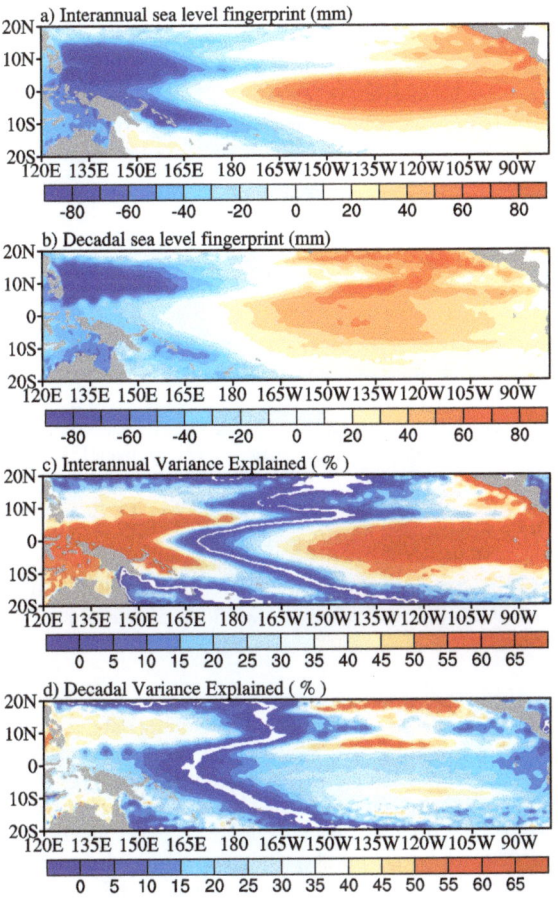

Figure 12. (a) Interannual sea level fingerprint (mm) (b) decadal sea level fingerprint (mm), (c) the explained variances of the interannual sea level associated with the ENSO, (d) the explained variances of the decadal sea level related to the PDO.

In addition, we used Equation (1) to calculate the decadal sea level fingerprint as the regression of sea level with respect to DCI, which is equivalent to the regression coefficient a2 in Equation (2) (closely related to PDO). For the decadal sea level fingerprint, the most significant region of the regression result is located in the northwest TPO, namely the eastern Philippines, with a maximum of −80 mm/unit DCI. In addition, the regions with more significant variations include the eastern and northeastern regions of the Pacific Ocean. The north American coast ranges between 40–70 mm/unit DCI, and the "seesaw" structure is no longer apparent compared with the interannual sea level fingerprint (Figure 12b). We used Equation (1) to calculate the contribution of PDO-related sea level variations to the interannual to decadal altimeter sea level variations (Figure 12d), which strongly corresponds to the decadal sea level fingerprint. The regions in the narrow and long latitudes with high contributions contribute more than 60%, and the contribution in the northwestern region is between 35–50%. In comparison, it contributes less than 40% in all other regions, indicating that PDO can better explain the interannual to decadal sea level variations only in the northwest and northeast TPO.

From the perspective of interannual sea level fingerprint, decadal sea level fingerprint, and their respective contributions to sea level variations, the combined effect of ENSO and

PDO can explain 100% of the interannual to decadal sea level variations in the northwest and East TPO. For the coast of north America, the contribution of ENSO and PDO can explain more than 90%. In the southwestern TPO, the joint action of ENSO and PDO has little impact on the interannual to decadal sea level variations in this region. The low-frequency sea level variations in this region are dominated by the steric effect and other processes (Figure 6a, Figure 7a, Figure 9a, Figure 10a).

5. Summary and Discussion

5.1. Discussion

The quantitative analysis results of the influence of wind field and heat steric factors on sea level variations have many interannual scales and a relative lack of decadal scales. The length and quality of the data are the key factors that affect the quantitative analysis of sea level variations in the TPO [29]. The satellite altimeter data, thermohaline data, and model data used in this paper span only 27 years, which is insufficient to study decadal sea level variations. Secondly, substantially different sources of wind field data and temperature and salt data are also reasons for the above problems. In addition, the dynamic sea level simulated by the 1.5-layer reduced gravity model is only driven by the wind field. The simulation is not the real dynamic sea level due to the lack of real terrain and stratification. Therefore, it is necessary for a high-precision three-dimensional ocean numerical model to quantitatively compare the effect of different wind forcing in future research. Finally, although the method estimates the magnitude of interannual to decadal sea level variations associated with ENSO, the regression was based on a limited time frame spanning only a few ENSO events. With the continuous growth of time series, the regression amplitude could be improved.

5.2. Summary

In this paper, we used satellite altimeter data, temperature and salinity data, and the 1.5-layer reduced gravity model to analyze the characteristics of the sea level temporal and spatial variations in the TPO and quantitatively assess the contribution of the steric effect and the dynamic process to sea level variations at different time scales. At the same time, we also discuss the impact of wind forcing in different regions on sea level variations in the TPO. Based on a multiple regression model, we quantified the relative contribution of interannual variations related to ENSO and decadal variations related to PDO to low-frequency sea level variations in the TPO. The sea level variations in the TPO exhibit significant multi-timescale variations [15–18]. Seasonally, the meridional mean sea level of TPO produces negative anomalies from January to May and positive anomalies from June to December. The steric sea level variations are the same as the satellite altimeter sea level variations. In contrast, the dynamic sea level propagates westward from the eastern Pacific Ocean, distinct from the altimeter sea level. In terms of the spatial distribution, the sea level variations are most significant in the long and narrow latitude zone from 170°W eastward between 5° and 15°N, with a variance range of 30–50 cm^2, explaining 40–60% of the monthly variations. Secondly, it is also more evident in the southeastern and southwestern regions of the Tropical Pacific 30–50 cm^2. The seasonal sea level variations range from 10–40 cm^2 and explained 35–45% of the monthly mean sea level variations. The steric effect and the dynamic process together influences the seasonal sea level variations in the TPO. Among them, the steric effect is dominant, and the contribution in most regions exceeds 70%. The dynamic process mainly affects the regions on both sides of the equator and the southwestern TPO. The contribution of this region is about 55–85%. Concurrently, the dynamic sea level variations are induced by the combined effect of the local wind forcing and the subtropical north Pacific Ocean wind forcing, of which the local wind forcing plays a leading role.

On the interannual time scales, the eastern and western TPO exhibits a seesaw spatial pattern during ENSO. During the El Niño period, the sea level in the eastern Pacific Ocean presented positive anomalies, and the sea level in the western Pacific Ocean presented negative anomalies. The opposite is the case during the La Niña period. The sea level anomalies are consistent with the altimeter sea level anomalies in changeable amplitude and phases. The dynamic process mainly affects the interannual sea level variations in the central and eastern TPO through the westward propagation of Rossby waves. At the same time, there are significant spatial differences in the interannual sea level variations in the TPO. The interannual variations are most significant in the northwest, southwest, and central and eastern regions of TPO, with a variance of more than 40 cm^2. The interannual sea level variations vary from month to month, while the contribution of the mean sea level variations is as high as 45–60%. The steric effect and the dynamic process strongly explains the interannual sea level variations in the TPO. Among them, the steric effect contributes relatively higher, up to 95% near the equator, to the interannual sea level variations in the TPO. The influence of the dynamic process is mainly concentrated in the northwest, northeast, and southern regions of the TPO, and its contribution is about 55–80%. The local wind forcing in the TPO is the dominant influence on interannual sea level variations, and the contribution of wind forcing in other regions is small or even negative.

On the decadal time scales, sea level variations in the TPO are mainly controlled by large-scale ocean–atmosphere processes such as PDO and the decadal variations of ENSO [29]. The region with the most dramatic decadal sea level variations in the TPO is located in the northwestern region of the eastern Philippines, and its variance exceeds 40 cm^2. This area can explain 25–45% of the monthly mean sea level variations. The regions with the greatest contribution to the monthly mean sea level variations are in the southeast and northwest, 50–70%. The steric effect and the dynamic process explained 100% of the decadal sea level variations in most regions of the TPO. The region near the equator mainly contributed more than 85% of the steric effect. The dynamic process primarily acts on the southwest and northwest regions of the TPO, contributing only about 55–70% in others. The local wind forcing in the TPO is the leading role of decadal sea level variations. The contribution of wind forcing in other regions is small or even negative.

The most significant regions with interannual and decadal sea level fingerprints in the TPO are located in the western and central-eastern regions of TPO, with their maximum reaching 80 mm/unit ICI and 80 mm/unit DCI. The combined effect of ENSO and PDO explains 100% of the interannual to decadal sea level variations in the northwestern and eastern TPO. For the coast of north America, ENSO and PDO explained more than 90%. In the southwestern TPO, the combined effect of ENSO and PDO has minimal effect on the interannual to decadal sea level variations. The process that dominated this region's low-frequency sea level variations is mainly the steric effect and other processes.

Author Contributions: J.L. proposed the idea. J.W. collected data and made the analysis. J.L. and J.Y. wrote the manuscript. W.T. and Y.L. helped the data preparation and participated in discussions on the results. All authors have read and agreed to the published version of the manuscript.

Funding: This work was funded by National Natural Science Foundation of China (42006021, 41806039, 42076233, 41806004).

Institutional Review Board Statement: Not applicable.

Informed Consent Statement: Not applicable.

Data Availability Statement: All the data links can be found in Section 2.1.

Acknowledgments: The satellite altimeter data is maintained and provided by Archiving, Validation, and Interpretation of Satellite Oceanographic center, with support from the AVISO (ftp://ftp-access.aviso.altimetry.fr/climatology, accessed on 22 September 2021). We acknowledge the anonymous reviewers. The authors would like to express their gratitude to EditSprings (https://www.editsprings.cn/, accessed on 22 September 2021) for the expert linguistic services provided.

Conflicts of Interest: The authors declare they have no conflict of interest.

References

1. Hallegatte, S.; Green, C.; Nicholls, R.J.; Corfee-Morlot, J. Future flood losses in major coastal cities. *Nat. Clim. Chang.* **2013**, *3*, 802–806. [CrossRef]
2. Wong, P.P.; Losada, I.J.; Gattuso, J.P.; Hinkel, J.; Khattabi, A.; McInnes, K.L.; Saito, Y.; Sallenger, A. Climate Change 2014—Impacts, Adaptation, and Vulnerability. Part A: Global and Sectoral Aspects. Contribution of Work-ing Group II to the Fifth Assessment Report of the Intergovernmental Panel on Climate Change. In *Coastal Systems and Low–Lying Areas*; Field, C.B., Barros, V.R., Dokken, D.J., Mach, K.J., Mastrandrea, M.D., Bilir, T.E., Chatterjee, M., Ebi, K.L., Estrada, Y.O., Genova, R.C., et al., Eds.; Cambridge University Press: Cambridge, UK, 2014; pp. 361–409.
3. Church, J.A.; White, N.J. Sea-level rise from the late 19th to the early 21st century. *Surv. Geophys.* **2011**, *32*, 585–602. [CrossRef]
4. Becker, M.; Meyssignac, B.; Letetrel, C.; Llovel, W.; Cazenave, A.; Delcroix, T. Sea level variations at tropical Pacific islands since 1950. *Glob. Planet. Chang.* **2011**, *80*, 85–98. [CrossRef]
5. Zhang, X.B.; Church, J.A. Sea level trends, interannual and decadal variability in the Pacific Ocean. *Geophys. Res. Lett.* **2012**, *39*. [CrossRef]
6. Merrifield, M.A.; Mathew, E.M. Regional sea level trends due to a Pacific trade wind intensification. *Geophys. Res. Lett.* **2011**, *38*. [CrossRef]
7. Qiu, B.; Chen, S.; Wu, L.; Kida, S. Wind- versus eddy-forced regional sea level trends and variability in the North Pacific Ocean. *J. Clim.* **2015**, *28*, 1561–1577. [CrossRef]
8. Axel, M.; Mikael, C.; David, A.T.H.; Thomas, S. Ordovician and Silurian sea–water chemistry, sea level, and climate: A synopsis. *Palaeogeogr. Palaeoclimatol. Palaeoecol.* **2010**, *296*, 389–413. [CrossRef]
9. Merrifield, M.A.; Thompson, P.R.; Lander, M. Multidecadal sea level anomalies and trends in the western tropical Pacific. *Geophys. Res. Lett.* **2012**, *39*. [CrossRef]
10. Chen, D.; Cane, M.A.; Zebiak, S.E.; Kaplan, A. The impact of sea level data assimilation on the Lamont Model Prediction of the 1997/98 El Niño. *Geophys. Res. Lett.* **1998**, *25*, 2837–2840. [CrossRef]
11. Chambers, D.P.; Chen, J.; Nerem, R.S.; Tapley, B.D. Interannual mean sea level change and the Earth's water mass budget. *Geophys. Res. Lett.* **2000**, *27*, 3073–3076. [CrossRef]
12. Christopher, G.P.; Rui, M.P. Buoyancy-driven interannual sea level changes in the southeast tropical pacific. *Geophys. Res. Lett.* **2012**, *39*. [CrossRef]
13. Li, X.; Xu, J.; Cai, R.S. Trends of sea level rise in the South China Sea during the 1990s: An altimetry result. *Chin. Sci. Bull.* **2013**. [CrossRef]
14. Cazenave, A.; Llovel, W. Contemporary Sea Level Rise. *Annu. Rev. Mar. Sci.* **2010**, *2*, 145–173. [CrossRef]
15. Anny, C.; Frédérique, R. Sea level and climate: Measurements and causes of changes. *Wiley Interdiscip. Rev. Clim. Chang.* **2011**, *2*, 647–662. [CrossRef]
16. Merrifield, M.A. A Shift in Western Tropical Pacific Sea Level Trends during the 1990s. *J. Clim.* **2011**, *24*, 4126–4138. [CrossRef]
17. Han, W.Q.; Meehl, G.A.; Hu, A.X.; Alexander, M.A.; Yamagata, T.; Yuan, D.L.; Ishii, M.; Pegion, P.; Zheng, J.; Hamlington, B.D.; et al. Intensification of decadal and multi-decadal sea level variability in the western tropical Pacific during recent decades. *Clim. Dyn.* **2014**, *43*, 1357–1379. [CrossRef]
18. Hu, S.J.; Fu, D.Y.; Han, Z.L.; Ye, H. Density, Demography, and Influential Environmental Factors on Overwintering Populations of Sogatella furcifera (Hemiptera: Delphacidae) in Southern Yunnan, China. *J. Insect Sci.* **2015**, *15*, 58. [CrossRef]
19. Lu, Q.; Zuo, J.C.; Wu, L.J. Low-frequency variation in sea level in the tropical Pacific. *Haiyang Xuebao* **2017**, *39*, 43–52. [CrossRef]
20. Lu, Q.; Zuo, J.C.; Li, Y.F.; Chen, M.X. Interannual sea level variability in the tropical Pacific Ocean from 1993 to 2006. *Glob. Planet. Chang.* **2013**, *107*, 70–81. [CrossRef]
21. Chen, J.L.; Shum, C.K.; Wilson, C.R.; Chambers, D.P.; Tapley, B.D. Seasonal sea level change from TOPEX/Poseidon observation and thermal contribution. *J. Geod.* **2000**, *73*, 638–647. [CrossRef]
22. Yu, H.L. The characteristic of Northwest Pacific sea level variation, and influencing factor. *J. Ocean. Univ. China* **2013**, *43*, 9–20. [CrossRef]
23. Li, X.H.; Wei, G.J.; Shao, L.; Liu, Y.; Liang, X.R.; Jian, Z.M.; Sun, M.; Wang, P.X. Geochemical and Nd isotopic variations in sediments of the South China Sea: A response to Cenozoic tectonism in SE Asia. *Earth Planet. Sci. Lett.* **2003**, *211*, 207–220. [CrossRef]
24. Li, J.; Tan, W.; Chen, M.X.; Luo, F.Y.; Liu, Y.; Fu, Q.J.; Li, B.T. An extreme sea level event along the northwest coast of the South China sea in 2011–2012. *Cont. Shelf Res.* **2020**, *196*, 104073. [CrossRef]
25. Busalacchi, A.J.; Mcphaden, M.J.; Picaut, J.; Springer, S.R. Sensitivity of wind-driven tropical Pacific Ocean simulations on seasonal and interannual time scales. *J. Mar. Syst.* **1990**, *1*, 119–154. [CrossRef]
26. Sergey, V.; Vinogradov, R.M.; Ponte, P.H.; Carl, W. The mean seasonal cycle in sea level estimated from a data-constrained general circulation model. *J. Geophys. Res. Ocean.* **2008**, *113*. [CrossRef]
27. Roberts, C.D.; Calvert, D.; Dunstone, N.; Hermanson, L.; Palmer, M.D.; Smith, D. On the Drivers and Predictability of Seasonal-to-Interannual Variations in Regional Sea Level. *J. Clim.* **2016**, *29*, 7565–7585. [CrossRef]
28. Chen, X.; Qiu, B.; Cheng, X.H.; Qi, Y.Q.; Du, Y. Intra-seasonal variability of Pacific-origin sea level anomalies around the Philippine Archipelago. *J. Oceanogr.* **2015**, *71*, 239–249. [CrossRef]

29. Chen, M.X.; Zuo, C.S.; Zhang, W.H.; Jia, Y.R.; Lyu, X.F. Research progress of inter-annual and multi-decadal sea level variability in tropical Pacific Ocean. *J. Oceanogr.* **2017**, *45*, 249–255. [CrossRef]
30. Chang, Y.T.; Du, L.; Zhang, S.W.; Huang, P.F. Sea level variations in the tropical Pacific Ocean during two types of recent El Niño events. *Glob. Planet. Chang.* **2013**, *108*, 119–127. [CrossRef]
31. Landerer, F.W.; Jungclaus, J.H.; Marotzke, J. El Niño–Southern Oscillation signals in sea level, surface mass redistribution, and degree-two geoid coefficients. *J. Geophys. Res. Ocean.* **2008**, *113*, C08014. [CrossRef]
32. Wang, G.Q.; Jin, J.L.; Wan, S.C.; Zhai, R. Variation of sea level rise along the China coastal area during the past decades. In Proceedings of the 2015 Academic Annual Meeting of China Water Conservancy Society, Nanjing, China, 26–28 October 2015; Volume II.
33. Li, L.H.; Zhou, L.H.; Lian, H.J. Sea-level Rise Forecast Model. *J. Comput.* **2011**, *6*, 2120–2126. [CrossRef]
34. Han, W.Q.; Gerald, A.; Meehl, A.; Detlef, S.; Aixue, H.; Benjamin, H.; Jessica, K.; Hindumathi, P.; Philip, T. Spatial Patterns of Sea Level Variability Associated with Natural Internal Climate Modes. *Surv. Geophys.* **2017**, *38*, 221–254. [CrossRef]
35. Meng, L.S.; Zhuang, W.; Zhang, W.W.; Ditri, A.; Yan, X.H. Decadal Sea Level Variability in the Pacific Ocean: Origins and Climate Mode Contributions. *J. Atmos. Ocean. Technol.* **2019**, *36*, 689–698. [CrossRef]
36. Duan, J.; Li, Y.L.; Zhang, L.; Wang, F. Impacts of the Indian Ocean Dipole on Sea Level and Gyre Circulation of the Western Tropical Pacific Ocean. *J. Clim.* **2020**, *33*, 4207–4228. [CrossRef]
37. Merrifield, M.; Kilonsky, B.; Nakahara, S. Interannual sea level changes in the tropical Pacific associated with ENSO. *Geophys. Res. Lett.* **1999**, *26*, 3317–3320. [CrossRef]
38. Nerem, R.S.; Chambers, D.P.; Choe, C.; Mitchum, G.T. Estimating Mean Sea Level Change from the TOPEX and Jason Altimeter Missions. *Mar. Geod.* **2010**, *33*, 435–446. [CrossRef]
39. Long, X.Y.; Widlansky, M.J.; Schloesser, F.; Thompson, P.R.; Annamalai, H.; Merrifield, M.A.; Yoon, H. Higher Sea Levels at Hawaii Caused by Strong El Niño and Weak Trade Winds. *J. Clim.* **2020**, *33*, 3037–3059. [CrossRef]
40. Fu, L.; Qiu, B. Low-frequency variability of the North Pacific Ocean: The roles of boundary- and wind-driven baroclinic Rossby waves. *J. Geophys. Res. Ocean.* **2002**, *107*, 3220. [CrossRef]
41. Zhuang, W.; Qiu, B.; Du, Y. Low-frequency western Pacific Ocean sea level and circulation changes due to the connectivity of the Philippine archipelago. *J. Geophys. Res.* **2013**, *118*, 6759–6773. [CrossRef]
42. Qiu, B.; Chen, S. Interannual-to-Decadal Variability in the Bifurcation of the North Equatorial Current off the Philippines. *J. Phys. Oceanogr.* **2010**, *40*, 2525–2538. [CrossRef]
43. Mantua, N.J.; Hare, S.R.; Zhang, Y.; Wallace, J.M.; Francis, R.C. A Pacific interdecadal climate oscillation with impacts on salmon production. *Bull. Am. Meteorol. Soc.* **1997**, *78*, 1069–1079. [CrossRef]
44. Wu, B.Y.; Wang, J. Winter Arctic oscillation, Siberian high and East Asian winter monsoon. *Geophys. Res. Lett.* **2002**, *29*, 1897. [CrossRef]
45. Feng, W.; Zhong, M.; Xu, H.Z. Sea level variation in the South China Sea inferred from satellite gravity, altimetry, and oceanographic data. *Sci. China Earth Sci.* **2012**, *55*, 1696–1701. [CrossRef]
46. Peng, D.J.; Palanisamy, H.; Cazenave, A.; Meyssignac, B. Interannual Sea Level Variations in the South China Sea Over 1950–2009. *Mar. Geod.* **2013**, *36*, 164–182. [CrossRef]
47. Wu, L.L.; Wang, X.L.L.; Yang, F. Historical wave height trends in the South and East China Seas, 1911-2010. *J. Geophys. Res. Ocean.* **2014**, *119*, 4399–4409. [CrossRef]
48. Zhou, J.; Li, P.L.; Yu, H.L. Characteristics and mechanisms of sea surface height in the South China Sea. *Glob. Planet. Chang.* **2012**, *88*, 20–31. [CrossRef]
49. Deng, W.F.; Wei, G.J.; Xie, L.H.; Ke, T.; Wang, Z.B.; Zeng, T.; Liu, Y. Variations in the Pacific Decadal Oscillation since 1853 in a coral record from the northern South China Sea. *J. Geophys. Res. Ocean.* **2013**, *118*, 2358–2366. [CrossRef]
50. Han, S.L. Estimation of extreme sea levels along the Bangladesh coast due to storm surge and sea level rise using EEMD and EVA. *J. Geophys. Res. Ocean.* **2013**, *118*, 4273–4285. [CrossRef]
51. Church, J.A.; White, N.J.; Hunter, J.R. Sea-level rise at tropical Pacific and Indian Ocean islands. *Glob. Planet. Chang.* **2006**, *53*, 155–168. [CrossRef]
52. Nidheesh, A.G.; Lengaigne, M.; Vialard, J.; Izumo, T.; Unnikrishnan, A.S.; Meyssignac, B.; Hamlington, B.; Boyer Montegut, C. Robustness of observation-based decadal sea level variability in the Indo-Pacific Ocean. *Geophys. Res. Lett.* **2017**, *44*, 7391–7400. [CrossRef]
53. Moon, J.; Song, Y.T.; Lee, H.K. PDO and ENSO modulations intensified decadal sea level variability in the tropical Pacific. *J. Geophys. Res. Ocean.* **2015**, *120*, 8229–8237. [CrossRef]
54. Deepa, J.S.; Gnanaseelan, C.; Mohapatra, S.; Chowdary, J.S.; Karmakar, A.; Kakatkar, R.; Parekh, A. The Tropical Indian Ocean decadal sea level response to the Pacific Decadal Oscillation forcing. *Clim. Dyn.* **2019**, *52*, 5045–5058. [CrossRef]
55. Deepa, J.S.; Gnanaseelan, C.; Parekh, A. The sea level variability and its projections over the Indo-Pacific Ocean in CMIP5 models. *Clim. Dyn.* **2021**, *57*, 173–193. [CrossRef]
56. Piecuch, C.G.; Thompson, P.R.; Ponte, R.M.; Merrifield, M.A.; Hamlington, B.D. What Caused Recent Shifts in Tropical Pacific Decadal Sea-Level Trends? *J. Geophys. Res. Ocean.* **2019**, *124*, 7575–7590. [CrossRef]
57. Hu, Z.Z.; Kumar, A.; Ren, H.L.; Wang, H.; L'Heureux, M.; Jin, F.F. Weakened Interannual Variability in the Tropical Pacific Ocean since 2000. *J. Clim.* **2013**, *26*, 2601–2613. [CrossRef]

58. Thomson, R.E.; Tabata, S. Steric height trends at Ocean Station PAPA in the northeast Pacific Ocean. *Mar. Geod.* **1987**, *11*, 103–113. [CrossRef]
59. Gharineiat, Z.; Deng, X.L. Description and assessment of regional sea-level trends and variability from altimetry and tide gauges at the northern Australian coast. *Adv. Space Res.* **2018**, *61*, S0273117718301984. [CrossRef]
60. Nidheesh, A.G.; Lengaigne, M.; Vialard, J.; Unnikrishnan, A.S.; Dayan, H. Decadal and long-term sea level variability in the tropical Indo-Pacific Ocean. *Clim. Dyn.* **2013**, *41*, 381–402. [CrossRef]
61. Wyrtki, K. Investigation of the El Nino Phenomenon in the Pacific Ocean. *Environ. Conserv.* **1975**, *2*, 281. [CrossRef]
62. Li, S.; Zhao, J.; Li, Y.; Qu, P. Study on Decadal Variability of Sea Surface Height in the Tropical Pacific Ocean. *Adv. Mar. Sci.* **2008**. [CrossRef]
63. Zhu, Z.N.; Zhu, X.H.; Guo, X.Y. Coastal tomographic mapping of nonlinear tidal currents and residual currents. *Cont. Shelf Res.* **2017**, *143*, 219–227. [CrossRef]
64. Palanisamy, H.; Cazenave, A.; Delcroix, T.; Meyssignac, B. Spatial trend patterns in the Pacific Ocean sea level during the altimetry era: The contribution of thermocline depth change and internal climate variability. *Ocean. Dyn.* **2015**, *65*, 341–356. [CrossRef]
65. Matthew, J.W.; Axel, T.; Karl, S.; Shayne, M.; Niklas, S.; Matthew, H.E.; Matthieu, L.; Wenju, C. Changes in South Pacific rainfall bands in a warming climate. *Nat. Clim. Chang.* **2013**, *3*, 417–423. [CrossRef]
66. Matthew, J.; Widlansky, A.T.; Wenju, C. Future extreme sea level seesaws in the tropical Pacific. *Sci. Adv.* **2015**, *1*, e1500560. [CrossRef]
67. Gaël, A.; Thierry, D. Interannual sea level changes and associated mass transports in the tropical Pacific from TOPEX/Poseidon data and linear model results (1964–1999). *J. Geophys. Res. Ocean.* **2002**, *107*, 17-1–17-22. [CrossRef]
68. Nerem, R.S.; Chambers, D.P.; Leuliette, E.W.; Mitchum, G.T.; Giese, B.S. Variations in global mean sea level associated with the 1997–1998 ENSO event: Implications for measuring long-term sea level change. *Geophys. Res. Lett.* **1999**, *26*, 3005–3008. [CrossRef]

Article

Monitoring Meteorological Drought in Southern China Using Remote Sensing Data

Li Liu [1,2,3], Ran Huang [4], Jiefeng Cheng [5], Weiwei Liu [6], Yan Chen [1,2,3], Qi Shao [1,2,3], Dingding Duan [7], Pengliang Wei [1,2,3], Yuanyuan Chen [8] and Jingfeng Huang [1,2,3,*]

[1] Institute of Applied Remote Sensing and Information Technology, Zhejiang University, Hangzhou 310058, China; 21714122@zju.edu.cn (L.L.); chen_yan0309@zju.edu.cn (Y.C.); 11914070@zju.edu.cn (Q.S.); weipengliang@zju.edu.cn (P.W.)
[2] Key Laboratory of Agricultural Remote Sensing and Information Systems, Zhejiang University, Hangzhou 310058, China
[3] Key Laboratory of Environment Remediation and Ecological Health, Ministry of Education, College of Natural Resources and Environmental Science, Zhejiang University, Hangzhou 310058, China
[4] School of Automation, Hangzhou Dianzi University, Hangzhou 310018, China; huang@hdu.edu.cn
[5] Zhejiang Geopher Spatial Planning Technology Co., Ltd., Hangzhou 310000, China; papainho@163.com
[6] Department of Geography and Spatial Information Techniques, Ningbo University, Ningbo 315211, China; liuweiwei@nbu.edu.cn
[7] Institute of Agricultural Resources and Regional Planning, Chinese Academy of Agricultural Sciences, Beijing 100081, China; duandingding@caas.cn
[8] College of Environment, Zhejiang University of Technology, Hangzhou 310000, China; yychen91@zjut.edu.cn
* Correspondence: hjf@zju.edu.cn; Tel.: +86-139-5717-1636

Citation: Liu, L.; Huang, R.; Cheng, J.; Liu, W.; Chen, Y.; Shao, Q.; Duan, D.; Wei, P.; Chen, Y.; Huang, J. Monitoring Meteorological Drought in Southern China Using Remote Sensing Data. *Remote Sens.* 2021, 13, 3858. https://doi.org/10.3390/rs13193858

Academic Editors: Baojie He, Ayyoob Sharifi, Chi Feng and Jun Yang

Received: 11 August 2021
Accepted: 24 September 2021
Published: 27 September 2021

Publisher's Note: MDPI stays neutral with regard to jurisdictional claims in published maps and institutional affiliations.

Copyright: © 2021 by the authors. Licensee MDPI, Basel, Switzerland. This article is an open access article distributed under the terms and conditions of the Creative Commons Attribution (CC BY) license (https://creativecommons.org/licenses/by/4.0/).

Abstract: Severe meteorological drought is generally considered to lead to crop damage and loss. In this study, we created a new standard value by averaging the values distributed in the middle 30–70% instead of the traditional mean value, and we proposed a new index calculation method named Normalized Indices (NI) for meteorological drought monitoring after normalized processing. The TRMM-derived precipitation data, GLDAS-derived soil moisture data, and MODIS-derived vegetation condition data from 2003 to 2019 were used, and we compared the NI with commonly used Condition Indices (CI) and Anomalies Percentage (AP). Taking the mid-to-lower reaches of the Yangtze River (MLRYR) as an example, the drought monitoring results for paddy rice and winter wheat showed that (1) NI can monitor well the relative changes in real precipitation/soil moisture/vegetation conditions in both arid and humid regions, while meteorological drought was overestimated with CI and AP, and (2) due to the monitoring results of NI, the well-known drought event that occurred in the MLRYR from August to October 2019 had a much less severe impact on vegetation than expected. In contrast, precipitation deficiency induced an increase in sunshine and adequate heat resources, which improved crop growth in 78.8% of the area. This study discusses some restrictions of CI and AP and suggests that the new NI index calculation provides better meteorological drought monitoring in the MLRYR, thus offering a new approach for future drought monitoring studies.

Keywords: meteorological drought; drought impact; paddy rice; winter wheat

1. Introduction

Drought, rainstorms, typhoons, high-temperature-induced damage, low temperature chilling injuries, and hailstorms have occurred frequently around the world in recent years. These meteorological disasters have a negative impact on normal socioeconomic development [1–4]. Drought is one of the most devastating natural disasters [5], especially in areas that rely heavily on rain-fed subsistence agriculture. Drought-induced famine seriously affects human survival and agricultural production [6–8].

After vegetation indices were developed in the 1980s, the Normalized Difference Vegetation Index (NDVI) was used to effectively monitor rainfall and drought and to estimate the impact of weather on crops and pastures in nonhomogeneous areas [9–12]. The problem is that, in addition to the weather influence, the difference in vegetation levels in these areas is also related to the differences between geographical resources (climate, soil, vegetation types, and terrain). For eliminating that portion of the NDVI, Kogan [13] calculated with Advanced Very High Resolution Radiometer (AVHRR) data the largest and lowest NDVI values during 1984–1987 for each of the 52 weeks of the year and for each pixel of Sudan. The maximum and minimum NDVI were used as the criteria for estimating the upper (favorable weather) and lower (unfavorable weather) limits of the ecosystem resources [14,15]. The difference between the maximum and minimum NDVI time series is due to weather variation. For enhancing the weather-related signal in NDVI values, the Vegetation Condition Index (VCI) was developed. The results showed that VCI was linearly positively correlated with precipitation. It was not sufficiently comprehensive to monitor drought only by the decline in NDVI, but the research proposed a generalized global meteorological disaster monitoring method based on the remote sensing index, so disaster monitoring achieved development from point to surface [16]. Similar to the VCI algorithm, various drought evaluation indexes based on different meteorological factors appeared gradually.

In 1995, the Temperature Condition Index (TCI) was developed by Kogan to estimate the maximum/minimum of the temperature envelope, which was used to determine temperature-related vegetation stress in addition to stress caused by excess rain [17]. High temperatures in the middle of the season indicate unfavorable or drought conditions, while low temperatures indicate mostly favorable conditions. Based on the Tropical Rainfall Measuring Mission (TRMM) precipitation data, Rhee et al. [18] proposed Scaled TRMM, which has the same calculation method as the VCI, while in 2013, Zhang and Jia [19] proposed the Soil Moisture Condition Index (SMCI) based on Advanced Microwave Scanning Radiometer for EOS (AMSR-E)-derived soil moisture. Over a long period of time, a variety of remote sensing drought monitoring indices have been developed for assessing meteorological drought, agricultural drought, and hydrological drought based on these Condition Indices (CI, such as VCI, TCI, PCI, and SMCI), some of which are shown in Table 1 [20–26].

The "Classification of Meteorological Drought" implemented in China on 1 November 2006, is the first national standard for monitoring meteorological drought disasters. It specifies the indicator, percentage of precipitation anomalies, which represents the changes in precipitation in a certain period compared with the average precipitation of all years. This indicator is used in daily business by the departments of the China Meteorological Administration, and it can assess monthly, seasonal, and annual drought events. The anomalies of soil moisture, vegetation, and temperature are also widely used in many studies [27–34], and they are collectively referred to as the Anomalies Percentage (AP).

Table 1. Summary of typical studies based on Condition Indices for drought monitoring.

Reference	Region and Year	Indices (Optimal Index Displayed in Bold)	Main Conclusion and Correlation between Index and Precipitation/Crop Yield
Kogan [13]	Sudan, Africa (1984–1987)	NDVI/**VCI**	VCI was first proposed and was positively correlated with precipitation.
Kogan [17]	the United States (1985–1993)	VCI/**TCI**	TCI was first proposed; the combination of VCI and TCI was the basis for VHI.
Rhee, Im, and Carbone [18]	North Carolina/South Carolina/Arizona/New Mexico (2000–2009)	scaled LST/scaled TRMM/scaled NDVI/scaled NMDI/scaled NDWI/scaled NDDI/VHI/**SDCI**/Z-Index	PCI was first proposed; SDCI performed better than existing indices such as NDVI and VHI and was positively correlated with crop yield.
Zhang and Jia [19]	Northern China (2003–2010)	PCI/SMCI/TCI/VCI/PSMCI/PTCI/SMTCI/**MIDI**	SMCI was first proposed; MIDI was the optimum in monitoring short-term drought, especially for meteorological drought across northern China.
Du, et al. [35]	Shandong, China (2013–2017)	PCI/TCI/VCI/**SDI**/SPI	SDI was positively correlated with precipitation and crop yield. VCI/SDI/TCI were all negatively correlated with drought affected crop area.
Zhang, et al. [36]	Hubei, Yunnan, Hebei Provinces, China (1981–2011)	PCI/SMCI/VCI/**PADI**/PDSI/SPI	Compared with the correlation with precipitation, soil moisture and vegetation data alone, PADI correlated well with wheat yield loss.
Liu, et al. [37]	Shandong, China (2013–2017)	PCI/SMCI/TCI/VCI/**MCDIs**/SPI/SPEI/MI	MCDIs is positively correlated with SPI-1 and MI. MCDI-1 was suitable to monitor meteorological drought and MCDI-9 was a good indicator for agricultural drought.
Wei, et al. [38]	Southwestern China (2001–2019)	PCI/SMCI/TCI/**OMDI**/SPI/SPEI	There is a significant positive correlation between OMDI and grain yield as well as between OMDI and NPP in most areas of China.
Wei, et al. [39]	Northwest China (2001–2019)	PCI/SMCI/TCI/VCI/**RSDEI**/SPEI	RSDEI had a strong correlation with NPP and crop yield except in some western parts of the study area.

Reviewing past studies, CI, AP, and synthetic indices based on them, have been widely used in existing drought monitoring, but there are few studies on their drought monitoring effects in southern China. In addition, the monitored results of these drought indices were usually validated by observed precipitation or statistical crop yield data. The conclusion was that the more severe the meteorological drought, the more severe the crop yield reduction (Table 1). However, contradictory phenomena are often overlooked in areas with abundant precipitation. Therefore, it is necessary to propose more effective drought monitoring methods in areas with abundant precipitation. Thus, there were three main objectives in this study: (1) to explore the applicability of the CI and AP for meteorological drought monitoring in southern China; (2) to propose a new index calculation approach, Normalized Indices (NI), for meteorological drought monitoring in southern China; and (3) to study the actual relationship between meteorological drought and crop health, such as paddy rice (*Oryza sativa* L.) and winter wheat (*Triticum aestivum* L.). The study developed a new drought index calculation method and provides a novel approach for future drought monitoring studies.

2. Study Area and Data

2.1. Study Area

The study area is located in the mid-to-lower reaches of the Yangtze River (MLRYR), extending from 24.5° N to 35.1° N and 108.4° E to 121.9° E (Figure 1). The area covers five administrative provincial units: Jiangsu, Anhui, Hubei, Jiangxi, and Hunan. While a single-cropped rice cultivation system is dominant in Jiangsu, Anhui, and Hubei Provinces, paddy rice is mainly cropped in rotation with winter wheat; a double-cropped rice cultivation system is practiced in Jiangxi and Hunan Provinces. The area has a subtropical monsoon climate with warm temperatures and abundant precipitation (Figure 2). From August to October, when crops mature, the East Asian Summer Monsoon retreats southward. Droughts and floods happen easily in this season and have caused serious economic losses and environmental damage [40–43]. In addition, the catchment area of the Yangtze River is the most concentrated area of freshwater lakes in China. Most parts of the study area are relatively flat and low-lying, including the famous Poyang and Dongting Lakes [44,45].

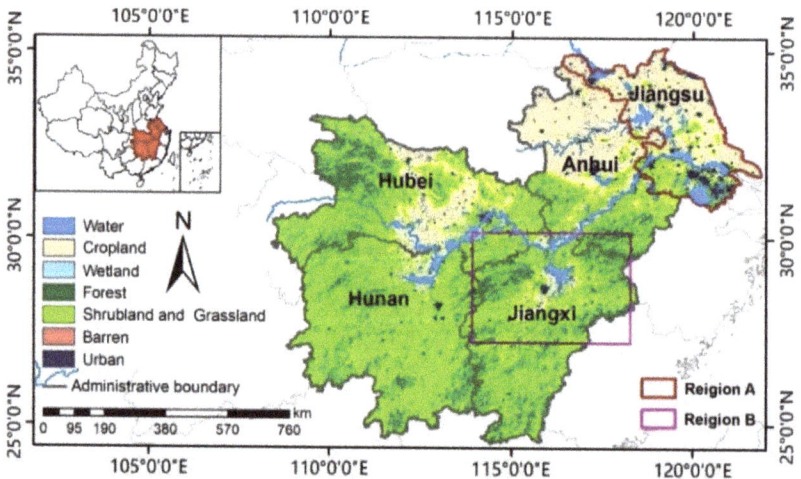

Figure 1. The study area.

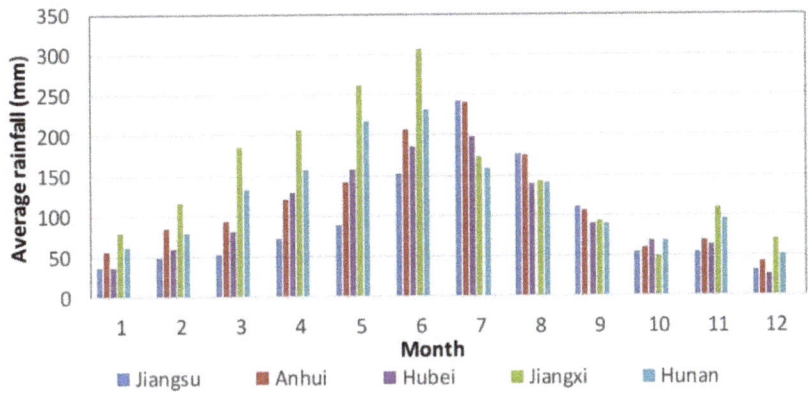

Figure 2. Average rainfall from 2003 to 2019 in study area.

2.2. Data

To achieve drought evolution process monitoring, long-term precipitation, root zone soil moisture, and vegetation data were integrated. Additionally, distribution maps of winter wheat and paddy rice were used to explore the impact of drought on crops. The crop yield data were also calculated for validation purposes. The data and related information used in this study are shown in Table 2.

Table 2. Data and related information used in the study.

Data	Source	Study Year	Temporal Resolution	Spatial Resolution
Precipitation	TRMM3B42/ TRMM3B43	2003–2019	8 days/month	0.25°
Soil Moisture	GLDAS-2.1	2003–2019	8 days/month	0.25°
Vegetation	MOD09A1/ MYD09A1	2003–2019	8 days/month	500 m
Cropland	MCD12Q1	2013	year	500 m
Wheat map	Decision Tree Classification	2011–2015	year	500 m
Rice map	PhenoRice	2011–2015	year	500 m
Growth stage	CMDSC	2011–2015	-	-
Yield	JMIC	2003–2019	year	County level

2.2.1. TRMM Data

The Tropical Rainfall Measuring Mission (TRMM), a joint project of the National Aeronautics and Space Administration (NASA) of the USA and the Japan Aerospace Exploration Agency (JAXA), was launched in November 1997 [46]. For this study, daily 3B42 precipitation data and monthly 3B43 precipitation data were used at a spatial resolution of 0.25°. The eight-day precipitation data were generated through temporal averaging of the daily 3B42 precipitation data. The precipitation data were preprocessed and downloaded on the Google Earth Engine (GEE) platform.

2.2.2. GLDAS Data

Root zone soil moisture is important and relatively stable compared with the surface soil moisture because the surface soil moisture is sensitive to other environmental variables (e.g., temperature) that drive atmospheric evaporative demand. The Global Land Data Assimilation System version 2 (GLDAS-2) has two components: one forced entirely with Princeton meteorological forcing data (GLDAS-2.0) and the other forced with a combination of model and observation-based forcing datasets (GLDAS-2.1) [47,48]. The three-hourly GLDAS-2.1 Noah Land Surface Model L4 product at 0.25° resolution from 2003 to 2019 was used to generate the eight-day root zone soil moisture data through temporal averaging, which were preprocessed and downloaded on the GEE platform.

2.2.3. MODIS Data

The 500 m, eight-day composite surface reflectance products (MOD09A1 and MYD09A1) of the Terra and Aqua satellites from 2003 to 2019 were downloaded from NASA's Level 1 and Atmosphere Archive and Distribution System (LAADS) (26 February 2020: https://ladsweb.modaps.eosdis.nasa.gov/search/). With the data processing method combination of EVI2_BLUE_MYO [49], the processing procedures mainly included image mosaicking, subsetting, spectral indices calculation, data quality labeling, cloudy pixel removal, interpolation of vegetation index images, image stacking, and Savitzky–Golay smoothing [50], all of which were implemented using Python v.3.7 programming language.

2.2.4. Land Cover Data

The distribution of winter wheat in Jiangsu Province from 2011 to 2015 came from Chen [51], and the spatial resolution had been resampled from 250 m to 500 m. The distribution of rice from 2011 to 2015 was obtained by the PhenoRice algorithm, with a resolution of 500 m [49,52]. Both maps are based on decision tree classification, combined with the phenology information of crops, with accuracies greater than 90%. The 500 m MODIS Land Cover Type products (MCD12Q1) of 2013 were downloaded from LAADS. Land_Cover_Type_1 was selected from datasets of land cover type products. The types of land cover had been merged from the original 17 categories to form 6 categories for use as a base map; the results are shown in Figure 1.

2.2.5. Other Data

Yield data of paddy rice were provided by the Jiangsu Meteorological Information Centre (JMIC) of China, including the statistical area and yield data of 72 counties in Jiangsu Province (Region A) from 2003 to 2019. The growth stage data of field observations from 2003 to 2015 were downloaded from the China Meteorological Data Service Centre (CMDSC, 25 June 2018: http://data.cma.cn/). The entire growing season of winter wheat was divided into two stages: Wheat Stage 1 (from sowing to the end of the regreening period—late October of the previous year to late February) and Wheat Stage 2 (from jointing period to maturity—early March to early June). The growing season of paddy rice was also divided into two stages: Rice Stage 1 (from transplanting to the end of jointing period—mid-June to late July) and Rice Stage 2 (from booting period to maturity—early August to mid-October).

3. Methodology

3.1. Calculation of the Condition Indices and Anomalies Percentage

PCI, SMCI, and VCI, calculated using TRMM, GLDAS, and MODIS data, respectively, are collectively called the Condition Indices (CI) and are computed as follows:

$$CI_i = \frac{F_i - F_{min}}{F_{max} - F_{min}} \quad (1)$$

where F_i, F_{max}, and F_{min} are the pixel values of precipitation (or root zone soil moisture or EVI2) and its maximum and minimum values, respectively. CI_i varies from 0 to 1, but a value of 0.5 is usually set as the threshold to monitor anomalous events. When CI_i equals 0.5, it is not difficult to obtain

$$F_i = Standard_{CI} = \frac{F_{max} + F_{min}}{2} \quad (2)$$

Take PCI as an example. During a meteorological drought with low precipitation, the PCI is close to or equal to 0, while during flooding conditions it is close to 1. If the PCI is less than 0.5, it means the precipitation is less than Standard$_{PCI}$.

The Precipitation Anomalies Percentage (PAP), Soil Moisture Anomalies Percentage (SMAP), and Vegetation Anomalies Percentage (VAP), collectively referred to here as the Anomalies Percentage (AP), are computed as follows:

$$AP_i = \frac{F_i - \overline{F}}{\overline{F}} \times 100\% \qquad (3)$$

where F_i and $\overline{F}$ are the pixel values of precipitation (or root zone soil moisture or EVI2), and the mean value is computed as follows:

$$Standard_{AP} = \overline{F} = \frac{F_1 + F_2 + \cdots + F_n}{n} \qquad (4)$$

Taking PAP as an example, in the ideal condition of a meteorological drought with low precipitation, PAP is close to or equal to −100%, while during flooding conditions the PAP is close to positive infinity. When PAP is less than 0, it means that the precipitation is less than Standard$_{PAP}$; that is, less than the average precipitation over the years.

3.2. Principle and Construction of the Normalized Indices

When using multiyear RS data to monitor drought, whether CI or AP, the purpose of the calculation is to compare with a standard value to judge the degree of drought or moisture. Therefore, this standard value needs to be typical and can represent the normal level of the pixel over a long period of time; thus, we proposed a new index calculation method named Normalized Indices (NI), the development of which is shown in Figure 3.

Name of calculation method	Formula	Standard value	Indices
Anomalies Percentage	$\frac{F_i - \overline{F}}{\overline{F}} \times 100\%$	$\overline{F} = \frac{F_1 + F_1 + \cdots + F_n}{n}$	PAP/SMAP/VAP
	① Development ①: Replace $\overline{F}$ with $\overline{F}'$		
Enhanced Anomalies Percentage	$\frac{F_i - \overline{F}'}{\overline{F}'} \times 100\%$	$\overline{F}'$	EPAP/ESMAP/EVAP
	② Development ②: Normalized calculation		
Normalized Indices	$\frac{F_i - \overline{F}'}{F_i + \overline{F}'}$	$\overline{F}'$	NPI/NSMI/NVI

Figure 3. The development of Normalized Indices. Note: $\overline{F}$ is the mean value of precipitation (or root zone soil moisture or vegetation index) over many years. $\overline{F}'$ is calculated by arranging the precipitation value of a single pixel over many years, from small to large, and averaging the values distributed in the middle 30–70% (40% in total).

Because the extreme values are added to the calculation of Standard$_{PCI}$ and Standard$_{PAP}$, they cannot represent well the real normal level of the pixels. Based on the AP calculation method, we use $\overline{F}'$ instead of $\overline{F}$ to obtain the calculation formula of Enhanced Anomalies Percentage (EAP):

$$EAP_i = \frac{F_i - \overline{F}'}{\overline{F}'} \times 100\% \qquad (5)$$

$\overline{F}'$ is calculated by arranging the precipitation value (or soil moisture/vegetation index) of a single pixel for many years, from small to large, and averaging the values distributed in the middle 30–70% (40% in total). However, when F_i exceeds twice $\overline{F}$ (or $\overline{F}'$), PAP and EPAP are greater than 100%. Since there is no upper limit for PAP and EPAP under ideal conditions, the modified Normalized Indices (NI) is proposed to monitor changes in precipitation (Normalized Precipitation Index, NPI), soil moisture (Normalized Soil Moisture Index, NSMI), and crop growth status (Normalized Vegetation Index, NVI), which is defined as follows:

$$NI_i = \frac{F_i - \overline{F}'}{F_i + \overline{F}'} \tag{6}$$

NI_i varies from −1 to 1, and the value of 0 is set as the threshold for monitoring the anomalous change:

$$Standard_{EAP} = Standard_{NI} = \overline{F}' \tag{7}$$

Take the multiyear precipitation events of typical pixels in the study area as an example (Figure 4). The area where Pixel 2 is located experienced extraordinary rainstorm events from 2003 to 2019, but Pixel 1 did not. Due to the small difference in precipitation over the years for Pixel 1, $Standard_{PCI}$, $Standard_{PAP}$, and $Standard_{NPI}$ are not very different; they are all close to the normal level. For Pixel 2, because the extreme maximum value was added to the calculation, the $Standard_{PCI}$ is much higher than the pixel values of normal years. Precipitation for all years was less than $Standard_{PCI}$, except in the year when the maximum occurred. As a result, PCI-based algorithms monitor different degrees of meteorological drought in the subsequent 16 years, which is completely inconsistent with the facts. The AP also has the same problem with CI, but the degree is relatively minor. In contrast, $Standard_{NI}$ is typical and can represent the normal level of the pixel over a long period of time.

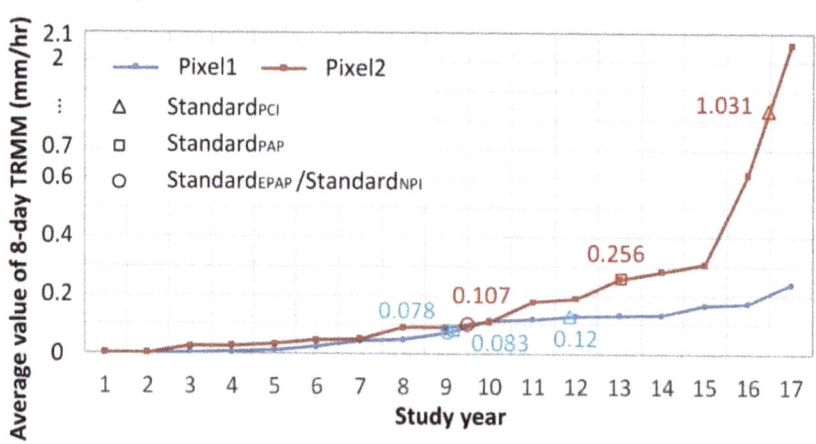

Figure 4. Time series curve of eight-day average TRMM data of typical pixels in the study area from 2003 to 2019. The row and column number of Pixel 1 is (36,50), shown in blue; Pixel 2, which encounters an abnormal rainy event with coordinates of (24,76), is given in red. Note: Study years sort by TRMM value from small to large.

3.3. Differences in Monitoring Effects of Different Indices

Take precipitation as an example. Assume that the precipitation range (true value) is between 0 and 10, where 0 means no precipitation, 10 means the maximum precipitation recorded in the history of all regions, and 5 is the normal precipitation in an ordinary semiarid and semihumid region. Among them, the omitted year (ellipsis) pixel value in the first column (Figure 5(a1,b1,c1)) is the same as in Year n. Taking Figure 5(b1) as an example, all of them are 0.25. Pixel 1 represents normal pixels that show no extreme drought or extraordinary rainstorm event occurring, or that show places where both have occurred with similar severity in all monitoring years; Pixel 2 represents only severe drought events that occurred in a certain year; Pixel 3 represents only severe humid events that occurred (such as a sudden increase in precipitation, sudden irrigation, dry land becoming paddy field, etc.).

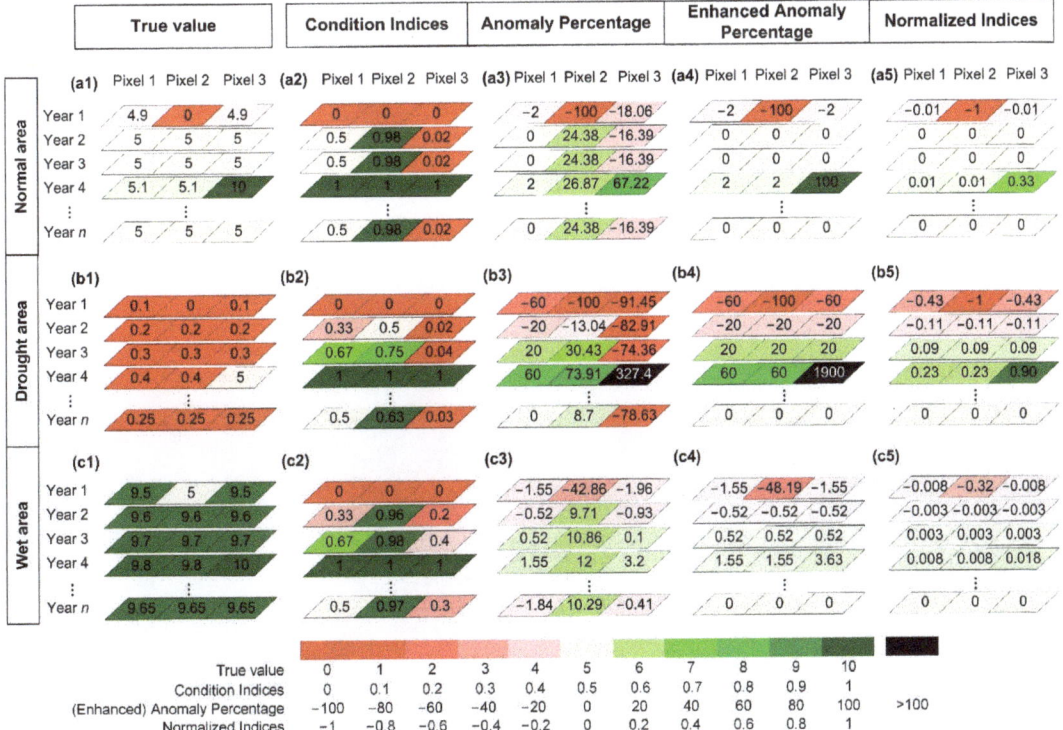

Figure 5. The simulation of different indices calculated in (a1–a5) normal regions, (b1–b5) arid regions, and (c1–c5) humid regions, using the data of various pixels for many years. Pixel 1 represents normal pixels; Pixel 2 represents pixels where only severe drought events occurred; Pixel 3 represents pixels where only severe flood events occurred.

The monitoring results of the CI (Figure 5, column 2) have two problems: (1) Once an extreme precipitation event occurs in one year, drought overestimation is likely to occur in other years. Compared with a1 and a2, Pixel 1 is a normal pixel, and the degree of drought and flood is more consistent, so the CI is relatively symmetrical; Pixel 2 only shows a severe water shortage in Year 1, which makes the CI of other nonextreme years generally larger; Pixel 3 has an extreme precipitation event in Year 4, which makes the CI of other nonextreme years generally small. In other words, for years (Year 2, Year 3, Year n, etc.) when precipitation is normal (the pixel value is 5 in a normal region), the monitoring results of CI show that Pixel 1 is consistent with the actual situation, Pixel 2 is wetter, and Pixel 3 has severe drought overestimation, compared with the actual situation (Figure 5a2). The results of b2 and c2 have the same problem as a2. This is because the Standard$_{PCI}$ of all pixels is very different due to the extreme value being added to the calculation, as shown in Figure 5, resulting in the same true value of pixels in normal years, but the precipitation status shown by PCI is very different. (2) Due to the calculation method of the CI, there are always the values of 0 (extreme drought) and 1 (extraordinary precipitation) for each pixel, regardless of whether real extreme events occurred. It is easy to monitor extreme abnormalities (a1—Pixel 1 and a2—Pixel 1) even if the true values of pixels are similar.

NI does not have this problem (Figure 5, column 5). In regions with similar daily conditions, the same true value will have very similar monitoring results (a5/b5/c5—Years 2 and 3); for regions with different moisture conditions, the same true value will have different monitoring results, such as all the pixels with a true value of 5 in a5, b5, and c5. The true value of 5 in a1, b1, and c1 has completely different meanings: it is the normal rainfall in normal (semiarid and semihumid) regions (a1); it means high rainfall in arid regions (b1); it means low rainfall in a humid region (c1). NI can monitor the relative changes of real precipitation (or soil moisture or vegetation conditions) of pixels in different regions. It changes from −1 to 1, which is convenient for mapping. However, the legend display of NI is not symmetrical, as shown in Table 3. The main advantages and disadvantages of CI, AP, EAP, and NI are summarized in Table 4.

Table 3. Legend meaning of Normalized Indices.

n (×Standard)	Label	n (×Standard)	Label	n (×Standard)	Label
0	−1	1	0	2	0.333
0.1	−0.818	1.1	0.048		
0.2	−0.667	1.2	0.091	3	0.5
0.3	−0.538	1.3	0.130		
0.4	−0.429	1.4	0.167	4	0.6
0.5	−0.333	1.5	0.20		
0.6	−0.250	1.6	0.231	10	0.818
0.7	−0.176	1.7	0.259		
0.8	−0.111	1.8	0.286	100	0.980
0.9	−0.053	1.9	0.310		
1	0	2	0.333	MAX	≈1

Table 4. The advantages and disadvantages of Condition Indices, Anomalies Percentage, Enhanced Anomalies Percentage, and Normalized Indices.

Index	Advantages	Disadvantages
CI	(1) CI is accurate in places where both drought and flood have occurred with similar severity. (2) The legend display is symmetrical.	(1) Once extreme precipitation event occurs in one year, drought overestimation is likely to occur in other years and vice versa. (2) There are always the values of 0 (drought) and 1 (precipitation) for each pixel, regardless of whether the real extreme events occur.
AP	(1) AP can well present the distance between the current value and the average value.	(1) The same as point (1) of CI to a lesser degree. (2) There is no upper limit under ideal conditions.
EAP	(1) EAP can monitor the relative changes of real situation of pixels in both arid and humid regions.	(1) There is no upper limit under ideal conditions.
NI	(1) NI does not have the limitations of above indices, and can monitor the relative changes of real precipitation (or soil moisture or vegetation conditions) of pixels in both arid and humid regions.	(1) The legend display is not symmetrical.

The monitoring results of the AP (Figure 5, column 3) have the following problems: (1) The first problem of CI, but the degree is relatively minor. When extreme events occur in certain years of the pixel, for other years with normal pixel values (Year 2, Year 3, Year n, etc.), the monitoring results using AP will be wetter (a3,b3,c3—Pixel 2) or drier (a3,b3,c3—Pixel 3) than the actual situation. (2) When the pixel value exceeds twice $\overline{F}$ (or $\overline{F}'$), the AP will be greater than 100%. The monitoring results of the EAP (Figure 5, column 4) do not have the first problem of the AP, but the second problem persists.

3.4. Validation of Study Results

The *Yearbook of Meteorological Disasters in China* and crop yield data from 2003 to 2019 were used to validate the study results. The drought and flood events recorded in the disaster yearbook are a summary of the meteorological observation data of China's meteorological departments at all levels and of the on-site monitoring results of meteorological stations. The main resource for drought and flood disaster analysis is precipitation data from field observations. In addition, when a severe drought is encountered, there will be records related to the state of soil moisture and crops, facilitating a comprehensive validation of remote sensing monitoring results. We also conducted a field survey in Jiangxi Province (Region B) in 2019 as a supplement, to validate the RS monitoring results.

Meteorological disasters, insects, diseases, and nutrients can all affect crop health and yield variation, but the meteorological factor is usually the main factor in crop monitoring of a large region. In this study, we used the changes in vegetation index to monitor the health of crops. The correlation between NVI and crop yield was used to validate the vegetation index monitoring results. Except for the Yield Anomalies Percentage (YAP) and Normalized Yield Index (NYI), the Standardized Variable of Yield (SVY) of each county [35] was also used to monitor the variation of crop yield, which is calculated as follows:

$$SVY_i = \frac{Y_i - \overline{Y}}{\sigma} \times 100\% \qquad (8)$$

where Y_i is the crop yield in i year of one county, $\overline{Y}$ is the average, and σ is the standard deviation of crop yield from 2003 to 2019.

4. Results

4.1. Application and Results Validation of Different Indices

4.1.1. Temporal Differences in PCI, PAP, EPAP, and NPI

Region A experienced continuous rainy weather from 6 August to 18 September 2014. The province's average precipitation was 60% higher than in the same period in normal years, which has been rare in recent years. This included a number of heavy rainstorms, sometimes accompanied by typhoons, which caused water to accumulate in farmland and crops to fail. From mid-June to late July 2014, there was a severe precipitation reduction, and the precipitation in December was also much lower than in previous years (as recorded in the *Yearbook of Meteorological Disasters in China*).

Compared with the actual results, PCI was significantly lower than the actual precipitation, so the rainy weather from 6 August to 18 September could not be monitored (Figure 6). This was because, compared with normal pixels that had no extreme events, pixels with extraordinary rainstorm events occurred and would have larger values of Standard$_{PCI}$, as shown in Figure 6—Pixel 2, resulting in a lower PCI value for normal years (Figure 5(a2,b2,c2)—Pixel 3). The results of PCI would suggest more severe drought events than the actual situation. Compared with PCI, the trend changes in precipitation monitored by PAP were more realistic. EPAP tended to have a value greater than the upper limit of the map display (far greater than 100%), which was not conducive to statistics and display. The NPI could monitor well the abnormal events of precipitation in the long-term series. The monitoring of the start and end time of the abnormal event was also more accurate and in line with the actual situation.

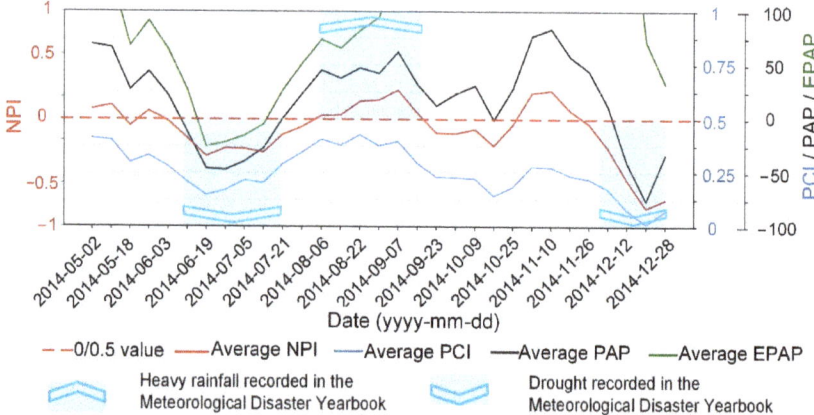

Figure 6. Average crop pixel values of PCI, PAP, EPAP, and NPI in Region A from May 2014 to December 2014, while PCI, PAP, EPAP, and NPI were updated every eight days and are marked in different colors; 0.5 is the threshold of PCI, whereas the other indices use 0.

4.1.2. Spatial Differences in Normalized Indices and Condition Indices

Regional changes and intensity changes in precipitation in the entire study area monitored by NPI from July to October were highly consistent with the drought and heavy rain events recorded in the yearbook. The changes in soil moisture monitored by NSMI were also in good agreement with changes in precipitation (NPI) (Figure 7a). The following *Yearbook* records are introduced in chronological order (months): (1) There were torrential rains and floods in the study area in July 2014, while a moderate to severe meteorological drought occurred in north-central Jiangsu, northwest Anhui, and central Hubei in the same month. (2) From 7 to 31 August, there were continuous low temperatures and rainy weather in the entire study area. The temperature in most areas was 2–3 °C lower than normal, and the sunshine hours were 60–80 h fewer than normal. Among them, the sunshine hours in Jiangsu Province were the lowest since 1961. The continuous low temperature and inadequate illumination in August caused damage to vegetation growth and decreased NVI (Figure 7a). (3) Jiangsu Province experienced continuous rainy weather from 1 to 18 September; Anhui Province's average precipitation was 32 days from 1 August to 30 September, the most in the same period since 1961; there were continuous rainy days in most parts of Hubei Province from 8 to 19 September. (4) However, drought occurred in central and southern Jiangxi from mid-September to early November and obvious meteorological droughts occurred in southern and eastern Hunan from mid-September to late October. From 16 to 21 October, there was continuous rain in western Hubei; from 27 to 30 October, there was continuous heavy rainfall in the MLRYR. Heavy to extreme rain occurred in some regions, which adversely affected crop growth (2014 *Yearbook of Meteorological Disasters in China*).

Compared with the meteorological observation results, the precipitation events monitored by PCI from July to October 2014 were generally small in scope and low in intensity, as shown in the purple circles of Figure 7. The results of soil moisture distribution monitored by SMCI were quite different from the PCI results. For example, except for Hubei Province, August showed continuous rainy weather with little sunshine; the PCI monitoring result was that the precipitation was relatively low, while the soil moisture monitored by the SMCI was obviously humid. This is mainly because the precipitation event in August had a long duration and wide range, but the overall intensity was not large. The total monthly rainfall was not high in the same month of all years, so the PCI monitoring result was drought (as shown in Figure 7b), which did not match the actual situation. For regions that suffered heavy rainfall events, the monitoring results of the Condition Indices would reflect severe drought overestimation, in which the error of precipitation (PCI) would be greater than that of soil moisture and vegetation (SMCI and VCI).

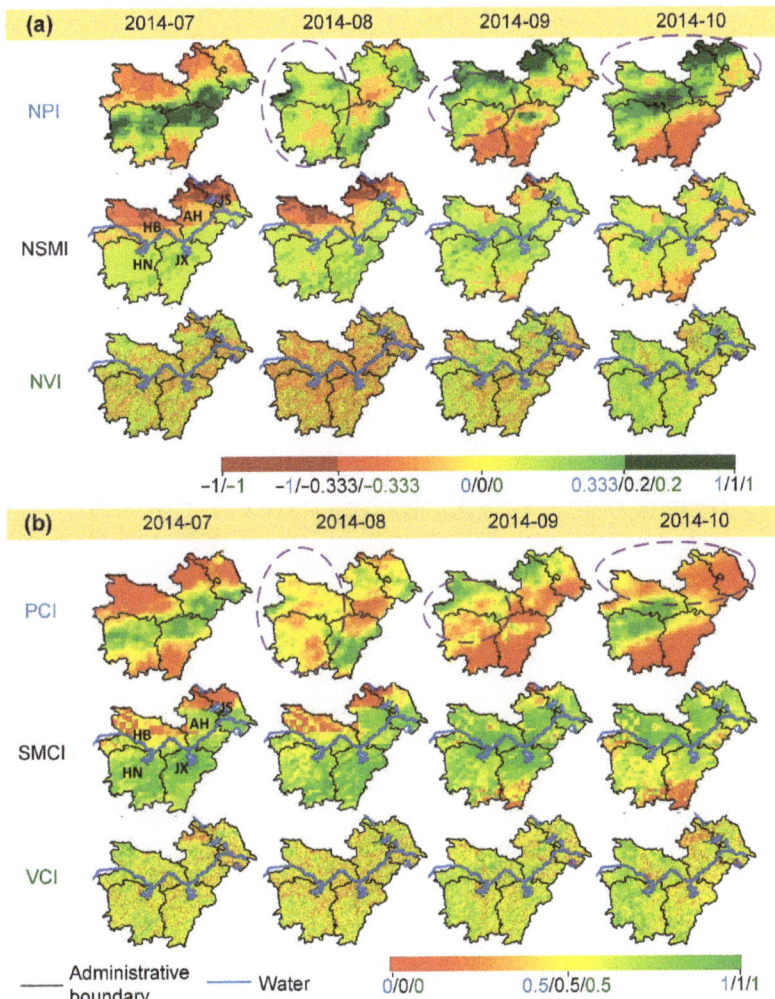

Figure 7. Spatial evolution of NPI/NSMI/NVI (**a**) and PCI/SMCI/VCI (**b**) in Jiangsu, Anhui, Hubei, Jiangxi, and Hunan Provinces, China. The results are organized on a timeline from July 2014 to October 2014 (the main growth season of rice). All indices were updated monthly.

4.2. Multiyear Drought Monitoring Based on Normalized Indices

4.2.1. Temporal Evolution of NPI, NSMI, and NVI

The PCI from 2011 to 2015 was below 0.5, indicating a four-year meteorological drought event that was obviously inconsistent with the facts; it did not match the excessive precipitation events recorded in the *Yearbook* (Figure 8a). The NPI could better monitor the drought and precipitation events recorded in the *Yearbook* than PCI (Figure 8b). The difference between VCI and NVI was smaller than that between PCI and VCI because the weather-related part of EVI2 affected by weather changes (precipitation, drought, high temperature, etc.) was relatively small [16], leading to more accurate results for VCI. However, the occasional excessive rainfall increases the value of Standard$_{PCI}$, making the PCI of normal years relatively small and causing the overestimation of meteorological drought.

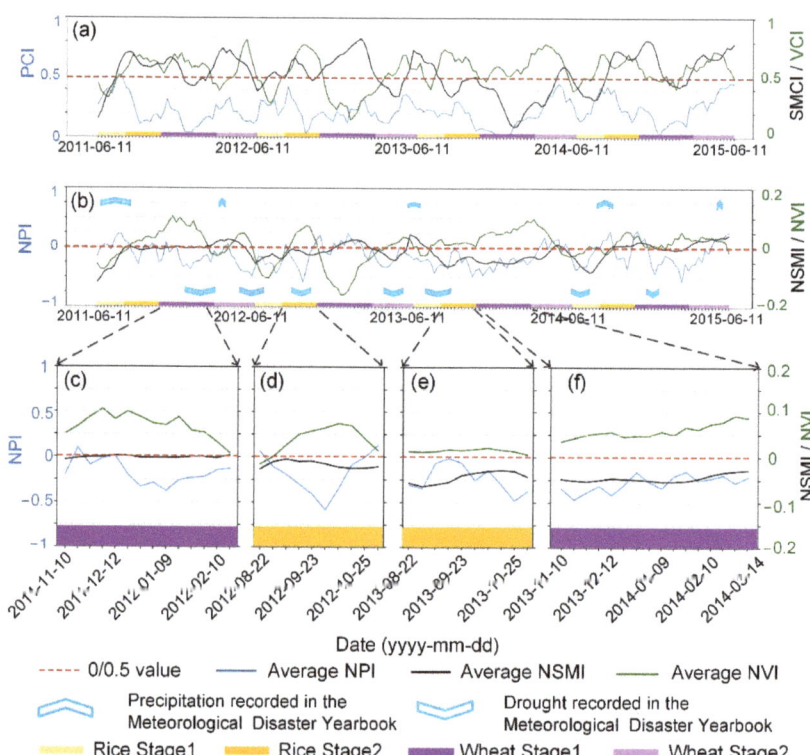

Figure 8. Average values of wheat and rice pixels of PCI/SMCI/VCI (**a**) and NPI/NSMI/NVI (**b**). (**c**–**f**) NI temporal evolution of four typical meteorological drought events in detail. The values of 0.5 and 0 are the threshold of CI and NI, respectively; meteorological drought is indicated when PCI is less than 0.5 or NPI is less than 0. Wheat Stage 1 means the sowing to regreening period of wheat; Wheat Stage 2 means the jointing to maturity; Rice Stage 1 means the transplanting to the jointing period of rice; Rice Stage 2 means the booting to maturity (see Section 2.2.5). All indices were updated every eight days and are marked in different colors.

It is especially worth noting that when the NPI was less than 0, the NVI (Figure 8b–f) was greater than 0; in other words, when meteorological drought (precipitation lower than the normal level) occurred, the crops grow better in Rice Stage 2 and wheat growing seasons in the MLRYR. This is because, in arid regions that rely on precipitation for irrigation, water is the main factor affecting crop health. However, in the MLRYR, which has abundant precipitation and numerous rivers and lakes, the continuous rainy weather is usually accompanied by reduced illumination and lower temperatures, which are not conducive to crop growth. In contrast, the meteorological drought means adequate illumination in the MLRYR, so crops grow better. However, during the wheat sowing period or rice transplanting period, severe meteorological drought will affect the growth and survival of seedlings, which causes serious damage to crops.

4.2.2. Spatial Evolution of Drought in 2019

In the postmonsoon (August–October) season of 2019 [53], there was great public concern about the severe drought event in the MLRYR, so we carried out a week-long field survey in Region B of Jiangxi Province in late October 2019. We visited the Agricultural Meteorological Center, surveyed a total of 180 rice samples (evenly distributed in the main rice-growing areas), and interviewed 12 rice growers to understand the evolution of drought (Figure 9).

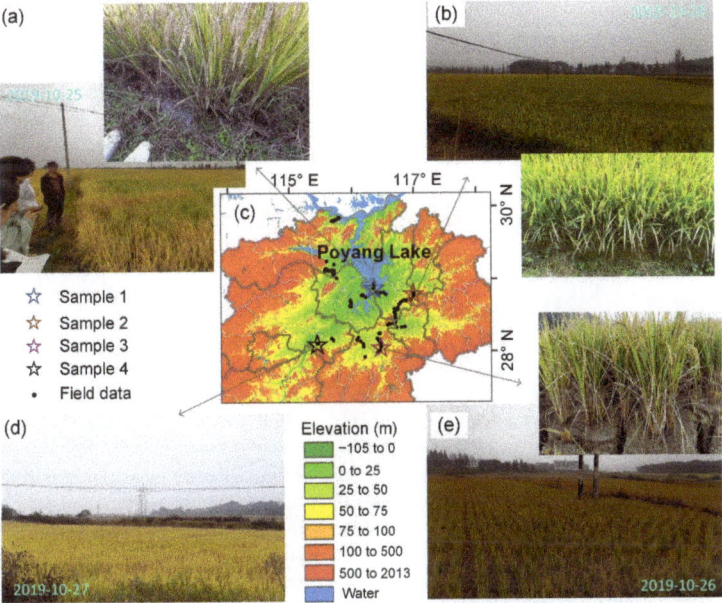

Figure 9. Field survey and ground geotagged photos of paddy rice with different health conditions in Region B in late October 2019. (**c**) Topography of Region B and the distribution of field survey points; (**a**,**b**,**d**) rice paddies with irrigation; (**e**) rice paddies without irrigation.

The survey results showed that Region B had abundant precipitation before August, and some regions had more than 40 consecutive days of precipitation before 14 July. Starting in late July, most areas of Region B had more than 100 consecutive days without precipitation. The growth conditions of rice were roughly divided into three types: (1) In most areas, due to the large water storage capacity of the reservoir and the good irrigation system, the soil moisture was normal or slightly lower than in previous years. The growth of rice was not significantly affected (Figure 9b,d) and the estimated yield had not changed obviously from previous years. (2) In the area close to Poyang Lake, water could be seen in the fields (Figure 9a). Due to the abundant sunshine from August to October, a slight increase in yield was expected. (3) In the small area with higher altitudes or poorer irrigation conditions, the soil moisture was obviously lower, reducing the yield by about 50% (Figure 9e) or even resulting in no harvest.

The NPI-based precipitation monitoring results showed severe meteorological drought (Figure 10). The soil moisture of most pixels in the entire study area was lower than that of the same period by about 20%. Since the soil moisture in the MLRYR was high in normal years, the reduction in soil moisture in most parts of the study area did not cause serious damage to vegetation growth. The NVI of the entire study area increased from −0.015 in July to 0.012 in October month by month, indicating that the vegetation growth showed a tendency to improve.

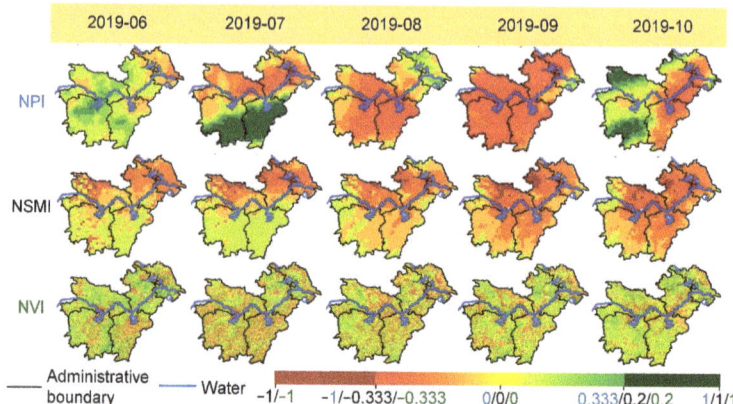

Figure 10. Spatial evolution of NPI, NSMI, and NVI in study area from June to October 2019 (growth season of rice). All indices were updated monthly.

We used changes in the vegetation index to monitor the crop health and the correlation between NVI and crop yield to validate the vegetation index monitoring results. The NPI and NVI of the main growing season of rice in Region A were averaged, and the results (Figure 11a,b) showed that the meteorological drought in the east was more serious but that the EVI2 had increased compared with previous years because there were more lakes in the western part of Region A. Compared with the average value of the entire rice growing season, the NVI of the harvest period was more consistent with the spatial variation of rice yields (Figure 11c,f). The effect of increasing production in the west was more obvious than in the east and, compared with YAP (Figure 11d) and SVY (Figure 11e), the spatial variation of NYI (Figure 11f) was more consistent with NVI. Both the vegetation index and the rice yield were negatively correlated with precipitation (Figure 12). With the decrease in precipitation, the vegetation index of 78.8% of pixels increased, while the yield of 97.1% of pixels increased.

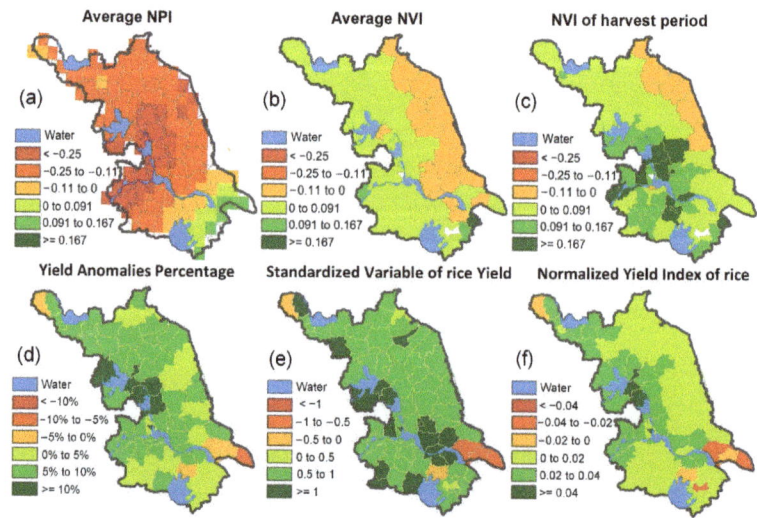

Figure 11. Spatial evolution of NPI, NVI, and rice yield change of Region A in growth season of rice in 2019.

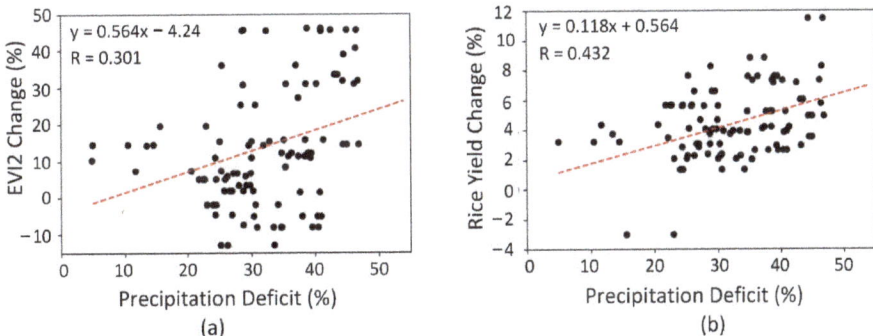

Figure 12. Scatter plot and correlation analysis between precipitation deficit, vegetation index (**a**), and rice yield (**b**) of region A in 2019.

5. Discussion

Most of our understanding of drought is based on remote sensing data, using calculated RS drought indices to monitor the conditions of precipitation, soil moisture, temperature, and vegetation. Each factor was assigned different weights based on empirical analysis, principal component analysis (PCA), kernel entropy component analysis (KECA), spatial principal component analysis (SPCA), and other methods [23,35,54,55]. New composite drought indices were then formed and used to monitor meteorological drought, agricultural drought, or hydrological drought in different regions. When using RS data for many years in drought monitoring, the purpose of the calculation is to compare the current state with a standard value, judging the drought or moisture degree in the region at a certain time. Due to the calculation principle of existing Condition Indices and Anomalies Percentage, drought overestimation occurs easily, especially in regions with abundant precipitation. However, the $Standard_{NI}$ can better represent the normal level of a region, which makes the results of precipitation, soil moisture, and vegetation changes monitored by Normalized Indices more consistent with the actual situation. Based on Normalized Indices, we realized the accurate monitoring of meteorological drought events in the study area over many years.

In addition, we studied the well-known meteorological drought event that occurred in the MLRYR from August to October 2019 and found that it had a much less severe impact on vegetation than expected. Meteorological or climatological drought is defined simply in terms of the magnitude and duration of a precipitation shortfall (16 August 2020: https://www.ametsoc.org/index.cfm/ams/about-ams/ams-statements/statements-of-the-ams-in-force/drought/). When a severe meteorological drought event occurs, people generally associate it with damaged crops and reduced yields. This phenomenon is evident in arid regions that rely solely on precipitation irrigation, which is also the focus of most studies. However, the fact that meteorological drought induces an increase in the vegetation index and crop yield is readily overlooked due to the lack of systematic studies. Our investigations have found that in southern China the water demand of crops can be satisfied by irrigation when meteorological drought occurs. The irrigation sources include lakes, reservoirs, pond water storage, and underground pumping. At the same time, a reduction in rainfall means an increase in illumination and adequate heat resources, so crops grow better. Anderson et al. [56] suggested that a finer crop model was needed that could consider moisture and temperature extremes during critical phenological stages of crop growth. We should also give more attention to the illumination. In addition, in accordance with the definition of meteorological drought as only a shortage of rainfall, it is not recommended to add on other factors such as soil moisture or vegetation; this would cause adverse effects because precipitation is not simply positively correlated with the vegetation index.

Agricultural drought links meteorological drought characteristics to agricultural impacts, associating precipitation shortages most immediately with higher evapotranspiration levels and soil moisture deficits. Our results prove that severe precipitation deficiency and meteorological drought do not necessarily lead to agricultural drought. Even when precipitation is the main water input for crops, in some regions a statistically weak relationship between precipitation and yield loss may be seen [36,57], while the timing of the precipitation is also an important factor. Therefore, it is not sufficient to use only measured precipitation to assist in the construction of agricultural drought monitoring models (such as determining the weight of each component) or for validating the monitoring results. Our results can provide some new ideas for the construction of agricultural insurance models.

Despite the good performance at capturing drought impacts, some key limitations exist when using Normalized Indices, as seen in this study.

The first limitation was introduced by the computation method. When calculating Condition Indices or Anomalies Percentage using soil moisture or vegetation index data over many years, the types of land cover are not distinguished, which is also one of the sources of error. For example, there was one area where a certain pixel had been represented as dry land for many years and only one year where it was used as a paddy field. When calculating CI or AP, if this single "paddy field" year was not removed using land cover data, other years would have appeared as having low soil moisture and severe drought, which was inconsistent with the real situation (Pixel 3 of Figure 5(a1,a2)). Land cover changes, such as crops to trees or farmland to ponds, will also increase the error in drought monitoring. Using Normalized Indices can avoid calculation errors, but the land use inconsistency of time series can also cause errors in judgment. Therefore, the use of multiyear land cover data to filter the time series in index calculations can improve the accuracy of disaster identification. Another limitation is the data requirement. The spatial resolution of the data used in this study was too low, especially for soil moisture. It is necessary to use the same resolution for soil moisture data as for the vegetation index. Soil moisture is a factor that directly impacts vegetation health, and the response of different vegetation changes in soil moisture varies. Adding high-spatiotemporal-resolution soil moisture data will be the basis for the high-precision monitoring of agricultural drought in the future.

6. Conclusions

According to the calculation principle of commonly used RS drought indices, and for achieving more accurate drought monitoring, we proposed a new index calculation method, referred to as Normalized Indices or NI. TRMM precipitation, GLDAS soil moisture, and MODIS reflection datasets were used to calculate drought indices. The disaster events recorded in the *Yearbook of Meteorological Disasters in China*, field survey data, and statistical crop yield data were used to validate the monitoring results of paddy rice and winter wheat. Through the simulation of different types of moisture conditions and multiyear drought monitoring of the study area, the monitoring results showed:

- NI can monitor well the relative changes in real precipitation/soil moisture/vegetation conditions, in both arid and humid regions, while meteorological drought is easily overestimated with CI in areas with abundant precipitation;
- The error of precipitation (PCI) is greater than that of soil moisture and vegetation (SMCI and VCI), the same as AP;
- The well-known drought event that occurred in the MLRYR from August to October 2019 had a much less severe impact on vegetation than expected. In contrast, the precipitation deficiency induced an increase in sunshine and adequate heat resources, which improved crop growth in most areas.

This study shows some restrictions and shortcomings of recognized CI and AP, and it proposes a new index calculation method of NI to better monitor meteorological drought in the MLRYR of China, providing a new method for future drought monitoring studies.

Author Contributions: Conceptualization and resources, J.H.; methodology and software, L.L.; validation, Y.C. (Yan Chen) and P.W.; formal analysis, Y.C. (Yuanyuan Chen); investigation, W.L. and D.D.; data curation, Y.C. (Yan Chen); writing—original draft preparation, L.L.; writing—review and editing, J.C.; visualization, Q.S.; supervision, R.H.; project administration, J.H.; funding acquisition, J.H. All authors have read and agreed to the published version of the manuscript.

Funding: This research was funded by "National Key R&D Program of China", grant number 2017YFD0300402-3 and Eramus+ Programme of the European Union (Grant NO. 598838-EPP-1-2018-EL-EPPKA2-CBHE-JP).

Institutional Review Board Statement: Not applicable.

Informed Consent Statement: Not applicable.

Acknowledgments: We would like to thank the Jiangsu Meteorological Bureau of China for the data support. The editors' and anonymous reviewers' valuable comments and advice that were significant for improving this manuscript are also much appreciated.

Conflicts of Interest: The authors declare no conflict of interest.

Abbreviations

AMSR-E	Advanced Microwave Scanning Radiometer for Earth Observing System
AP	Anomalies Percentage
AVHRR	Advanced Very High Resolution Radiometer
CI	Condition Indices
CMDSC	China Meteorological Data Service Centre
EAP	Enhanced Anomalies Percentage
EPAP	Enhanced Precipitation Anomalies Percentage
ESMAP	Enhanced Soil Moisture Anomalies Percentage
EVAP	Enhanced Vegetation Anomalies Percentage
EVI2	2-band Enhanced Vegetation Index
GEE	Google Earth Engine
GLDAS	Global Land Data Assimilation System
JAXA	Japan Aerospace Exploration Agency
JMIC	Jiangsu Meteorological Information Centre
KECA	Kernel Entropy Component Analysis
LAADS	NASA's Level 1 and Atmosphere Archive and Distribution System
MCDIs	Composite Drought Indices based on multivariable linear regression
MI	Moisture Index
MIDI	Microwave Integrated Drought Index
MLRYR	Mid-to-Lower Reaches of the Yangtze River
MODIS	Moderate-resolution Imaging Spectroradiometer
NASA	National Aeronautics and Space Administration
NDDI	Normalized Difference Drought Index
NDVI	Normalized Difference Vegetation Index
NDWI	Normalized Difference Water Index
NI	Normalized Indices
NMDI	Normalized Multiband Drought Index
NPI	Normalized Precipitation Index
NPP	Net Primary Productivity
NSMI	Normalized Soil Moisture Index
NVI	Normalized Vegetation Index
NYI	Normalized Yield Index
OMDI	Optimized Meteorological Drought Index
PADI	Process-based Accumulated Drought Index
PAP	Precipitation Anomalies Percentage
PCA	Principal Component Analysis
PCI	Precipitation Condition Index
PDSI	Palmer Drought Severity Index

PR	Precipitation Radar
PSMCI	TRMM Precipitation and Soil Moisture Condition Index
PTCI	TRMM Precipitation and Temperature Condition Index
RS	Remote Sensing
RSDEI	Remote Sensing Drought Evaluation Index
SDCI	Scaled Drought Condition Index
SDI	Synthesized Drought Index
SMAP	Soil Moisture Anomalies Percentage
SMCI	Soil Moisture Condition Index
SMTCI	Soil Moisture and Temperature Condition Index
SPCA	Spatial Principal Component Analysis
SPEI	Standardized Precipitation Evapotranspiration Index
SPI	standardized precipitation index
SVY	Standardized Variable of crop Yield
TCI	Temperature Condition Index
TMI	TRMM Microwave Imager
TRMM	Tropical Rainfall Measuring Mission
VAP	Vegetation Anomalies Percentage
VCI	Vegetation Condition Index
VIRS	Visible and Infrared Scanner
YAP	Yield Anomalies Percentage

References

1. Halwatura, D.; McIntyre, N.; Lechner, A.; Arnold, S. Capability of meteorological drought indices for detecting soil moisture droughts. *J. Hydrol. Reg. Stud.* **2017**, *12*, 396–412. [CrossRef]
2. Lai, C.; Zhong, R.; Wang, Z.; Wu, X.; Chen, X.; Wang, P.; Lian, Y. Monitoring hydrological drought using long-term satellite-based precipitation data. *Sci. Total. Environ.* **2019**, *649*, 1198–1208. [CrossRef]
3. Dou, Y.; Huang, R.; Mansaray, L.R.; Huang, J. Mapping high temperature damaged area of paddy rice along the Yangtze River using Moderate Resolution Imaging Spectroradiometer data. *Int. J. Remote Sens.* **2019**, *41*, 471–486. [CrossRef]
4. Sun, S.; Li, Q.; Li, J.; Wang, G.; Zhou, S.; Chai, R.; Hua, W.; Deng, P.; Wang, J.; Lou, W. Revisiting the evolution of the 2009–2011 meteorological drought over Southwest China. *J. Hydrol.* **2019**, *568*, 385–402. [CrossRef]
5. Javed, T.; Li, Y.; Rashid, S.; Li, F.; Hu, Q.; Feng, H.; Chen, X.; Ahmad, S.; Liu, F.; Pulatov, B. Performance and relationship of four different agricultural drought indices for drought monitoring in China's mainland using remote sensing data. *Sci. Total. Environ.* **2021**, *759*, 143530. [CrossRef]
6. Agutu, N.; Awange, J.; Zerihun, A.; Ndehedehe, C.E.; Kuhn, M.; Fukuda, Y. Assessing multi-satellite remote sensing, reanalysis, and land surface models' products in characterizing agricultural drought in East Africa. *Remote Sens. Environ.* **2017**, *194*, 287–302. [CrossRef]
7. Yao, N.; Li, Y.; Sun, C. Effects of changing climate on reference crop evapotranspiration over 1961–2013 in Xinjiang, China. *Theor. Appl. Clim.* **2018**, *131*, 349–362. [CrossRef]
8. Zhao, M.; Huang, S.; Huang, Q.; Wang, H.; Leng, G.; Xie, Y. Assessing socio-economic drought evolution characteristics and their possible meteorological driving force. *Geomat. Nat. Hazards Risk* **2019**, *10*, 1084–1101. [CrossRef]
9. Tucker, C.J.; Townshend, J.R.; Goff, T.E. African Land-Cover Classification Using Satellite Data. *Science* **1985**, *227*, 369–375. [CrossRef]
10. Hielkema, J.U.; Prince, S.D.; Astle, W.L. Rainfall and vegetation monitoring in the Savanna Zone of the Democratic Republic of Sudan using the NOAA Advanced Very High Resolution Radiometer. *Int. J. Remote Sens.* **1986**, *7*, 1499–1513. [CrossRef]
11. Malingreau, J.-P. Global vegetation dynamics: satellite observations over Asia. *Int. J. Remote Sens.* **1986**, *7*, 1121–1146. [CrossRef]
12. Justice, C.O.; Townshend, J.R.G.; Holben, B.N.; Tucker, C.J. Analysis of the phenology of global vegetation using meteorological satellite data. *Int. J. Remote Sens.* **1985**, *6*, 1271–1318. [CrossRef]
13. Kogan, F.N. Remote sensing of weather impacts on vegetation in non-homogeneous areas. *Int. J. Remote Sens.* **1990**, *11*, 1405–1419. [CrossRef]
14. Liu, W.T.; Kogan, F.N. Monitoring regional drought using the Vegetation Condition Index. *Int. J. Remote Sens.* **1996**, *17*, 2761–2782. [CrossRef]
15. Kogan, F.N. Global Drought Watch from Space. *Bull. Am. Meteorol. Soc.* **1997**, *78*, 621–636. [CrossRef]
16. Kogan, F.N. Droughts of the Late 1980s in the United States as Derived from NOAA Polar-Orbiting Satellite Data. *Bull. Am. Meteorol. Soc.* **1995**, *76*, 655–668. [CrossRef]
17. Kogan, F. Application of vegetation index and brightness temperature for drought detection. *Adv. Space Res.* **1995**, *15*, 91–100. [CrossRef]
18. Rhee, J.; Im, J.; Carbone, G. Monitoring agricultural drought for arid and humid regions using multi-sensor remote sensing data. *Remote Sens. Environ.* **2010**, *114*, 2875–2887. [CrossRef]

19. Zhang, A.; Jia, G. Monitoring meteorological drought in semiarid regions using multi-sensor microwave remote sensing data. *Remote Sens. Environ.* **2013**, *134*, 12–23. [CrossRef]
20. Ghaleb, F.; Mario, M.; Sandra, A.N. Regional Landsat-Based Drought Monitoring from 1982 to 2014. *Climate* **2015**, *3*, 563–577. [CrossRef]
21. Jiao, W.; Wang, L.; Novick, K.A.; Chang, Q. A new station-enabled multi-sensor integrated index for drought monitoring. *J. Hydrol.* **2019**, *574*, 169–180. [CrossRef]
22. Jiao, W.; Tian, C.; Chang, Q.; Novick, K.A.; Wang, L. A new multi-sensor integrated index for drought monitoring. *Agric. For. Meteorol.* **2019**, *268*, 74–85. [CrossRef]
23. Hao, Z.; Singh, V.P. Drought characterization from a multivariate perspective: A review. *J. Hydrol.* **2015**, *527*, 668–678. [CrossRef]
24. Najem, S.; Al Bitar, A.; Faour, G.; Jarlan, L.; Mhawej, M.; Fadel, A.; Zribi, M. Drought Assessment using Micro-Wave Timeseries of Precipitation and Soil Moisture Over the Mena Region. In Proceedings of the 2020 Mediterranean and Middle-East Geoscience and Remote Sensing Symposium (M2GARSS), Tunis, Tunisia, 9–11 March 2020; pp. 289–292. [CrossRef]
25. Wei, W.; Pang, S.; Wang, X.; Zhou, L.; Xie, B.; Zhou, J.; Li, C. Temperature Vegetation Precipitation Dryness Index (TVPDI)-based dryness-wetness monitoring in China. *Remote Sens. Environ.* **2020**, *248*, 111957. [CrossRef]
26. Zhang, Q.; Yu, H.; Sun, P.; Singh, V.P.; Shi, P. Multisource data based agricultural drought monitoring and agricultural loss in China. *Glob. Planet. Chang.* **2019**, *172*, 298–306. [CrossRef]
27. Qian, W.; Ai, Y.; Leung, J.C.-H.; Zhang, B. Anomaly-based synoptic analysis and model product application for 2020 summer southern China rainfall events. *Atmos. Res.* **2021**, *258*, 105631. [CrossRef]
28. Li, X.; Huang, W.-R. How long should the pre-existing climatic water balance be considered when capturing short-term wetness and dryness over China by using SPEI? *Sci. Total. Environ.* **2021**, *786*, 147575. [CrossRef]
29. Sgroi, L.C.; Lovino, M.A.; Berbery, E.H.; Müller, G.V. Characteristics of droughts in Argentina's core crop region. *Hydrol. Earth Syst. Sci.* **2021**, *25*, 2475–2490. [CrossRef]
30. Wu, M.; Vico, G.; Manzoni, S.; Cai, Z.; Bassiouni, M.; Tian, F.; Zhang, J.; Ye, K.; Messori, G. Early Growing Season Anomalies in Vegetation Activity Determine the Large-Scale Climate-Vegetation Coupling in Europe. *J. Geophys. Res. Biogeosciences* **2021**, *126*. [CrossRef]
31. Parinussa, R.M.; Wang, G.; Liu, Y.; Lou, D.; Hagan, D.F.T.; Zhan, M.; Su, B.; Jiang, T. Improved surface soil moisture anomalies from Fengyun-3B over the Jiangxi province of the People's Republic of China. *Int. J. Remote Sens.* **2018**, *39*, 8950–8962. [CrossRef]
32. Otkin, J.A.; Zhong, Y.; Lorenz, D.; Anderson, M.C.; Hain, C. Exploring seasonal and regional relationships between the Evaporative Stress Index and surface weather and soil moisture anomalies across the United States. *Hydrol. Earth Syst. Sci.* **2018**, *22*, 5373–5386. [CrossRef]
33. Lorenz, D.J.; Otkin, J.A.; Svoboda, M.; Hain, C.R.; Anderson, M.C.; Zhong, Y. Predicting U.S. Drought Monitor States Using Precipitation, Soil Moisture, and Evapotranspiration Anomalies. Part I: Development of a Nondiscrete USDM Index. *J. Hydrometeorol.* **2017**, *18*, 1943–1962. [CrossRef]
34. Lorenz, D.J.; Otkin, J.A.; Svoboda, M.; Hain, C.R.; Anderson, M.C.; Zhong, Y. Predicting the U.S. Drought Monitor Using Precipitation, Soil Moisture, and Evapotranspiration Anomalies. Part II: Intraseasonal Drought Intensification Forecasts. *J. Hydrometeorol.* **2017**, *18*, 1963–1982. [CrossRef]
35. DU, L.; Tian, Q.; Yu, T.; Meng, Q.; Jancsó, T.; Udvardy, P.; Huang, Y. A comprehensive drought monitoring method integrating MODIS and TRMM data. *Int. J. Appl. Earth Obs. Geoinf.* **2013**, *23*, 245–253. [CrossRef]
36. Zhang, X.; Chen, N.; Li, J.; Chen, Z.; Niyogi, D. Multi-sensor integrated framework and index for agricultural drought monitoring. *Remote Sens. Environ.* **2017**, *188*, 141–163. [CrossRef]
37. Liu, Q.; Zhang, S.; Zhang, H.; Bai, Y.; Zhang, J. Monitoring drought using composite drought indices based on remote sensing. *Sci. Total. Environ.* **2020**, *711*, 134585. [CrossRef] [PubMed]
38. Wei, W.; Zhang, J.; Zhou, J.; Zhou, L.; Xie, B.; Li, C. Monitoring drought dynamics in China using Optimized Meteorological Drought Index (OMDI) based on remote sensing data sets. *J. Environ. Manag.* **2021**, *292*, 112733. [CrossRef] [PubMed]
39. Wei, W.; Zhang, H.Y.; Zhou, J.J.; Zhou, L.; Xie, B.B.; Li, C.H. Drought monitoring in arid and semi-arid region based on mul-ti-satellite datasets in northwest, China. *Environ. Sci. Pollut.* **2021**, *28*, 51556–51574. [CrossRef] [PubMed]
40. Niu, N.; Li, J. Interannual variability of autumn precipitation over South China and its relation to atmospheric circulation and SST anomalies. *Adv. Atmos. Sci.* **2008**, *25*, 117–125. [CrossRef]
41. Zhang, W.; Jin, F.-F.; Turner, A. Increasing autumn drought over southern China associated with ENSO regime shift. *Geophys. Res. Lett.* **2014**, *41*, 4020–4026. [CrossRef]
42. Zhang, W.; Jin, F.-F.; Zhao, J.-X.; Qi, L.; Ren, H.-L. The Possible Influence of a Nonconventional El Niño on the Severe Autumn Drought of 2009 in Southwest China. *J. Clim.* **2013**, *26*, 8392–8405. [CrossRef]
43. Wang, P.; Tam, C.-Y.; Xu, K. El Niño–East Asian monsoon teleconnection and its diversity in CMIP5 models. *Clim. Dyn.* **2019**, *53*, 6417–6435. [CrossRef]
44. Pei, F.; Zhou, Y.; Xia, Y. Application of Normalized Difference Vegetation Index (NDVI) for the Detection of Extreme Precipitation Change. *Forests* **2021**, *12*, 594. [CrossRef]
45. Zhang, D.Q.; Chen, L.J. Possible mechanisms for persistent anomalous rainfall over the middle and lower reaches of Yangtze River in winter 2018/2019. *Int. J. Climatol.* **2021**. [CrossRef]

46. Chen, Y.; Huang, J.; Sheng, S.; Mansaray, L.R.; Liu, Z.; Wu, H.; Wang, X. A new downscaling-integration framework for high-resolution monthly precipitation estimates: Combining rain gauge observations, satellite-derived precipitation data and geographical ancillary data. *Remote Sens. Environ.* **2018**, *214*, 154–172. [CrossRef]
47. Rodell, M.; Houser, P.R.; Jambor, U.; Gottschalck, J.; Mitchell, K.; Meng, C.-J.; Arsenault, K.; Cosgrove, B.; Radakovich, J.; Bosilovich, M.; et al. The Global Land Data Assimilation System. *Bull. Am. Meteorol. Soc.* **2004**, *85*, 381–394. [CrossRef]
48. Chen, Y.; Yang, K.; Qin, J.; Zhao, L.; Tang, W.; Han, M. Evaluation of AMSR-E retrievals and GLDAS simulations against observations of a soil moisture network on the central Tibetan Plateau. *J. Geophys. Res. Atmos.* **2013**, *118*, 4466–4475. [CrossRef]
49. Liu, L.; Huang, J.; Xiong, Q.; Zhang, H.; Song, P.; Huang, Y.; Dou, Y.; Wang, X. Optimal MODIS data processing for accurate multi-year paddy rice area mapping in China. *GIScience Remote Sens.* **2020**, *57*, 687–703. [CrossRef]
50. Chen, Y.; Lu, D.; Moran, E.; Batistella, M.; Dutra, L.V.; Sanches, I.D.; da Silva, R.F.B.; Huang, J.; Luiz, A.J.B.; de Oliveira, M.A.F. Mapping croplands, cropping patterns, and crop types using MODIS time-series data. *Int. J. Appl. Earth Obs. Geoinf.* **2018**, *69*, 133–147. [CrossRef]
51. Chen, Y.Y. Risk Assessment and Monitoring of Winter Wheat Waterlogging Combining Ground-Based Observations and Satel-lite-Derived Data. Ph.D. Thesis, Zhejiang University, Hangzhou, China, 2018. (In Chinese).
52. Boschetti, M.; Busetto, L.; Manfron, G.; Laborte, A.; Asilo, S.; Pazhanivelan, S.; Nelson, A. PhenoRice: A method for automatic extraction of spatio-temporal information on rice crops using satellite data time series. *Remote Sens. Environ.* **2017**, *194*, 347–365. [CrossRef]
53. Xu, K.; Miao, H.; Liu, B.; Tam, C.; Wang, W. Aggravation of Record-Breaking Drought over the Mid-to-Lower Reaches of the Yangtze River in the Post-monsoon Season of 2019 by Anomalous Indo-Pacific Oceanic Conditions. *Geophys. Res. Lett.* **2020**, *47*. [CrossRef]
54. Rajsekhar, D.; Singh, V.P.; Mishra, A.K. Multivariate drought index: An information theory based approach for integrated drought assessment. *J. Hydrol.* **2015**, *526*, 164–182. [CrossRef]
55. Cardil, A.; Vega-Garcia, C.; Ascoli, D.; Molina-Terren, D.; Silva, C.; Rodrigues, M. How does drought impact burned area in Mediterranean vegetation communities? *Sci. Total. Environ.* **2019**, *693*, 133603. [CrossRef] [PubMed]
56. Anderson, M.; Zolin, C.A.; Sentelhas, P.C.; Hain, C.R.; Semmens, K.; Yilmaz, M.T.; Gao, F.; Otkin, J.; Tetrault, R. The Evaporative Stress Index as an indicator of agricultural drought in Brazil: An assessment based on crop yield impacts. *Remote Sens. Environ.* **2016**, *174*, 82–99. [CrossRef]
57. Niyogi, D.; Liu, X.; Andresen, J.; Song, Y.; Jain, A.K.; Kellner, O.; Takle, E.S.; Doering, O.C. Crop models capture the impacts of climate variability on corn yield. *Geophys. Res. Lett.* **2015**, *42*, 3356–3363. [CrossRef]

Article

The Grain for Green Program Intensifies Trade-Offs between Ecosystem Services in Midwestern Shanxi, China

Baoan Hu [1], Zhijie Zhang [2], Hairong Han [1,*], Zuzheng Li [3], Xiaoqin Cheng [1], Fengfeng Kang [1] and Huifeng Wu [1]

1. School of Ecology and Nature Conservation, Beijing Forestry University, Beijing 100083, China; baoanhu@bjfu.edu.cn (B.H.); cloud2014@bjfu.edu.cn (X.C.); phoonkong@bjfu.edu.cn (F.K.); huifengwu@bjfu.edu.cn (H.W.)
2. Hohhot Meteorological Bureau, Hohhot 010020, China; zhzhj1982@163.com
3. State Key Laboratory of Urban and Regional Ecology, Research Center for Eco-Environmental Sciences, Chinese Academy of Sciences, Beijing 100085, China; zuzhengli@rcees.ac.cn
* Correspondence: hanhr6015@bjfu.edu.cn; Tel.: +86-010-62336015

Citation: Hu, B.; Zhang, Z.; Han, H.; Li, Z.; Cheng, X.; Kang, F.; Wu, H. The Grain for Green Program Intensifies Trade-Offs between Ecosystem Services in Midwestern Shanxi, China. *Remote Sens.* **2021**, *13*, 3966. https://doi.org/10.3390/rs13193966

Academic Editors: Baojie He, Ayyoob Sharifi, Chi Feng and Jun Yang

Received: 30 August 2021
Accepted: 30 September 2021
Published: 3 October 2021

Publisher's Note: MDPI stays neutral with regard to jurisdictional claims in published maps and institutional affiliations.

Copyright: © 2021 by the authors. Licensee MDPI, Basel, Switzerland. This article is an open access article distributed under the terms and conditions of the Creative Commons Attribution (CC BY) license (https:// creativecommons.org/licenses/by/ 4.0/).

Abstract: Ecological engineering is a widely used strategy to address environmental degradation and enhance human well-being. A quantitative assessment of the impacts of ecological engineering on ecosystem services (ESs) is a prerequisite for designing inclusive and sustainable engineering programs. In order to strengthen national ecological security, the Chinese government has implemented the world's largest ecological project since 1999, the Grain for Green Program (GFGP). We used a professional model to evaluate the key ESs in Lvliang City. Scenario analysis was used to quantify the contribution of the GFGP to changes in ESs and the impacts of trade-offs/synergy. We used spatial regression to identify the main drivers of ES trade-offs. We found that: (1) From 2000 to 2018, the contribution rates of the GFGP to changes in carbon storage (CS), habitat quality (HQ), water yield (WY), and soil conservation (SC) were 140.92%, 155.59%, −454.48%, and 92.96%, respectively. GFGP compensated for the negative impacts of external environmental pressure on CS and HQ, and significantly improved CS, HQ, and SC, but at the expense of WY. (2) The GFGP promotes the synergistic development of CS, HQ, and SC, and also intensifies the trade-off relationships between WY and CS, WY and HQ, and WY and SC. (3) Land use change and urbanization are significantly positively correlated with the WY–CS, WY–HQ, and WY–SC trade-offs, while increases in NDVI helped alleviate these trade-offs. (4) Geographically weighted regression explained 90.8%, 94.2%, and 88.2% of the WY–CS, WY–HQ, and WY–SC trade-offs, respectively. We suggest that the ESs' benefits from the GFGP can be maximized by controlling the intensity of land use change, optimizing the development of urbanization, and improving the effectiveness of afforestation. This general method of quantifying the impact of ecological engineering on ESs can act as a reference for future ecological restoration plans and decision-making in China and across the world.

Keywords: Grain for Green Program; ecosystem services trade-offs; scenario analysis; spatial regression; Midwestern Shanxi

1. Introduction

Ecosystem services (ESs) refer to all the benefits that human beings obtain directly or indirectly from the natural ecosystem to meet and maintain their living needs [1,2]. The Millennium Ecosystem Assessment (MEA) divides ESs into four basic types, including regulating services (e.g., soil conservation and carbon storage), provisioning services (e.g., water and timber), supporting services (e.g., biodiversity conservation and nutrient cycling), and cultural services (e.g., forest recreation) [2,3]. Human development patterns over the last few centuries have detrimentally affected the health and resilience of natural ecosystems [3,4]. Declines in ESs have been observed at global and regional scales, and these declines pose a significant threat to human well-being [2,5]. Ecological engineering is a widely adopted countermeasure that attempts to mitigate the contradiction between human development and ecosystem protection [6,7]. Ecological engineering aims to increase the sustainable supply of ESs by repairing or improving ecosystem functioning [8,9]. At present, global investment in the development of ecological engineering amounts to billions of dollars per year [9], and quantitative assessment of the effects of ecological engineering on ESs has attracted the attention of many managers, research organizations, and researchers [10].

ESs are good indicators for evaluating the ecological benefits of ecological engineering, as they effectively connect human well-being and the natural environment [1,11]. ES trade-offs occur when the increase of a certain ES is at the cost of reducing another ES [2,12]. Therefore, revealing the influencing factors of ES trade-offs is crucial to maintaining the sustainable supply of multiple ESs [12,13]. The frequent conversion between land use types caused by high-intensity human activities is the main cause of ES declines [14,15]. Rebuilding the ecological functioning of degraded ecosystems by changing land use patterns and intensity of use is the main aim of most ecological projects [8,16], which will have strong impacts on the supply and trade-offs between ESs [17,18]. Ecological engineering that unilaterally promotes a single ES makes it difficult to maximize ecological benefits [19], and may even negatively affect ecosystem functioning and cause other services to decline [20,21].

Ecological degradation is one of the main reasons for the increasing frequency of natural disasters [22]. In order to achieve carbon neutrality and strengthen national ecological security, the Chinese government has implemented the world's largest ecological project since 1999: the Grain for Green Program (GFGP) [23]. With the implementation of the GFGP, vegetation cover has increased significantly [24] and various ESs, such as biodiversity and climate regulation, have been significantly improved [25]. However, large-scale planting of non-native vegetation not only leads to a significant increase in water consumption and evapotranspiration [26], which aggravates the potential conflict between regional ecosystem functioning and human demand for water resources [27], but also further challenges the achievement of balance between green and grain land, especially in arid regions [28,29]. This has rendered uncertain the sustainability of the ecological benefits of the GFGP. Therefore, quantitative assessment of the impacts of ecological engineering on ESs and analysis of the dominant factors driving ES trade-offs are prerequisites for the design of inclusive and sustainable ecological engineering [30].

In the context of rapid socio-economic development, most studies have confirmed that the GFGP can improve ESs and change the relationship between ecosystem support services and regulation services [31,32]. However, this change is influenced by multiple factors, such as natural, anthropogenic, climatic, and socio-economic factors [33]. There are few studies that quantify the contribution rate of ecological engineering to changes in ESs and the impacts of ecological engineering on the relationship between different ESs. In this study, we focus on Lvliang City, Shanxi Province, an area typical of the GFGP. This region has serious soil erosion and is a typical ecologically fragile zone. Our specific objectives are: (1) to quantify the contribution rate of the GFGP to changes in ESs; (2) to analyze the impacts of the GFGP on the trade-offs and synergy between ESs; and (3) to identify the factors influencing the trade-offs between ESs and put forward suggestions for promoting the inclusive and sustainable development of the GFGP. This research should serve as a reference for future ecological engineering projects in China and around the world.

2. Materials and Methods

2.1. Study Area

Lvliang City is located in the east-central region of China's Loess Plateau and the western region of Shanxi Province, and has an area of about 21,100 km^2 (Figure 1). Lvliang City has a continental monsoon climate with four distinct seasons, synchronized rain and heat, and sufficient sunlight. The average annual temperature is between 0.4 °C and 12.2 °C, and the average annual precipitation is between 438 and 588 mm. The elevation of the study area ranges from 561 to 2806 m a.s.l., with high terrain in the middle of the study area and lower terrain on the edges (Figure 1). Vegetation cover in the mountains of the central and eastern regions is relatively high, and human activities and industrial development are mainly concentrated in the southeastern plains; the western loess hilly regions have broken terrain, barren soil, and sparse vegetation [34,35]. Due to the low coverage rate of surface vegetation coupled with the landform type of prevalent ravines, the area has serious soil erosion and is typically an ecologically fragile area [34]. In recent years, because of the GFGP, the vegetation coverage rate in this area has increased significantly, the functions of various ecosystems such as climate regulation and soil conservation have improved significantly, and ESs have, accordingly, changed significantly [24,35].

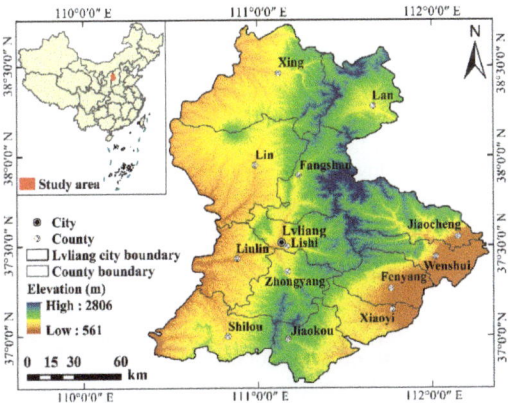

Figure 1. Location and elevation of study area.

2.2. Data Sources and Descriptions

In this study, we used multi-source data products, such as land use, meteorology, soil, and digital elevation models, to evaluate ESs. Detailed descriptions and data sources are shown in Table 1. In ArcGIS 10.2, all data are converted to the same projected coordinate system (WGS_1984_UTM_Zone_49N), and the "Resample" tool is used to unify the raster data resolution to 30 m.

Table 1. Description and sources of data used to evaluate ESs.

Data	Data Format	Data Description	Data Sources
Land use maps	Raster (30 m)	Land use maps interpreted from Landsat TM/ETM/OLI images. Land use types are classified into seven categories: farmland, forest, grassland, shrub land, water body, construction land, and unused land.	Data Center for Resources and Environmental Sciences, Chinese Academy of Sciences (http://www.resdc.cn/ (accessed on 16 March 2021))
Digital Elevation Model	Raster (30 m)	Elevation data.	Geospatial Data Cloud (http://www.gscloud.cn (accessed on 16 March 2021))
Meteorological data	Raster (1 km)	Including monthly average temperature and precipitation, annual average temperature and precipitation, and potential evapotranspiration.	National Earth System Science Data Center (http://www.geodata.cn/ (accessed on 16 March 2021))
Soil properties	Raster (1 km)	Including soil texture, topsoil sand fraction, topsoil silt fraction, topsoil clay fraction, topsoil organic carbon, root restricting layer depth, and plant available water content.	Harmonized World Soil database (http://www.iiasa.ac.at/Research/LUC/External-World-soil-database/HTML/ (accessed on 16 March 2021))

Table 1. Cont.

Data	Data Format	Data Description	Data Sources
Evapotranspiration coefficient (Kc)	Excel format	Plant evapotranspiration for different land use types.	Food and Agriculture Organization of the United Nations (FAO) (http://www.fao.org/3/X0490E/x0490e0b.htm (accessed on 16 March 2021))
Watershed boundary	Shapefile	Digital watershed atlas.	HydroSHEDS (http://hydrosheds.org/ (accessed on 16 March 2021))

2.3. Quantifying Ecosystem Services

The InVEST model is used to quantify four key ESs: water yield (WY), soil conservation (SC), habitat quality (HQ), and carbon storage (CS). WY is calculated based on the difference between annual average precipitation and actual annual evapotranspiration [29,36]. SC refers to the erosion control ability of the ecosystem to prevent soil loss and the ability to store and maintain sediment [36,37]. The sediment delivery ratio module calculates soil conservation services based on the difference between potential (under extremely degraded conditions without vegetation cover) and actual (under current land cover and management conditions) soil loss [19,36,37]. HQ refers to the ability to provide resources and environmental conditions for the survival and development of species or populations, which depends on the abundance of natural resources [36,38]. The habitat quality module calculates the HQ according to the habitat suitability of each land use type, the impact distance and weight of threat factors, and the sensitivity of each land use type to threat factors [36,38]. Through previous studies [38–41], we determined the impact distance and weight of threat factors, the habitat suitability of each land use type, and the sensitivity parameters to each threat factor (Tables S1 and S2). The carbon module quantifies CS using previous local research on the carbon density of different land use types [42–44] (Table S3). To avoid the influence of abnormal climate fluctuations in a single year, we selected the average rainfall and temperature from 2000 to 2018 as the general results from the study area [45,46]. Table 2 provide greater detail on the process of assessment of each ES.

Table 2. Methods for quantifying ESs.

ESs	Methods	Mathematical Expression
WY	InVEST model water yield module	$WY_x = (1 - AET_x/P_x) \times P_x$ WY_x: annual water yield for each grid cell; AET_x: annual actual evapotranspiration for pixel x; P_x: annual precipitation on pixel x; Biophysical coefficients of model input are shown in Table S3.
SC	InVEST model sediment delivery ratio module	$SC = R \times K \times LS \times (1 - C \times P)$ SC: soil conservation; R: rainfall erosion factor; K: soil erosion factor; LS: slope length and gradient factor; C: vegetation cover factor; P: support practice factor. R and K are calculated to refer to the method of Yang et al. [32] and Zhang et al. [37]. We assigned C and P values according to existing literature [17,36,39] (Table S3).
HQ	InVEST model habitat quality module	$HQ = H_j \times \left[1 - \left(\frac{D_{xj}^Z}{D_{xj}^Z + K^Z}\right)\right]$ HQ: habitat quality; H_j: habitat suitability for habitat type j; D_{xj}: degree of habitat degradation in pixel x that is in habitat type j; K: half-saturation constant; Z: default parameter of the normalized constant model.
CS	InVEST model carbon module	$CS = C_a + C_b + C_s + C_d$ CS: carbon storage; C_a, C_b, C_s, and C_d are carbon densities in aboveground biomass, belowground biomass, soil, and dead matter, respectively, for each land use type.

2.4. Calculation of Trade-Offs Between Ecosystem Services

Correlation analysis is an effective tool to identify relationships between pairs of ESs, with significant negative correlations representing trade-offs and positive correlations representing synergies [47]. The size of the Pearson correlation coefficient indicates the strength of the trade-off and synergy relationships [47]. Obviously, this method ignores the difference in the geographical space of the change rate of the ES trade-offs. The root mean squared error (*RMSE*) quantifies the average difference between the standard deviation of a single ES and the average ES's standard deviation [47,48]. The dispersion degree of the standard deviation of distance of average ESs is described, and reflects the difference in the geographical space of the change rate of the ES's trade-offs [49,50]. Therefore, this study uses the *RMSE* to quantify the trade-offs between ESs. To eliminate the influence of ES unit differences, we first standardize the value of each ES.

$$ES_i = \frac{ES_{i,obs} - ES_{i,min}}{ES_{i,max} - ES_{i,min}}. \tag{1}$$

where ES_i is the standardized value; $ES_{i,obs}$ is the raw value; and $ES_{i,min}$ and $ES_{i,max}$ are the minimum and maximum values of the i ESs, respectively. *RMSE* is calculated as follows:

$$RMSE = \sqrt{\frac{1}{n-1}\sum_{i=1}^{n}(ES_i - \overline{ES})^2} \tag{2}$$

where $\overline{ES}$ is the expected value of n kinds of ESs. In two dimensions, *RMSE* represents the distance from the coordinate point to the diagonal, and the relative position of the coordinate point represents the relative benefit of a certain ecosystem service [49]. Lu et al. [48] and Luo et al. [50] provide detailed instructions and procedures for the calculation of such trade-offs.

2.5. Actual Land Use Changes and Scenarios

The local administrative department of the GFGP provided vector data for the implementation area of the GFGP in Lvliang City as of the end of 2018. We set up a scenario where the GFGP was not implemented and quantified the impact of the GFGP on regional ESs by comparing this alternative scenario with the actual scenario.

(1) Actual scenario: we evaluated ESs before (2000) and after (2018) the implementation of the GFGP based on actual land use. By comparing ESs in 2000 and 2018, we can understand actual changes of ESs under the implementation of the GFGP.

(2) Alternative scenarios where the GFGP was not implemented (2018S): this is a simulated scenario. We assume that during the period 2000–2018, the actual GFGP implementation area did not implement the GFGP; that is, the land use types remained, unchanged, at their state in 2000, while the land use types in other regions were consistent with actual changes. By comparing ESs in 2018S and 2018, we were able to quantify the impact of the GFGP on ESs.

Terrain fragmentation due to soil erosion is the main cause of ecological degradation in the Loess Plateau [51]. The design and implementation of the GFGP on the sub-watershed scale to carry out comprehensive control of soil erosion has achieved good results [19,52]. In addition, as a physical geographical unit, the sub-watershed scale can more accurately reflect biophysical characteristics [19]. Therefore, we obtained the average value of each ES at the sub-watershed scale through the Zonal Statistics tool in ArcGIS 10.2. At the sub-watershed scale, the impact of GFGP on ESs was quantified and the trade-offs among ESs and their influencing factors were analyzed.

2.6. Geographically Weighted Regression Model

Previous studies have confirmed that there are obvious geographical differences in ESs [53,54]. It is difficult for the classic global regression to reflect the differences in the relationship between ES trade-offs and influencing factors in geographic space, and not fully reflect actual local processes [55]. Geographically weighted regression (GWR) obtains local coefficients by minimizing residuals, taking into account differences in the spatial variation in the relationship between ES trade-offs and influencing factors, which improves the reliability of the model [56].

ESs and their trade-off relationships are affected by factors such as land use [57], climate [58], vegetation [19], geomorphology [59], and urbanization [60]. We selected eight influencing factors to include in our model: the dynamic degree of comprehensive land use change (LUD; Refer to Li et al. [61] method for calculation), annual average temperature (TEM), NDVI, annual average precipitation (PRE), percentage of construction land (CON), elevation (DEM), slope (SLO), and potential evapotranspiration (PET) (Table S4 provides detailed calculation methods or sources). To avoid the influence of multicollinearity, all factors were tested for multicollinearity in SPSS 21, and the factors with VIF greater than five were eliminated (Table S5). These preliminary analyses left us with LUD, NDVI, PRE, DEM, and CON as independent variables and the trade-off relationships between ESs as dependent variables for our GWR model. The lower the AIC_c value of the model output, the more concise the model and the more reliable the regression estimation. The higher adjusted R^2 indicates a higher explanatory power and a better fit [56]. The mathematical expression of the model is as follows:

$$y_i = \beta_0(u_i, v_i) + \sum_{j=1}^{p} \beta_j(u_i, v_i) x_{ij} + \varepsilon_i, \ i \in \{1, 2, \ldots, n\} \tag{3}$$

where y is the dependent variable; (u_i, v_i) is the spatial location of the i-th sample; $\beta_0(u_i, v_i)$ is the intercept; p is the number of influencing factors; x_{ij} represents the independent variables; $\beta_j(u_i, v_i)$ is the estimated coefficient of the i-th sample for the j-th driving factors; and ε_i is the error term.

3. Results

3.1. Land Use Change

Figure 2 shows the land use patterns in 2000, 2018, and 2018S (the scenario if the GFGP were not implemented). Compared with 2000, the area of farmland decreased by 28.90% in 2018, and the area of construction land, forest, grassland, and shrub land increased by 259.31%, 13.7%, 23.98%, and 4.51%, respectively (Table 3). The area of farmland and forest under the 2018S scenario decreased by 5.21% and 10.92%, respectively, and the area of grassland and shrub land increased by 4.20% and 0.87%, respectively (Table 3). Our results show that the GFGP led to a decrease of 23.69% in the area of farmland, and an increase of 24.62%, 19.78%, and 3.64% in the area of forest, grassland, and shrub land, respectively (Table 3), and thus was the main driving force for the significant increase in regional vegetation cover.

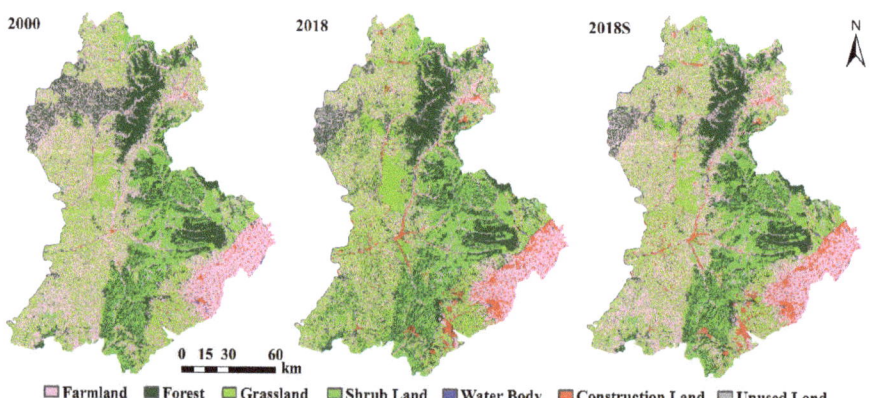

Figure 2. Land use patterns in 2000, 2018, and 2018S. 2018S: the scenario if the GFGP were not implemented.

Table 3. Land use changes from 2000 to 2018 and 2018S. 2018S: the scenario if the GFGP were not implemented.

	Land Use Types	Farmland	Forest	Grassland	Shrub Land	Water Body	Construction Land	Unused Land
2018	Change area (km²)	−2515.04	527.15	1214.05	140.63	−24.39	656.97	0.62
	Change ratio (%)	−28.90%	13.70%	23.98%	4.51%	−17.36%	259.31%	87.06%
2018S	Change area (km²)	−453.05	−420.01	212.61	27.23	−24.39	656.97	0.64
	Change ratio (%)	−5.21%	−10.92%	4.20%	0.87%	−17.36%	259.31%	90.36%
	Effect of GFGP on land use change (%)	−23.69%	24.62%	19.78%	3.64%	0	0	−3.30%

3.2. ESs Change

Figure 3 shows the spatial pattern of ESs in 2000, 2018, and 2018S (the scenario if the GFGP were not implemented). The spatial pattern of CS is consistent with land use, and high-value areas are distributed in mountainous regions with higher forest cover (Figure 3). Compared with 2000, the average CS in 2018 and 2018S increased by 15.47 (Mg/ha) and decreased by 6.33 (Mg/ha), respectively. During the study period, the contribution rate of the GFGP to CS changes was 140.92% (Figure 3). The central and eastern areas are dominated by forest and grassland, with high HQ, while in the western loess hilly region and the southeastern plains, HQ is relatively low (Figure 3). Compared with 2000, the average HQ in 2018 and 2018S increased by 0.035 and decreased by 0.019, respectively. During the study period, the contribution rate of the GFGP to HQ changes was 155.59% (Figure 3). WY was high in the center of the study area and low in the outer regions (Figure 3). Compared with 2000, the average WY in 2018 and 2018S increased by 0.79 (mm) and 4.36 (mm), respectively. During the study period, the GFGP had a significant negative impact on WY, with a contribution rate of −454.48% (Figure 3). The central and eastern regions had high SC values, while the southeast and western regions had relatively low SC (Figure 3). Compared with 2000, the average SC in 2018 and 2018S increased by 0.947 (ton/ha) and 0.067 (ton/ha), respectively. During the study period, the contribution rate of the GFGP to SC changes was 92.96% (Figure 3). In general, the implementation of the GFGP from 2000 to 2018 compensated for the negative impacts of external environmental pressures on CS and HQ, and significantly improved CS, HQ, and SC; however, this improvement came at the expense of WY.

3.3. Trade-Offs Between ESs

The correlation between changes in ESs from 2000 to 2018 and 2018S (the scenarios if the GFGP were not implemented) was analyzed at the sub-watershed scale. CS, HQ, and SC have a significant synergistic relationship, and there is a significant trade-off between these ESs and WY (Table 4). In addition, the correlation coefficients (including positive and negative correlations) between paired ESs in the actual scenario are larger than those in the alternative scenario if the GFGP were not implemented (Table 4). This indicates that the GFGP has intensified the trade-offs and synergies between ESs.

We visualized the WY-CS, WY-HQ, and WY-SC trade-offs using root mean squared error (*RMSE*), and our results show that the west and southeast are the high value areas of the trade-offs (Figure 4). Average tradeoff values of WY-CS, WY-HQ, and WY-SC are 0.051, 0.050, and 0.016, respectively, in the actual scenario, and the average tradeoff values of WY-CS, WY-HQ, and WY-SC are 0.028, 0.030, and 0.014, respectively, in the alternative scenario if the GFGP were not implemented (Figure 4). This indicates that the implementation of the GFGP strengthens the trade-offs between WY-CS, WY-HQ, and WY-SC.

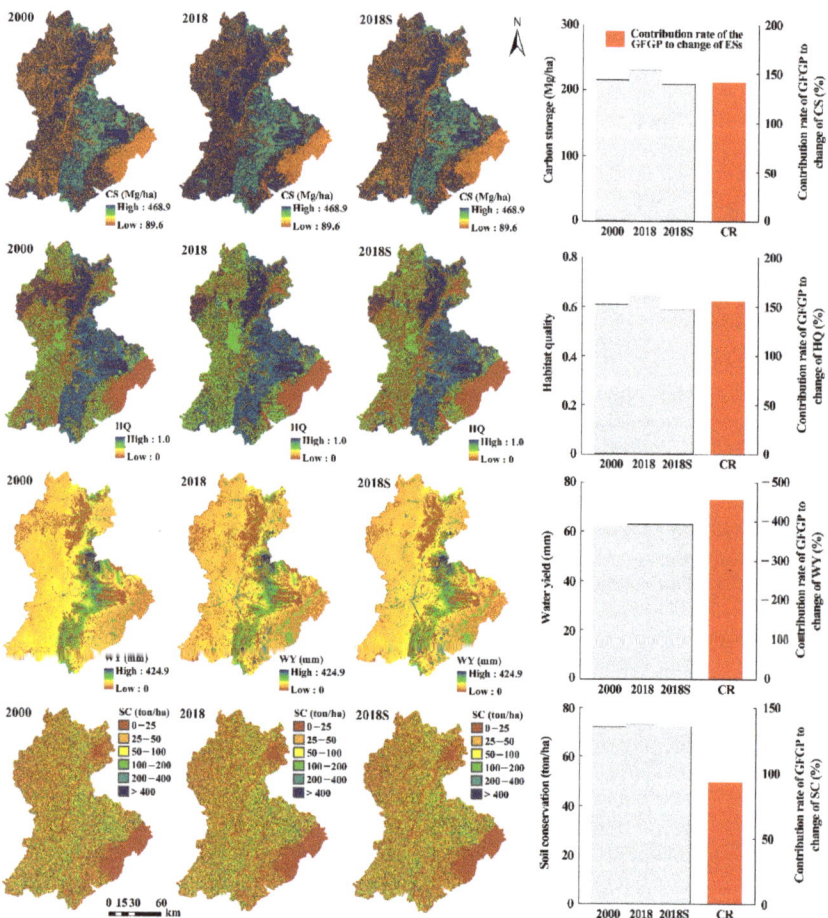

Figure 3. Spatial pattern of ESs in 2000, 2018, and 2018S. 2018S: the scenario if the GFGP were not implemented. The bar chart on the right represents the average value of ESs in 2000, 2018, and 2018S and the contribution rate of the GFGP to the changes in ESs. Abbreviations: CS: carbon storage; WY: water yield; SC: soil conservation; HQ: habitat quality.

Table 4. Pearson's correlation analysis between changes in ecosystem services.

N = 181	CS2018	HQ2018	SC2018	WY2018	CS2018S	HQ2018S	SC2018S	WY2018S
CS2018	1							
HQ2018	0.920 **	1						
SC2018	0.835 **	0.889 **	1					
WY2018	−0.804 **	−0.898 **	−0.641 **	1				
CS2018S					1			
HQ2018S					0.684 **	1		
SC2018S					0.384 **	0.397 **	1	
WY2018S					−0.645 **	−0.878 **	−0.075	1

CS2018 (CS2018S), HQ2018 (HQ2018S), SC2018 (SC2018S), and WY2018 (WY2018S), respectively, indicate changes in carbon storage, habitat quality, soil conservation, and water yield in 2018 (scenarios with and without the implementation of the GFGP) relative to 2000; N represents the number of sub-watersheds; ** indicates significance at the $p < 0.01$ level.

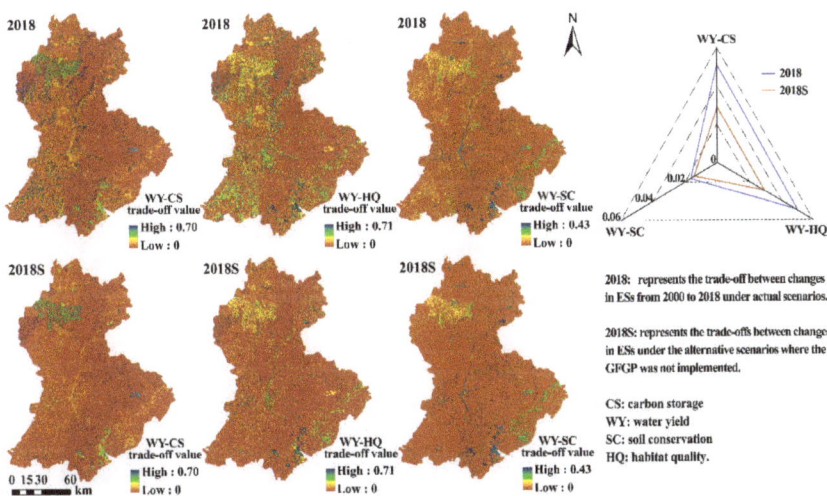

Figure 4. The spatial distributions of trade-offs between ESs. The radar graph on the right represents the average values of trade-offs between ESs.

3.4. FACTORS Influencing ESs Trade-Offs

We built a GWR model to explore the geospatial relationship between ES trade-offs and the factors that influence them. Compared with OLS, the adjusted R^2 of the GWR model was greater, and the AIC_c value decreased significantly (Table 5), indicating that the GWR results have higher explanatory power and can more accurately reflect the processes at play.

Table 5. Model fit metrics for ordinary least squares (OLS) regression and GWR.

ES Trade-Offs	Fit Metrics	Model	
		OLS	GWR
WY-CS	R^2 (adjust)	0.837	0.908
	AIC_c	194.958	128.907
WY-HQ	R^2 (adjust)	0.901	0.942
	AIC_c	104.279	48.576
WY-SC	R^2 (adjust)	0.721	0.882
	AIC_c	291.957	182.843

Abbreviations: CS: carbon storage; WY: water yield; SC: soil conservation; HQ: habitat quality.

The correlation coefficient of the GWR model reflects the spatial non-stationary response of ES trade-offs to influencing factors (Figure 5). LUD is significantly positively correlated with WY-CS, WY-HQ, and WY-SC trade-offs, and the correlation coefficient is relatively high in the northeast (Figure 5 (a1–a3), Table 6). NDVI is significantly negatively correlated with WY-CS, WY-HQ, and WY-SC trade-offs, and the correlation coefficient is relatively high in the southwest (Figure 5 (b1–b3), Table 6). PRE is positively correlated with WY-CS and WY-HQ trade-offs, but negatively correlated with the WY-SC trade-off, and the correlation coefficient is high in the southeast (Figure 5 (c1–c3), Table 6). DEM is negatively correlated with the WY-HQ trade-off, but positively correlated with WY-CS and WY-SC trade-offs, and the correlation coefficient is larger in the north (Figure 5 (d1–d3), Table 6). CON is negatively correlated with the WY-CS trade-off, but significantly positively correlated with WY-HQ and WY-SC trade-offs, and the correlation coefficient is larger in the south (Figure 5 (e1–e3), Table 6).

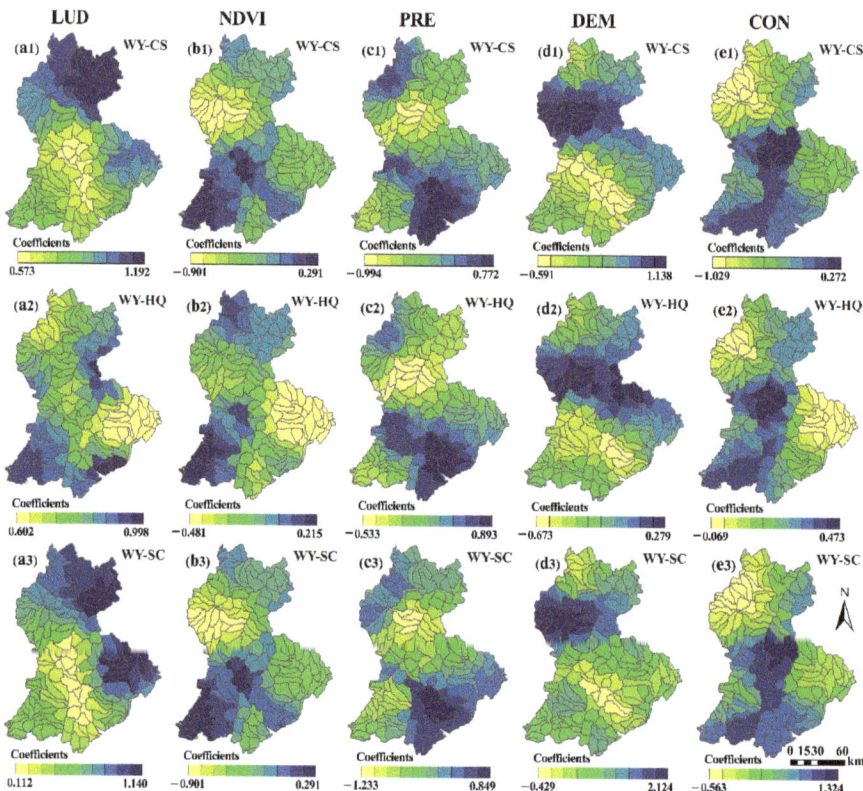

Figure 5. GWR coefficients between ES trade-offs and their influencing factors. Abbreviations: CS: carbon storage; WY: water yield; SC: soil conservation; HQ: habitat quality; LUD: dynamic degree of comprehensive land use change; NDVI: Normalized Difference Vegetation Index; PRE: precipitation; DEM: elevation; CON: percentage of construction land.

Table 6. Mean statistics of GWR coefficients between ES trade-offs and influencing factors.

ESs Trade-Offs	LUD	NDVI	PRE	DEM	CON
WY-CS	0.888	−0.036	0.044	0.070	−0.143
WY-HQ	0.794	−0.052	0.120	−0.126	0.206
WY-SC	0.595	−0.054	−0.010	0.424	0.619

Abbreviations: CS: carbon storage; WY: water yield; SC: soil conservation; HQ: habitat quality; LUD: dynamic degree of comprehensive land use change; NDVI: Normalized Difference Vegetation Index; PRE: precipitation; DEM: elevation; CON: percentage of construction land.

Local R^2 maps describe spatial differences in model goodness of fit, ranging between 0 and 1. Our results show that the selected five influencing factors are closely related to ES trade-offs, explaining 90.8%, 94.2%, and 88.2% of the WY-CS, WY-HQ, and WY-SC trade-offs, respectively (Figure 6, Table 5). In general, LUD, CON, and NDVI are the most important driving factors of ES trade-offs, and they are significantly positively correlated with LUD and CON while being negatively correlated with NDVI (Figure 6, Table 6). This means that increasing vegetation cover, controlling the intensity of land use change, and optimizing the development of urbanization are effective ways to alleviate the trade-offs between ESs and realize the synergistic promotion of multiple ESs.

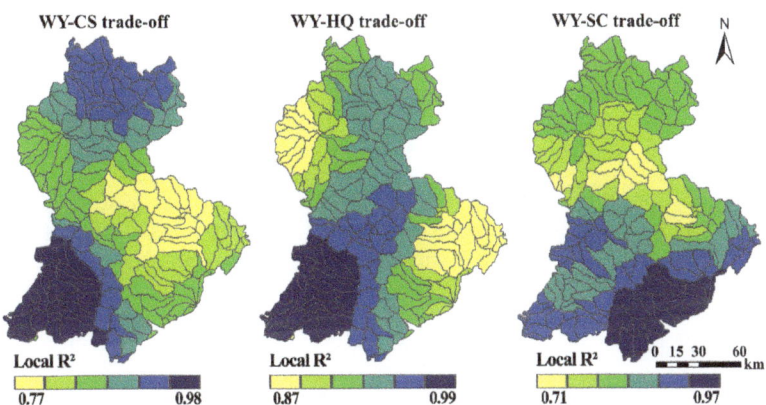

Figure 6. Spatial patterns of local R^2 from GWR between ES trade-offs and influencing factors. Abbreviations: CS: carbon storage; WY: water yield; SC: soil conservation; HQ: habitat quality.

4. Discussion

4.1. Effects of the GFGP on ESs

The GFGP is a successful program for coping with environmental degradation and increasing the supply of ESs [31]. Although the quality of the regional ecological environment has been greatly improved [17,62], realizing the coordinated development of multiple ESs is still a key consideration in optimizing GFGP policies. Our results show that the implementation of the GFGP significantly increased forest, grassland, and shrub land area (Table 3), and vegetation cover increased significantly [24], leading to significant increases in CS, HQ, and SC. This indicates that the GFGP promoted the synergistic relationship among CS, HQ, and SC. However, CS and HQ, under an alternative scenario where the GFGP was not implemented, significantly decreased (Figure 3), indicating that the GFGP effectively compensated for the negative impacts of external environmental pressures on CS and HQ. By comparing the actual scenario in 2018 with the alternative scenario, we found that the GFGP is the most important driving force for the increase in SC, with a contribution rate of 92.96% (Figure 3). In addition, our results show that the GFGP has had a significant negative impact on WY, with a contribution rate of −454.48% (Figure 3). This is mainly due to the large-scale planting of non-native vegetation, which leads to a significant increase in water consumption and evapotranspiration [27,63]. Our research confirms that the GFGP produces significant ecological benefits while also exacerbating regional water resource conflicts, which is consistent with previous studies [26,27,51,64]. However, unlike previous studies, we quantified the contribution rate of the GFGP to ES changes and the impact on the trade-offs/synergies between ESs, providing a more direct reference for alleviating regional water resource conflicts and realizing the synergistic promotion of multiple ESs.

4.2. Suggestions on the Inclusiveness and Sustainable Development of the GFGP

Identifying the dominant factors influencing the trade-offs between ESs is critical to formulating an inclusive and sustainable plan for the GFGP. There are obvious spatial differences in the relationships between ES trade-offs and their influencing factors [65,66], and classical global regression models did not fully reflect the relationships between the two in geographic space [59]. The local coefficients obtained by the GWR model by minimizing the residuals reflect the spatial non-stationary relationships between them [56], effectively overcoming the problems with classic regression models. We used the GWR model to explore the spatially non-stationary relationship between ES trade-offs and their influencing factors, and our results show that LUD, CON, and NDVI are the most important driving factors for ES trade-offs (Figure 5, Table 6). The WY–CS, WY–HQ, and WY–SC trade-offs were significantly positively correlated with LUD and CON, but negatively correlated with NDVI (Figure 5 (a1–a3, b1–b3, e1–e3), Table 6).

Land use change and urbanization are the main drivers of declines in CS, HQ, and SC [57,67], and also have negative impacts on the water conservation capacity of ecosystems [68]. However, land use change and urbanization also reduced the evapotranspiration of surface vegetation to a certain degree [69], and their impacts on precipitation at smaller timescales are also limited [70]. Therefore, increases in LUD and CON intensify the tradeoffs between WY–CS, WY–HQ, and WY–SC. NDVI is the most direct manifestation of the effectiveness of afforestation [71]. The GFGP is the main driver of the increase in regional NDVI [24], which not only improves CS, HQ, and SC but also improves the water conservation capacity of the ecosystem [72,73]. Therefore, increasing NDVI helps to alleviate the trade-offs between WY–CS, WY–HQ, and WY–SC.

The correlation coefficient of the GWR model reflects the spatial non-stationary response of ES trade-offs to their influencing factors (Figure 5). In the northeast and south of the study area, urbanization developed rapidly and the intensity of human activity was high (Figure 2), so the correlation coefficient between LUD, CON, and ES trade-offs was relatively large (Figure 5 (a1–a3, e1–e3)). In the southwestern region, the terrain is rugged and vegetation is relatively scarce [34], so the correlation coefficient between NDVI and ESs trade-offs is relatively high (Figure 5 (b1–b3)). Therefore, controlling LUD and CON in the northeast and south, and increasing vegetation cover in the southwest, is essential to alleviate the WY–CS, WY–HQ, and WY–SC trade-offs.

In summary, we propose that future engineering projects should take into account the geospatial relationships between ES trade-offs and their influencing factors. By controlling the intensity of land use change, optimizing the development of urbanization, and improving the effectiveness of afforestation, the inclusive and sustainable development of the regional GFGP can be realized.

4.3. Uncertainties and Limitations

Our research provides a direct and flexible method to quantify the impacts of the GFGP on ESs, but it still has certain limitations. First, changing ESs is a complex process driven by factors such as nature, human activities, and climate change [17,74]. It is very difficult to completely quantify the impact of the GFGP on ESs. Our study used the average climate parameters from 2000 to 2018. Although this method is widely adopted [45,46], climate change during the research period will certainly have had an impact on ESs. Second, the input parameters of the model evaluation are taken from previous studies, but due to the limitations of our data sources, quality, and availability, we did not verify the results of the ES evaluations. Third, because of the limited availability of data, we had to ignore some details of the GFGP, such as tree species selection and configuration, vegetation management methods, etc., although these practices certainly could have a strong impact on ESs [71]. These problems may introduce some uncertainty into our model results. Therefore, it is necessary to obtain long-term positioning observation data and conduct more detailed research on the impacts of the GFGP.

5. Conclusions

Based on scenario analysis, we quantified the impacts of the GFGP on changes in ESs in Lvliang City, a typical ecologically fragile area, and analyzed the main forces driving ES trade-offs through spatial regression. Our research shows that the GFGP compensated for the negative impacts of external environmental pressures on CS and HQ, and significantly improved CS, HQ, and soil conservation (SC), but this improvement came at the expense of water yield (WY). While the GFGP promotes the synergistic development of CS, HQ, and SC, it also intensifies the trade-off relationships between these services and WY. Land use change and urbanization are significantly positively correlated with the trade-offs between WY–CS, WY–HQ, and WY–SC, while increasing NDVI helped to alleviate these trade-offs. Therefore, controlling the intensity of land use change, optimizing the development of urbanization, and improving the effectiveness of afforestation are effective ways to realize the inclusive and sustainable development of the GFGP. The general methods used in this study to quantify the impacts of ecological engineering on ESs can provide a reference for future ecological restoration plans and decision-making in China and around the world.

Supplementary Materials: The following are available online at https://www.mdpi.com/article/10.3390/rs13193966/s1, Table S1: Threats and their maximum distance of influence and weights. Table S2: The sensitivity of habitat types to each threat. Table S3: Table of biophysical coefficients for InVEST. Table S4: Description of variables selected in this study. Table S5: Multicollinearity test among influencing factors.

Author Contributions: Conceptualization, Methodology, and Writing—original draft, B.H.; Formal analysis, Z.Z.; Validation and Resources, H.H.; Software and Data curation, Z.L.; Project administration, X.C.; Supervision, F.K.; Visualization, H.W. All authors have read and agreed to the published version of the manuscript.

Funding: This research was supported by the National Key Research and Development Program of China (No. 2019YFA0607304).

Institutional Review Board Statement: Not applicable.

Informed Consent Statement: Not applicable.

Data Availability Statement: Publicly available datasets were analyzed in this study. Data sources and access links are detailed in Table 1 and Table S4.

Acknowledgments: We sincerely thank the reviewers and editors for their efforts to improve the quality of our paper. We would like to express our gratitude to those who participated in the manuscript revisions.

Conflicts of Interest: The authors declare no conflict of interest.

References

1. Costanza, R.; d'Arge, R.; De Groot, R.; Farber, S.; Grasso, M.; Hannon, B.; Limburg, K.; Naeem, S.; O'Neill, R.V.; Paruelo, J.; et al. The value of the world's ecosystem services and natural capital. *Nature* **1997**, *387*, 253–260. [CrossRef]
2. Carpenter, S.R.; Mooney, H.A.; Agard, J.; Capistrano, D.; Defries, R.S.; Díaz, S.; Dietz, T.; Duraiappah, A.K.; Oteng-Yeboah, A.; Pereira, H.M.; et al. Science for managing ecosystem services: Beyond the millennium ecosystem assessment. *Proc. Natl. Acad. Sci. USA* **2009**, *106*, 1305–1312. [CrossRef] [PubMed]
3. Costanza, R.; De Groot, R.; Sutton, P.; van der Ploeg, S.; Anderson, S.J.; Kubiszewski, I.; Farber, S.; Turner, R.K. Changes in the global value of ecosystem services. *Glob. Environ. Chang.* **2014**, *26*, 152–158. [CrossRef]
4. Xiang, H.; Wang, Z.; Mao, D.; Zhang, J.; Zhao, D.; Zeng, Y.; Wu, F. Surface mining caused multiple ecosystem service losses in China. *J. Environ. Manag.* **2021**, *290*, 112618. [CrossRef] [PubMed]
5. Newbold, T.; Hudson, L.N.; Hill, S.L.L.; Contu, S.; Lysenko, I.; Senior, R.A.; Börger, L.; Bennett, D.J.; Choimes, A.; Collen, B.; et al. Global effects of land use on local terrestrial biodiversity. *Nature* **2015**, *520*, 45–50. [CrossRef]
6. Wunder, S.; Brouwer, R.; Engel, S.; Ezzine-de-Blas, D.; Muradian, R.; Pascual, U.; Pinto, R. From principles to practice in paying for nature's services. *Nat. Sustain.* **2018**, *1*, 145–150. [CrossRef]
7. Zeng, J.; Chen, T.; Yao, X.; Chen, W. Do protected areas improve ecosystem services? A case study of Hoh Xil Nature Reserve in Qinghai-Tibetan Plateau. *Remote Sens.* **2020**, *12*, 471. [CrossRef]
8. Benayas, J.M.R.; Newton, A.C.; Diaz, A.; Bullock, J.M. Enhancement of biodiversity and ecosystem services by ecological restoration: A meta-analysis. *Science* **2009**, *325*, 1121–1124. [CrossRef]
9. Salzman, J.; Bennett, G.; Carroll, N.; Goldstein, A.; Jenkins, M. The global status and trends of Payments for Ecosystem Services. *Nat. Sustain.* **2018**, *1*, 136–144. [CrossRef]
10. Costanza, R.; De Groot, R.; Braat, L.; Kubiszewski, I.; Fioramonti, L.; Sutton, P.; Farber, S.; Grasso, M. Twenty years of ecosystem services: How far have we come and how far do we still need to go? *Ecosyst. Serv.* **2017**, *28*, 1–16. [CrossRef]
11. Gao, J. Editorial for the Special Issue "Ecosystem Services with Remote Sensing". *Remote Sens.* **2020**, *12*, 2191. [CrossRef]
12. Turkelboom, F.; Leone, M.; Jacobs, S.; Kelemen, E.; García-Llorente, M.; Baró, F.; Termansen, M.; Barton, D.N.; Berry, P.; Stange, E.; et al. When we cannot have it all: Ecosystem services trade-offs in the context of spatial planning. *Ecosyst. Serv.* **2018**, *29*, 566–578. [CrossRef]
13. Qian, D.; Du, Y.; Li, Q.; Guo, X.; Cao, G. Alpine grassland management based on ecosystem service relationships on the southern slopes of the Qilian Mountains, China. *J. Environ. Manag.* **2021**, *288*, 112447. [CrossRef] [PubMed]
14. Li, S.; Zhang, Y.; Wang, Z.; Li, L. Mapping human influence intensity in the Tibetan Plateau for conservation of ecological service functions. *Ecosyst. Serv.* **2018**, *30*, 276–286. [CrossRef]

15. Schirpke, U.; Tscholl, S.; Tasser, E. Spatio-temporal changes in ecosystem service values: Effects of land-use changes from past to future (1860–2100). *J. Environ. Manag.* **2020**, *272*, 111068. [CrossRef]
16. Grafius, D.R.; Corstanje, R.; Warren, P.H.; Evans, K.L.; Hancock, S.; Harris, J.A. The impact of land use/land cover scale on modelling urban ecosystem services. *Landsc. Ecol.* **2016**, *31*, 1509–1522. [CrossRef]
17. Liu, Y.; Lü, Y.; Fu, B.; Harris, P.; Wu, L. Quantifying the spatio-temporal drivers of planned vegetation restoration on ecosystem services at a regional scale. *Sci. Total. Environ.* **2019**, *650*, 1029–1040. [CrossRef]
18. Zheng, H.; Li, Y.; Robinson, B.E.; Liu, G.; Ma, D.; Wang, F.; Lu, F.; Ouyang, Z.; Daily, G.C. Using ecosystem service trade-offs to inform water conservation policies and management practices. *Front. Ecol. Environ.* **2016**, *14*, 527–532. [CrossRef]
19. Feng, Q.; Zhao, W.; Hu, X.; Liu, Y.; Daryanto, S.; Cherubini, F. Trading-off ecosystem services for better ecological restoration: A case study in the Loess Plateau of China. *J. Clean. Prod.* **2020**, *257*, 120469. [CrossRef]
20. Divinsky, I.; Becker, N.; Kutiel, P. Ecosystem service tradeoff between grazing intensity and other services-A case study in Karei-Deshe experimental cattle range in northern Israel. *Ecosyst. Serv.* **2017**, *24*, 16–27. [CrossRef]
21. Peng, J.; Hu, X.; Wang, X.; Meersmans, J.; Liu, Y.; Qiu, S. Simulating the impact of Grain-for-Green Programme on ecosystem services trade-offs in Northwestern Yunnan, China. *Ecosyst. Serv.* **2019**, *39*, 100998. [CrossRef]
22. Cai, W.; Borlace, S.; Lengaigne, M.; Rensch, P.V.; Collins, M.; Vecchi, G.; Timmermann, A.; Santoso, A.; McPhaden, M.J.; Wu, L.; et al. Increasing frequency of extreme El Niño events due to greenhouse warming. *Nat. Clim. Chang.* **2014**, *4*, 111–116. [CrossRef]
23. Geng, Q.; Ren, Q.; Yan, H.; Li, L.; Zhao, X.; Mu, X.; Wu, P.; Yu, Q. Target areas for harmonizing the Grain for Green Programme in China's Loess Plateau. *Land Degrad. Dev.* **2019**, *31*, 325–333. [CrossRef]
24. Zheng, K.; Wei, J.; Pei, J.; Cheng, H.; Zhang, X.; Huang, F.; Li, F.; Ye, J. Impacts of climate change and human activities on grassland vegetation variation in the Chinese Loess Plateau. *Sci. Total. Environ.* **2019**, *660*, 236–244. [CrossRef]
25. Hou, Y.; Lü, Y.; Chen, W.; Fu, B. Temporal variation and spatial scale dependency of ecosystem service interactions: A case study on the central Loess Plateau of China. *Landsc. Ecol.* **2017**, *32*, 1201–1217. [CrossRef]
26. Wen, X.; Théau, J. Spatiotemporal analysis of water-related ecosystem services under ecological restoration scenarios: A case study in northern Shaanxi, China. *Sci. Total. Environ.* **2020**, *720*, 137477. [CrossRef] [PubMed]
27. Feng, X.; Fu, B.; Piao, S.; Wang, S.; Ciais, P.; Zeng, Z.; Lü, Y.; Zeng, Y.; Li, Y.; Jiang, X.; et al. Revegetation in China's Loess Plateau is approaching sustainable water resource limits. *Nat. Clim. Chang.* **2016**, *6*, 1019–1022. [CrossRef]
28. Chen, Y.; Wang, K.; Lin, Y.; Shi, W.; Song, Y.; He, X. Balancing green and grain trade. *Nat. Geosci.* **2015**, *8*, 739–741. [CrossRef]
29. Yang, S.; Bai, Y.; Alatalo, J.M.; Wang, H.; Jiang, B.; Liu, G.; Chen, G. Spatio-temporal changes in water-related ecosystem services provision and trade-offs with food production. *J. Clean. Prod.* **2021**, *286*, 125316. [CrossRef]
30. Mandle, L.; Shields-Estrada, A.; Chaplin-Kramer, R.; Mitchell, M.G.E.; Bremer, L.L.; Gourevitch, J.D.; Hawthorne, P.; Johnson, J.A.; Robinson, B.E.; Smith, J.R.; et al. Increasing decision relevance of ecosystem service science. *Nat. Sustain.* **2021**, *4*, 161–169. [CrossRef]
31. Ouyang, Z.; Zheng, H.; Xiao, Y.; Polasky, S.; Liu, J.; Xu, W.; Wang, Q.; Zhang, L.; Xiao, Y.; Rao, E.; et al. Improvements in ecosystem services from investments in natural capital. *Science* **2016**, *352*, 1455–1459. [CrossRef]
32. Yang, S.; Zhao, W.; Liu, Y.; Wang, S.; Wang, J.; Zhai, R. Influence of land use change on the ecosystem service trade-offs in the ecological restoration area: Dynamics and scenarios in the Yanhe watershed, China. *Sci. Total. Environ.* **2018**, *644*, 556–566. [CrossRef]
33. Peng, J.; Tian, L.; Zhang, Z.; Zhao, Y.; Green, S.M.; Quine, T.A.; Liu, H.; Meersmans, J. Distinguishing the impacts of land use and climate change on ecosystem services in a karst landscape in China. *Ecosyst. Serv.* **2020**, *46*, 101199. [CrossRef]
34. Sun, C.; Li, X.; Zhang, W.; Chen, W.; Wang, J. Evaluation of ecological security in poverty-stricken region of Lüliang Mountain based on the remote sensing image. *China Environ. Sci.* **2019**, *39*, 5352–5360.
35. Li, J.; Wang, Y. Spatial coupling characteristics of eco-environment quality and economic poverty in Lüliang area. *Chin. J. Appl. Ecol.* **2014**, *25*, 1715–1724.
36. Sharp, H.T.; Tallis, H.T.; Ricketts, T.; Guerry, A.D.; Wood, S.A.; Chaplin-Kramer, R.; Nelson, E.; Ennaanay, D.; Wolny, S.; Olwero, N.; et al. InVEST 3.8.0 User's Guide. The Natural Capital Project: Stanford University, University of Minnesota, The Nature Conservancy, and World Wildlife Fund. 2020. Available online: http://releases.naturalcapitalproject.org/invest-userguide/latest/#supporting-tools (accessed on 26 November 2020).
37. Zhang, L.; Fu, B.; Lü, Y.; Zeng, Y. Balancing multiple ecosystem services in conservation priority setting. *Landsc. Ecol.* **2015**, *30*, 535–546. [CrossRef]
38. Liu, L.; Zhang, H.; Gao, Y.; Zhu, W.; Liu, X.; Xu, Q. Hotspot identification and interaction analyses of the provisioning of multiple ecosystem services: Case study of Shaanxi Province, China. *Ecol. Indic.* **2019**, *107*, 105566. [CrossRef]
39. Sun, X.; Lu, Z.; Li, F.; Crittenden, J.C. Analyzing spatio-temporal changes and trade-offs to support the supply of multiple ecosystem services in Beijing, China. *Ecol. Indic.* **2018**, *94*, 117–129. [CrossRef]
40. Liu, C.; Wang, C. Spatio-temporal evolution characteristics of habitat quality in the Loess Hilly Region based on land use change: A case study in Yuzhong county. *Acta Ecol. Sin.* **2018**, *38*, 7300–7311.
41. Zhou, L.; Tang, J.; Liu, X.; Dang, X.; Mu, H. Effects of urban expansion on habitat quality in densely populated areas on the Loess Plateau: A case study of Lanzhou, Xi'an-Xianyang and Taiyuan, China. *Chin. J. Appl. Ecol.* **2021**, *32*, 261–270.
42. Liang, Y.; Hashimoto, S.; Liu, L. Integrated assessment of land-use/land-cover dynamics on carbon storage services in the Loess Plateau of China from 1995 to 2050. *Ecol. Indic.* **2021**, *120*, 106939. [CrossRef]

43. Zhang, Y.; Shi, X.; Tang, Q. Carbon storage assessment in the upper reaches of the Fenhe River under different land use scenarios. *Acta Ecol. Sin.* **2021**, *41*, 360–373.
44. Tang, X.; Zhao, X.; Bai, Y.; Tang, Z.; Wang, W.; Zhao, Y.; Wan, H.; Xie, Z.; Shi, X.; Wu, B.; et al. Carbon pools in China's terrestrial ecosystems: New estimates based on an intensive field survey. *Proc. Natl. Acad. Sci. USA* **2018**, *115*, 4021–4026. [CrossRef] [PubMed]
45. Zheng, H.; Wang, L.; Peng, W.; Zhang, C.; Li, C.; Robinson, B.E.; Wu, X.; Kong, L.; Li, R.; Xiao, Y.; et al. Realizing the values of natural capital for inclusive, sustainable development: Informing China's new ecological development strategy. *Proc. Natl. Acad. Sci. USA* **2019**, *116*, 8623–8628. [CrossRef]
46. Fu, Q.; Li, B.; Hou, Y.; Bi, X.; Zhang, X. Effects of land use and climate change on ecosystem services in Central Asia's arid regions: A case study in Altay Prefecture, China. *Sci. Total. Environ.* **2017**, *607*, 633–646. [CrossRef] [PubMed]
47. Bradford, J.B.; D'Amato, A.W. Recognizing trade-offs in multi-objective land management. *Front. Ecol. Environ.* **2012**, *10*, 210–216. [CrossRef]
48. Lu, N.; Fu, B.; Jin, T.; Chang, R. Trade-off analyses of multiple ecosystem services by plantations along a precipitation gradient across Loess Plateau landscapes. *Landsc. Ecol.* **2014**, *29*, 1697–1708. [CrossRef]
49. Xu, J.; Chen, J.; Liu, Y. Partitioned responses of ecosystem services and their tradeoffs to human activities in the Belt and Road region. *J. Clean. Prod.* **2020**, *276*, 123205.
50. Luo, Y.; Lü, Y.; Fu, B.; Zhang, Q.; Li, T.; Hu, W.; Comber, A. Half century change of interactions among ecosystem services driven by ecological restoration: Quantification and policy implications at a watershed scale in the Chinese Loess Plateau. *Sci. Total. Environ.* **2019**, *651*, 2546–2557. [CrossRef]
51. Fu, B.; Wang, S.; Liu, Y.; Liu, J.; Liang, W.; Miao, C. Hydrogeomorphic ecosystem responses to natural and anthropogenic changes in the Loess Plateau of China. *Annu. Rev. Earth Planet. Sci.* **2017**, *45*, 223–243. [CrossRef]
52. Jiang, C.; Zhang, H.; Zhang, Z. Spatially explicit assessment of ecosystem services in China's Loess Plateau: Patterns, interactions, drivers, and implications. *Glob. Planet. Chang.* **2018**, *161*, 41–52. [CrossRef]
53. Wang, L.; Ma, S.; Jiang, J.; Zhao, Y.; Zhang, J. Spatiotemporal Variation in Ecosystem Services and Their Drivers among Different Landscape Heterogeneity Units and Terrain Gradients in the Southern Hill and Mountain Belt, China. *Remote Sens.* **2021**, *13*, 1375. [CrossRef]
54. Ahmed, M.A.; Abd-Elrahman, A.; Escobedo, F.J.; Cropper, W.P.; Martin, T.A.; Timilsina, N. Spatially-explicit modeling of multi-scale drivers of aboveground forest biomass and water yield in watersheds of the Southeastern United States. *J. Environ. Manag.* **2017**, *199*, 158–171. [CrossRef] [PubMed]
55. Zhang, Z.; Liu, Y.; Wang, Y.; Liu, Y.; Zhang, Y.; Zhang, Y. What factors affect the synergy and tradeoff between ecosystem services, and how, from a geospatial perspective? *J. Clean. Prod.* **2020**, *257*, 120454. [CrossRef]
56. Fotheringham, A.S.; Yang, W.; Kang, W. Multiscale geographically weighted regression (MGWR). *Ann. Am. Assoc. Geogr.* **2017**, *107*, 1247–1265. [CrossRef]
57. Clerici, N.; Cote-Navarro, F.; Escobedo, F.J.; Rubiano, K.; Villegas, J.C. Spatio-temporal and cumulative effects of land use-land cover and climate change on two ecosystem services in the Colombian Andes. *Sci. Total. Environ.* **2019**, *685*, 1181–1192. [CrossRef] [PubMed]
58. Sannigrahi, S.; Zhang, Q.; Pilla, F.; Joshi, P.K.; Basu, B.; Keesstra, S.; Roy, P.S.; Wang, Y.; Sutton, P.C.; Chakraborti, S.; et al. Responses of ecosystem services to natural and anthropogenic forcings: A spatial regression based assessment in the world's largest mangrove ecosystem. *Sci. Total. Environ.* **2020**, *715*, 137004. [CrossRef]
59. Gao, J.; Zuo, L.; Liu, W. Environmental determinants impacting the spatial heterogeneity of karst ecosystem services in Southwest China. *Land Degrad. Dev.* **2020**, *32*, 1718–1731. [CrossRef]
60. Li, S.; He, Y.; Xu, H.; Zhu, C.; Dong, B.; Lin, Y.; Si, B.; Deng, J.; Wang, K. Impacts of Urban Expansion Forms on Ecosystem Services in Urban Agglomerations: A Case Study of Shanghai-Hangzhou Bay Urban Agglomeration. *Remote Sens.* **2021**, *13*, 1908. [CrossRef]
61. Li, B.; Wang, W. Trade-offs and synergies in ecosystem services for the Yinchuan Basin in China. *Ecol. Indic.* **2018**, *84*, 837–846. [CrossRef]
62. Guo, S.; Han, X.; Li, H.; Wang, T.; Tong, X.; Ren, G.; Feng, Y.; Yang, G. Evaluation of soil quality along two revegetation chronosequences on the Loess Hilly Region of China. *Sci. Total. Environ.* **2018**, *633*, 808–815. [CrossRef] [PubMed]
63. Yang, L.; Che, L.; Wei, W.; Yu, Y.; Zhang, H. Comparison of deep soil moisture in two re-vegetation watersheds in semi-arid regions. *J. Hydrol.* **2014**, *513*, 314–321. [CrossRef]
64. Jia, X.; Fu, B.; Feng, X.; Hou, G.; Liu, Y.; Wang, X. The tradeoff and synergy between ecosystem services in the Grain-for-Green areas in Northern Shaanxi, China. *Ecol. Indic.* **2014**, *43*, 103–113. [CrossRef]
65. García, A.M.; Santé, I.; Loureiro, X.; Miranda, D. Green infrastructure spatial planning considering ecosystem services assessment and trade-off analysis. Application at landscape scale in Galicia region (NW Spain). *Ecosyst. Serv.* **2020**, *43*, 101115. [CrossRef]
66. Liang, J.; Li, S.; Li, X.; Li, X.; Liu, Q.; Meng, Q.; Lin, A.; Li, J. Trade-off analyses and optimization of water-related ecosystem services (WRESs) based on land use change in a typical agricultural watershed, southern China. *J. Clean. Prod.* **2021**, *279*, 123851. [CrossRef]
67. Wang, X.; Yan, F.; Su, F. Impacts of Urbanization on the Ecosystem Services in the Guangdong-Hong Kong-Macao Greater Bay Area, China. *Remote Sens.* **2020**, *12*, 3269. [CrossRef]

68. Zhang, Y.; Liu, Y.; Zhang, Y.; Liu, Y.; Zhang, G.; Chen, Y. On the spatial relationship between ecosystem services and urbanization: A case study in Wuhan, China. *Sci. Total. Environ.* **2018**, *637–638*, 780–790. [CrossRef]
69. Gao, J.; Li, F.; Gao, H.; Zhou, C.; Zhang, X. The impact of land-use change on water-related ecosystem services: A study of the Guishui River Basin, Beijing, China. *J. Clean. Prod.* **2017**, *163*, S148–S155. [CrossRef]
70. Lang, Y.; Song, W.; Zhang, Y. Responses of the water-yield ecosystem service to climate and land use change in Sancha River Basin, China. *Phys. Chem. Earth* **2017**, *101*, 102–111. [CrossRef]
71. Wu, X.; Wang, S.; Fu, B.; Liu, J. Spatial variation and influencing factors of the effectiveness of afforestation in China's Loess Plateau. *Sci. Total. Environ.* **2021**, *771*, 144904. [CrossRef]
72. Wen, X.; Deng, X.; Zhang, F. Scale effects of vegetation restoration on soil and water conservation in a semi-arid region in China: Resources conservation and sustainable management. *Resour. Conserv. Recycl.* **2019**, *151*, 104474. [CrossRef]
73. Zhou, G.; Wei, X.; Chen, X.; Zhou, P.; Liu, X.; Xiao, Y.; Sun, G.; Scott, D.F.; Zhou, S.; Han, L.; et al. Global pattern for the effect of climate and land cover on water yield. *Nat. Commun.* **2015**, *6*, 5918. [CrossRef] [PubMed]
74. He, Y.; Kuang, Y.; Zhao, Y.; Ruan, Z. Spatial Correlation between Ecosystem Services and Human Disturbances: A Case Study of the Guangdong–Hong Kong–Macao Greater Bay Area, China. *Remote Sens.* **2021**, *13*, 1174. [CrossRef]

Article

Increased Ecosystem Carbon Storage between 2001 and 2019 in the Northeastern Margin of the Qinghai-Tibet Plateau

Peijie Wei [1,2], Shengyun Chen [1,3,4,*], Minghui Wu [1,2], Yinglan Jia [1,2], Haojie Xu [5] and Deming Liu [4]

1 Cryosphere and Eco-Environment Research Station of Shule River Headwaters,
  State Key Laboratory of Cryospheric Science, Northwest Institute of Eco-Environment and Resources,
  Chinese Academy of Sciences, Lanzhou 730000, China; weipeijie19@mails.ucas.ac.cn (P.W.);
  wumh2017@lzb.ac.cn (M.W.); jiayinglan20@mails.ucas.edu.cn (Y.J.)
2 University of Chinese Academy of Sciences, Beijing 100049, China
3 School of Geographical Sciences, Academy of Plateau Science and Sustainability, Qinghai Normal University,
  Xining 810008, China
4 Long-Term National Scientific Research Base of the Qilian Mountain National Park, Xining 810000, China;
  demingliu2021@163.com
5 State Key Laboratory of Grassland Agroecosystems, College of Pastoral Agriculture Science and Technology,
  Lanzhou University, Lanzhou 730020, China; xuhaojie@lzu.edu.cn
* Correspondence: sychen@lzb.ac.cn

Abstract: Global alpine ecosystems contain a large amount of carbon, which is sensitive to global change. Changes to alpine carbon sources and sinks have implications for carbon and climate feedback processes. To date, few studies have quantified the spatial-temporal variations in ecosystem carbon storage and its response to global change in the alpine regions of the Qinghai-Tibet Plateau (QTP). Ecosystem carbon storage in the northeastern QTP between 2001 and 2019 was simulated and systematically analyzed using the Integrated Valuation of Ecosystem Services and Tradeoffs (InVEST) model. Furthermore, the Hurst exponent was obtained and used as an input to perform an analysis of the future dynamic consistency of ecosystem carbon storage. Our study results demonstrated that: (1) regression between the normalized difference vegetation index (NDVI) and biomass (coefficient of determination (R^2) = 0.974, $p < 0.001$), and between NDVI and soil organic carbon density (SOCD) ($R^2 = 0.810$, $p < 0.001$) were valid; (2) the spatial distribution of ecosystem carbon storage decreased from the southeast to the northwest; (3) ecosystem carbon storage increased by 13.69% between 2001 and 2019, and the significant increases mainly occurred in the low-altitude regions; (4) climate and land use (LULC) changes caused increases in ecosystem carbon storage of 4.39 Tg C and 2.25 Tg C from 2001 to 2019, respectively; and (5) the future trend of ecosystem carbon storage in 92.73% of the study area shows high inconsistency but that in 7.27% was consistent. This study reveals that climate and LULC changes have positive effects on ecosystem carbon storage in the alpine regions of the QTP, which will provide valuable information for the formulation of eco-environmental policies and sustainable development.

Keywords: InVEST model; global change; ecosystem carbon storage; Hurst exponent; Qinghai-Tibet Plateau

1. Introduction

Ecosystem carbon storage is recognized as a key indicator of ecosystem function because it is closely related to the climate regulation and productivity of terrestrial ecosystems [1,2]. Ecosystem carbon storage refers to the cumulative amount of carbon stored in terrestrial ecosystems [3]. The maintenance of ecosystem carbon storage is one of the hotspots of common concern around the world [4,5]. Quantitative studies of ecosystem carbon storage can provide a theoretical basis for the integrated management of natural ecosystems and the sustainable utilization of natural resources [5,6].

The quantitative methods of ecosystem carbon storage include field surveys, model simulations, and remote sensing [7]. As one of the assessment methods, modeling is becoming increasingly prominent because it can conduct evaluations of carbon storage at different scales, including global [8], national [9], and regional [10]. Many models have been used to evaluate carbon storage, such as Vector Autoregression (VAR) [11], High Accuracy Surface Modeling (HASM) [12], Century [13], Biome-BGC [14], General Ensemble Biogeochemical Modeling System (GEMS) [4,15], and InVEST [16,17]. The InVEST model provides new technology for conducting spatial expression, dynamic analysis, and quantitative evaluations of ecosystem service function [18]. More importantly, the InVEST model can easily be used to assess the impacts of climate and LULC changes on ecosystem carbon storage.

The InVEST model has been used in the United States, Tanzania, Indonesia, and China [7,19,20]. However, most of these studies have focused on the carbon pool of single ecosystems, and there are few systematic studies on the carbon storage of terrestrial ecosystems [7]. Previous studies have shown that climate and LULC changes are the major factors affecting ecosystem carbon storage [21–24]. Generally, climate change controls the balance between carbon inputs from plant productivity and carbon outputs from soil carbon decomposition and alters ecosystem carbon storage [25]. LULC change is an important process that affects carbon storage: changes in LULC from one type to another are usually accompanied by a large amount of carbon exchange [23,26]. LULC change can alter the carbon cycle process by changing the ecosystem's structure (species composition, biomass) and function (energy balance, biodiversity, and the cycle of carbon, nitrogen, and water.) [23].

The QTP is the highest and largest plateau in the world, with an average elevation of more than 4000 m [22,27]. The soil has been reported to have accumulated plentiful soil organic matter, and its carbon density is obviously higher than that at similar latitudes, mainly due to its relatively low temperature and very high altitude, and thus the QTP has been regarded as a huge carbon pool [22,28,29]. Some studies of the QTP have evaluated carbon storage by using the InVEST model, but the spatial-temporal pattern of ecosystem carbon storage is still uncertain due to studies with low statistical power and insufficient sample sizes [30]. In addition, due to its unique geographic and ecological conditions, the alpine ecosystem of the QTP is very fragile and sensitive to climate change [31]. In the past 50 years, the air temperature of the QTP has increased by 0.2 °C per decade, roughly twice the observed rate of global warming [22], and annual precipitation has increased at the rate of 0.91 mm per year (1961–2007) [32]. Rapid warming and wetting undoubtedly shape the structures and processes of the ecosystem, which, in turn, lead to dramatic changes in the carbon cycle. Because climate warming tends to boost both plant production and soil respiration, there is some uncertainty about the change trend of ecosystem carbon [33,34]. It should be noted that the global warming trend will continue, according to the Fifth Assessment Report of the Intergovernmental Panel on Climate Change (IPCC) [35], which may further modify the dynamics of ecosystem carbon storage and amplify the uncertainty regarding the QTP. Hence, it has become urgent to analyze the spatial-temporal patterns of ecosystem carbon storage in the QTP more accurately, assess the impact of global change on ecosystem carbon storage, and detect the dynamic consistency of ecosystem carbon in the future.

Currently, studies on the spatial-temporal patterns of ecosystem carbon storage and their response to global change in the Shule River Basin on the northeastern margin of the QTP are especially rare. Taking the upstream regions of the Shule River Basin as a case study, the main goals of this research were: (1) to clarify and analyze the spatial-temporal patterns of ecosystem carbon storage and its change trend; (2) to explore the response characteristics of ecosystem carbon storage to climate and LULC changes; and (3) to detect the consistency of dynamic of ecosystem carbon storage in the future. This study has important guiding significance for the rational planning of environmental protection and the formulation of relevant policies in alpine regions.

2. Materials and Methods

2.1. Study Area and Climate Conditions

The Shule River, known as the "natural water tower" and "lifeline" of herders and farmers, is the second-largest inland river in the Hexi Corridor of China [36]. The upstream regions of the Shule River Basin (96.2°–99.0°E, 38.2°–40.0°N) lies in a mountainous area with abundant precipitation, which is the catchment area of the main stream of the Shule River Basin, and covers an area of about 1.38×10^4 km^2 (Figure 1). The altitude ranges from 1900 to 5733 m and gradually decreases from the edge to the middle region. Further, dozens of glaciers lie in this area [37]. The study area has a typical continental climate [38,39], with a low mean annual air temperature, little precipitation, and high actual evaporation [36,37]. The mean annual air temperature and annual precipitation from 1990 to 2019 were −5.24 °C and 201.64 mm, respectively (Figure 2). During this period, the mean annual air temperature and annual precipitation increased at the rate of 0.03 °C and 4.70 mm per year ($p < 0.01$), respectively. The change in climate in the study area showed a trend towards warmer and wetter conditions. In addition, the dominant vegetation types are alpine swamp meadows, alpine steppes, and alpine meadows [40].

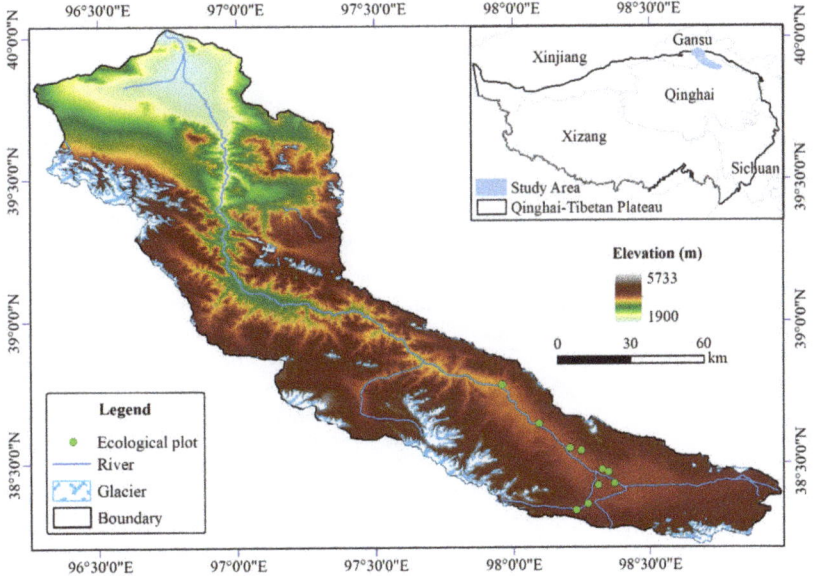

Figure 1. Location of the study area and ecological plots.

2.2. Data Source and Processing

Meteorological data. The meteorological station data from around China from 1990–2019 were provided by the China Meteorological Data Service Center (http://data.cma.cn/ (accessed on 7 October 2019)). Original air temperature and precipitation data were interpolated into grid data using the spatial interpolation tool of geographic information systems.

LULC data. The Moderate Resolution Imaging Spectroradiometer (MODIS) Land Cover Type (MCD12Q1) Version 6 product provides global LULC types at yearly intervals (2001–2019). The classification method of the International Geosphere Biosphere Program (IGBP) was used in this study. The product was provided by NASA Earth Science Data Systems (https://search.earthdata.nasa.gov/search (accessed on 16 October 2020)).

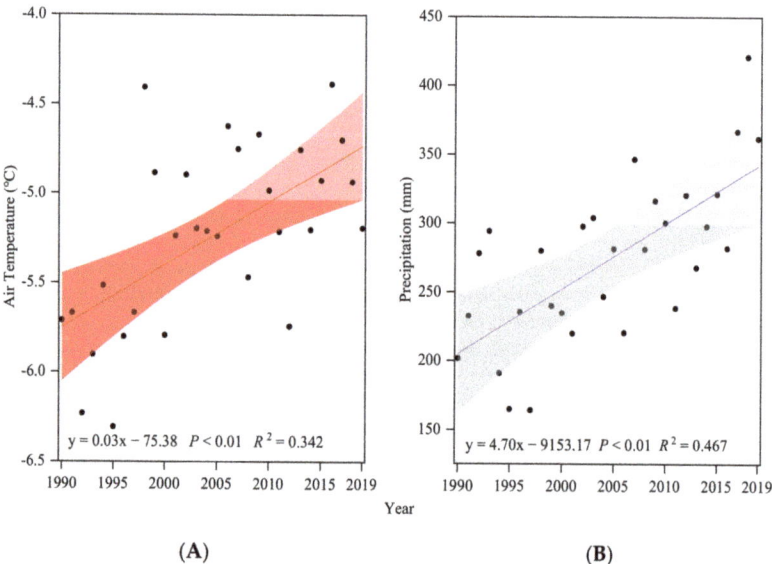

Figure 2. Temporal dynamics of the mean annual air temperature (**A**) and annual precipitation (**B**) in 1990–2019. These dots are data points for each year, and the shaded areas represent the 99.0% confidence intervals of the regression equation.

NDVI data. Based on the SPOT/VEGETATION NDVI satellite remote sensing data, datasets of monthly NDVI were generated using the maximum value synthesis method. The NDVI values for the growth seasons (May to September) from 2000 to 2019 were obtained for this study. The datasets were provided by the Resource and Environmental Science and Data Center (http://www.resdc.cn/ (accessed on 21 November 2020)).

Ecosystem carbon parameters. Biomass and SOCD data were collected in 2000. The biomass data were obtained from the Carbon Dioxide Information Analysis Center (CDIAC), and SOCD data were provided by the European Soil Data Centre (ESDAC) (0–30 cm). Because biological carbon is relatively stable and small, it was obtained from field surveys and the literature [7,41–43]. In addition, the root-to-shoot ratios (*RSRs*) of different LULC types were collected from the IPCC 2006 national greenhouse gas emission inventory [44].

Field survey data. The LULC type of the ecological plots is grassland. Sampling occurred during the growth seasons of 2011–2019. The locations of the ecological plots can be found in Figure 1. We collected about 374 biomass samples (including aboveground biomass (AGB) and belowground biomass (BGB)) and 585 soil organic carbon (SOC) samples. In detail, each 50 × 50 cm quadrat was selected randomly in each ecological plot, and the living AGB was harvested. Previous studies reported that more than 90% of the BGB in alpine grassland is concentrated in the top 30 cm of the soil [38,45]. Hence, we collected BGB from each quadrat at a depth of 0–30 cm using soil cores 4.8 cm in diameter. Soil samples were crumbled by hand, rocks were removed, and samples were then passed through a sieve with a 0.2 cm pore size, then cleaned repeatedly to obtain the BGB. The AGB and BGB were weighed (0.01 g) after drying (24 h at 80 °C), and the carbon contents were converted by a ratio of 0.475 [44].

Samples for SOC were collected from each quadrat and were split into 0–10, 10–20, and 20–30 cm sections, then packed in plastic bags and brought to the laboratory. The dichromate oxidation method (Walkley–Black procedure) was used to measure the SOC [46]. In addition, volumetric soil samples for each depth were sampled using a cutting ring (volume of 100 g/cm^3) and dried at 105 °C to determine the bulk density. The volume of rock fragments (i.e., coarser than 2 mm) was measured by submerging moist rock fragments and recording the volume of displaced water [38]. The soil organic carbon density (SOCD, unit: kg/m^2) was calculated using Equation (1) [38,47]:

$$SOCD = \sum_{i=1}^{n} h_i \times BD_i \times SOC_i \times (1 - C_i)/100 \quad (1)$$

where h_i, BD_i, SOC_i, and C_i are soil thickness (cm), bulk density (g/cm^3), SOC (g/kg), and the volume percentage of soil particles >2 mm at layer i, respectively.

In addition, all grid data were resampled with a spatial resolution of 500 m and projected using the World Geodetic System 1984.

2.3. Methods

2.3.1. Carbon Module

The carbon storage module in the InVEST model was used to estimate the ecosystem carbon storage in a specific region. This module consists of four carbon pools: aboveground biomass carbon, belowground biomass carbon, SOC, and humus carbon [1]. Aboveground biomass includes living plant materials above the soil level (such as leaves, bark, branches, trunks, etc.); belowground biomass refers to the living root systems of the aboveground plants. In contrast, soil organic matter comprises the organic component of soil, whereas humus is derived from leaf litter and the wood of standing and lying trees. The LULC types and their carbon density are the basic parameters of the module [18]. In the InVEST 3.2.0 User's Guide, more detailed descriptions can be found [48]. The calculation formulae are shown in Equations (2) and (3):

$$C_{ecosystem} = C_{above} + C_{below} + C_{soil} + C_{humus} \quad (2)$$

$$C_{zone} = \sum_{i=1}^{n} C_i \times A_i \quad (3)$$

where $C_{ecosystem}$, C_{above}, C_{below}, C_{soil}, and C_{humus} are the ecosystem carbon, aboveground biomass carbon, belowground biomass carbon, SOC, and humus carbon, respectively; C_i is the carbon density of the LULC type i; and A_i is the area of LULC type i.

2.3.2. Parameter Inversion of the Key Carbon Pool

Reviewing previous studies, we found that four main methods are used to estimate carbon density: literature reviews, field measurements, empirical modeling, and remote sensing. Due to differences in the estimation methods, the results may vary greatly [30]. Some studies used MODIS-NDVI products to establish the relationship between biomass and SOCD and demonstrated the scientific adaptability of this approach [30,49–51]. In this study, we used the spatial analysis tool of ArcGIS to acquire various values (including NDVI, biomass carbon, and SOCD) for each LULC type based on the dataset of 2000. Jan Joseph et al. [52] thought that the carbon density of built-up land and water bodies is zero, and thus the urban and built-up land, permanent snow and ice, and other water bodies were not considered in our study. In addition, nonirrigated farmland and bare land were also not considered due to their anomalous carbon density data. In accordance with the requirements of the fit of strong correlation, two quadratic curve equations, namely Biomass–NDVI and SOCD–NDVI, were constructed (Table 1). The key parameters (including biomass and SOCD) were calculated for 2001–2019 according to the regression equations and NDVI. Additionally, the biomass of each LULC type was divided into AGB and BGB according to the *RSRs*.

In addition, the performance of regression equations can be assessed by comparing the simulated and observed values, and systematic quantification of their performance accuracy can be determined by the coefficient of determination (R^2) (ranging from 0 to 1), p-value (p), and root mean squared errors (*RMSE*). Among these, the R^2 describes the decrease in collinearity between the measured and simulated data [39,53]. A higher value of R^2 represents less error variance, and values above 0.5 are regarded as acceptable. RMSE is one of the statistical indicators of error. It is commonly accepted that the lower the RMSE, the better the model performance [54]. All R^2 values in the text refer to the adjusted R^2.

Table 1. The regression equations of biomass–NDVI and SOCD–NDVI.

Regression Equations	R^2	p
Biomass = −45.64 × NDVI² + 32.08 × NDVI − 1.27	0.974	0.001
SOCD = −471.71 × NDVI² + 289.33 × NDVI + 5.67	0.810	0.001

2.3.3. Hypotheses of Climate and LULC Changes

Generally, changes in ecosystem carbon storage in the region can be attributed to changes in the LULC type and carbon density [49]. To some extent, the change in carbon density is a direct indicator that reflects the effect of climate change on ecosystem carbon storage. To further explore the main factor affecting the change in ecosystem carbon storage, we took 2001 as the control year, and three hypotheses were designed regarding 2019 values: actual condition, LULC change only, and climate change only (Figure 3). This process is described as follows [49]:

1. The change in ecosystem carbon storage under actual condition is described as ΔC, and the formula is expressed in Equation (4):

$$\Delta C = \sum_{i=1}^{n}(A_{i2}D_{i2} - A_{i1}D_{i1}) \tag{4}$$

where A_{i1} and A_{i2} are the area of LULC type i before and after the change, respectively; D_{i1} and D_{i2} are the ecosystem carbon densities of LULC type i before and after the change, respectively.

2. The change in ecosystem carbon storage caused by climate change only can be expressed by Equation (5):

$$\Delta C_D = \sum_{i=1}^{n} A_{i1}(D_{i2} - D_{i1}) \tag{5}$$

3. The ecosystem carbon density of each LULC type is constant, and thus the change in ecosystem carbon storage is caused by LULC change only, which can be expressed as Equation (6):

$$\Delta C_L = \sum_{i=1}^{n} D_{i1}(A_{i2} - A_{i1}) \tag{6}$$

4. According to all three hypotheses, the contribution of climate and LULC change to the change in the ecosystem carbon storage can be described by Equations (7) and (8):

$$R_L = \frac{\Delta C_L}{\Delta C_L + \Delta C_D} \times 100\% \tag{7}$$

$$R_D = \frac{\Delta C_D}{\Delta C_L + \Delta C_D} \times 100\% \tag{8}$$

where R_L and R_D are the contributions of LULC and climate changes to the change in ecosystem carbon storage, respectively.

Figure 3. Simulation of ecosystem carbon storage under different hypotheses.

2.3.4. Linear Regression Analysis

Linear regression analysis was used to detect the spatial change in the research objects (including meteorological elements and ecosystem carbon storage) with time-series data. The slope of the linear regression is considered to be the best index for quantifying the change trend of the research object during the study period. The slope can be calculated by Equation (9):

$$slope = \frac{n \times \sum_{i=1}^{n} i \times B_i - (\sum_{i=1}^{n} i)(\sum_{i=1}^{n} B_i)}{n \times \sum_{i=1}^{n} i^2 - (\sum_{i=1}^{n} i)^2} \qquad (9)$$

where n is the number of years in the study period, i is the serial number of the year, and B_i is the value of research object in the year i. Positive and negative values of the slope refer to positive and negative trends, respectively.

The correlation coefficient (r) of the linear regression can be used to test the significance of the change trend. It is expressed in Equation (10):

$$r = \frac{cov(i, B_i)}{\sqrt{var(i)var(B_i)}} \qquad (10)$$

where cov and var are the covariance and variance functions, respectively. When $p < 0.01$, the change trend is extremely significant; when $0.01 < p < 0.05$, the change trend is significant; and when $p > 0.05$, the change trend is non-significant.

2.3.5. Hurst Exponent

The Hurst exponent is a classic method for detecting long memory in time series, which was proposed by the hydrologist H.E Hurst in 1951 [55]. R/S analysis is a superior and well-known method used for estimating Hurst exponent and was introduced by Mandelbrot [55,56]. In this study, we used this method to test the consistency of the future dynamics of ecosystem carbon. The main calculation procedures are shown in Equations (11)–(15) [56]:

1. Divide the time series $\{\xi(\tau)\}$ ($\tau = 1, 2, \ldots, n$) into τ subseries $x(t)$, and, for each subseries, $t = 1, \ldots, \tau$.
2. Define the mean sequence of the time series:

$$\xi_\tau = \frac{1}{\tau}\sum_{t=1}^{\tau} x(t), \ \tau = 1, 2, \ldots, n \qquad (11)$$

3. Calculate the cumulative deviation:

$$X(t, \tau) = \sum_{u=1}^{t} (\xi(u) - (\xi)_\tau), \ 1 \leq t \leq \tau \tag{12}$$

4. Create the range sequence:

$$R(\tau) = \max_{1 \leq t \leq \tau} X(t, \tau) - \min_{1 \leq t \leq \tau} X(t, \tau), \ \tau = 1, 2, \ldots, n \tag{13}$$

5. Create the standard deviation sequence:

$$S(\tau) = \left(\frac{1}{\tau} \sum_{t=1}^{\tau} (\xi(t) - (\xi)_\tau)^2 \right)^{1/2}, \ \tau = 1, 2, \ldots, n \tag{14}$$

6. Rescale the range:

$$\frac{R(\tau)}{S(\tau)} = (c\tau)^H \tag{15}$$

The value of the Hurst exponent ranges from 0 to 1, according to Hurst [57] and Mandelbrot [58]. When the value is equal to 0.5, this indicates that the time series is a random series without consistency (i.e., the change trend of the time series in the future would not be related with that in the study period); when the value is greater than 0.5, it refers to the consistency of the time series (i.e., the change trend of the time series in the future is the same as that in the study period, with the greater value for the more consistency); and when the value is less than 0.5, which indicates the inconsistency of the time series in the future, with theless value for the more inconsistency.

3. Results

3.1. Verification of Key Carbon Pool Parameters

To evaluate the performance of the regression equations, we analyzed the relationships between the observed values from the field survey and the simulated values (including AGB, BGB, and SOCD) for 2011–2019 (Figure 4). The R^2, p, and $RMSE$ were used to evaluate the agreement between the observed and simulated variables. The results indicated that the simulated values were sufficiently consistent with the measured values ($p < 0.01$), with correlation coefficients of 0.901, 0.925, and 0.866, respectively. Additionally, the $RMSE$ values were relatively small.

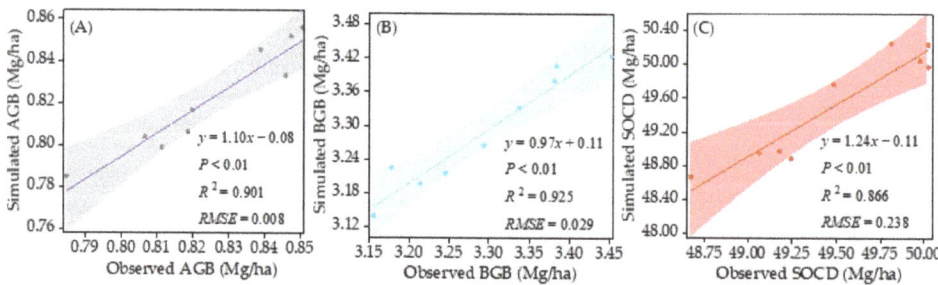

Figure 4. Comparisons of the simulated and observed aboveground biomass (AGB) (**A**), belowground biomass (BGB) (**B**), and soil organic carbon density (SOCD) (**C**) of grassland in 2011–2019. These lines are the fitting lines of the regression equation, and the shaded areas represent the 99.0% confidence intervals of the regression equations.

3.2. Climate Change

The spatial distribution of mean air annual temperature and annual precipitation could be characterized by large regional features. Specifically, the spatial variation range in the mean annual air temperature was -17.35 to 6.20 °C, gradually increasing from southeast to northwest (Figure 5A), while annual precipitation varied from 100.17 to 533.24 mm, diminishing from the southeast to northwest (Figure 5B). In addition, as shown in Figure 5C, significant warming mainly occurred in parts of the northwest ($p < 0.05$) from 2001 to 2019, and the rest of the study area also experienced obvious increases ($p > 0.05$). Figure 5D shows the changes in annual precipitation during the study period, which clearly indicate an extremely significant increase in precipitation across the whole study area ($p < 0.01$).

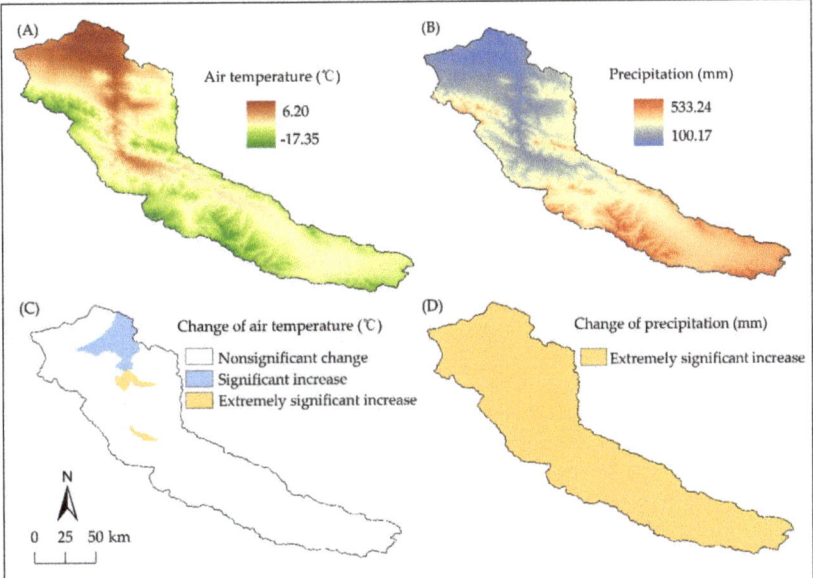

Figure 5. Distributions of the mean annual air temperature (**A**), and annual precipitation (**B**), and changes in the mean annual air temperature (**C**) and annual precipitation (**D**) in 2001–2019.

3.3. Distribution of LULC and Change in LULC

In this study area, grassland is the major LULC type, encompassing more than 44.74% of the regional area. This LULC type is mostly located at the east of the northwest and in low-altitude regions of the southeast (Figure 6). The second most common LULC type is desert, accounting for about 41.40%, mainly distributed in the west of the northwest and in high-altitude regions of the southeast. Further, cropland, permanent snow and ice, permanent wetland, and built-up land accounted for about 4.72%, 4.04%, 2.73%, and 2.37%, respectively. In the marginal high-altitude regions, the main LULC types are desert and permanent snow and ice. Cropland and permanent wetland are located in the northwest of the study area, while built-up land is scattered throughout the study area.

In Table 2, the rows and columns represent the area of the six LULC types in 2001 and 2019, respectively. The values of the main diagonal represent the area of each LULC type that persisted in 2019, and the off-diagonal values show the converted area of each LULC type from 2001 to 2019. In fact, only the desert showed a diminishing trend in 2019 compared with 2001, shrinking by 1492.86 km². Compared with 2001, the area increased by 1174.33 km² for grassland and by 128.11, 97.28, 60.64, and 32.49 km² for permanent snow and ice, cropland, permanent wetland, and built-up land in 2019, respectively (Table 2). Hence, the area of grassland increased the most during the study period (78.66%), followed by permanent snow and ice (8.58%). In practice, there was a decline in desert, which can be mainly attributed to the transition from desert to grassland, permanent snow and ice, cropland, and permanent wetland: 691.85, 300.59, 322.26, and 151.05 km², respectively. Additionally, a small part of the grassland was also converted to built-up land (17.50 km²) and permanent snow and ice (11.04 km²).

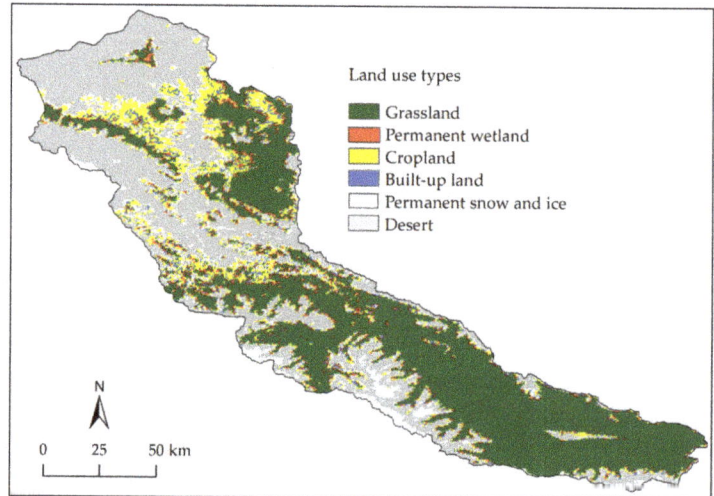

Figure 6. Geographical distribution of LULC types during the study period.

Table 2. Conversion of LULC types between 2001 and 2019 (unit: km²).

	LULC Type	2019					
		GL	PW	CL	BUL	PSI	DL
2001	GL	5438.18	51.72	48.7	17.5	11.04	15.84
	PW	168.09	132.44	18.8	9.37	6.42	10.36
	CL	237.64	35.98	246.47	26.89	16.07	38.81
	BUL	121.88	19.41	24.69	115.75	7.83	20.62
	PSI	99.67	15.52	38.22	13.77	279.62	46.66
	BA	691.85	151.05	322.26	159.39	300.59	4830.95

Note: Grassland—GL; permanent wetland—PW; cropland—CL; built-up land—BUL; permanent snow and ice—PSI; Desert land—DL

3.4. Differences in Ecosystem Carbon among LULC Types

The ecosystem carbon density in different LULC types is shown in Figure 7A. The highest amount of ecosystem carbon density was found in grassland, reaching 52.80 Mg/ha, followed by permanent wetland and cropland (about 44.83 and 42.00 Mg/ha, respectively), whereas the ecosystem carbon density of desert was the lowest, up to 29.87 Mg/ha. Compared with 2001, the ecosystem carbon density of grasslands, permanent wetland, cropland, and desert increased by 4.14, 7.90, 6.71, and 2.15 Mg/ha in 2019, respectively; the permanent wetland increased the most.

In terms of ecosystem carbon storage, grassland had the highest, reaching 32.48 Tg C, followed by desert (up to 16.87 Tg C) (Figure 7B). In contrast, the ecosystem carbon storage of permanent wetland and cropland was relatively small, up to just 1.81 and 2.86 Tg C, respectively. Ecosystem carbon storage in 2001 and 2019 was 50.56 and 57.49 Tg C, respectively, implying that ecosystem carbon storage increased during this period. Specifically, the ecosystem carbon storage of grassland increased the most (8.69 Tg C), followed by permanent wetland (0.61 Tg C) and cropland (0.88 Tg C), while ecosystem carbon storage in deserts decreased (−3.26 Tg C).

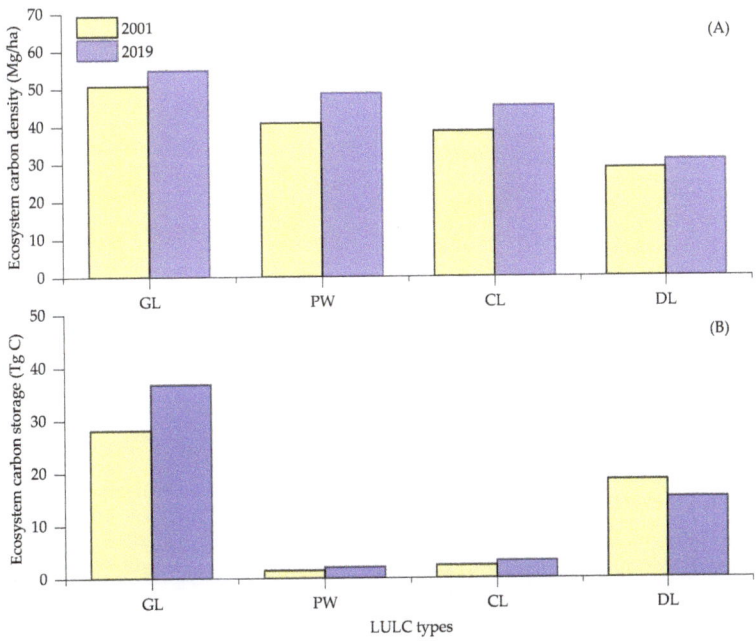

Figure 7. The ecosystem carbon density (**A**) and ecosystem carbon storage (**B**) of various LULC types.

3.5. Spatial Distribution of and Change in Ecosystem Carbon Storage

The ecosystem carbon storage exhibited strong spatial heterogeneity in the study area, and the spatial distribution of the ecosystem carbon storage was basically consistent throughout the study period, i.e., decreasing from southeast to northwest (Figure 8). In detail, the lowest value of ecosystem carbon storage was for water bodies (close to 0), where were mainly distributed in marginal high-altitude regions. The highest value of ecosystem carbon storage occurred in the east of the northwest and in low-altitude regions of the southeast.

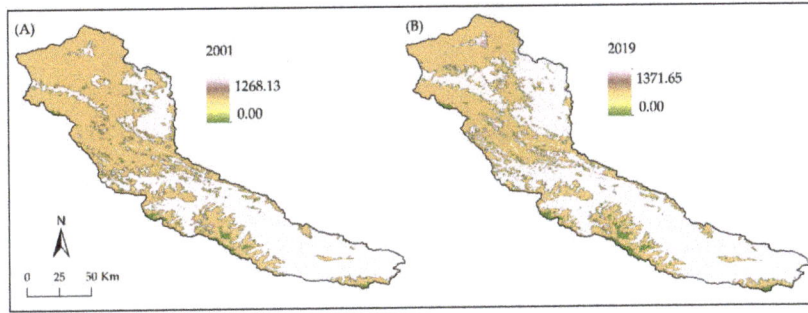

Figure 8. Distribution of ecosystem carbon storage in 2001(**A**) and 2019 (**B**).

The change trend of ecosystem carbon storage of each pixel is shown in Figure 9, indicating that the linear trend of ecosystem carbon storage showed distinct spatial differences from 2001 to 2019. In sum, more than 51.48% of the study area experienced extremely significant ($p < 0.01$) or significant ($p < 0.05$) increases in ecosystem carbon storage, mainly distributed in the east of the northwest and low-altitude regions of the southeast, while 6.11% of the area had extremely significant or significant decreases; these locations were scattered throughout the study area. In contrast, the area with non-significant changes in ecosystem carbon storage made up about 42.41%, mainly concentrated in the northwest and high-altitude regions of the southeast.

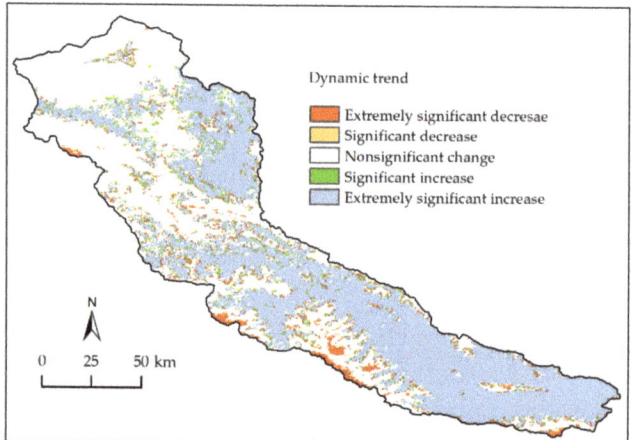

Figure 9. Changes in ecosystem carbon storage between 2001 and 2019.

3.6. Impacts of Changes in Climate and LULC on Ecosystem Carbon Storage

Under actual conditions, the ecosystem carbon storage increased by 6.92 Tg C between 2001 and 2019 (Table 3). Under the hypothesis with LULC change, the ecosystem carbon storage increased or decreased with LULC change, depending on the type of conversion. However, the amplitude of the increase was greater than the amplitude of the decrease, leading to a net increase of 2.25 Tg C in total ecosystem carbon storage. The major increase in ecosystem carbon storage was caused by land conversion from desert to grassland (691.85 km^2), cropland (322.26 km^2), and permanent wetland (151.05 km^2). In contrast, land conversion from grassland, cropland, and permanent wetland to desert caused only minor variations in ecosystem carbon storage. Under the hypothesis with climate change, the total ecosystem carbon storage increased by 4.39 Tg C in 2019 compared with 2001, an increase of 8.69%, which can be attributed to the increase in carbon density promoted by climate warming and wetting.

Table 3. Ecosystem carbon storage under the different hypotheses (unit: Tg C).

LULC Type	Control Year	Actual Condition	Hypothesis with LUCC Change	Hypothesis with Climate Change
	2001	2019	2019	2019
GL	28.14	36.82	34.04	35.57
PW	1.51	2.12	1.78	1.68
CL	2.42	3.30	2.81	2.57
DL	18.50	15.24	14.18	15.14
Regional carbon storage	50.56	57.49	52.81	54.95

When the three hypotheses were compared, we found that the contribution of climate change to the change in total regional ecosystem carbon storage was 66.10%, whereas LULC change only accounted for 33.90%. This reveals the fact that the impact of climate change on the total ecosystem carbon storage was far greater than that of LULC change. Further, under the hypothesis with LULC change, the ecosystem carbon density per pixel was lower than actual conditions (Figure 10A,B). The main regions with a reduction in ecosystem carbon storage were distributed in the northwest under the hypothesis with climate change (Figure 10C), compared with actual conditions.

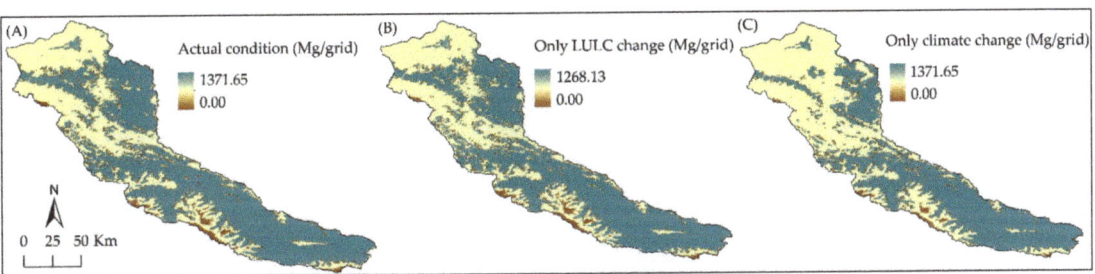

Figure 10. Ecosystem carbon storage under actual condition (**A**), the hypotheses with LULC change (**B**), and the hypothesis with climate change (**C**) in 2019.

3.7. Future Trends of Ecosystem Carbon Storage

The Hurst exponent of the ecosystem carbon storage time series in the study area distinctly increased from southeast to northwest (Figure 11), but the Hurst exponent in most of the study area, accounting for about 92.73%, was lower than 0.5, indicating that the trend in ecosystem carbon storage in the future is highly inconsistent. The trend for ecosystem carbon storage was consistent with the future in only 7.27% of the study area; these areas were scattered throughout the region.

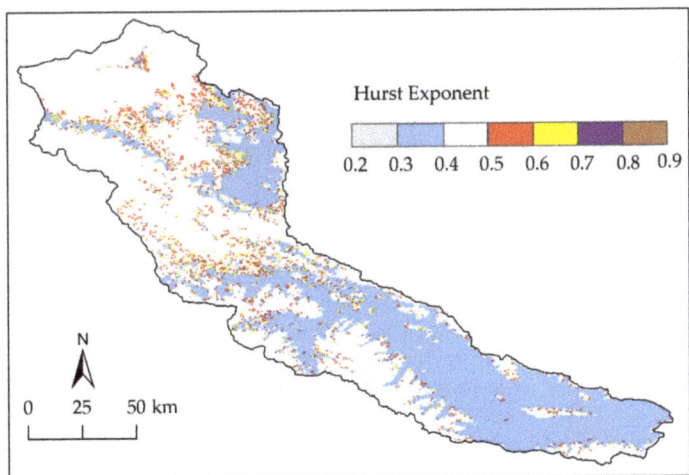

Figure 11. Distribution of Hurst exponent for the ecosystem carbon storage time series.

4. Discussion

4.1. Temporal Dynamics and Factors Influencing Ecosystem Carbon Storage

The QTP makes an important contribution to the global carbon pool and carbon cycle, and this ecosystem's carbon dynamics can mitigate or enhance the impact on atmospheric CO_2 and global warming [59]. In this context, some scholars have explored the temporal dynamics of ecosystem carbon storage in the QTP. Among them, Zhao et al. [49] found that the ecosystem carbon storage of the QTP exhibited overall growth in 2001–2010. The Qinghai Lake Basin (QLB), located in the northeastern margin of the QTP, is the largest saltwater lake in China and is an important wetland globally. Li et al. [7] evaluated the temporal dynamics of the ecosystem carbon storage of the QLB in 1990–2015, indicating that the ecosystem carbon storage increased by 1.60 Tg C. Located in the hinterland of the QTP, the Three-River Headwaters Region (TRHR) is the source of the Yangtze River, the Yellow River, and the Lancang River, and it is also an extremely sensitive area of China's ecological environment and the initiating area of climate change. Zhang et al. [60] discovered that the ecosystem carbon storage in the TRHR fluctuated upward in 2000–2010. In this study, we found that ecosystem carbon storage obviously increased in the upstream regions of the Shule River Basin between 2019 and 2001, which is consistent with the results of the abovementioned studies. This may be attributed to the increase in vegetation productively caused by climate and LULC changes. On the one hand, over the past few decades, the study area has experienced warming and wetting. Meanwhile, a significant portion of it is underlaid by permafrost, and with climate warming, permafrost thawing is likely to occur. These changes have induced changes in the ecosystem carbon processes, such as carbon gains attributed to stimulated vegetation productivity and carbon losses from thawing permafrost; the balance of these fluxes depends on the feedback of permafrost to warming [30,61]. In fact, the increase in carbon inputs from vegetation caused by climate change was greater than the loss of carbon, resulting in an obvious increase in ecosystem carbon storage during the study period. Further, the increase in ecosystem carbon density in the desert was the least, and deserts were largely replaced by other LULC types, resulting in the ecosystem carbon storage of deserts decreasing by 3.26 Tg C from 2001 to 2019. However, the increase in ecosystem carbon storage of other LULC types (which increased by 10.18 Tg C) is enough to offset the corresponding decrease in deserts; most of the increased ecosystem carbon storage is due to grassland. Interestingly, compared with other LULC types, although the ecosystem carbon density of permanent wetland increased the most (7.90 Mg/ha) in 2019 compared with 2001, its carbon storage increased the least, up to just 0.61 Tg C, mainly attributed to its relatively small area.

Through analysis of the three hypotheses, we found that the contribution of climate change to the change in ecosystem carbon storage was 66.10%, while the contribution of LULC change was 33.90%. Thus, climate change is more important than LULC change in affecting the regional ecosystem carbon storage, which has been confirmed by previous studies [49]. The main reason is that the scale of LULC change is small, and the conversion patterns between LULC types can cause both positive and negative effects on ecosystem carbon storage, whereas climate change can directly alter the inputs and outputs of ecosystem carbon storage.

4.2. Spatial Distribution and Factors Affecting Ecosystem Carbon Storage

The spatial distribution of ecosystem carbon storage in the study area was basically consistent over the study period. Under the integrated impact of climate and LULC changes, ecosystem carbon storage exhibited strong spatial heterogeneity, decreasing from southeast to northwest. The lowest values were mainly located in the marginal high-altitude regions. Because the mean annual air temperature of the high-altitude regions is lower than 0 °C, the LULC type is dominated by permanent snow and ice, so it is difficult for vegetation to survive, and the ecosystem carbon storage is close to 0. In contrast, the low-altitude regions can provide more suitable water and temperature environments for vegetation growth, and the ecosystem carbon storage in these areas is higher. In particular, the highest ecosystem carbon storage was found in the east of the northwest and in low-altitude altitude regions of the southeast due to their relatively low air temperature and favorable precipitation. Plants in these regions grow well and create higher primary productivity, which is beneficial for carbon accumulation, whereas the opposite is true in the other parts of the northwest.

In 2001–2019, the precipitation in the whole study area increased significantly, and air temperature showed an obvious increasing trend. Previous studies [62–64] have demonstrated that warming and wetting can stimulate plant growth, thus promoting gross and net primary production. Further, desert has been largely replaced by grassland. Hence, more than 51.48% of the study area experienced significant increases in ecosystem carbon storage in 2001–2019. Additionally, only 6.11% of the study area experienced significant decreases in ecosystem carbon storage, which can be attributed to the conversion of a small part of the desert and grassland into built-up land between 2001 and 2019. It is worth mentioning that the inconsistent dynamic trend of ecosystem carbon storage in the future may be related to climate change and the intensification of human activities such as urbanization and overgrazing.

4.3. Uncertainties and Limitations

Firstly, the InVEST model can reliably estimate ecosystem carbon storage in alpine regions, but the model does not consider the process of carbon sequestration and release [49]. Secondly, the parameter inversion of the regression model springs from remote sensing data, which may bring some uncertainty to estimates of carbon storage. Thirdly, *RSRs* usually change under the influence of grazing and climate change [59]. The *RSRs* come from the IPCC in this study, which may increase the uncertainty of the simulation. Further, due to the defects of the InVEST model, we assessed the impact of climate change on ecosystem carbon storage by controlling the change in carbon density. Finally, detecting the future trend of ecosystem carbon storage using R/S analysis was a great advance in this study, but the time of the expected dynamic trends in the future could not be determined.

5. Conclusions

We analyzed the spatial-temporal patterns and change trends of ecosystem carbon storage and explored the impact of climate and LULC changes on ecosystem carbon storage between 2001 and 2019. Under the integrated influence of climate and LULC changes, ecosystem carbon storage increased from 50.56 Tg C in 2001 to 57.49 Tg C in 2019, an increase of 13.69%. Between 2001 and 2019, more than 51.48% of the ecosystem carbon storage in the study area increased significantly, and regions with significant decreases accounted for 6.11%. Additionally, the spatial distribution of ecosystem carbon storage was heterogeneous, i.e., it decreased from southeast to northwest. By varying three hypotheses, we found that climate change has a dominant impact on the change in ecosystem carbon storage, while LULC change has a relatively small impact. The future trend of ecosystem carbon storage may be dominated by inconsistency. Evaluations of ecosystem carbon storage in alpine areas are of great significance for decision-making regarding ecological environmental protection and sustainable development.

Author Contributions: Conceptualization, P.W., M.W., S.C.; visualization, P.W., Y.J.; methodology, Y.J., H.X.; resources, P.W., H.X. and D.L.; formal analysis, P.W., M.W.; data curation, S.C., H.X.; writing—original draft preparation, P.W.; writing—review and editing, P.W., M.W., S.C. and H.X.; S.C.; funding acquisition, S.C.; supervision, S.C.; project administration. All authors have read and agreed to the published version of the manuscript.

Funding: This research was funded by the National Science Foundation of China (41871064, 41690142), Qinghai Key R&D and Transformation Program (2020-SF-146), the Freedom Project of the State Key Laboratory of Cryospheric Science, Northwest Institute of Eco-Environment and Resources, Chinese Academy of Sciences (SKLCS-ZZ-2021) and Qinghai Province High-level Innovative "Thousand Talents" Program.

Data Availability Statement: Not applicable.

Conflicts of Interest: The authors declare no conflict of interest.

References

1. Zhao, M.; He, Z.; Du, J.; Chen, L.; Lin, P.; Fang, S. Assessing the effects of ecological engineering on carbon storage by linking the CA-Markov and InVEST models. *Ecol. Indic.* **2019**, *98*, 29–38. [CrossRef]
2. Houghton, R.A. Revised estimates of the annual net flux of carbon to the atmosphere from changes in land use and land management 1850–2000. *Tellus B Chem. Phys. Meteorol.* **2003**, *55*, 378–390.
3. He, C.; Zhang, D.; Huang, Q.; Zhao, Y. Assessing the potential impacts of urban expansion on regional carbon storage by linking the LUSD-urban and InVEST models. *Environ. Model. Softw.* **2016**, *75*, 44–58. [CrossRef]

4. Zhao, S.; Liu, S.; Yin, R.; Li, Z.; Deng, Y.; Tan, K.; Deng, X.; Rothstein, D.; Qi, J. Quantifying terrestrial ecosystem carbon dynamics in the Jinsha watershed, upper Yangtze, China from 1975 to 2000. *Environ. Manag.* **2010**, *45*, 466–475. [CrossRef]
5. Chen, D.; Deng, X.; Jin, G.; Samie, A.; Li, Z. Land-use-change induced dynamics of carbon stocks of the terrestrial ecosystem in Pakistan. *Phys. Chem. Earth Parts A/B/C* **2017**, *101*, 13–20. [CrossRef]
6. Upadhyay, T.P.; Sankhayan, P.L.; Solberg, B. A review of carbon sequestration dynamics in the Himalayan region as a function of land-use change and forest/soil degradation with special reference to Nepal. *Agric. Ecosyst. Environ.* **2005**, *105*, 449–465. [CrossRef]
7. Li, J.; Gong, J.; Guldmann, J.; Li, S.; Zhu, J.C. Dynamics in the Northeastern Qinghai–Tibetan Plateau from 1990 to 2030 Using Landsat Land Use/Cover Change Data. *Remote Sens.* **2020**, *12*, 528. [CrossRef]
8. Quéré, C.L.; Peters, G.P.; Andres, R.J.; Andrew, R.M.; Boden, T.A.; Ciais, P.; Friedlingstein, P.; Houghton, R.A.; Marland, G.; Moriarty, R.; et al. Global carbon budget 2013. *Earth Syst. Sci. Data* **2014**, *6*, 235–263. [CrossRef]
9. Fang, J.; Chen, A.; Peng, C.; Zhao, S.; Ci, L. Changes in forest biomass carbon storage in China between 1949 and 1998. *Science* **2001**, *292*, 2320–2322. [CrossRef] [PubMed]
10. Tao, Y.; Li, F.; Liu, X.; Zhao, D.; Sun, X.; Xu, L. Variation in ecosystem services across an urbanization gradient: A study of terrestrial carbon stocks from Changzhou, China. *Ecol. Model.* **2015**, *318*, 210–216. [CrossRef]
11. Wu, S.; Li, Y.; Yu, D.; Zhou, L.; Zhou, W.; Guo, Y.; Wang, X.; Dai, L. Analysis of factors that influence forest vegetation carbon storage by using the VAR model: A case study in Shaanxi Province. *Acta Ecol. Sin.* **2015**, *35*, 196–203.
12. Zhao, M.; Yue, T.; Zhao, N.; Sun, X.; Zhang, X. Combining LPJ-GUESS and HASM to simulate the spatial distribution of forest vegetation carbon stock in China. *J. Geogr. Sci.* **2014**, *24*, 249–268. [CrossRef]
13. Parton, W.J.; Scurlock, J.M.O.; Ojima, D.S.; Gilmanov, T.G.; Scholes, R.J.; Schimel, D.S.; Kirchner, T.; Menaut, J.C.; Seastedt, T.; Moya, E.G.; et al. Observations and modeling of biomass and soil organic matter dynamics for the grassland biome worldwide. *Glob. Biogeochem. Cycles* **1993**, *7*, 785–809. [CrossRef]
14. Running, S.W.; Hunt, E.R., Jr. *Generalization of a Forest Ecosystem Process Model for Other Biomes, Biome-BGC, and an Application for Global-Scale Models*; Academic Press, Inc.: San Diego, CA, USA, 1993.
15. Zhao, S.; Liu, S.; Sohl, T.; Young, C.; Werner, J. Land use and carbon dynamics in the southeastern United States from 1992 to 2050. *Environ. Res. Lett.* **2013**, *8*, 044022. [CrossRef]
16. Zhang, Y.; Zhang, X.; Chen, Z.; Wang, W.; Chen, D. Research on the spatiotemporal variation of carbon storage in coastal zone ecosystem of Jiangsu based on InVEST Model. *Res. Soil Water Conserv.* **2016**, *23*, 100–105.
17. Liang, Y.; Liu, L.; Huang, J. Integrating the SD-CLUE-S and InVEST models into assessment of oasis carbon storage in northwestern China. *PLoS ONE* **2017**, *12*, e0172494. [CrossRef]
18. Liu, S.; Hu, N.; Zhang, J.; Lv, Z. Spatiotemporal change of carbon storage in the Loess Plateau of northern Shaanxi, based on the InVEST Model. *Sci. Cold Arid. Reg.* **2018**, *10*, 240–250.
19. Huang, C.; Yang, J.; Zhang, W. Development of ecosystem services evaluation models: Research progress. *Chin. J. Ecol.* **2013**, *32*, 3360–3367.
20. Wu, Z.; Chen, X.; Liu, B.; Chu, J.; Peng, L. Research progress and application of InVEST model. *Chin. J. Trop. Agric.* **2013**, *33*, 58–62.
21. Feng, W. Change Characteristics of Soil Organic Carbon in Shiyang River Basin and Its Response to Climate and Land Use Changes. Master's Thesis, Northwest Normal University, Lanzhou, China, 2020.
22. Chen, H.; Zhu, Q.; Peng, C.; Wu, N.; Wang, Y.; Fang, X.; Gao, Y.; Zhu, D.; Yang, G.; Tian, J.; et al. The impacts of climate change and human activities on biogeochemical cycles on the Qinghai-Tibetan Plateau. *Glob. Chang. Biol.* **2013**, *19*, 2940–2955. [CrossRef]
23. Chen, G.; Tian, H. Land use/cover change effects on carbon cycling in terrestrial ecosystems. *Chin. J. Plant Ecol.* **2007**, *31*, 189–204.
24. Hayes, D.J.; McGuire, D.A.; Kicklighter, D.W.; Burnside, T.J.; Melillo, J.M. The effects of land cover and land use change on the contemporary carbon balance of the arctic and boreal terrestrial ecosystems of northern Eurasia. In *Eurasian Arctic Land Cover and Land Use in a Changing Climate*; Gutman, G., Reissell, A., Eds.; Springer: Berlin, Germany, 2010; pp. 109–136.
25. Wang, X.; Li, Y.; Gong, X.; Niu, Y.; Chen, Y.; Shi, X.; Li, W. Storage, pattern and driving factors of soil organic carbon in an ecologically fragile zone of northern China. *Geoderma* **2019**, *343*, 155–165. [CrossRef]
26. Noble, I.R.; Bolin, B.; Ravindranath, N.H.; Verardo, D.J.; Dokken, D.J. Land use, land use change, and forestry. *Environ. Conserv.* **2000**, *28*, 284–293.
27. Baumann, F.; HE, J.; Schmidt, K.; Kuhn, P.; Scholten, T. Pedogenesis, permafrost, and soil moisture as controlling factors for soil nitrogen and carbon contents across the Tibetan Plateau. *Glob. Chang. Biol.* **2009**, *15*, 3001–3017. [CrossRef]
28. Zhang, Y.; Tang, Y.; Jiang, J.; Yang, Y. Characterizing the dynamics of soil organic carbon in grasslands on the Qinghai-Tibetan Plateau. *Sci. China Ser. D Earth Sci.* **2007**, *50*, 113–120. [CrossRef]
29. Wang, S.; Zhou, C. Estimating soil carbon reservoir of terrestrial ecosystem in China. *Geo. Res.* **1999**, *18*, 349–356.
30. Liu, S.; Sun, Y.; Dong, Y.; Zhao, H.; Dong, S.; Zhao, S.; Beazley, R. The spatio-temporal patterns of the topsoil organic carbon density and its influencing factors based on different estimation models in the grassland of Qinghai-Tibet Plateau. *PLoS ONE* **2019**, *14*, e0225952. [CrossRef] [PubMed]
31. Klein, J.A.; Harte, J.; Zhao, X. Experimental warming causes large and rapid species loss, dampened by simulated grazing, on the Tibetan Plateau. *Ecol. Lett.* **2004**, *7*, 1170–1179. [CrossRef]

32. Li, L.; Yang, S.; Wang, Z.; Zhu, X.; Tang, H. Evidence of warming and wetting climate over the Qinghai-Tibet Plateau. *Arct. Antarct. Alp. Res.* **2010**, *42*, 449–457. [CrossRef]
33. Zhang, Y.; Qi, W.; Zhou, C.; Ding, M.; Liu, L.; Gao, J.; Bai, W.; Wang, Z.; Zheng, D. Spatial and temporal variability in the net primary production of alpine grassland on the Tibetan Plateau since 1982. *J. Geogr. Sci.* **2014**, *24*, 269–287. [CrossRef]
34. Zhuang, Q.; He, J.; Lu, Y.; Ji, L.; Xiao, J.; Luo, T. Carbon dynamics of terrestrial ecosystems on the Tibetan Plateau during the 20th century: An analysis with a process-based biogeochemical model. *Glob. Ecol. Biogeogr.* **2010**, *19*, 649–662. [CrossRef]
35. Peng, L.; Lin, Y.; Chen, G.; Lien, W. Climate change impact on spatiotemporal hotspots of hydrologic ecosystem services: A case study of Chinan catchment, Taiwan. *Water* **2019**, *11*, 867. [CrossRef]
36. Chen, S.; Liu, W.; Qin, X.; Liu, Y.; Zhang, T.; Chen, K.; Hu, F.; Ren, J.; Qin, D. Response characteristics of vegetation and soil environment to permafrost degradation in the upstream regions of the Shule River Basin. *Environ. Res. Lett.* **2012**, *7*, 045406. [CrossRef]
37. Sheng, Y.; Li, J.; Wu, J.-C.; Ye, B.-S.; Wang, J. Distribution patterns of permafrost in the upper area of Shule River with the application of GIS technique. *J. China Univ. Min. Technol.* **2010**, *39*, 32–39.
38. Liu, W.; Chen, S.; Qin, X.; Baumann, F.; Scholten, T.; Zhou, Z.; Sun, W.; Zhang, T.; Ren, J.; Qin, D. Storage, patterns, and control of soil organic carbon and nitrogen in the northeastern margin of the Qinghai–Tibetan Plateau. *Environ. Res. Lett.* **2012**, *7*, 035401. [CrossRef]
39. Wei, P.; Chen, S.; Wu, M.; Deng, Y.; Xu, H.; Jia, Y.; Liu, F. Using the InVEST Model to Assess the Impacts of Climate and Land Use Changes on Water Yield in the Upstream Regions of the Shule River Basin. *Water* **2021**, *13*, 1250. [CrossRef]
40. Liu, W.; Chen, S.; Liang, J.; Qin, X.; Kang, S.; Ren, J.; Qin, D. The effect of decreasing permafrost stability on ecosystem carbon in the northeastern margin of the Qinghai–Tibet Plateau. *Sci. Rep.* **2018**, *8*, 1–10. [CrossRef]
41. Chuai, X.; Huang, X.; Zheng, Z.; Zhang, M.; Liao, Q.; Lai, L.; Lu, J. Land use change and its influence on carbon storage of terrestrial ecosystems in Jiangsu Province. *Resour. Sci.* **2011**, *33*, 1932–1939.
42. Huang, M.; Ji, J.; Cao, M.; Li, K. Modeling study of vegetation shoot and root biomass in China. *Acta Ecol. Sin.* **2006**, *26*, 4156–4163.
43. Xie, X.; Sun, B.; Zhou, H.; Li, Z. Soil carbon stocks and their influencing factors under native vegetations in China. *Acta Pedol. Sin.* **2004**, *41*, 699–705.
44. Eggleston, H.S.; Buendia, L.; Miwa, K.; Ngara, T.; Tanabe, K. *IPCC Guidelines for National Greenhouse Gas Inventories*; Institute for Global Environmental Strategies: Hayama, Japan, 2006.
45. Yang, Y.; Fang, J.; Ji, C.; Han, W. Above-and belowground biomass allocation in Tibetan grasslands. *J. Veg. Sci.* **2009**, *20*, 177–184. [CrossRef]
46. Nelson, D.W.; Sommers, L.E. Total carbon, organic carbon, and organic matter. Methods of soil analysis. *Chem. Microbiol. Prop.* **1983**, *09*, 539–579.
47. Liu, W.; Chen, S.; Zhao, Q.; Sun, Z.; Ren, J.; Qin, D. Variation and control of soil organic carbon and other nutrients in permafrost regions on central Qinghai-Tibetan Plateau. *Environ. Res. Lett.* **2014**, *9*, 114013. [CrossRef]
48. Sharp, R.; Tallis, H.; Ricketts, T.; Guerry, A.; Wood, S.; Chaplin-Kramer, R.; Nelson, E.; Ennaanay, D.; Wolny, S.; Olwero, N.; et al. *Invest Version 3.2. 0 User's Guide. The Natural Capital Project*; Stanford University: Stanford, CA, USA, 2015.
49. Zhao, Z.; Liu, G.; Mou, N.; Xie, Y.; Xu, Z.; Li, Y. Assessment of carbon storage and its influencing factors in Qinghai-Tibet Plateau. *Sustainability* **2018**, *10*, 1864. [CrossRef]
50. Dai, E.; Zhai, R.; Ge, Q.; Wu, X. Detecting the storage and change on topsoil organic carbon in grasslands of Inner Mongolia from 1980s to 2010s. *J. Geogr. Sci.* **2014**, *24*, 1035–1046. [CrossRef]
51. Jia, Y.; Guo, N.; Huang, L.; Jia, J. Ananlyses on MODIS-NDVI index saturation in northwest China. *Plateau Meteorol.* **2008**, *27*, 896–903.
52. Dida, J.J.V.; Tiburan, C.; Tsutsumida, N.; Saizen, I. Carbon Stock Estimation of Selected Watersheds in Laguna, Philippines Using InVEST. *Philipp. J. Sci.* **2021**, *150*, 501–513.
53. Van Liew, M.W.; Arnold, J.G.; Garbrecht, J.D. Hydrologic simulation on agricultural watersheds: Choosing between two models. *Trans. ASAE* **2003**, *46*, 1539. [CrossRef]
54. Moriasi, D.N.; Arnold, J.G.; Van Liew, M.W.; Bingner, R.L.; Harmel, R.D.; Veith, T.L. Model evaluation guidelines for systematic quantification of accuracy in watershed simulations. *Trans. ASABE* **2007**, *50*, 885–900. [CrossRef]
55. Sánchez Granero, M.A.; Trinidad Segovia, J.E.; Pérez, J.G. Some comments on Hurst exponent and the long memory processes on capital markets. *Phys. A Stat. Mech. Its Appl.* **2008**, *387*, 5543–5551. [CrossRef]
56. Peng, J.; Liu, Z.; Liu, Y.; Wu, J.; Han, Y. Trend analysis of vegetation dynamics in Qinghai-Tibet Plateau using Hurst Exponent. *Ecol. Indic.* **2012**, *14*, 28–39. [CrossRef]
57. Hurst, H.E. Long-term storage capacity of reservoirs. *Trans. Am. Soc. Civ. Eng.* **1951**, *116*, 770–799. [CrossRef]
58. Mandelbrot, B.B.; Wallis, J.R. Robustness of the rescaled range R/S in the measurement of noncyclic long run statistical dependence. *Water Resour. Res.* **1969**, *5*, 967–988. [CrossRef]
59. Dai, L.; Ke, X.; Guo, X.; Du, Y.; Zhang, F.; Li, Y.; Li, Q.; Lin, L.; Peng, C.; Shu, K.; et al. Responses of biomass allocation across two vegetation types to climate fluctuations in the northern Qinghai–Tibet Plateau. *Ecol. Evol.* **2019**, *9*, 6105–6115. [CrossRef] [PubMed]
60. Zhang, J.; Liu, C.; Hao, H.; Sun, L.; Qiao, Q.; Wang, H. Spatial-temporal changes of carbon storage and carbon sink of grassland ecosystem in the Three-River Headwaters Region based on MODIS GPP/NPP data. *Ecol. Environ. Sci.* **2015**, *24*, 8–13.

61. Ding, J.; Chen, L.; Ji, C.; Hugelius, G.; Li, Y.; Liu, L.; Qin, S.; Zhang, B.; Yang, G.; Li, F.; et al. Decadal soil carbon accumulation across Tibetan permafrost regions. *Nat. Geosci.* **2017**, *10*, 420–424. [CrossRef]
62. Piao, S.; Fang, J.; He, J. Variations in vegetation net primary production in the Qinghai-Xizang Plateau, China, from 1982 to 1999. *Clim. Chang.* **2006**, *74*, 253–267. [CrossRef]
63. Zhao, M.S.; Running, S.W. Drought-induced reduction in global terrestrial net primary production from 2000 through 2009. *Science* **2010**, *329*, 940–943. [CrossRef]
64. Wang, S.; Duan, J.; Xu, G.; Wang, Y.; Zhang, Z.; Rui, Y.; Luo, C.; Xu, B.; Zhu, X.; Chang, X.; et al. Effects of warming and grazing on soil N availability, species composition, and ANPP in an alpine meadow. *Ecology* **2012**, *93*, 2365–2376. [CrossRef]

Article

Comparison of Total Column and Surface Mixing Ratio of Carbon Monoxide Derived from the TROPOMI/Sentinel-5 Precursor with In-Situ Measurements from Extensive Ground-Based Network over South Korea

Ukkyo Jeong [1,2] and Hyunkee Hong [3,*]

1 Earth System Science Interdisciplinary Center, University of Maryland, College Park, MD 20740, USA; ukkyo.jeong@nasa.gov
2 NASA Goddard Space Flight Center, Greenbelt, MD 20771, USA
3 National Institute of Environmental Research, Seogu Hwangyong-ro 42, Incheon 22689, Korea
* Correspondence: wanju77@korea.kr

Citation: Jeong, U.; Hong, H. Comparison of Total Column and Surface Mixing Ratio of Carbon Monoxide Derived from the TROPOMI/Sentinel-5 Precursor with In-Situ Measurements from Extensive Ground-Based Network over South Korea. *Remote Sens.* **2021**, *13*, 3987. https://doi.org/10.3390/rs13193987

Academic Editors: Baojie He, Ayyoob Sharifi, Chi Feng and Jun Yang

Received: 20 August 2021
Accepted: 26 September 2021
Published: 5 October 2021

Publisher's Note: MDPI stays neutral with regard to jurisdictional claims in published maps and institutional affiliations.

Copyright: © 2021 by the authors. Licensee MDPI, Basel, Switzerland. This article is an open access article distributed under the terms and conditions of the Creative Commons Attribution (CC BY) license (https://creativecommons.org/licenses/by/4.0/).

Abstract: Atmospheric carbon monoxide (CO) significantly impacts climate change and human health, and has become the focus of increased air quality and climate research. Since 2018, the Troposphere Monitoring Instrument (TROPOMI) has provided total column amounts of CO ($C_{TROPOMI}$) with a high spatial resolution to monitor atmospheric CO. This study compared and assessed the accuracy of $C_{TROPOMI}$ measurements using surface in-situ measurements (S_{KME}) obtained from an extensive ground-based network over South Korea, where CO level is persistently affected by both local emissions and trans-boundary transport. Our analysis reveals that the TROPOMI effectively detected major emission sources of CO over South Korea and efficiently complemented the spatial coverage of the ground-based network. In general, the correlations between $C_{TROPOMI}$ and S_{KME} were lower than those for NO_2 reported in a previous study, and this discrepancy was partly attributed to the lower spatiotemporal variability. Moreover, vertical CO profiles were sampled from the ECMWF CAMS reanalysis data (EAC4) to convert $C_{TROPOMI}$ to surface mixing ratios ($S_{TROPOMI}$). $S_{TROPOMI}$ showed a significant underestimation compared with S_{KME} by approximately 40%, with a moderate correlation of approximately 0.51. The low biases of $S_{TROPOMI}$ were more significant during the winter season, which was mainly attributed to the underestimation of the EAC4 CO at the surface. This study can contribute to the assessment of satellite and model data for monitoring surface air quality and greenhouse gas emissions.

Keywords: carbon monoxide; TROPOMI; surface mixing ratio; Korea; EAC4; climate; air quality

1. Introduction

Major sources of atmospheric carbon monoxide (CO) include the incomplete combustion of fossil fuels, biomass burning, and the oxidation of methane and non-methane hydrocarbons, predominantly activated by the hydroxyl radical (OH). CO is removed by photochemical oxidation, which consumes OH during the process [1,2], thus affecting the atmospheric cleansing capacity [2] and lifetime of methane (CH_4) [3,4]. In addition, this reaction produces greenhouse gases such as carbon dioxide (CO_2) and tropospheric ozone (O_3); therefore, CO is regulated by worldwide air quality standards and is designated a significant greenhouse gas with a radiative forcing of 0.23 W m^{-2} [5]. The lifetime of CO varies from weeks to months [6], which is long enough to persist through horizontal and vertical transport but too short to be well mixed globally. Owing to the moderate lifetime of CO, it is frequently utilized as a tracer for the propagation of pollution [7,8]. For these reasons, the Monitoring Atmospheric Composition and Climate (MACC) project of the Global Monitoring for Environment and Security (GMES) program prioritized CO as an important chemical species for air quality and climate studies [9].

Nadir-viewing passive sensors provide global distributions of CO retrievals from either near-infrared or thermal-infrared (TIR) radiances. Since the first measurement of CO during four flights of the space shuttle between 1981 and 1999 [10], the measurement of pollution in the troposphere (MOPPIT) has provided decades of global CO retrievals since 2000 from the 1-0 CO absorption band at 4.7 μm [11]. These TIR measurements are sensitive to CO in the middle troposphere and depend on the spectral resolution and thermal contrast in the lower troposphere. The Atmospheric Infrared Sounder [12] onboard the Aqua launched in 2002, the Tropospheric Emission Spectrometer (TES) [13] onboard the Aura launched in 2004, and the Infrared Atmospheric Sounding Interferometer [14] onboard the Meteorological Operational (METOP) also utilize this TIR absorption band of CO.

For clear atmospheric conditions, the shortwave-infrared (SWIR) earth-radiances near the first overtone 2-0 absorption band of CO (between 2.30–2.39 μm) is negligibly affected by scattering in the atmosphere but is dominated by atmospheric absorption and surface reflectance. Therefore, SWIR measurements are sensitive to the total column amount of CO along the light path, making CO retrievals using these wavelengths suitable for detecting emission sources of CO. In addition to the more recent progress of the MOPITT using its SWIR measurements [15], the Scanning Imaging Absorption Spectrometer for Atmospheric Chartography [16] on the Envisat satellite has provided continuous time series of global CO SWIR measurements since 2002. Worden et al. (2010) combined the TIR and SWIR measurements of MOPITT to retrieve global CO trends and assessed its theoretical information content, which showed increased retrieval sensitivity near the surface compared with those using a single band [17]. In October 2017, the TROPOspheric Monitoring Instrument (TROPOMI) onboard the Sentinel-5 Precursor (S5P) of the European Space Agency (ESA) was launched and continues to measure CO using SWIR radiances with higher spatial resolution and better radiometric performance [18]. Moreover, the TROPOMI allows for the detection of weak regional sources, such as individual wildfires, from its daily overpasses.

The retrieval sensitivity of CO near the surface is critical for the operational use of satellite data for air quality and climate applications, as its emissions and major chemical interactions occur within the boundary layer. In addition to the nadir-viewing sensors, instruments on solar occultation satellites such as the Atmospheric Chemistry Experiment-Fourier Transform Spectrometer (ACE-FTS) on board SCISAT [19–21], or of a limb viewing geometry including the Michelson Interferometer for Passive Atmospheric Sounding (MI-PAS)/ENVISAT [22] and Microwave Limb Sounder (MLS) /Aura [23] provide informative CO profile retrievals. However, CO retrievals from these sensors are limited by their lower horizontal resolution and coverage compared with those from the nadir viewing instruments; therefore, they are not suitable for accurately identifying regional emissions. Retrievals using both SWIR and TIR radiances show promising results with high sensitivity near surfaces [17]; however, to the best of our knowledge, these retrieval data are not currently available as an operational product. To overcome the limitations of satellite products, previous studies have combined model simulations and column retrievals from satellites to derive surface concentrations of aerosols [24] and to trace gases [25].

Zhang et al. (2020) reported that the annual mean values of the MOPITT CO over Asia decreased significantly at a rate of 0.58 ± 0.15% per year from 2003 to 2017 and associated this decrease with reduced biomass burning over southeast Asia during the spring season [26]. Similar results were reported by Buchholz et al. (2021), who demonstrated a decreasing global CO trend of approximately 0.5% per year between 2002 and 2018 based on MOPITT data. They also attributed the significant decline in CO over Northeast China from 2002 to 2018 to improvements in combustion efficiency [27]. Zheng et al. (2018) suggested that decreased CO emissions in China from four primary sectors (iron and steel industries, residential sources, gasoline-powered vehicles, and construction materials industries) could be responsible for 76% of the inversion-based trend of east Asian CO emissions [28]. Kang et al. (2019) estimated that the anthropogenic contribution of CO decreased to approximately 94% from 2001 to 2011 over east China [29]. Figure 1 shows

the mean total column amounts of CO over east and southeast Asia for 2019 from the
TROPOMI, which were binned to a 0.05° × 0.05° horizontal grid. To calculate the average
values, the CO data with a quality flag greater than or equal to 0.5, were sampled. As
shown in this figure, significant amounts of CO prevailed over east China throughout
2019, which also affected downwind regions, including the Korean peninsula [30]. Jeong
and Hong (2021) derived surface-level NO_2 by combining the TROPOMI and reanalysis
data to assess long-term exposure for epidemiological studies [25]. They compared the
estimated NO_2 with an extensive ground-based network over South Korea managed by the
Korean Ministry of Environment (KME). To the best of our knowledge, only a few studies
have compared satellite-retrieved and ground in-situ CO measurements, despite their
significance for assessments of satellite retrievals [31,32]. This study is a follow-up study
of [25], which aimed to compare and assess CO products of the TROPOMI ($C_{TROPOMI}$)
for complementing surface measurements using an extensive ground-based network over
South Korea, and thereby to contribute to the improvement of our understanding of the air
quality impacts of CO and provide a guideline for climate studies.

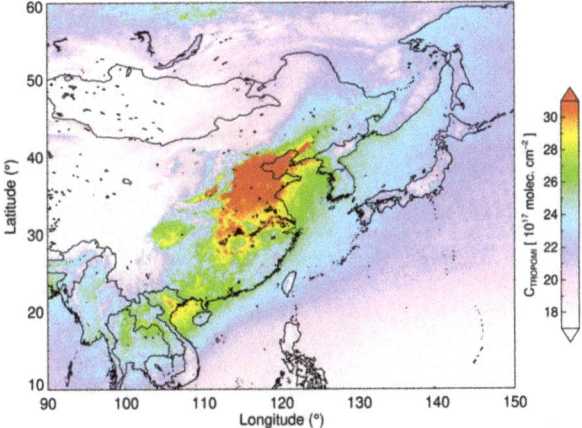

Figure 1. Average CO total column amounts for 2019 from TROPOMI binned to a 0.05° × 0.05°
horizontal grid over east and southeast Asia. TROPOMI CO data with quality flags ≥0.5 were used
to calculate the average values.

2. Data

2.1. TROPOMI Total Column Density of CO

The TROPOMI is the unique payload of the S5P satellite mission and has measured
reflected solar light by the Earth using two spectrometer modules since 2017: one covering
the ultraviolet–visible (270–495 nm) and near-infrared (675–775 nm) spectra and the other
covering the SWIR between 2305 and 2385 nm. The SWIR spectrometer was developed
by Surrey Satellite Technology Limited, United Kingdom, and has a spectral resolution
of approximately 0.25 nm with a sampling resolution of approximately 0.1 nm. The
TROPOMI also measures the Sun directly through the irradiance port and internal diffuser
for calibration [18,33].

The SWIR measurements of the TROPOMI feed the Shortwave Infrared CO Retrieval
(SICOR) algorithm to retrieve total CO column amounts and effective cloud parameters
(i.e., cloud optical thickness and cloud center height) [34,35]. The SICOR algorithm is
based on the SCIAMACHY heritage [36] and is improved for cloudy and aerosol-loaded
atmospheres. The inversion utilizes a profile-scaling method based on monthly averaged
vertical profiles of CO from the global chemistry transport model, version 5 (TM5) [37].
Moreover, it generates vertically integrated columns of CO with an averaging kernel for
each retrieval [35], which are tested extensively using SCIAMACHY measurements and

cover the TROPOMI spectral range with a similar spectral resolution [38]. The SICOR algorithm consists of two steps. In the first step, the SICOR algorithm retrieves the total amount of CH_4 from the TROPOMI radiances between 2315 and 2324 nm to filter optically thick clouds and aerosols assuming a non-scattering atmosphere. A full-physics algorithm retrieves $C_{TROPOMI}$ in the second step from radiances between 2324 and 2338 nm. The CH_4 retrievals from the first step were used to derive the effective cloud parameters at this stage. One of the merits of the SICOR algorithm is that it provides reliable retrievals for cloudy conditions because the sensitivity of the measurement to the CO above the cloud is utilized to retrieve $C_{TROPOMI}$ assuming a certain vertical profile shape from the TM5 [35]. In addition, the high reflectance of the cloud enhances retrieval sensitivity.

Borsdorff et al. (2018) compared $C_{TROPOMI}$ with the European Center for Medium-Range Weather Forecasts (ECMWF)/Integrated Forecasting System (IFS) products of the Copernicus Atmosphere Monitoring Service (CAMS), which assimilates IASI and MOPITT observations of CO [35,39]. Both CO observations show a marginal mean difference of $3.2 \pm 5.5\%$ with a Pearson correlation coefficient (r) of 0.97. Martínez-Alonso et al. (2020) compared $C_{TROPOMI}$ to the MOPITT and airborne (ATom, Atmospheric Tomography mission) datasets, which showed excellent agreement with a mean bias of less than 3.73% [40]. $C_{TROPOMI}$ also showed good agreement with ground-based Total Carbon Column Observing Network (TCCON) measurements, with a mean bias of about 6.2 ppb [41]. In general, the accuracy and precision of the CO data product meets the level 2 user requirements: within an accuracy of <15% and a precision with $\leq$10%.

2.2. Surface Network of CO Measurements

The KME has monitored particulate matter, NO_2, CO, O_3, and SO_2 since the 2000s from extensive surface air quality monitoring stations in South Korea. In 2019, 569 stations measured the surface mixing ratios of CO. These stations are predominately situated at ambient locations in urban and rural areas far from major roadways and typically deployed on the roofs of public buildings with fewer than five stories. To monitor roadside air quality, several stations (41 in 2019) are situated near major roads with a height of approximately 2.5 m above the ground level. The KME measures CO mixing ratios based on a nondispersive method using CO analyzers (model 3008, Dasibi Environmental Corp.; US Environmental Protection Agency reference method RFCA-0488-067) with a lower detection limit of 0.1 ppm and response time of 120 s. Linearity of the detector is better than 1%, and span drift is about ±1% for 24 h and ±2% for one week. Instruments are inspected monthly. The standard inspection procedure consisted of a two-step process: first, abnormal samples were screened based on the conditions of the instrument (i.e., calibration, inspection, or malfunction). Next, data exceeding the normal range or rate of change were screened [42,43]. Five minutes of temporal resolution of the raw data was averaged hourly after the quality assurance procedures and then reported to the public [42].

2.3. ECMWF Atmospheric Composition Reanalysis 4

The 4th generation global CAMS reanalysis data of the ECMWF (EAC4) assimilates the total column CO, tropospheric column NO_2, aerosol optical depth, and total column/profiles of O_3 from satellite retrievals to furnish the three-dimensional fields of these species [44]. The EAC4 covers the period from 2003 with a three-hour temporal resolution and a horizontal resolution of approximately 80 km ($0.75° \times 0.75°$) at 60 vertical model grids. The EAC4 assimilates the MOPITT TIR total column CO (TCCO, Version 6) retrievals that are sensitive to those in the mid and upper troposphere [45]. We sampled the vertical profile shape of the CO from EAC4 to convert $C_{TROPOMI}$ to a surface-mixing ratio ($S_{TROPOMI}$) for comparison with the surface measurements (S_{KME}). Table 1 summarizes the measurement parameters for CO used in this study, obtained from different sources.

Table 1. Descriptions of different parameters of CO from TROPOMI, surface measurements, and reanalysis data (EAC4).

Acronym	Definition
$C_{TROPOMI}$	Total vertical column density of CO from TROPOMI
C_{EAC4}	Total vertical column density of CO from EAC4
S_{KME}	Surface mixing ratio of CO from ground network of Korea Ministry of Environment
$S_{TROPOMI}$	Surface mixing ratio of CO converted from $C_{TROPOMI}$
S_{EAC4}	Surface mixing ratio of CO from EAC4

3. Results

3.1. Comparison of Spatial Distributions of CO from TROPOMI and Ground Network

The mean values of $C_{TROPOMI}$ for 2019 over South Korea are shown in Figure 2 and were binned to a comparable resolution of the TROPOMI (0.05° × 0.05° horizontal grid). Panel (a) of Figure 2 represents South Korea, and panels (b) to (d) focus on the most significant emission areas of CO in the domain of panel (a). In general, high $C_{TROPOMI}$ values were observed in eastern South Korea, where the low $C_{TROPOMI}$ vales in Figure 2a were predominately observed over mountainous areas. Figure 2b depicts the values over the Seoul metropolitan area, where more than half of the Korean population (~26 million) is distributed. As shown in this figure, the TROPOMI clearly indicate high values of the $C_{TROPOMI}$ over Seoul, Incheon, and active ironworks in Dangjin. One of the largest industrial complexes in Gwangyang and the ironworks in Pohang resulted in a significant CO burden, as shown in Figure 2c,d, respectively. Large amounts of $C_{TROPOMI}$ over the western sea of the Korean peninsula are likely associated with trans-boundary transport from East China [30] (also see Figure 1).

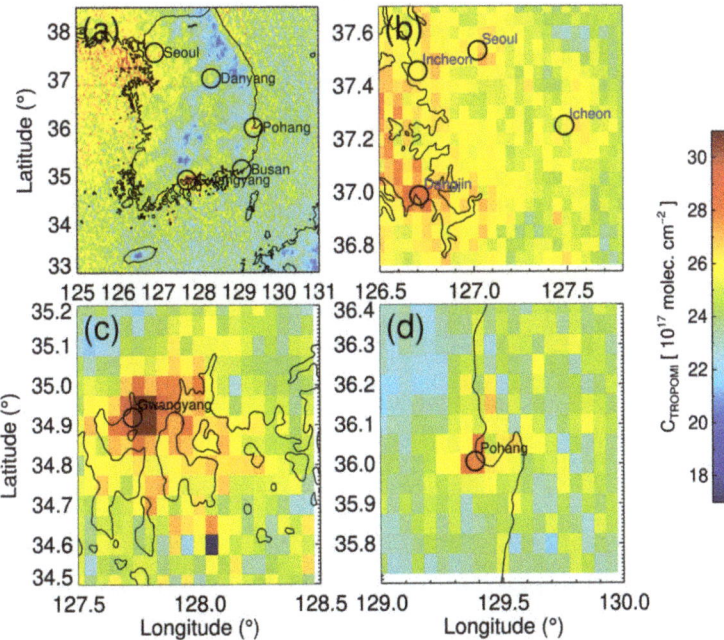

Figure 2. Average TROPOMI CO total columns for 2019 binned to a 0.05° × 0.05° horizontal grid over (**a**) South Korea, (**b**) Seoul metropolitan area, (**c**) Gwangyang, and (**d**) Pohang. TROPOMI CO data with quality flags ≥0.5 were used to calculate the average values to avoid optically thick cloud and aerosol contaminations. The black circles in panel (**a**) indicate major CO sources in South Korea.

The annual mean value of the CO surface mixing ratio measured by the KME network in South Korea is shown in Figure 3. The domains of the panels in Figure 3 are the same as those shown in Figure 2. As shown in this figure, the KME network was densely distributed over highly populated areas, particularly in cities situated in the Seoul metropolitan area (Figure 3b). Such strategic distribution of the ground-based network is efficient for monitoring NO_2, which is predominantly emitted from transportation in South Korea [25]. However, these network spatial distributions are not optimal for monitoring CO, as this compound is predominately emitted from industrial activities. As indicated in Figures 2 and 3, a vast number of stations over the Seoul metropolitan area demonstrate the spatial distribution of $C_{TROPOMI}$ (indicated by the comparison of Figures 2b and 3b), whereas the sparse distribution of surface measurements detected limited areas of the emission sources (as shown by comparing the lower panels of Figure 2 with those of Figure 3). Satellite retrievals, such as TROPOMI, can efficiently complement such limitations of ground-based networks.

The circles and squares in Figure 2 indicate ambient and roadside monitoring stations, respectively; Jeong and Hong (2021) reported significantly higher values of NO_2 from the roadside stations than the nearby ambient monitoring sites [25]. By comparing the values of the circles and squares in Figure 2b, we determined that unlike NO_2, the CO mixing ratios measured at roadside stations did not show significant differences from those at ambient monitoring stations. This is likely due to the relatively longer lifetime of CO; the emitted burden of CO remains in the atmosphere for a sufficient period to be well-mixed within a boundary layer over the Seoul metropolitan areas. Therefore, $C_{TROPOMI}$ is likely to experience less horizontal heterogeneity within its footprint but is more closely related to boundary layer height.

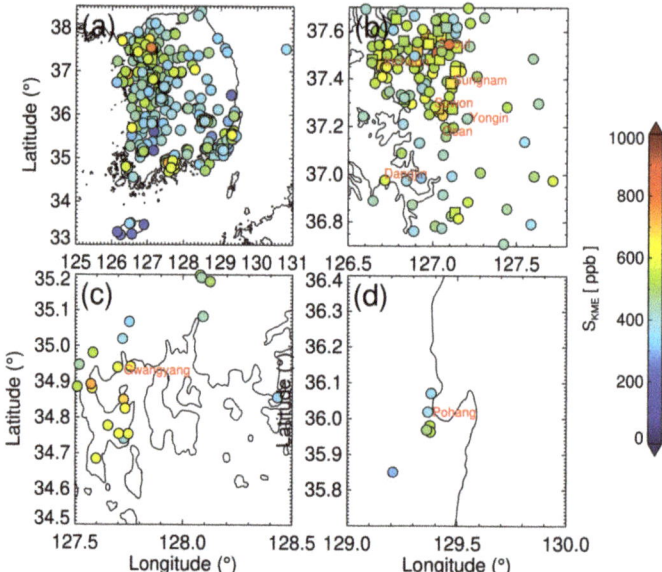

Figure 3. Mean values of surface CO mixing ratio from ground air-quality monitoring network of Korea Ministry of Environment in 2019. Panel (**a**) depicts the values over South Korea, and panels (**b**–**d**) show large emission sources in domain (**a**). Panel (**b**) represents the Seoul metropolitan area, and panels (**c**,**d**) indicate industrial complexes in Gwangyang and ironworks in Pohang, respectively. The squares within these panels indicate air quality monitoring stations on the side of roads with heavy traffic and the circles represent ambient air quality monitoring sites.

Figure 4a compares the annual mean values of $C_{TROPOMI}$ and S_{KME} over the KME stations. $C_{TROPOMI}$ values within ±0.025° from each ground station were averaged for spatial collocation. Note that the $C_{TROPOMI}$ and S_{KME} are not linearly comparable due to the spatiotemporal variabilities of vertical profile. However, as a major fraction of CO supposed to be distributed within the mixing layer, we expect such comparison may provide a primitive but basic assessment of the satellite retrievals before converting the $C_{TROPOMI}$ to surface mixing ratio for direct comparison. The green circles and red rectangles in Figure 4a represent the ambient urban/rural monitoring stations and roadside stations, respectively. The RMSE denotes the root-mean-square error, and the MBE represents the mean bias error. For a similar comparison for NO_2, the annual mean values of the TROPOMI and surface measurements show a high correlation (r = 0.84), particularly over the ambient monitoring sites (r = 0.88) [25]. However, the correlation between $C_{TROPOMI}$ and S_{KME} was lower (r = 0.37), partly attributed to the lower variability of CO compared to that of NO_2. As discussed, the comparison for the roadside monitoring stations did not show a notable difference from that of the ambient sites (revealed by comparing the green circles and red squares in Figure 4a). Spatiotemporally coincident samples ($C_{TROPOMI}$ within ±0.025° of the KME stations and S_{KME} within ±30 min of the TROPOMI overpass time) of $C_{TROPOMI}$ and S_{KME} in 2019 are compared in Figure 4b, and show a slightly lower correlation (r = 0.33) than that in panel (a).

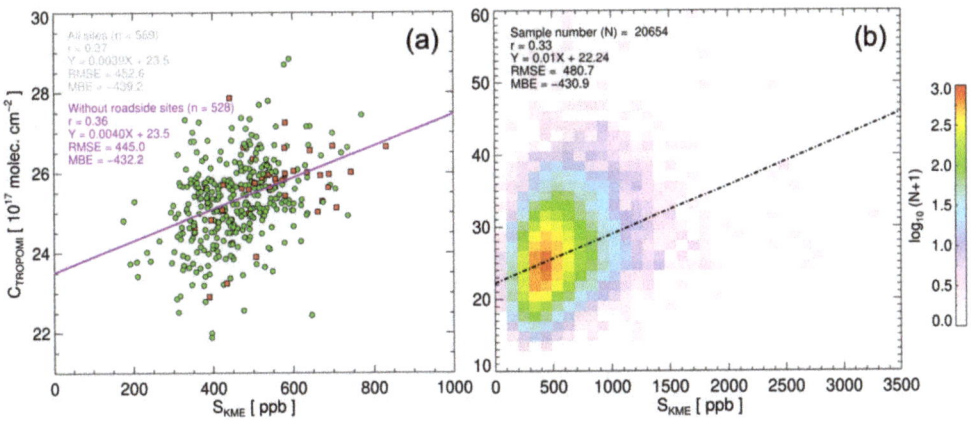

Figure 4. Comparison of total column CO from TROPOMI ($C_{TROPOMI}$) and in-situ surface mixing ratio from KME network (S_{KME}) over South Korea in 2019. Panel (**a**) compares annual mean values at each station, and panel (**b**) compares all collocated samples. Green circles and red rectangles in panel (**a**) indicate ambient and roadside monitoring stations, respectively. The RMSE stands for root-mean-squared-error, and the MBE denotes mean-bias-error.

The correlation coefficients between $C_{TROPOMI}$ and S_{KME} at each KME station in 2019 are shown in Figure 5. In general, a higher correlation appeared near the emission sources of CO owing to the higher retrieval sensitivity of the TROPOMI and variability of CO. For NO_2, the correlations between the TROPOMI retrievals and KME measurements over the roadside stations (squares) were significantly lower than those over the ambient stations (circles) because of their higher spatiotemporal variability near the source areas [25]. Such differences were not observed for CO, as shown in this figure, which was attributed to its relatively longer lifetime, as shown in Figure 3.

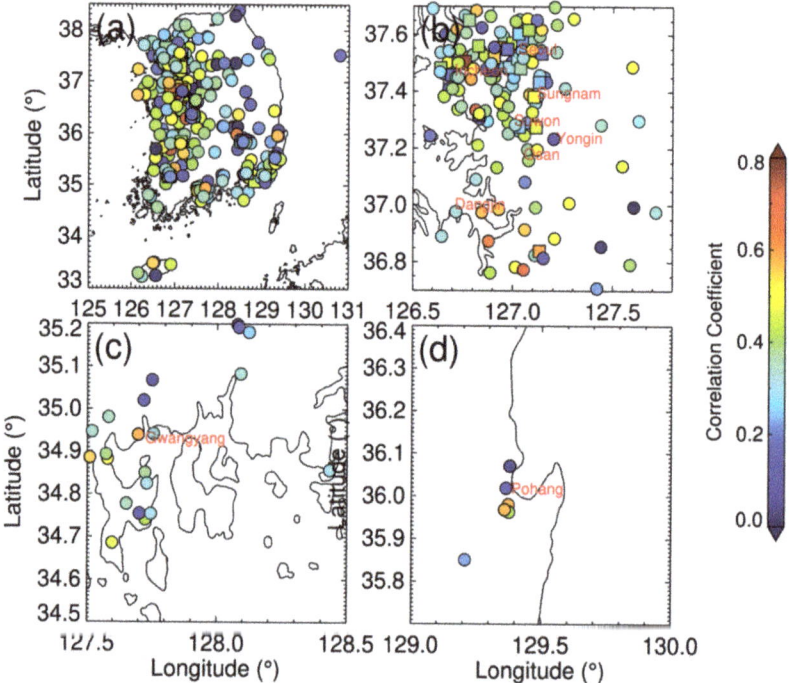

Figure 5. Correlation coefficients between $C_{TROPOMI}$ and S_{KME} at the KME monitoring stations over (**a**) South Korea, (**b**) Seoul metropolitan area, (**c**) industrial complexes in Gwangyang, and (**d**) ironworks in Pohang in 2019. Circles and squares represent ambient air-quality monitoring sites and roadside air-quality monitoring stations, respectively.

The TROPOMI retrievals utilize the profile-scaling method based on monthly averaged vertical profiles of CO from the TM5 [35,37], thus the ratio of S_{KME} to $C_{TROPOMI}$ is assumed to be higher near strong emission sources: the CO mixing ratio at the surface of these areas is likely higher than that at ambient (well-mixed) stations. The ratios over the KME stations are shown in Figure 6 and were significantly high near emission sources in South Korea. The ratios along the coastal line were highly variable at each station, and could likely be attributed to complex boundary layer processes occurring over these areas (see Figure 6c,d). A similar complexity was observed for NO_2 [25], which again emphasizes the importance of intensive field campaigns combined with model simulations over these areas (e.g., ozone water-land environmental transition study [46]).

3.2. Estimation of CO Surface Mixing Ratio from TROPOMI and CAMS Reanalysis Data

To derive surface air quality from satellite data, Jeong and Hong (2021) utilized the ratio of surface mixing ratios to total column amounts from the EAC4, which are multiplied by $C_{TROPOMI}$ as follows [25]:

$$S_{TROPOMI} = \frac{S_{EAC4}}{C_{EAC4}} C_{TROPOMI} \quad (1)$$

where $S_{TROPOMI}$ is the estimated surface CO mixing ratio from $C_{TROPOMI}$, and S_{EAC4} and C_{EAC4} are the surface mixing ratio and total column amount of CO from EAC4, respectively. As the CAMS model system (for EAC4) and TM5 (for $C_{TROPOMI}$) utilizes the same chemical mechanism, which is a modified and extended version of the CB05 [47,48], we expected the biases that arose from the different averaging kernels to be minimized. Furthermore, some of the systematic biases (e.g., emission inventory) could be canceled out because the ratio of S_{EAC4} to C_{EAC4} was relatively more accurate than their absolute values.

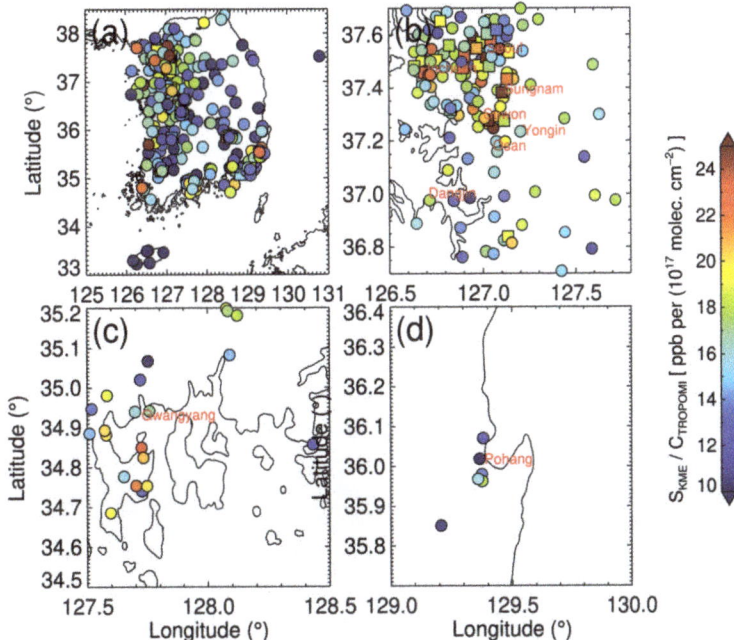

Figure 6. Ratio of surface mixing ratio to total column density for CO at the KME monitoring stations over (**a**) South Korea, (**b**) Seoul metropolitan area, (**c**) industrial complexes in Gwangyang, and (**d**) ironworks in Pohang in 2019. Circles and squares represent ambient air-quality monitoring sites and roadside air-quality monitoring stations, respectively.

Figure 7 shows the annual statistics of the CO vertical profiles from the EAC4 of longitudes from 125° to 131° and latitudes from 33° to 39°. The black line with circles depicts the mean values of the CO mixing ratio at each layer, the dark gray area indicates the standard deviation (±σ), and the light gray area shows the data range (minimum and maximum values) at each level. The mean values with a ±σ of S_{KME} for entire stations are indicated by the red circle with an error bar (466 ± 218 ppb), which was significantly higher than that of the EAC4 (193 ± 95 ppb). This difference is partially attributable to the spatial coverage of the KME network; most of the ground stations are located near urban areas or large emission sources, whereas the EAC4 values in this figure were calculated from data over the entire target region of South Korea. The mean (±σ) values of $C_{TROPOMI}$ over this target domain and over the KME stations were 24.8 (±3.6) × 10^{17} molec. cm^{-2} and 25.3 (±4.5) × 10^{17} molec. cm^{-2}, respectively. The horizontal heterogeneity of CO within a TROPOMI pixel is relatively small due to its moderate lifetime; therefore, the spatial coverage of the KME stations does not fully explain the difference between S_{KME} and S_{EAC4}. Turquety et al. (2008) [49] compared the Laboratoire de Météorologie Dynamique, zoom; version 4 (LMDz) and Interactive Chemistry and Aerosols; version 2 (INCA) model simulations [50,51] to the Measurement of Ozone and Water Vapor on Airbus In-Service Aircraft (MOZAIC) aircraft-based in-situ profiles [52] over Asia, and reported relatively lower biases of CO from the model in the troposphere, suggesting the underestimation of CO emissions. The uncertainties of CO emissions over South Korea in the EAC4 may have propagated errors in the CO vertical profiles, particularly near the surface, which could have affected the difference between S_{KME} and S_{EAC4}.

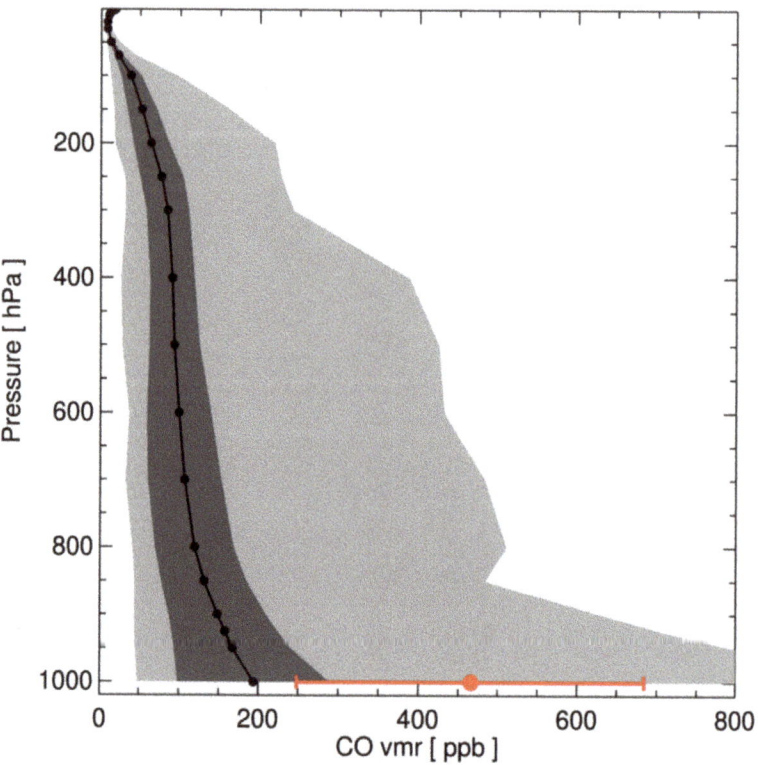

Figure 7. Statistics of CO vertical profiles from ECMWF CAMS reanalysis data (EAC4). The black line with circles indicates the mean values at each level, the dark gray area indicates ±one standard deviation, and the light gray area shows the minimum and maximum values. Profiles were sampled at longitudes from 125° to 131° and latitudes from 33° to 39° in 2019. The red circle denotes the mean CO value from all stations of the KME network for 2019, and the red line presents its ±one standard deviation.

The surface mixing ratios of CO from EAC4 and derived from TROPOMI were compared with the KME measurements in Figure 8. A comparison between Figures 4 and 8 reveals that the correlations of the CO surface mixing ratios between the different sources showed a higher correlation (r = 0.48–0.51) than that between S_{KME} and $C_{TROPOMI}$. Moreover, $S_{TROPOMI}$ shows a slightly higher correlation with S_{KME} than that between S_{EAC4} and S_{KME} with a statistical significance (z-score of about 4.04). The slope of the regression line between the $S_{TROPOMI}$ and S_{KME} also shows slightly better consistency than that between the S_{EAC4} and S_{KME} (t-value of about 6.9). Accordingly, the RMSE and MBE values between the $S_{TROPOMI}$ and S_{KME} were lower than those between the S_{EAC4} and S_{KME}, which quantifies the benefit of using TROPOMI to derive the surface CO mixing ratio. However, such agreement between the $S_{TROPOMI}$ and S_{KME} was lower than that for NO_2 using the identical technique over the same spatiotemporal domain [25], and the low bias of $S_{TROPOMI}$ (MBE = −187.6 ppb) compared to S_{KME} was still significant with respect to the average value of S_{KME} (466 ± 218 ppb) which is discussed at following figures.

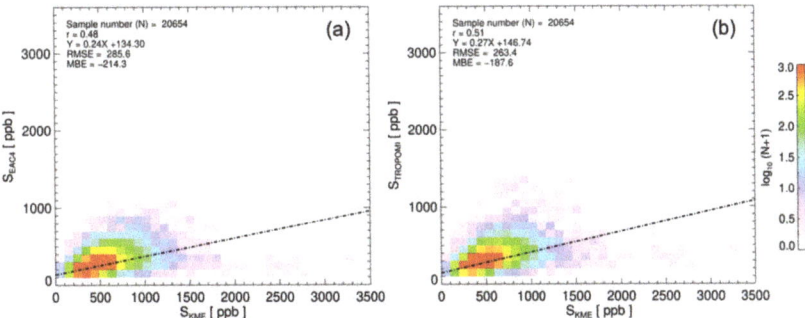

Figure 8. (**a**) Comparison of surface CO mixing ratios from ECMWF CAMS reanalysis data (S_{EAC4}) and measured from KME stations (S_{KME}) over South Korea for 2019. Panel (**b**) compares the ratios estimated from TROPOMI ($S_{TROPOMI}$) and S_{KME} during the same period.

Monthly mean values of $C_{TROPOMI}$ (red circles) and C_{EAC4} (green squares) are shown in Figure 9a. The dark colors indicate the mean values over the KME stations, and the lighter colors depict those over the entire domain of South Korea (longitudes from 125° to 131° and latitudes from 33° to 39°). Over the KME stations, the mean C_{EAC4} values were lower than those of $C_{TROPOMI}$ throughout the year by approximately 10% (2.6×10^{17} molec. cm^{-2}). This could be partly attributed to the lower spatial resolution of the EAC4 (i.e., approximately 80 km) compared to that of the TROPOMI (i.e., approximately 7 km), as the greater collocated pixels of the EAC4 for each site may contain a greater fraction of background areas around the KME stations. The average values of C_{EAC4} and $C_{TROPOMI}$ over broader and identical spatial domains experienced these sampling issues to a lesser degree, as demonstrated by the light colors in Figure 9a. The mean values of C_{EAC4} over the entire target domain were 7% lower than those of $C_{TROPOMI}$. Regarding similar comparisons, Borsdorff et al. (2018) reported biases of approximately ±15% depending on the region (see Figure 2 of [35]), and we expected that these biases were within the uncertainty ranges of $C_{TROPOMI}$ and C_{EAC4}. In addition, the monthly variations over the KME stations showed an excellent correlation (r = 0.98), as shown in Figure 9a.

As shown in Figures 7 and 8, significant underestimations of S_{EAC4} (by approximately 46%) and $S_{TROPOMI}$ (by approximately 40%) compared to S_{KME} were also observed in the monthly mean values of these parameters throughout the year (Figure 9b). The black diamonds in Figure 9b depict the monthly mean values of S_{KME} for 24-h measurements, and the blue squares indicate those for the TROPOMI overpass time. In general, the S_{KME} values were high in winter and low in summer, despite the $C_{TROPOMI}$ peak observed in March (revealed by comparison of Figure 9a,b). The high $C_{TROPOMI}$ values in March were likely associated with active biomass burning over southeast Asia, whereas the S_{KME} peak in January was attributed to the stable boundary layer during this period. The mean S_{KME} values from 24-h samples and from the TROPOMI overpass time showed slight differences of approximately 6–10% in spring and winter but comparable values in summer, which was attributed to diurnal boundary layer development. Note that the NO_2 from the KME at the TROPOMI overpass time was consistently lower by approximately 23% than the 24-h mean values because of a combination of its chemical processes and boundary layer development during the daytime [25]. The monthly mean S_{EAC4} values showed generally similar tendencies (r = 0.89), but with significantly low biases throughout the year, particularly in winter. Such relatively low biases of S_{EAC4} resulted in similar degrees of underestimation of $S_{TROPOMI}$, as presented in Figure 9b.

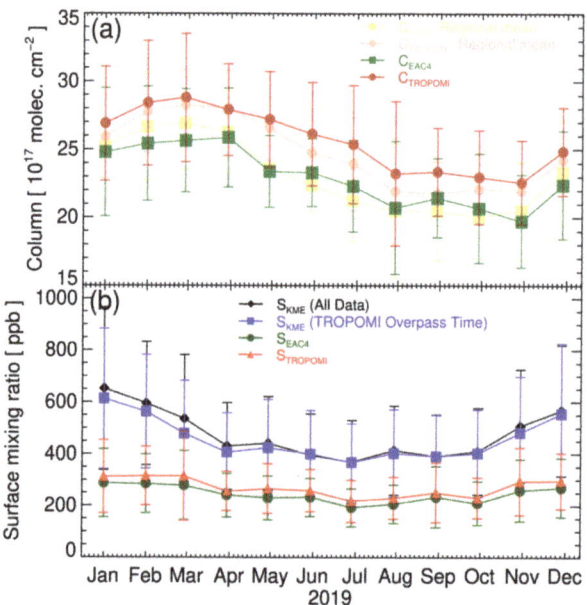

Figure 9. (a) Monthly variations of total column CO from ECMWF CAMS reanalysis data (C_{EAC4}; green square) and TROPOMI ($C_{TROPOMI}$; red circle) in 2019. Dark colors of this panel depict their mean values over the KME stations, and light colors indicate the mean values over the entire South Korean domain (125° to 131° longitude and 33° to 39° latitude). Panel (**b**) presents monthly variations in the surface CO mixing ratio from the KME stations (black diamond: 24-h average, blue square: at TROPOMI overpass time, approximately 13:00 local time), EAC4 (green circle), and the TROPOMI (red triangle).

4. Summary and Discussions

This study aimed to assess $C_{TROPOMI}$ using an extensive ground-based network over South Korea to derive the surface mixing ratio of CO over the globe, key information in understanding its role in the regional air quality climate. Our analysis reveals that the CO concentration over South Korea is persistently affected by both local emissions and trans-boundary transport, emphasizing the importance of satellite-based remote sensing over the region. The TROPOMI accurately detected major sources of CO over South Korea (e.g., Seoul, Dangjin, Pohang, and Gwangyang), complementing the spatial coverage of ground-based networks. In general, the correlations between $C_{TROPOMI}$ and S_{KME} (r = 0.33 for all coincident samples, r = 0.37 for annual mean values at each site) were lower than those for NO_2 reported in a previous study [25], and this observation was partly attributed to the lower spatiotemporal variability. Moreover, higher correlations were observed near the emission sources. We utilized vertical profiles from EAC4 to convert the total column amounts of CO from TROPOMI to the surface mixing ratio. This converted $S_{TROPOMI}$ was directly compared to S_{KME}, which showed a significant underestimation of approximately 40%, with a moderate correlation of approximately 0.51. The relatively low biases of $S_{TROPOMI}$ were more significant in winter and were associated with the underestimation of S_{EAC4}.

Turquety et al. (2008) also reported a significant underestimation of CO (by approximately 49% below 850 hPa) from the LMDz-INCA model compared to the MOZAIC aircraft measurements over highly polluted areas in Bangkok, Thailand. They suggested that part of this underestimation could be attributed to the relatively low horizontal resolution of the model (i.e., 3.75° in longitude 2.5° in latitude) [49], which may not accurately resolve highly polluted areas. Moreover, Khan et al. (2017) also reported a significantly low bias of the MOPITT CO compared to the KME measurement in Seoul, South Korea, by approximately 35%, with a low correlation of 0.28 [53]. To the best of our knowledge, the factors affecting the low biases of the surface CO mixing ratio from satellites and models over this region remain uncertain. Intensive field campaigns combining various chemistry models of high spatial resolution (comparable to that of the TROPOMI) and in-situ profile measurements (e.g., from aircraft or unmanned aerial systems) may help to better understand these discrepancies. Moreover, multi-band retrievals of CO using both SWIR and TIR [17] may also help to detect the surface burden of CO more efficiently over a broader region. One of the important merits of this study is that this method is applicable to other regions (e.g., other Asian or developing countries, where in-situ measurements are sparse) as the EAC4 and TROPOMI provides relatively uniform quality over globe. However, comparison studies between the satellite retrievals and surface measurements are essential for broader regions to understand uncertainties in the assumed CO profiles and emissions.

Author Contributions: Conceptualization, methodology, formal analysis, investigation, resources, data curation, writing, and editing by U.J. Software, validation, resources, writing and editing, visualization, and funding acquisition by H.H. All authors have read and agreed to the published version of the manuscript.

Funding: This research was supported by the National Institute of Environmental Research (NIER) of the Ministry of Environment, Republic of Korea (grant no. NIER-2021-01-01-052).

Institutional Review Board Statement: Not applicable.

Informed Consent Statement: Not applicable.

Data Availability Statement: The TROPOMI data are available at https://scihub.copernicus.eu/ (accessed on 29 September 2019), and the EAC4 data are from https://ads.atmosphere.copernicus.eu (accessed on 29 September 2019). We obtained KME ground measurements from https://www.airkorea.or.kr (accessed on 29 September 2019).

Acknowledgments: The authors appreciate ESA and NASA for providing the TROPOMI and EAC4 data and thank the Korean Ministry of Environment for the in-situ CO data.

Conflicts of Interest: The authors declare no conflict of interest. The funder had no role in the design of the study; collection, analyses, or interpretation of data; writing of the manuscript; or in the decision to publish the results.

References

1. Spivakovsky, C.; Logan, J.; Montzka, S.; Balkanski, Y.; Foreman-Fowler, M.; Jones, D.; Horowitz, L.; Fusco, A.; Brenninkmeijer, C.; Prather, M.; et al. Three-dimensional climatological distribution of tropospheric OH: Update and evaluation. *J. Geophys. Res. Atmos.* **2000**, *105*, 8931–8980. [CrossRef]
2. Lelieveld, J.; Gromov, S.; Pozzer, A.; Taraborrelli, D. Global tropospheric hydroxyl distribution, budget and reactivity. *Atmos. Chem. Phys.* **2016**, *16*, 12477–12493. [CrossRef]
3. Prather, M.J. Lifetimes and time scales in atmospheric chemistry. *Philos. Trans. R. Soc.* **2007**, *A365*, 1705–1726. [CrossRef]
4. Gaubert, B.; Worden, H.M.; Arellano, A.F.J.; Emmons, L.K.; Tilmes, S.; Barr'e, J.; Martinez-Alonso, S.; Vitt, F.; Anderson, J.L.; Alkemade, F.; et al. Chemical feedback from decreasing carbon monoxide emissions. *Geophys. Res. Lett.* **2017**, *44*, 9985–9995. [CrossRef]
5. Myhre, G.; Shindell, D.; Bréon, F.-M.; Collins, W.; Fuglestvedt, J.; Huang, J.; Koch, D.; Lamarque, J.-F.; Lee, D.; Mendoza, B.; et al. Anthropogenic and Natural Radiative Forcing. In *Contribution of Working Group I to the Fifth Assessment Report of the Intergovernmental Panel on Climate Change*; Cambridge University Press: Cambridge, UK, 2014; pp. 659–740.
6. Holloway, T.; Levy II, H.; Kasibhatla, P. Global distribution of carbon monoxide. *J. Geophys. Res.* **2000**, *105*, 12123–12147. [CrossRef]
7. Heald, C.; Jacob, D.; Fiore, A.; Emmons, L.; Gille, J.; Deeter, M.; Warner, J.; Edwards, D.; Crawford, J.; Hamlin, A.; et al. Asian outflow and trans-Pacific transport of carbon monoxide and ozone pollution: An integrated satellite, aircraft, and model perspective. *J. Geophys. Res. Atmos.* **2003**, *108*, 4804. [CrossRef]
8. Gloudemans, A.M.S.; Krol, M.C.; Meirink, J.F.; de Laat, A.T.J.; van der Werf, G.R.; Schrijver, G.R.; van den Broek, M.M.P.; Aben, I. Evidence for long-range transport of Carbon Monoxide in the Southern Hemisphere from SCIAMACHY observations. *Geophys. Res. Lett.* **2006**, *33*, L16807. [CrossRef]

9. Hollingsworth, A.; Engelen, R.J.; Benedetti, A.; Boucher, O.; Chevallier, F.; Dethof, A.; Elbern, H.; Eskes, H.; Flemming, J. Toward a Monitoring and Forecasting System For Atmospheric Composition: The GEMS Project. *Bull. Am. Meteor. Soc.* **2008**, *89*, 1147. [CrossRef]
10. Reichle, H.G., Jr.; Connors, V.S. The mass of CO in the atmosphere during October 1984, April 1994, and October 1994. *J. Atmos. Sci.* **1999**, *56*, 307. [CrossRef]
11. Deeter, M.N.; Emmons, L.K.; Francis, G.L.; Edwards, D.P.; Gille, J.C.; Warner, J.X.; Khattatov, B.; Ziskin, D.; Lamarque, J.-F.; Ho, S.-P.; et al. Operational carbon monoxide retrieval algorithm and selected results for the MOPITT instrument. *J. Geophys. Res.* **2003**, *108*, 4399. [CrossRef]
12. McMillan, W.W.; Barnet, C.; Strow, L.; Chahine, M.T.; McCourt, M.L.; Warner, J.X.; Novelli, P.C.; Korontzi, S.; Maddy, E.S.; Datta, S. Daily global maps of carbon monoxide from NASA's Atmospheric Infrared Sounder. *Geophys. Res. Lett.* **2005**, *32*, L11801. [CrossRef]
13. Rinsland, C.P.; Luo, M.; Logan, J.A.; Beer, R.; Worden, H.; Kulawik, S.S.; Rider, D.; Osterman, G.; Gunson, M.; Eldering, A.; et al. Measurements of carbon monoxide distributions by the tropospheric emission spectrometer instrument onboard the Aura Spacecraft: Overview of analysis approach and examples of initial results. *Geophys. Res. Lett.* **2006**, *33*, L22806. [CrossRef]
14. Turquety, S.; Hadji-Lazaro, J.; Clerbaux, C.; Hauglustaine, D.A.; Clough, S.A.; Cassé, V.; Schlüssel, P.; Mégie, G. Operational trace gas retrieval algorithm for the Infrared Atmospheric Sounding Interferometer. *J. Geophys. Res.* **2004**, *109*, D21301. [CrossRef]
15. Deeter, M.N.; Edwards, D.P.; Gille, J.C.; Drummond, J.R. CO retrievals based on MOPITT near-infrared observations. *J. Geophys. Res.* **2009**, *114*, D04303. [CrossRef]
16. Bovensmann, H.; Burrows, J.P.; Buchwitz, M.; Frerick, J.; Noël, S.; Rozanov, V. SCIAMACHY: Mission objectives and measurement modes. *J. Atmos. Sci.* **1999**, *56*, 127–150. [CrossRef]
17. Worden, H.M.; Deeter, M.N.; Edwards, D.P.; Gille, J.C.; Drummond, J.R.; Nédélec, P. Observations of near-surface carbon monoxide from space using MOPITT multispectral retrievals. *J. Geophys. Res.* **2010**, *115*, D18314. [CrossRef]
18. Veefkind, J.P.; Aben, I.; McMullan, K.; Förster, H.; de Vries, J.; Otter, G.; Claas, J.; Eskes, H.J.; de Haan, J.F.; Kleipool, Q.; et al. Tropical on the ESA Sentinel-5 Precursor: A GMES mission for global observations of the atmospheric composition for climate and air quality applications. *Remote Sens. Environ.* **2012**, *120*, 70. [CrossRef]
19. Bernath, P.F.; McElroy, C.T.; Abrams, M.C.; Boone, C.D.; Butler, M.; Camy-Peyret, C.; Carleer, M.; Clerbaux, C.; Coheur, P.-F.; Colin, R.; et al. Atmospheric Chemistry Experiment (ACE): Mission overview. *Geophys. Res. Lett.* **2005**, *32*, L15S01. [CrossRef]
20. Clerbaux, C.; Coheur, P.-F.; Hurtmans, D.; Barret, B.; Carleer, M.; Colin, R.; Semeniuk, K.; McConnell, J.C.; Boone, C.; Bernath, P. Carbon monoxide distribution from the ACE-FTS solar occultation measurements. *Geophys. Res. Lett.* **2005**, *32*, L16S01. [CrossRef]
21. Clerbaux, C.; George, M.; Turquety, S.; Walker, K.A.; Barret, B.; Bernath, P.; Boone, C.; Borsdorff, T.; Cammas, J.P.; Catoire, V.; et al. CO measurements from the ACE-FTS satellite instrument: Data analysis and validation using ground-based, airborne and spaceborne observations. *Atmos. Chem. Phys.* **2008**, *8*, 2569–2594. [CrossRef]
22. Funke, B.; López-Puertas, M.; Bermejo-Pantaleón, D.; von Clarmann, T.; Stiller, G.P.; Höpfner, M.; Grabowski, U.; Kaufmann, M. Analysis of nonlocal thermodynamic equilibrium CO 4.7 μm fundamental, isotopic, and hot band emissions measured by the Michelson Interferometer for Passive Atmospheric Sounding on Envisat. *J. Geophys. Res.* **2007**, *112*, D11305. [CrossRef]
23. Pumphrey, H.C.; Filipiak, M.J.; Livesey, N.J.; Schwartz, M.J.; Boone, C.; Walker, K.A.; Bernath, P.; Ricaud, P.; Barret, B.; Clerbaux, C.; et al. Waters, Validation of middle-atmosphere carbon monoxide retrievals from MLS on Aura. *J. Geophys. Res.* **2007**, *112*, D24S38.
24. Krishna, R.K.; Ghude, S.D.; Kumar, R.; Beig, G.; Kulkarni, R.; Nivdange, S.; Chate, D. Surface PM2.5: Estimate using satellite-derived aerosol optical depth over India. *Aerosol Air Qual. Res.* **2019**, *19*, 25–37. [CrossRef]
25. Jeong, U.; Hong, H. Assessment of tropospheric concentrations of NO_2 from the TROPOMI/Sentinel-5 Precursor for the estimation of long-term exposure to surface NO_2 over South Korea. *Remote Sens.* **2021**, *13*, 1877. [CrossRef]
26. Zhang, X.; Liu, J.; Han, H.; Zhang, Y.; Jiang, Z.; Wang, H.; Meng, L.; Li, Y.C.; Liu, Y. Satellite-observed variations and trends in carbon monoxide over Asia and their sensitivities to biomass burning. *Remote Sens.* **2020**, *12*, 830. [CrossRef]
27. Buchholz, R.R.; Worden, H.M.; Park, M.; Francis, G.; Deeter, M.N.; Edwards, D.P.; Emmons, L.K.; Gaubert, B.; Gille, J.; Martínez-Alonso, S.; et al. Air Pollution Trends Measured from Terra: CO and AOD over industrial fire-prone, and background regions. *Remote Sens. Environ.* **2021**, *256*, 112275. [CrossRef]
28. Zheng, B.; Chevallier, F.; Ciais, P.; Yin, Y.; Deeter, M.N.; Worden, H.M.; Wang, Y.; Zhang, Q.; He, K. Rapid decline in carbon monoxide emissions and export from East Asia between years 2005 and 2016. *Environ. Res. Lett.* **2018**, *13*, 44007. [CrossRef]
29. Kang, H.Q.; Zhu, B.; van der A, R.J.; Zhu, C.M.; de Leeuw, G.; Hou, X.W.; Gao, J.H. Natural and anthropogenic contributions to long-term variations of SO_2, NO_2, CO, and AOD over East China. *Atmos. Res.* **2019**, *215*, 284–293. [CrossRef]
30. Jeong, U.; Kim, J.; Lee, H.; Lee, Y.G. Assessing the effect of long-range pollutant transportation on air quality in Seoul using the conditional potential source contribution function method. *Atmos. Environ.* **2017**, *150*, 33–44. [CrossRef]
31. Lalitaporn, P.; Mekaumnuaychai, T. Satellite measurements of aerosol optical depth and carbon monoxide and comparison with ground data. *Environ. Monit. Assess.* **2020**, *192*, 369. [CrossRef]
32. Magro, C.; Nunes, L.; Gonçalves, O.C.; Neng, N.R.; Nogueira, J.M.F.; Rego, F.C.; Vieira, P. Atmospheric trends of CO and CH_4 from extreme wildfires in Portugal using Sentinel-5P TROPOMI level-2 data. *Fire* **2021**, *4*, 25. [CrossRef]
33. Van Hees, R.M.; Tol, P.J.J.; Cadot, S.; Krijger, M.; Persijn, S.T.; van Kempen, T.A.; Snel, R.; Aben, I.; Hoogeveen, R.W.M. Determination of the TROPOMI-SWIR instrument spectral response function. *Atmos. Meas. Tech.* **2018**, *11*, 3917–3933. [CrossRef]

34. Landgraf, J.; aan de Brugh, J.; Scheepmaker, R.; Borsdorff, T.; Hu, H.; Houweling, S.; Butz, A.; Aben, I.; Hasekamp, O. Carbon monoxide total column retrievals from TROPOMI shortwave infrared measurements. *Atmos. Meas. Tech.* **2016**, *9*, 4955–4975. [CrossRef]
35. Borsdorff, T.; de Brugh, J.A.; Hu, H.; Aben, I.; Hasekamp, O.; Landgraf, J. Measuring carbon monoxide with TROPOMI: First results and a comparison with ECMWF-IFS analysis data. *Geophys. Res. Lett.* **2018**, *45*, 2826–2832. [CrossRef]
36. Frankenberg, C.; Platt, U.; Wagner, T. Retrieval of CO from SCIAMACHY onboard ENVISAT: Detection of strongly polluted areas and seasonal patterns in global CO abundances. *Atmos. Chem. Phys.* **2005**, *4*, 8425. [CrossRef]
37. Krol, M.; Houweling, S.; Bregman, B.; van den Broek, M.; Segers, A.; van Velthoven, P.; Peters, W.; Dentener, F.; Bergamaschi, P. Two-way nested global chemistry-transport zoom model TM5: Algorithm and applications. *Atmos. Chem. Phys.* **2005**, *5*, 417–432. [CrossRef]
38. Borsdorff, T.; aan de Brugh, J.; Hu, H.; Nédélec, P.; Aben, I.; Landgraf, J. Carbon monoxide column retrieval for clear-sky and cloudy atmospheres: A full-mission data set from SCIAMACHY 2.3 µm reflectance measurements. *Atmos. Meas. Tech.* **2017**, *10*, 1769–1782. [CrossRef]
39. Inness, A.; Blechschmidt, A.-M.; Bouarar, I.; Chabrillat, S.; Crepulja, M.; Engelen, R.J.; Eskes, H.; Flemming, J.; Gaudel, A.; Hendrick, F.; et al. Data assimilation of satellite-retrieved ozone, carbon monoxide, and nitrogen dioxide with ECMWF's Composition-IFS. *Atmos. Chem. Phys.* **2015**, *15*, 5275–5303. [CrossRef]
40. Martínez-Alonso, S.; Deeter, M.; Worden, H.; Borsdorff, T.; Aben, I.; Commane, R.; Daube, B.; Francis, G.; George, M.; Landgraf, J.; et al. 1.5 years of TROPOMI CO measurements: Comparisons to MOPITT and Atom. *Atmos. Meas. Tech.* **2020**, *13*, 4841–4864. [CrossRef]
41. Borsdorff, T.; aan de Brugh, J.; Schneider, A.; Lorente, A.; Birk, M.; Wagner, G.; Kivi, R.; Hase, F.; Feist, D.G.; Sussmann, R.; et al. Improving the TROPOMI CO data product: Update the spectroscopic database and destriping of single orbits. *Atmos. Meas. Tech.* **2019**, *12*, 5443–5455. [CrossRef]
42. Air Korea. Available online: https://www.airkorea.or.kr (accessed on 1 January 2021).
43. National Institute of Environmental Research (NIER). *Annual Report of Air Quality in Korea*; Ministry of the Environment: Sejongsi, Korea, 2019.
44. Inness, A.; Ades, M.; Agustí-Panareda, A.; Barré, J.; Benedictow, A.; Blechschmidt, A.-M.; Dominguez, J.J.; Engelen, R.; Eskes, H.; Flemming, J.; et al. CAMS reanalysis of atmospheric composition. *Atmos. Chem. Phys.* **2019**, *19*, 3515–3556. [CrossRef]
45. Deeter, M.N.; Martínez-Alonso, S.; Edwards, D.P.; Emmons, L.K.; Gille, J.C.; Worden, H.M.; Sweeney, C.; Pittman, J.V.; Daube, B.C.; Wofsy, S.C. MOPITT Version 6 product: Algorithm enhancements and validation. *Atmos. Meas. Tech.* **2014**, *7*, 3623–3632. [CrossRef]
46. Sullivan, J.T.; Berkoff, T.; Gronoff, G.; Knepp, T.; Pippin, M.; Allen, D.; Twigg, L.; Swap, R.; Tzortziu, M.; Thompson, A.M.; et al. The Ozone Water–Land Environmental Transition Study: An innovative strategy for understanding chesapeake bay pollution events. *Bull. Am. Meteorol. Soc.* **2019**, *100*, 291–306. [CrossRef]
47. CAMx. *User's Guide: Comprehensive Air-Quality Model with Extensions, Version 5.40*; ENVIRON International Corporation: Novato, CA, USA, 2011. Available online: http://www.camx.com (accessed on 1 August 2021).
48. Huijnen, V.; Williams, J.; Van Weele, M.; Van Noije, T.; Krol, M.; Dentener, F.; Segers, A.; Houweling, S.; Peters, W.; De Laat, J.; et al. The global chemistry transport model TM5: Description and evaluation of the tropospheric chemistry version 3.0. *Geosci. Model. Dev.* **2010**, *3*, 445–473. [CrossRef]
49. Turquety, S.; Clerbaux, C.; Law, K.; Coheur, P.-F.; Cozic, A.; Szopa, S.; Hauglustaine, D.A.; Hadji-Lazaro, J.; Gloudemans, A.M.S.; Schrijver, H.; et al. CO emission and export from Asia: An Analysis Combining Complementary Satellite Measurements (MOPITT, SCIAMACHY, and ACE-FTS) with global modeling. *Chem. Phys.* **2008**, *8*, 5187–5204.
50. Hauglustaine, D.A.; Hourdin, F.; Walters, S.; Jourdain, L.; Filiberti, M.-A.; Larmarque, J.-F.; Holland, E.A. Interactive chemistry in the Laboratoire de Météorologie Dynamique general circulation model: Description and background tropospheric chemistry evaluation. *J. Geophys. Res.* **2004**, *109*, D04314. [CrossRef]
51. Folberth, G.; Hauglustaine, D.A.; Lathiére, J.; Brocheton, F. Impact of biogenic hydrocarbons on tropospheric chemistry: Results from a global chemistry-climate model. *Atmos. Chem. Phys.* **2006**, *6*, 2273–2319. [CrossRef]
52. Nedelec, P.; Cammas, J.-P.; Thouret, V.; Athier, G.; Cousin, J.-M.; Legrand, C.; Abonnel, C.; Lecoeur, F.; Cayez, G.; Marizy, C. An improved infrared carbon monoxide analyzer for routine measurements aboard commercial Airbus aircraft: Technical validation and first scientific results of the MOZAIC III program. *Atmos. Chem. Phys.* **2003**, *3*, 1551–1564. [CrossRef]
53. Khan, A.; Szulejko, J.E.; Bae, M.-S.; Shon, Z.H.; Sohn, J.-R.; Seo, J.W.; Jeon, E.-C.; Kim, K.-H. Long-term trend analysis of CO in the Yongsan district of Seoul, Korea, between 1987 and 2013. *Atmos. Pollut. Res.* **2017**, *8*, 988–996. [CrossRef]

Article

Global Runoff Signatures Changes and Their Response to Atmospheric Environment, GRACE Water Storage, and Dams

Sheng Yan [1,2,3], Jianyu Liu [1,2,3,*], Xihui Gu [4] and Dongdong Kong [4]

1. Laboratory of Critical Zone Evolution, School of Geography and Information Engineering, China University of Geosciences, Wuhan 430074, China; yansheng@cug.edu.cn
2. State Key Laboratory of Water Resources and Hydropower Engineering Science, Wuhan 430074, China
3. State Key Laboratory of Hydrology-Water Resources and Hydraulic Engineering, Nanjing Hydraulic Research Institute, Nanjing 210029, China
4. Department of Atmospheric Science, School of Environmental Studies, China University of Geosciences, Wuhan 430074, China; guxh@cug.edu.cn (X.G.); kongdongdong@cug.edu.cn (D.K.)
* Correspondence: liujy@cug.edu.cn

Abstract: Runoff signatures (RS), a special set of runoff indexes reflecting the hydrological process, have an important influence on many fields of both human and natural systems by flooding, drought, and available water resources. However, the global RS changes and their causes remain largely unknown. Here, we make a comprehensive investigation of RS changes and their response to total water storage anomalies (TWSA) from GRACE satellites, atmospheric circulation, and reservoir construction by using daily runoff data from 21,955 hydrological stations during 1975–2017. The global assessment shows that (1) in recent years, the global extreme flow signatures tend to decrease, while the low and average flow signatures are likely to increase in more regions; (2) the spatial patterns of trends are similar for different RS, suggesting that the runoff distribution tends to entirely upward in some regions, while downward in other regions; (3) the trends in RS are largely consistent with that in TWSA over most regions in North America and eastern South America during 1979–2017, indicating that the GRACE-based TWSA have great potential in hydrological monitoring and attribution; (4) atmospheric circulation change could partly explain the global spatiotemporal variation patterns of RS; (5) dams have important influences on reducing the high flow signature in the catchments including dams built during 1975–2017. This study provides a full picture of RS changes and their possible causes, which has important implications for water resources management and flood and drought disaster assessment.

Keywords: runoff signatures; GRACE satellites; atmospheric circulation; floods

1. Introduction

River runoff is a crucial link in the earth's water cycle and the most important component of available water resources; therefore, the accurate description of runoff characteristics is vital for hydrological risk assessment and water resources management [1]. Despite its importance in water resources management and flood and drought disaster assessments, unfortunately, the global spatial patterns of runoff signatures (RS) and its response to total water storage anomalies (TWSA), atmospheric circulation, and human activities remain largely unknown yet. This is primarily due to the lack of global observation data and the scarcity of indicators to comprehensively characterize runoff change [2]. Therefore, to better describe the process of runoff change, we introduce a special set of runoff indexes based on a synthetic set of daily runoff data that can fully reflect the change characteristics of river runoff, namely RS [3]. They are divided into five categories: low flow signature, high flow signature, average flow signature, flow dynamic signature, event frequency, and duration. We selected eight RS and show them in Table 1. The abbreviations used later for RS are also shown in Table 1.

Table 1. Summary of eight runoff signatures in the 10,044 stations.

Group	Signature	Definition	Unit	Median
Low flow	ZFR	Zero flow ratio	Unitless	0.00
Low flow	Q10	Daily flow at the 10th percentile	m^3/s	1.32
Low flow	Q50	Daily flow at the 50th percentile	m^3/s	3.91
High flow	Q99	Daily flow at the 99th percentile	m^3/s	49.48
Mean flow	Qm	Mean daily flow	m^3/s	7.77
Mean flow	Qw	Mean daily flow during winter (Dec.–Jan.–Feb.)	m^3/s	5.77
Mean flow	Qs	Mean daily flow during summer (Jun.–Jul.–Aug.)	m^3/s	7.84
Flow dynamics	Qstd	Standard deviation of daily streamflow	m^3/s	10.48

At present, most studies on runoff change analysis are limited to regional or national scales, such as these in China [4], United States [5], and Australia [3]. However, there are some limitations in the attribution investigation of runoff change at the regional scale: (1) runoff change is influenced by the local catchment microclimate and underlying surface conditions [6], and the results cannot represent the universal regular globally; (2) due to different research methods and different comparison periods, it is difficult to carry out comparative studies on large spatial scales. In recent years, a few studies have been conducted on the attribution of global runoff change [7–9]. However, due to the lack of observation data, researchers mostly use reconstructed data to analyze global runoff change [10,11]. These efforts have created conditions for the development of global runoff research, but the detection results are still affected by the uncertainty of simulated runoff data. In addition, these studies mostly take mean runoff as the research object, hence their results cannot comprehensively show the change characteristics of RS, such as extreme hydrological events, flood peak runoff and dry water runoff [3].

The large-scale river runoff change is the consequence of the complex interaction of climate conditions, human activity, vegetation, topography, total water storage, and other factors. Among them, total water storage, climate change, and human activities have relatively large variability, which is the main influencing factor of runoff change [12–14]. Therefore, we evaluate the RS variation trend from three aspects of TWSA, atmospheric circulation, and reservoir regulation. First, TWSA includes all the water components of the earth's continental regions and is an important indicator of global climate change [15]. At present, a few studies have already explored the response of floods to TWSA. For example, Reager et al. [16] point out that gradual changes in TWSA are a prerequisite for local flood potential. However, the response of RS to TWSA is not very clear yet. Therefore, our study can provide more supporting evidence for the causes of RS change from the perspective of TWSA change based on remote sensing satellite data. Second, atmospheric circulation strongly influences precipitation variability, thus affecting runoff changes [17,18]. Some studies have concentrated on the connection between large-scale atmospheric circulation change and catchment runoff processes or extreme runoff events, embedding basin runoff processes in the global atmospheric circulation context [19]. For example, changes in catchment runoff in North America, Australia, Africa, and elsewhere have been affected by the ENSO in large part [20,21]. However, the response of global runoff variation to large-scale atmospheric circulation has not been adequately studied, especially, the possible roles of geopotential height, horizontal wind, and water vapor flux in atmospheric circulation are still unclear yet. This study reveals the action process and underlying physical mechanism of atmospheric circulation on RS through spatiotemporal variation of atmospheric variables. Third, except TWSA and atmospheric circulation, human activities on the ground such as historical water and land management also influence surface runoff and hydrological extreme events [2]. For example, the construction and operation of dams would have an impact on runoff change. Additionally, the change of surface runoff will further affect the agriculture, natural environment, fishery industry, and infrastructure construction of the local river catchment [22]. However, due to the lack of reliable reservoir models and attribution methods, little is known about the effects of dam construction and operation on runoff. Therefore, our study evaluates the effect of dams on the changing trend of RS by applying field significance resampling methods at the global scale.

Overall, the RS changes are not investigated at the global scale, and the possible mechanisms of global RS changes remain largely unknown. Hence, the scientific questions that this study attempts to solve are as follows: (1) What is the temporal and spatial variation trend of the RS globally? (2) How does the RS respond to TWSA and atmospheric circulation? (3) What is the impact of dams on RS change?

2. Materials

2.1. Runoff Signatures Data

As mentioned in the introduction, due to the lack of hydrometric gauging stations, most studies on runoff change are limited to regional or national scales. To investigate the global RS changes and their causes, a set of daily runoff data from 21,955 gauging stations globally is synthesized. The specific data sources are shown in Table 2 below [23–25].

Table 2. Summary of station observations' sources.

Number	Source	Website or Reference
9180 stations	National Water Information System of the US; GAGES-II database	https://waterdata.usgs.gov/nwis; Falcone et al., 2010 (accessed on 4 August 2021)
4628 stations	Global Runoff Data Centre	http://grdc.bafg.de (accessed on 4 August 2021)
3029 stations	HidroWeb portal of the Brazilian Agência Nacional de Águas	http://www.snirh.gov.br/hidroweb (accessed on 4 August 2021)
2260 stations	EURO-FRIEND-Water	http://ne-friend.bafg.de (accessed on 4 August 2021)
1479 stations	Canada National Water Data Archive	https://www.canada.ca/en/environment-climate-change (accessed on 4 August 2021)
776 stations	Commonwealth Scientific and Industrial Research Organization (CSIRO); Australian Bureau of Meteorology	http://www.bom.gov.au/waterdata; Zhang et al., 2013 (accessed on 4 August 2021)
531 stations	Chilean Center for Climate and Resilience Research; CAMELS-CL	http://www.cr2.cl/recursos-y-publicaciones/bases-de-datos/datos-de-caudales; Alvarez-Garreton et al., 2018 (accessed on 4 August 2021)

Considering the runoff datasets were obtained from different sources, we carried out a series of criteria to control the quality of daily flow by referring to some mature data processing methods [26]. The relevant details of the standards used in this study are as follows:

1. The runoff data with more than 10 consecutive data are regarded as missing data [27].
2. For each station, the data in the year with missing observations more than 10% is discarded [28].
3. The station is deleted if the streamflow valid recording length is less than 10 years (not necessarily continuous) at a station during 1975–2017 (the period chosen for the research).

Finally, observations from a total of 10,044 gauging stations meet these requirements, namely dataset A. The global distribution of hydrological stations for dataset A is shown in Figure 1.

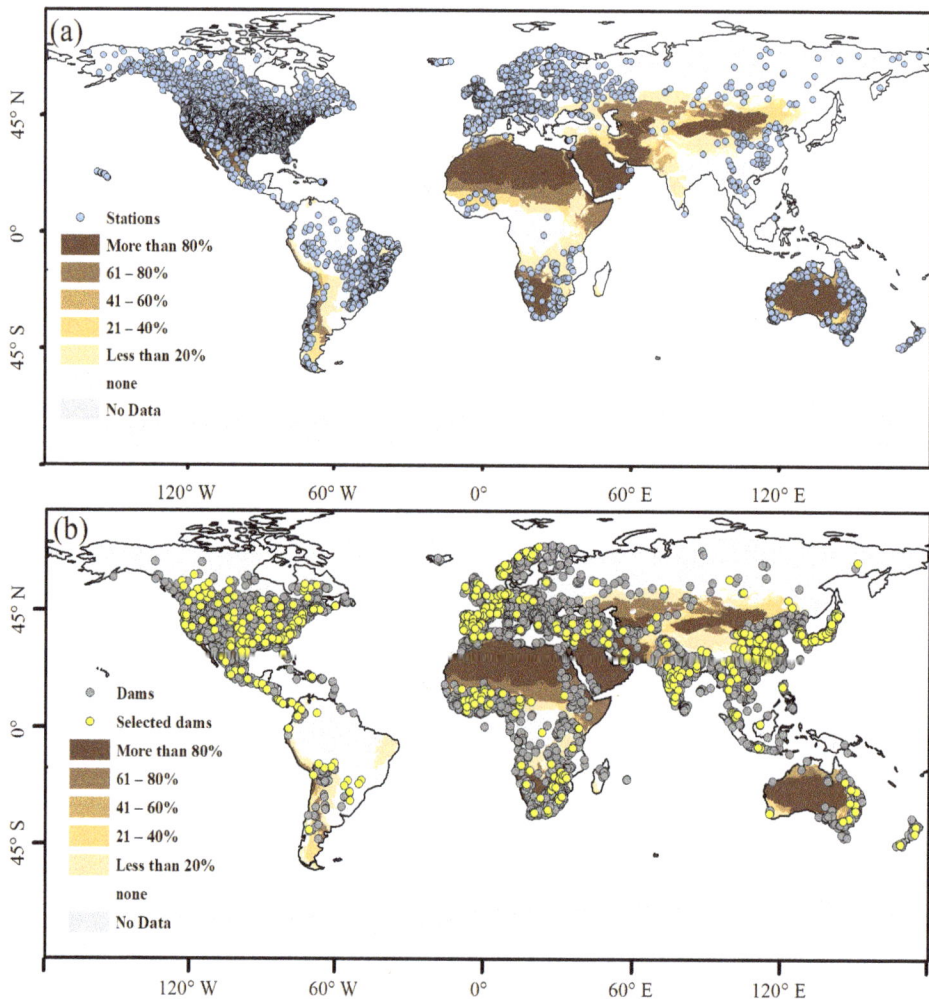

Figure 1. The spatial distribution of hydrological stations and dams. (**a**) The blue points indicate the selected the hydrological stations used in this study. (**b**) The global distribution of the dam is at the bottom. The grey points are the collected 7320 dams in the world, and the yellow points are selected 5182 dams located in the selected catchment boundaries from 1975 to 2017. Brown areas represent the arid regions of the world, and light and dark colors represent the percentage of arid land area.

2.2. GRACE Satellite Data

Global total water storage anomalies (TWSA) have been monitored by NASA's Gravity Recovery and Climate Experiment (GRACE) satellites via satellite gravimetry with unprecedented precision. GRACE data may be valuable for monitoring trends in the extreme runoff by providing information on both base flow stored in soil and groundwater and event flow driven by precipitation [29]. However, since the GRACE satellite was launched in 2002, the GRACE TWSA only covers the period of 2002–2017. Hence, we used a newly released reconstructed TWSA data based on GRACE observation, which covers the period of 1979–2019 with the spatial resolution of 0.5° [30]. Additionally, we assess the relationship between TWSA and RS changes by using the GRACE global monthly mass concentration blocks version RL05m from the Jet Propulsions Laboratory. To keep the same temporal resolutions, we used the common coverage periods of GRACE-based TWSA and RS in the analysis, i.e., 1979–2017. For the spatial resolutions, we also extracted the catchment-scale TWSA, and compared their trends with RS trends in the same catchment scale. The dataset could be downloaded (https://doi.org/10.5061/dryad.z612jm6bt, accessed on 4 August 2021).

2.3. Atmospheric Circulation Data

It is significant to evaluate the response of RS to atmospheric circulation. Fortunately, the Japan Meteorological Agency (JMA) has launched the Japanese global atmospheric reanalysis project, which is named the Japanese 55-year Reanalysis (JRA-55) [31]. In this study, the JRA-55 was used to explore the impacts of atmospheric circulation on RS. It covers a total of 62 years from 1958 to 2019. Monthly horizontal wind and geopotential height at 850, 500, and 300 hPa level with the 2.5° × 2.5° spatial resolution and the water vapor flux with a spatial resolution of 1.25° × 1.25° was chosen to carry on attribution analysis of RS change. The reanalysis dataset could be available on the NCAR/UCAR website (https://rda.ucar.edu/, accessed on 4 August 2021).

2.4. Dams Data

To evaluate the impact of dams on RS change globally, we collected a dam dataset with a relative complete variable from the Global Reservoir and Dam Database (GRanD) (http://wp.geog.mcgill.ca/hydrolab/grand/, accessed on 4 August 2021). This database includes 7320 dams with a height greater than 15 m or storage capacity larger than 0.1 km^3. It should be noted that this dataset was obtained voluntarily from various research groups around the world, which had different observation equipment, methods of data collection, and collation. Therefore, it is not realistic to provide a uniform description of the same standard for all dams globally [28]. In this dataset, there are 5182 dams built during 1975–2017. We determined the number of dams in each catchment according to the shapefiles of the catchment boundary. If there was one (or more) dam within the catchment boundary of a hydrological station, we regarded it as a dam-affected station. After this process, there were 193 dam-affected stations, named dataset A1. The remaining 9851 stations without dams were considered as dataset A2 ("no dams" group).

3. Methods

3.1. Trend Detection

The nonparametric Mann–Kendall test [32,33] was used for trend significance detection, which can exclude the interference of abnormal values and is also applicable to data with abnormal distribution or nonlinear trends. The null hypothesis of this test is that data are identically distributed and independent. At present, the Mann–Kendall test is widely used for trend examination, which can capture the overall trend of time series, including the detailed trend in recent years [9,27]. In this study, we detected spatial and temporal patterns of trends in eight RS and three large-scale environmental variables, that is, geopotential height, horizontal wind, and water vapor flux. In addition, we also examined the trends in TWSA measured by NASA's GRACE mission globally. The significance of the trend is set as the 0.05 confidence level.

3.2. Field Significance Resampling Methods

We evaluated the significance levels for the proportion of the stations showing significant trends by applying a field significance resampling procedure [34–36]. The specific resampling details of the method are summarized as follows:

1. Select a time series as the reference period for resampling, such as {1975, 1976, 1977, 1978, 1979, ..., 2009, 2010}, then randomly resample based on this reference to make the length of the new sequence unchanged and the order change. For example, {1980, 1996, 2003, 1975, 1986, ..., 2009, 1978}.

2. The time series obtained through resampling in step (1) corresponds to the observation value of RS in the corresponding year one by one for all stations to get a new resampled dataset [37].
3. Conduct the Mann–Kendall test for the time series obtained in step (2) at each station at the 0.05 significance level. Additionally, the percentages of stations with significant increase and decrease trends are calculated, respectively.
4. Repeat steps (1) through (3) 2000 times to obtain a dataset that can reflect the percentage distribution of stations with significant trends.
5. Calculate the 95th percentile in the dataset obtained in step (4), which represents the ratio of stations with significant trends. Additionally, the ratio of stations with significant trends in the reference observations is also calculated.
6. Compare the 95th percentile with the observed percentage value, if the latter is larger, it indicates that the observed percentage value is not generated randomly but is significant. That is, the no-change null hypothesis is rejected while the observed ratio value is outside the 90% confidence interval of the resampling distribution.

4. Results

4.1. Spatial Patterns of Trends in Runoff Signatures

The RS trends were examined by using the Mann–Kendall test based on dataset A records during 1975–2017 at the 0.05 significance level. As shown in Figure 2, zero flow ratio (ZFR) has a significantly increasing trend (SIT) in southern North America, eastern South America, southern Europe, and eastern Oceania, while it has a significantly decreasing trend (SDT) in northern North America, central Europe, southern Africa, and northwestern Oceania. The low flow signatures, Q10 and Q50, roughly share consistent spatial patterns, with a SIT in most areas of northern North America, southeastern South America, central and northern Europe, and northern Asia, while a SDT in eastern South America, southwestern Europe, central East Asia, and eastern Oceania. Notable is the high flow signature, Q99. Its trend change direction is generally opposite to ZFR. Additionally, Q99 shows an SIT in eastern and central North America, Western Europe, and central South America while showing a SDT in southwest North America, southern Europe, eastern South America, and eastern Oceania. Interestingly, the spatial patterns of average RS, Qm, Qw, and Qs are roughly consistent, with SIT in northeastern North America, northern Europe, central and southern South America while with SDT in southwestern North America, eastern South America, southern Europe, and eastern Oceania. However, Qs is slightly different, and the stations showing a SIT in most of northern Europe and northern Oceania are denser and more abundant. The flow dynamic signature, Qstd has a SIT in northeastern North America, Western Europe, and central South America, while it has a SDT in southwestern North America, southern Europe, eastern South America, southern Africa, Southeast Asia, and southeastern Oceania.

Generally, the extreme flow signatures tend to decrease in more stations. Overall, there are 18.2% and 9.9% stations with ZFR and Q99 showing SDT, respectively, while 14.5% and 6.7% of stations showing a SIT. For low and average flow signatures, however, the number of stations with a SIT is distinctly larger than that showing a SDT. Overall, the percentages of stations with Q10, Q50, Qs showing a SIT reach 22.6%, 17.7%, 15.4%, while that showing a SDT is only 13.8%, 12.3%, 8.4%, respectively. As for average RS, Qm, and Qw, the number of stations showing a SIT is similar to that showing a SDT.

In summary: (1) the extreme RS, including ZFR and Q99, show a SDT in more stations globally, implying that the flood risk tends to decrease in more regions over the past decades; (2) in addition, more stations show a SDT for Qstd, i.e., the standard deviation of streamflow is reducing, implying that the interannual variability of streamflow tends to be more stable; (3) the change direction (positive and negative) of runoff signatures, Q10, Q50, Q99, Qm is usually consistent, which suggests that the runoff distribution tends to entirely upward in some regions, while downward in other regions. It is worth mentioning that the spatial pattern of trend in ZFR is usually contrary to that of other RS, since different to other RS, lower ZFR (zero flow ratio) represents larger runoff.

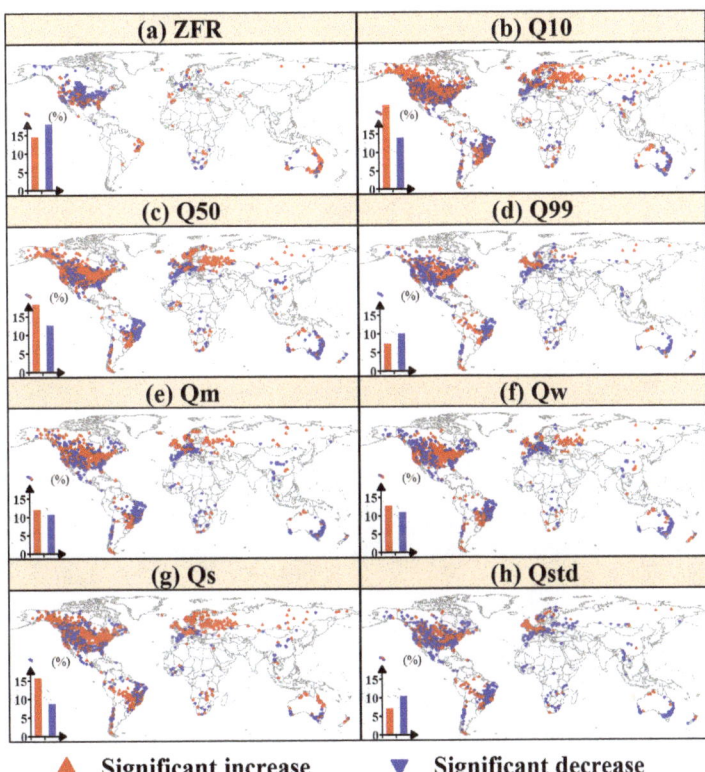

Figure 2. Spatial patterns of trends in eight runoff signatures based on the Mann-Kendall test at the 0.05 significance level (1975–2017). The red (blue) triangle represents a significant increase (decrease). The bar chart shows the percentages of stations with significant trends.

4.2. The Response of Runoff Signatures to Atmospheric Circulation

Investigation of atmospheric circulation change is an effective perspective to explain the causes and physical processes behind the RS trends [38]. Here, we explore the response of RS change to multiple atmospheric circulation indexes, including geopotential height, horizontal wind, and water vapor flux.

According to the different geopotential heights and horizontal wind in Figure 3, we can find that, in southwestern North America, the pressure has been increasing in recent years. Additionally, there is generation of a high-pressure center relative to the surrounding area and formation of an anticyclone at 850-hPa level with the strengthening of the prevailing downdraft. Meanwhile, due to the influence of the dry northeast trade winds from the inland, the water vapor flux is reducing, and then the precipitation is decreasing (Figure 4), thus resulting in the decreasing trend in most RS except for the ZFR. This is because a higher ZFR (zero flow ratio) means a smaller runoff, which is contrary to other RS. In contrast, in northern North America, where the prevailing westerlies are the main influence factor, it brings much warm water moisture from the northern Pacific. Compared with other regions, the increasing trend of air pressure is milder, and the vigorous updraft in the vertical structure promotes more water vapor convergence and condensation [39], which leads to the increasing trend in runoff in this region.

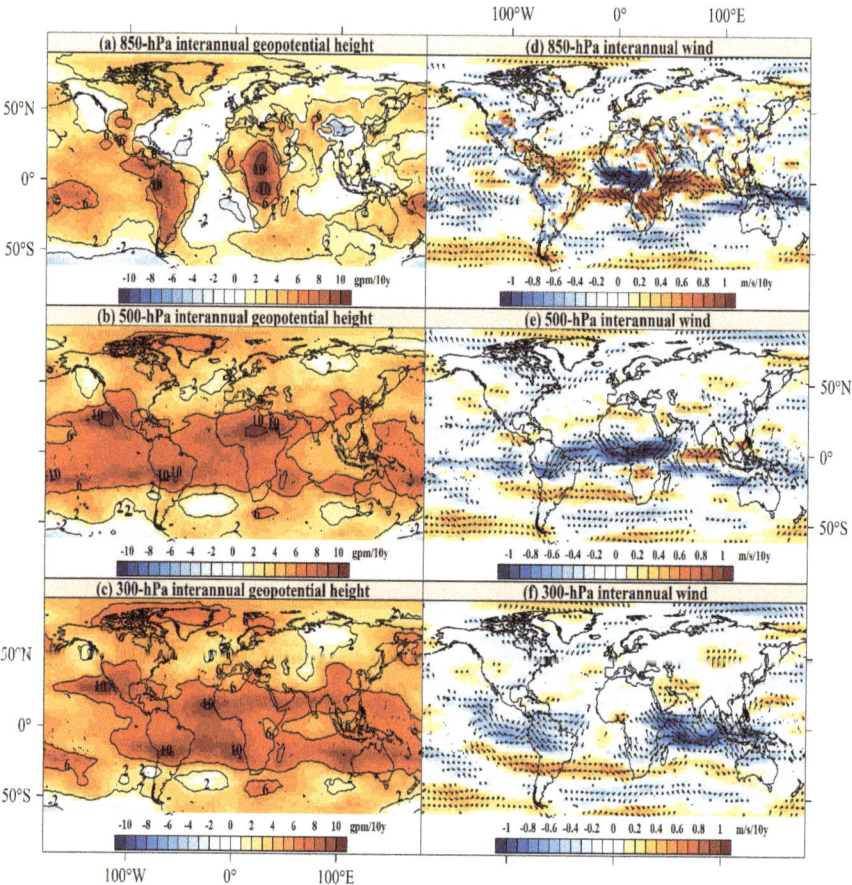

Figure 3. Spatial patterns of annual trends in atmospheric circulation. Trends in (**a**–**c**) geopotential height and (**d**–**f**) horizontal wind at 850 (top), 500 (middle), 300 (bottom) hPa based on the Mann-Kendall test. The red (blue) shades indicate the positive (negative) trends at the 0.05 level. The arrow represents the direction of the horizontal wind and its length represents the magnitude of significance. Blank areas indicate that the trend is not significant.

In eastern South America, it is mainly controlled by the intensive air pressure at different geopotential heights. Meanwhile, the air is becoming drier and sinks stronger. In addition, the dry west wind is enhanced from South America inland and brings more water vapor flux. However, wet northeasterly winds from the mid-Atlantic are weaker, and the corresponding water vapor flux is also decreasing. The increase of water vapor by the dry westerly wind is not enough to offset the decrease of water vapor by the wet northeast wind, which leads to the decrease of cloud formation and precipitation. As a result, decreasing runoff is observed in those regions. In central and southern South America, the trend change direction is just the opposite. As the region is mainly affected by the strengthening westerly wind from the southeast Pacific Ocean, which brings more warm and humid water vapor from the ocean. Thus, the precipitation is increasing, resulting in an increasing trend in the runoff.

In the whole European continent controlled by temperate climate (Temperate marine climate, Mediterranean climate), atmospheric circulation changes are relatively mild. Except for a slight increasing trend in air pressure, horizontal wind and water vapor flux changes are not particularly significant. However, we can still find out that the polar easterly winds at the 850 hPa level in northern Europe have a slightly enhanced trend, and the corresponding water vapor flux from the Arctic Ocean also shows an increasing trend as well as precipitation. Therefore, it causes an increasing flow in this region. In addition, we also find that the process of atmospheric circulation and the response of RS to it in some catchments of northern Asia is roughly the same as that of northern Europe.

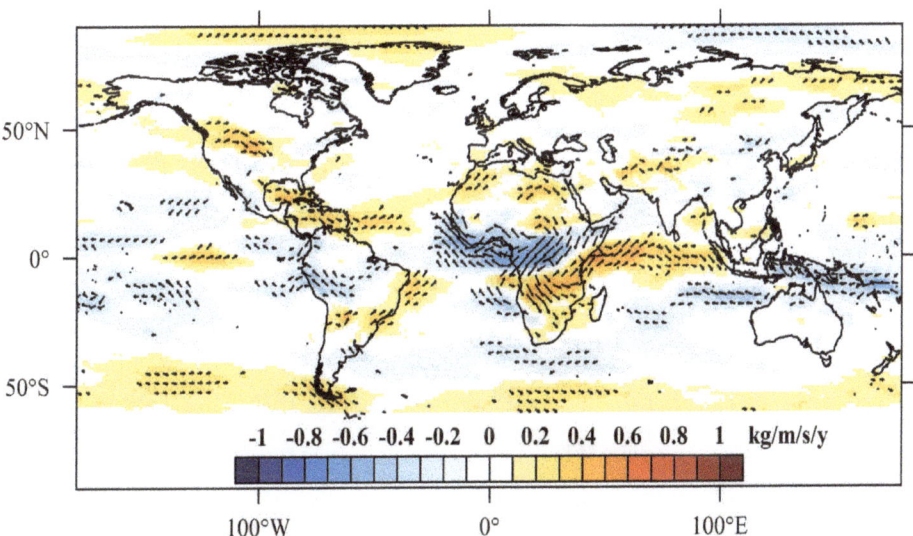

Figure 4. Spatial patterns of annual trends of the water vapor flux during 1975–2017 based on the Mann-Kendall test. The red (blue) shades represent the degree of positive (negative) trends at the 0.05 level. The arrow represents the direction of the water vapor flux and its length represents the magnitude of significance. Blank areas indicate that the trend is not significant.

In eastern Oceania, the air pressure at different geopotential heights shows an increasing trend, while the downdraft strengthens and the air becomes drier. At the same time, the humid southeast trade winds from the western Pacific are weakening along with the decrease of water vapor flux and precipitation, resulting in the decreasing trend in runoff in this region. In general, the consistent spatial and temporal patterns between atmospheric circulation variables and the change of RS are detected roughly, indicating that atmospheric circulation changes have a partial impact on the trend of RS [18].

4.3. The Response of Runoff Signatures to TWSA of GRACE Satellite

TWSA is a comprehensive reflection of regional precipitation, runoff, evapotranspiration, and groundwater, and it has already become an important parameter of global water cycle observation [29]. To explain the possible mechanisms of RS changes, we investigate the changing trend of RS from the perspective of the TWSA. We used the common coverage periods of GRACE-based TWSA and RS in the analysis, i.e., 1979–2017. In general, the spatial patterns of trend in TWSA are consistent with that in RS over most regions in North America and eastern South America from 1979 to 2017 (Figures 5 and 6). Specifically, the same SIT between RS and TWSA is found in central and western North America, northern Europe, and central South America. Meanwhile, consistent SDT is observed in southern North America, eastern and southwestern South America. This consistent appearance of positive/negative trends in most RS and TWSA indicates that gradual changes in catchment humidity influenced by a combination of climate and anthropogenic factors are a prerequisite for local extreme runoff potential [16,29]. In the process of the water cycle, if we assume that all the specific conditions in the basin are constant, runoff is generated only from precipitation. However, runoff change is not only affected by the pattern, duration, and location of precipitation, but also by the antecedent soil moisture and hydraulic characteristics of the watershed [40]. Hence, when TWSA is a decreasing trend in the catchment, a certain amount of precipitation will preferentially replenish the reduced water in the soil. Additionally, runoff will show a decreasing trend due to the decrease in the water supply. When TWSA shows an increasing trend, it indicates that the soil moisture is becoming more and more saturated and the precipitation required gradually decreases. In addition, runoff will be an increasing trend due to increased water supply. We assume, of course, that the other variables remain unchanged. Therefore, this provides evidence for the reason why the trend direction of TWSA is consistent with that in RS.

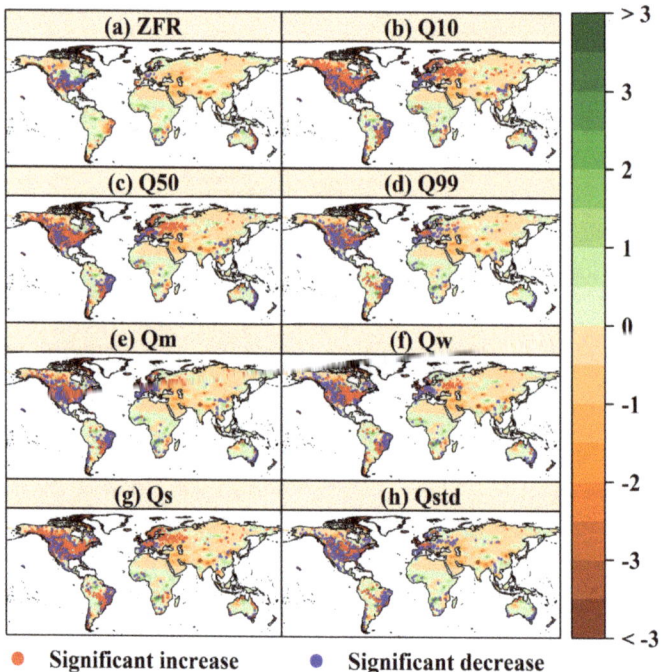

Figure 5. Spatial patterns of trends in eight runoff signatures, (**a**) ZFR, (**b**) Q10, (**c**) Q50, (**d**) Q99, (**e**) Qm, (**f**) Qw, (**g**) Qs, (**h**) Qstd based on the Mann–Kendall test at the 0.05 significance level, as well as trends in GRACE liquid water equivalent thickness in cm/year (1979–2017). The red circles indicate increasing trends, the blue represents decreasing trends. The green in the background shows an increasing trend in liquid water equivalent (basin water storage), and the orange shows a decreasing trend. The darker the color, the more obvious the changing trend.

To compare the trends in RS and TWSA in the same spatial resolution, we extracted the TWSA for each catchment, and investigated the trends at the catchment scale (Figure 6). Generally, the RS changes are consistent with TWSA, with more 60% of stations showing consistent trends. It is worth mentioning that the percentage for ZFR is less than 40%. This is because larger ZFR represents less streamflow, which is contrary to other RS. Hence, the trends in all RS are consistent with TWSA in most stations, implying that the impacts of TWSA on streamflow indeed exist and are not random.

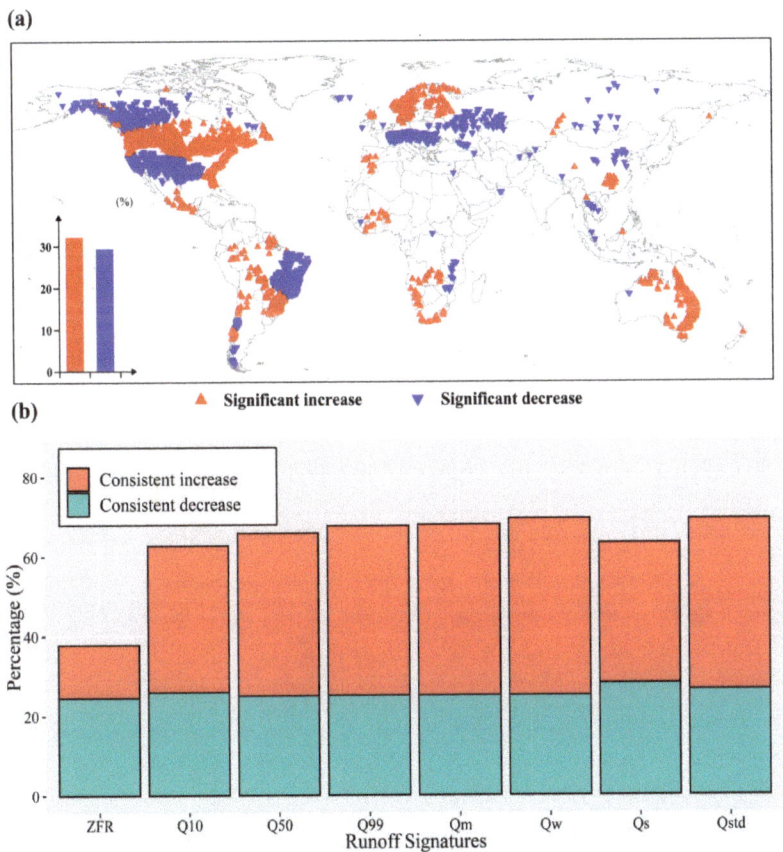

Figure 6. (a) Spatial patterns of trends in TWSA based on the Mann–Kendall test at the 0.05 significance level (1979~2017). The red (blue) triangle represents a significant increase (decrease). The bar chart shows the percentages of stations showing significant trends. (b) The percentage of stations with consistent increase/decrease in TWSA and eight RS. Red (blue) indicates the proportion of stations with consistent increase (decrease) in TWSA and RS.

4.4. The Influences of Dams on Runoff Signatures

To evaluate the differential impact of dams on positive and negative trends of RS, we applied the Mann–Kendall test combined with resampling to investigate whether the proportion of stations showing an increasing/decreasing RS is significantly based on dataset A1 (dam-affected stations) [34]. The percentage distributions of dam-affected stations with significant trends are shown in Figures 7 and 8. For low flow signatures, i.e., ZFR, Q10, and Q50, the percentage of stations with a SIT are close to that with a SDT. Specifically, the proportions of stations that show a SIT reach 1.55%, 13.98%, and 8.29%, while the percentages of stations with a SDT are 1.04%, 14.50%, and 11.91% for ZFR, Q10, and Q50, respectively. These low flow signatures show that the percentages of stations with a SIT/SDT are field significant and are inconsistent with the no-change null hypothesis. Additionally, this result indicates that the dams seem to have no material effect on the overall trend results of ZFR, Q10, and Q50.

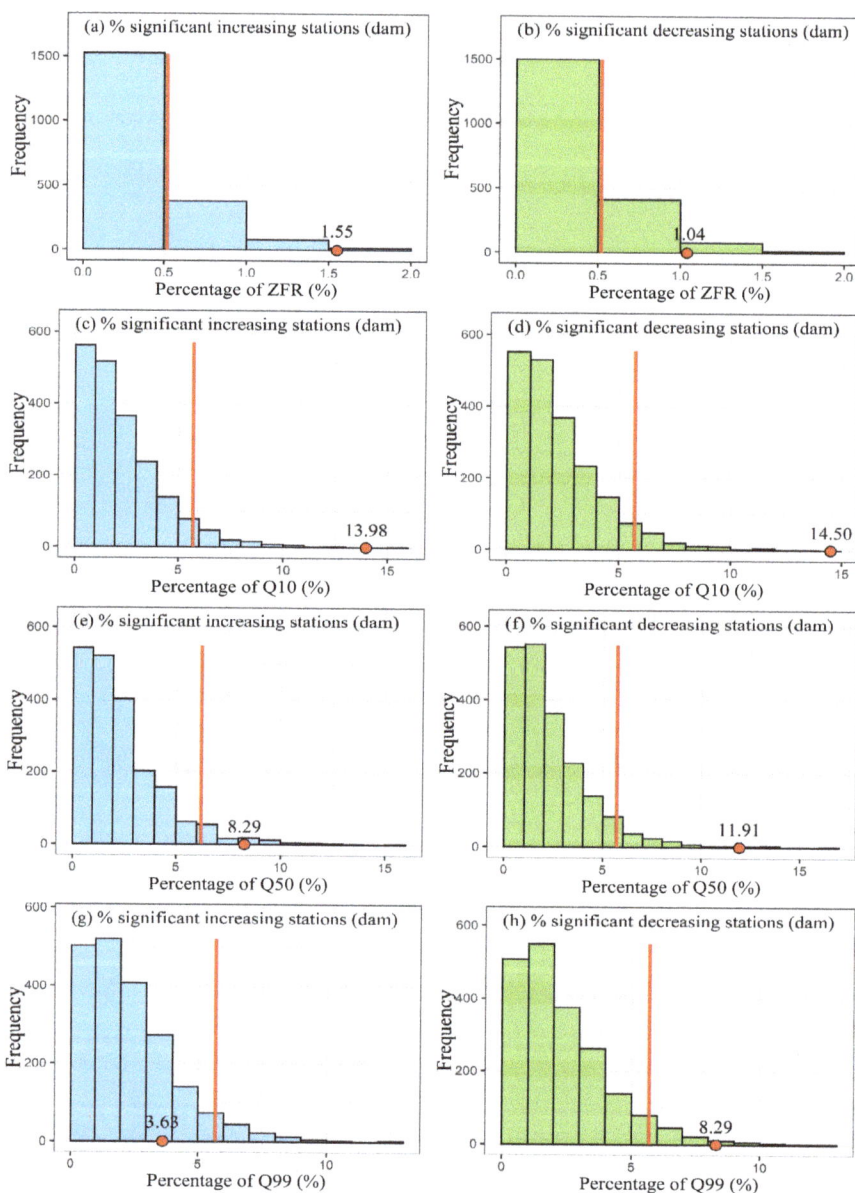

Figure 7. The proportion of stations with significant trends for 193 stations in the "dam" dataset on the four runoff signatures, i.e., (**a**,**b**) ZFR, (**c**,**d**) Q10, (**e**,**f**) Q50, (**g**,**h**) Q99, based on the Mann–Kendall test. Left panels indicate the results for the proportion of stations showing a significantly increasing trend, while right panels represent results for the proportion of stations showing a significantly decreasing trend. The histogram represents the distribution of proportion obtained from the 2000 moving-blocks field significance resampling procedure. The red dot indicates the observation value while the red line indicates the 95th percentile.

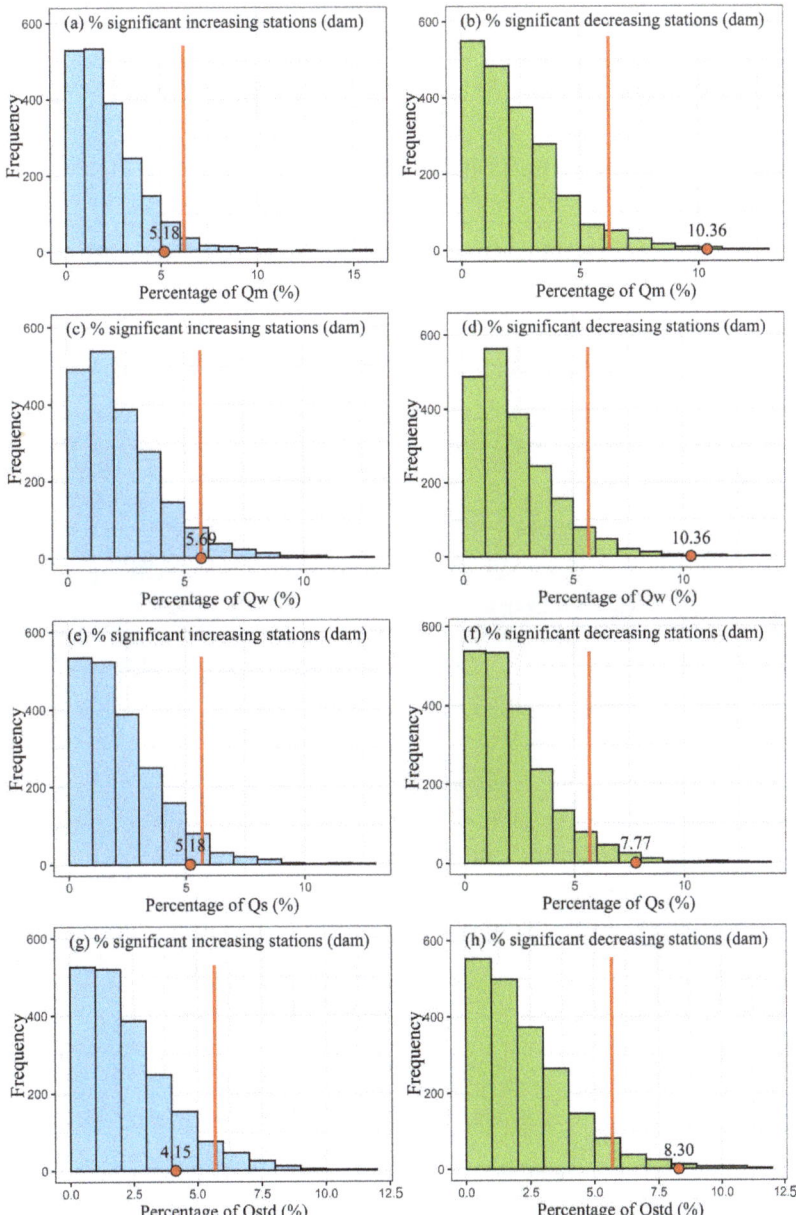

Figure 8. The same as Figure 7, but for the four runoff signatures, i.e., (**a**,**b**) Qm, (**c**,**d**) Qw, (**e**,**f**) Qs, (**g**,**h**) Qstd.

For high flow signature (Q99), the percentages of stations showing SDT are more than two times higher than that showing SIT. The proportion of stations with a SDT and SIT accounts for 8.29% and 3.63%, respectively. Similar results can be found in mean flow signatures (Qm and Qw) and flow dynamic signature (Qstd). Their percentages of stations showing a SDT are approximately twice as much as those showing a SIT. In particular, the percentages of stations showing a SDT reach 10.36%, 10.36%, and 8.30%, respectively, while the percentages of stations with a SIT are 5.18%, 5.69%, and 4.15%. Q99, Qm, Qw, and Qstd all show that the percentages of stations with a SIT are not the field significant and are consistent with the no-change null hypothesis, but the percentages of stations with a SDT are just the opposite. These results indicate that the increasing (decreasing) trend of Q99, Qm, Qw, and Qstd are (are not) caused by random changes, implying that dams have an important influence on reducing floods, mean runoff, and runoff variability.

5. Discussion

Although we collect global hydrological observation data as much as possible (Table 2), the conclusions in regions with fewer data are constrained (e.g., Asia) or muted (e.g., Africa). Consequently, the coverage of observations and the un-homogeneous of spatial distribution are still the limited factors of this study [2]. In fact, it is almost impossible to collect all hydrological stations across the globe with a regular distribution, particularly for some regions without rives and gauging stations. As shown in the new Figure 1, the regions without stations mainly distribute in the arid regions where no river and hydrological stations exist. At present, most studies on global hydrological change are still limited by uneven data distribution, such as Liu et al. [9], Do et al., [28], and Gudmundsson et al. [27]. Due to the diversity of instruments, collection and the limitations of documents, the GSIM dataset, with more than 30,000 stations, cannot guarantee its regular distribution across the world [41].

Over the past decades, the extreme RS in more stations shows a SDT globally, implying that both the flood and hydrological drought is more likely to decrease over the world. Generally, these results are consistent with previous regional research. For example, Hodgkins et al. [42] found that the flood in North America and Europe tends to decrease in more regions; Gudmundsson et al. [6] indicated that the runoff shows a decreasing trend in recent years in most regions of southern Europe. What is more, the change direction (positive and negative) of RS, Q10, Q50, Q99, Qm is usually consistent in the same station, implying that the runoff distribution tends to entirely upward in some regions, while downward in other regions, which agrees with the results of Gudmundsson et al. [27].

Atmospheric circulation has a significant impact on runoff variability, which has been widely verified in different regions, e.g., North America [43], Europe [44], and Australia [45]. Here, we assess the response of RS to atmospheric circulation on a global scale by focusing on precipitation as an intermediate variable. Additionally, our study finds that atmospheric circulation can explain the RS change in most regions. Nevertheless, atmospheric circulation is found to have a limited influence on RS changes in some regions due to the complexity of the hydrological process and its driving factors. The RS change driven by atmospheric circulation indicators such as geopotential height, horizontal wind, and water vapor flux is only a large-scale control factor [17]. However, except for the precipitation changes regulated by atmospheric circulation, the RS changes can be influenced by a series of other driving factors, such as soil moisture and groundwater (associated with TWSA) [46], and human activities (e.g., dams) [40]. Hence, we further investigate the response of RS to TWSA and dams to explore the possible mechanisms of global RS change.

GRACE is, to date, the first and only tool that is capable of global monitoring for TWSA. Besides, the Global Climate Observing System (GCOS) Steering Committee has been committed to treating TWSA as a new Essential Climate Variable (ECV) [30]. In consequence, it is reasonable and meaningful to explore the impacts of TWSA from GRACE on RS. Our results show that the trends in RS are generally consistent with that in TWSA over most regions in North America and eastern South America from 1979 to 2017, implying that the TWSA from the GRACE can largely explain spatial patterns of RS changes. These suggested that TWSA is an important factor affecting RS changes. On the one hand, larger TWSA always represents that more water storage (e.g., underground water and lake water) can recharge the river. On the other hand, the higher antecedent soil moisture accompanying larger TWSA favors a larger flooding and runoff. Additionally, previous studies suggested that the floods tend to decrease in the regions where TWSA and antecedent soil moisture showed decreasing trends [17,18]. Diffenbaugh et al. [47] found that California's severe TWSA shortage led to the reduction of soil moisture and the worst drought (low RS) on the west coast of the United States in centuries. Louise et al. [29] indicated that the broad shifts of TWSA with the soil moisture decreasing that occurs at the land surface and subsurface (TWSA change) is an important driving factor of flooding (high RS). In the dry soil moisture condition, the increase of precipitation does not have to translate into the increase of streamflow, and the reduced soil moisture and groundwater will reduce RS [14].

In addition, we also investigated the impacts of dams on RS by using the stations with dams in their catchments. Our results show that the percentages of stations showing SDT are almost twice more than that showing SIT for Q99, Qm, Qw, and Qstd. This suggests that dams play an important role in reducing the peak and mean value of streamflow, as well as the streamflow variability. Nonetheless, we also find that dams have limited impacts on the low flow signatures, ZFR, Q10, Q50. From a hydrological point of view, large dams are expected to have an important impact on flood flows; as, in many situations, dams are designed to lessen the floods risks and disasters [48,49].

6. Conclusions

This study, for the first time, assesses the spatiotemporal patterns of RS changes and their possible mechanisms on a global scale based on 10044 stations. Compared with previous studies restricted to regional scale or using only the specific runoff index, our research not only breaks through the regional limitations and eliminates the disturbance of the uncertainty of the simulation data but also introduces a complete set of RS indicators that could reflect the runoff characteristics. Therefore, this investigation provides a comprehensive picture of temporal and spatial characteristics of global RS, with important implications in global water resource management and flood and drought disaster assessment.

The extreme RS, ZFR, and Q99 show a SDT in more stations globally over the past decades, implying that the flood and the drought tend to decrease in more regions. In addition, more stations show a SDT for Qstd, implying that the interannual variability of streamflow tends to be more stable. Generally, the change direction (positive and negative) of most runoff signatures is usually consistent. This indicates that the runoff distribution tends to be entirely upward in some regions, while downward in other regions (consistently drier or wetter).

Through the temporal and spatial variations of geopotential height, horizontal wind, and water vapor flux globally, we reveal the active mechanism of atmospheric circulation on the RS change. Although these atmospheric circulation indicators are only the large-scale controlling factors, the atmospheric circulation can roughly explain the RS change in many regions. Additionally, this suggests that the global atmospheric circulation variation has an important impact on regional RS changes.

The spatial patterns of trends in RS agree well with that in TWSA from GRACE satellites over most regions in North America and eastern South America from 1979 to 2017. This suggests that GRACE satellites productions have great potential in simulation and attribution of hydrological change. In addition, the dams have important impacts on the peak, mean, and variability of runoff. In the dam-impacts stations, our results show that the percentages of stations showing SDT are almost twice more than that showing SIT for Q99, Qm, Qw, and Qstd.

Author Contributions: Conceptualization, J.L., D.K. and X.G.; methodology, S.Y.; software, S.Y.; validation, S.Y.; formal analysis, S.Y.; investigation, S.Y.; resources, J.L.; data curation, J.L. and S.Y.; writing—original draft preparation, S.Y.; writing—review and editing, J.L., D.K. and X.G.; visualization, S.Y.; supervision, J.L.; project administration, J.L.; funding acquisition, J.L. All authors have read and agreed to the published version of the manuscript.

Funding: This research was funded by the National Natural Science Foundation of China (42001042), Fundamental Research Funds for the Central Universities, China University of Geosciences (Wuhan) (grants G1323521106), Visiting Researcher Fund Program of the State Key Laboratory of Water Resources and Hydropower Engineering Science (2020SWG02), Opening funding of the State Key Laboratory of Hydrology-Water Resources and Hydraulic Engineering, Nanjing Hydraulic Research Institute (U2020nkms01), Opening funding of the State Key Laboratory of Loess and Quaternary Geology, Institute of Earth Environment, CAS (SKLLQG2018), the Fundamental Research Funds for the Central Universities, China University of Geosciences (Wuhan) (CUG2106351), Ministry of Education (GTYR202003). We acknowledge the World Climate Research Programme's Working Group on Coupled Modeling for CMIP6 simulations.

Institutional Review Board Statement: Not applicable.

Informed Consent Statement: Not applicable.

Data Availability Statement: The data is available from the website in Table 2.

Acknowledgments: We acknowledge Hylke E Beck and Yongqiang Zhang, who provided hydrological data and important suggestions.

Conflicts of Interest: The authors declare no conflict of interest.

References

1. Westerberg, I.K.; McMillan, H.K. Uncertainty in hydrological signatures. *Hydrol. Earth Syst. Sci.* **2015**, *19*, 3951–3968. [CrossRef]
2. Gudmundsson, L.; Boulange, J.; Do, H.X.; Gosling, S.N.; Grillakis, M.G.; Koutroulis, A.G.; Leonard, M.; Liu, J.G.; Schmied, H.M.; Papadimitriou, L.; et al. Globally observed trends in mean and extreme river flow attributed to climate change. *Science* **2021**, *371*, 1159–1162. [CrossRef]
3. Zhang, Y.Q.; Chiew, F.H.S.; Li, M.; Post, D. Predicting Runoff Signatures Using Regression and Hydrological Modeling Approaches. *Water Resour. Res.* **2018**, *54*, 7859–7878. [CrossRef]
4. Liu, J.Y.; Zhang, Q.; Singh, V.P.; Shi, P.J. Contribution of multiple climatic variables and human activities to streamflow changes across China. *J. Hydrol.* **2017**, *545*, 145–162. [CrossRef]
5. Ahn, K.H.; Merwade, V. Quantifying the relative impact of climate and human activities on streamflow. *J. Hydrol.* **2014**, *515*, 257–266. [CrossRef]
6. Gudmundsson, L.; Seneviratne, S.I.; Zhang, X.B. Anthropogenic climate change detected in European renewable freshwater resources. *Nat. Clim. Chang.* **2017**, *7*, 813–826. [CrossRef]
7. Berghuijs, W.R.; Larsen, J.R.; van Emmerik, T.H.M.; Woods, R.A. A Global Assessment of Runoff Sensitivity to Changes in Precipitation, Potential Evaporation, and Other Factors. *Water Resour. Res.* **2017**, *53*, 8475–8486. [CrossRef]
8. Berghuijs, W.R.; Woods, R.A. A simple framework to quantitatively describe monthly precipitation and temperature climatology. *Int. J. Climatol.* **2016**, *36*, 3161–3174. [CrossRef]
9. Liu, J.Y.; Zhang, Q.; Feng, S.Y.; Gu, X.H.; Singh, V.P.; Sun, P. Global Attribution of Runoff Variance Across Multiple Timescales. *J. Geophys. Res.-Atmos.* **2019**, *124*, 13962–13974. [CrossRef]
10. Marvel, K.; Cook, B.I.; Bonfils, C.J.W.; Durack, P.J.; Smerdon, J.E.; Williams, A.P. Twentieth-century hydroclimate changes consistent with human influence. *Nature* **2019**, *569*, 59–65. [CrossRef] [PubMed]
11. Douville, H.; Ribes, A.; Decharme, B.; Alkama, R.; Sheffield, J. Anthropogenic influence on multidecadal changes in reconstructed global evapotranspiration. *Nat. Clim. Chang.* **2013**, *3*, 59–62. [CrossRef]
12. Greve, P.; Gudmundsson, L.; Seneviratne, S.I. Regional scaling of annual mean precipitation and water availability with global temperature change. *Earth Syst. Dynam.* **2018**, *9*, 227–240. [CrossRef]
13. Haddeland, I.; Heinke, J.; Biemans, H.; Eisner, S.; Florke, M.; Hanasaki, N.; Konzmann, M.; Ludwig, F.; Masaki, Y.; Schewe, J.; et al. Global water resources affected by human interventions and climate change. *Proc. Natl. Acad. Sci. USA* **2014**, *111*, 3251–3256. [CrossRef]
14. Sharma, A.; Wasko, C.; Lettenmaier, D.P. If Precipitation Extremes Are Increasing, Why Aren't Floods? *Water Resour. Res.* **2018**, *54*, 8545–8551. [CrossRef]
15. Kusche, J.; Eicker, A.; Forootan, E.; Springer, A.; Longuevergne, L. Mapping probabilities of extreme continental water storage changes from space gravimetry. *Geophys. Res. Lett.* **2016**, *43*, 8026–8034. [CrossRef]
16. Reager, J.T.; Thomas, B.F.; Famiglietti, J.S. River basin flood potential inferred using GRACE gravity observations at several months lead time. *Nat. Geosci.* **2014**, *7*, 589–593. [CrossRef]
17. Liu, J.Y.; Zhang, Y.Q.; Yang, Y.T.; Gu, X.H.; Xiao, M.Z. Investigating Relationships Between Australian Flooding and Large-Scale Climate Indices and Possible Mechanism. *J. Geophys. Res.-Atmos.* **2018**, *123*, 8708–8723. [CrossRef]
18. Liu, J.Y.; Zhang, Y.Q. Multi-temporal clustering of continental floods and associated atmospheric circulations. *J. Hydrol.* **2017**, *555*, 744–759. [CrossRef]

19. Merz, B.; Aerts, J.; Arnbjerg-Nielsen, K.; Baldi, M.; Becker, A.; Bichet, A.; Bloschl, G.; Bouwer, L.M.; Brauer, A.; Cioffi, F.; et al. Floods and climate: Emerging perspectives for flood risk assessment and management. *Nat. Hazards Earth Syst. Sci.* **2014**, *14*, 1921–1942. [CrossRef]
20. Jain, S.; Lall, U. Floods in a changing climate: Does the past represent the future? *Water Resour. Res.* **2001**, *37*, 3193–3205. [CrossRef]
21. Ward, P.J.; Beets, W.; Bouwer, L.M.; Aerts, J.; Renssen, H. Sensitivity of river discharge to ENSO. *Geophys. Res. Lett.* **2010**, *37*, 6. [CrossRef]
22. Yun, X.B.; Tang, Q.H.; Wang, J.; Liu, X.C.; Zhang, Y.Q.; Lu, H.; Wang, Y.L.; Zhang, L.; Chen, D.L. Impacts of climate change and reservoir operation on streamflow and flood characteristics in the Lancang-Mekong River Basin. *J. Hydrol.* **2020**, *590*, 11. [CrossRef]
23. Zhang, Y.; Viney, N.; Frost, A.; Oke, A.; Brooks, M.; Chen, Y.; Campbell, N. *Collation of Australian Modeller's Streamflow Dataset for 780 Unregulated Australian Catchments*; CSIRO: Canberra, Australia, 2013.
24. Alvarez-Garreton, C.; Mendoza, P.A.; Boisier, J.P.; Addor, N.; Galleguillos, M.; Zambrano-Bigiarini, M.; Lara, A.; Puelma, C.; Cortes, G.; Garreaud, R.; et al. The CAMELS-CL dataset: Catchment attributes and meteorology for large sample studies-Chile dataset. *Hydrol. Earth Syst. Sci.* **2018**, *22*, 5817–5846. [CrossRef]
25. Falcone, J.A.; Carlisle, D.M.; Meador, W. GAGES (Geospatial Attributes of Gages for Evaluating Streamflow): A stream gage database for evaluating natural and altered flow conditions in the conterminous United States. *Ecology* **2010**, *91*, 621. [CrossRef]
26. Gudmundsson, L.; Greve, P.; Seneviratne, S.I. The sensitivity of water availability to changes in the aridity index and other factors-A probabilistic analysis in the Budyko space. *Geophys. Res. Lett.* **2016**, *43*, 6985–6994. [CrossRef]
27. Gudmundsson, L.; Leonard, M.; Do, H.X.; Westra, S.; Seneviratne, S.I. Observed Trends in Global Indicators of Mean and Extreme Streamflow. *Geophys. Res. Lett.* **2019**, *46*, 756–766. [CrossRef]
28. Do, H.X.; Westra, S.; Leonard, M. A global-scale investigation of trends in annual maximum streamflow. *J. Hydrol.* **2017**, *552*, 28–43. [CrossRef]
29. Slater, L.J.; Villarini, G. Recent trends in US flood risk. *Geophys. Res. Lett.* **2016**, *43*, 12428–12436. [CrossRef]
30. Li, F.P.; Kusche, J.; Chao, N.F.; Wang, Z.T.; Locher, A. Long-Term (1979–Present) Total Water Storage Anomalies Over the Global Land Derived by Reconstructing GRACE Data. *Geophys. Res. Lett.* **2021**, *48*, e2021GL093492. [CrossRef]
31. Kobayashi, S.; Ota, Y.; Harada, Y.; Ebita, A.; Moriya, M.; Onoda, H.; Onogi, K.; Kamahori, H.; Kobayashi, C.; Endo, H.; et al. The JRA-55 Reanalysis: General Specifications and Basic Characteristics. *J. Meteorol. Soc. Jpn.* **2015**, *93*, 5–48. [CrossRef]
32. Mann, H.B. Nonparametric test against trend. *Econometrica* **1945**, *13*, 245–259. [CrossRef]
33. Kendall, M.G. *Rank Correlation Measures*, 4th ed.; Charles Griffin: London, UK, 1975; p. 272.
34. Wilks, D.S. *Statistical Methods in the Atmospheric Sciences*; Academic Press: Cambridge, MA, USA, 2011; Volume 100.
35. Ishak, E.H.; Rahman, A.; Westra, S.; Sharma, A.; Kuczera, G. Evaluating the non-stationarity of Australian annual maximum flood. *J. Hydrol.* **2013**, *494*, 134–145. [CrossRef]
36. Westra, S.; Alexander, L.V.; Zwiers, F.W. Global Increasing Trends in Annual Maximum Daily Precipitation. *J. Clim.* **2013**, *26*, 3904–3918. [CrossRef]
37. Patton, A.; Politis, D.N.; White, H. Correction to "Automatic Block-Length Selection for the Dependent Bootstrap" by D. Politis and H. White. *Econom. Rev.* **2009**, *28*, 372–375. [CrossRef]
38. King, A.D.; Alexander, L.V.; Donat, M.G. Asymmetry in the response of eastern Australia extreme rainfall to low-frequency Pacific variability. *Geophys. Res. Lett.* **2013**, *40*, 2271–2277. [CrossRef]
39. Najibi, N.; Devineni, N.; Lu, M.Q.; Perdigao, R.A.P. Coupled flow accumulation and atmospheric blocking govern flood duration. *npj Clim. Atmos. Sci.* **2019**, *2*, 13. [CrossRef]
40. Johnson, F.; White, C.J.; van Dijk, A.; Ekstrom, M.; Evans, J.P.; Jakob, D.; Kiem, A.S.; Leonard, M.; Rouillard, A.; Westra, S. Natural hazards in Australia: Floods. *Clim. Chang.* **2016**, *139*, 21–35. [CrossRef]
41. Liu, J.Y.; You, Y.Y.; Zhang, Q.; Gu, X.H. Attribution of streamflow changes across the globe based on the Budyko framework. *Sci. Total Environ.* **2021**, *794*, 148662. [CrossRef] [PubMed]
42. Hodgkins, G.A.; Whitfield, P.H.; Burn, D.H.; Hannaford, J.; Renard, B.; Stahl, K.; Fleig, A.K.; Madsen, H.; Mediero, L.; Korhonen, J.; et al. Climate-driven variability in the occurrence of major floods across North America and Europe. *J. Hydrol.* **2017**, *552*, 704–717. [CrossRef]
43. Burn, D.H. Climatic influences on streamflow timing in the headwaters of the Mackenzie River Basin. *J. Hydrol.* **2008**, *352*, 225–238. [CrossRef]
44. Kingston, D.G.; Fleig, A.K.; Tallaksen, L.M.; Hannah, D.M. Ocean-Atmosphere Forcing of Summer Streamflow Drought in Great Britain. *J. Hydrometeorol.* **2013**, *14*, 331–344. [CrossRef]
45. Verdon, D.C.; Wyatt, A.M.; Kiem, A.S.; Franks, S.W. Multidecadal variability of rainfall and streamflow: Eastern Australia. *Water Resour. Res.* **2004**, *40*, 8. [CrossRef]
46. Barnett, T.P.; Pierce, D.W.; Hidalgo, H.G.; Bonfils, C.; Santer, B.D.; Das, T.; Bala, G.; Wood, A.W.; Nozawa, T.; Mirin, A.A.; et al. Human-induced changes in the hydrology of the western United States. *Science* **2008**, *319*, 1080–1083. [CrossRef] [PubMed]
47. Diffenbaugh, N.S.; Swain, D.L.; Touma, D. Anthropogenic warming has increased drought risk in California. *Proc. Natl. Acad. Sci. USA* **2015**, *112*, 3931–3936. [CrossRef] [PubMed]

48. FitzHugh, T.W.; Vogel, R.M. The impact of dams on flood flows in the United States. *River Res. Appl.* **2011**, *27*, 1192–1215. [CrossRef]
49. Jaramillo, F.; Destouni, G. Local flow regulation and irrigation raise global human water consumption and footprint. *Science* **2015**, *350*, 1248–1251. [CrossRef]

Article

Quantification of Natural and Anthropogenic Driving Forces of Vegetation Changes in the Three-River Headwater Region during 1982–2015 Based on Geographical Detector Model

Siqi Gao [1,2,3], Guotao Dong [3,4,*], Xiaohui Jiang [1,2], Tong Nie [1,2,3], Huijuan Yin [3] and Xinwei Guo [3]

1 Shaanxi Key Laboratory of Earth Surface System and Environmental Carrying Capacity, College of Urban and Environmental Sciences, Northwest University, Xi'an 710127, China; siqigao@stumail.nwu.edu.cn (S.G.); xhjiang@nwu.edu.cn (X.J.); nietong@stumail.nwu.edu.cn (T.N.)
2 Department of Physical Geography, College of Urban and Environmental Science, Northwest University, Xi'an 710127, China
3 Yellow River Conservancy Commission, Yellow River Institute of Hydraulic Research, Zhengzhou 450003, China; yinhuijuan@hky.yrcc.gov.cn (H.Y.); guoxinwei@hky.yrcc.gov.cn (X.G.)
4 Heihe Water Resources and Ecological Protection Research Center, Lanzhou 730030, China
* Correspondence: dongguotao@hhglj.yrcc.gov.cn; Tel.: +86-135-2306-3035

Citation: Gao, S.; Dong, G.; Jiang, X.; Nie, T.; Yin, H.; Guo, X. Quantification of Natural and Anthropogenic Driving Forces of Vegetation Changes in the Three-River Headwater Region during 1982–2015 Based on Geographical Detector Model. *Remote Sens.* 2021, 13, 4175. https://doi.org/10.3390/rs13204175

Academic Editors: Baojie He, Ayyoob Sharifi, Chi Feng and Jun Yang

Received: 3 August 2021
Accepted: 15 October 2021
Published: 19 October 2021

Publisher's Note: MDPI stays neutral with regard to jurisdictional claims in published maps and institutional affiliations.

Copyright: © 2021 by the authors. Licensee MDPI, Basel, Switzerland. This article is an open access article distributed under the terms and conditions of the Creative Commons Attribution (CC BY) license (https://creativecommons.org/licenses/by/4.0/).

Abstract: The three-river headwater region (TRHR) supplies the Yangtze, Yellow, and Lantsang rivers, and its ecological environment is fragile, hence it is important to study the surface vegetation cover status of the TRHR to facilitate its ecological conservation. The normalized difference vegetation index (NDVI) can reflect the cover status of surface vegetation. The aims of this study are to quantify the spatial heterogeneity of the NDVI, identify the main driving factors influencing the NDVI, and explore the interaction between these factors. To this end, we used the global inventory modeling and mapping studies (GIMMS)-NDVI data from the TRHR from 1982 to 2015 and included eight natural factors (namely slope, aspect, elevation, soil type, vegetation type, landform type, annual mean temperature, and annual precipitation) and three anthropogenic factors (gross domestic product (GDP), population density, and land use type), which we subjected to linear regression analysis, the Mann-Kendall statistical test, and moving *t*-test to analyze the spatial and temporal variability of the NDVI in the TRHR over 34 years, using a geographical detector model. Our results showed that the NDVI distribution of the TRHR was high in the southeast and low in the northwest. The change pattern exhibited an increasing trend in the west and north and a decreasing trend in the center and south; overall, the mean NDVI value from 1982 to 2015 has increased. Annual precipitation was the most important factor influencing the NDVI changes in the TRHR, and factors, such as annual mean temperature, vegetation type, and elevation, also explained the vegetation coverage status well. The influence of natural factors was generally stronger than that of anthropogenic factors. The NDVI factors had a synergistic effect, exhibiting mutual enhancement and nonlinear enhancement relationships. The results of this study provide insights into the ecological conservation of the TRHR and the ecological security and development of the middle and lower reaches.

Keywords: NDVI; spatiotemporal variation; driving factors; geographical detector; three-river headwater region

1. Introduction

The vegetation cover is an important component of surface ecosystems that connects the atmosphere, hydrosphere, pedosphere, and areas inhabited by humans [?]; thus, the study of regional vegetation cover is essential to regional ecological conservation. The normalized difference vegetation index (NDVI) can accurately reflect the status of surface vegetation cover, which is the best indicator of vegetation coverage and the most effective indicator for monitoring regional vegetation change and the ecological environment [? ?]. Regional vegetation coverage changes and their drivers have been studied at different scales, including globally [?], as well as in Central Asia [?], northern China [?], the Loess Plateau [?], the Qinghai–Tibet Plateau [?], the Yangtze River Basin [?], and the Amur-Heilongjiang River Basin [?], using the NDVI.

The geographical detector model proposed by Wang et al. [?] bridges the gap between the correlation analysis methods used in previous studies, and can quantify the spatial heterogeneity of vegetation and its driving factors, as well as the interaction between factors. This method has been successfully used to quantify the influence of driving factors on vegetation change. Zhao et al. [?] found that precipitation plays a crucial role in the growth of vegetation in northern China and even in other arid regions of the world. Yuan et al. [?] showed that vegetation exhibited significant spatial heterogeneity throughout the Heihe River Basin. Zhu et al. [?] found that land use types and precipitation were the main factors driving vegetation change in the middle reaches of the Heihe River Basin. Ran et al. [?] concluded that natural factors had a greater influence on vegetation than anthropogenic factors in northern Tibet. Zhang et al. [?] found that the influence of anthropogenic factors was greater than that of natural factors in the oasis-desert ecotone. Liu et al. [?] reported that precipitation was the main factor affecting the difference in the spatial distribution of the NDVI in the Qinghai–Tibet Plateau.

The ecological conservation of river sources is of vital importance to the ecological environment and the development of the middle and lower reaches of rivers. Located in the hinterland of Qinghai–Tibet Plateau, the TRHR is a natural ecological barrier in China with special alpine vegetation system and fragile ecological environment. Vegetation coverage plays an important role in its preservation. Studying the spatial and temporal variation characteristics of vegetation in alpine areas and its driving forces can better explain the environmental change process. Previous studies on the characteristics of vegetation change in the TRHR have had short time series and incomplete datasets, therefore the conclusions obtained are inconsistent [? ?]. There remains a gap in studying the temporal and spatial changes of vegetation in long time series, which cannot accurately reflect the distribution characteristics of vegetation in the TRHR at both temporal and spatial scales. Previous studies on the driving factors of vegetation change in the TRHR have mostly been limited to examining the effect of climatic factors, such as temperature and precipitation [? ? ?], and, hence, there remains a lack of research on the influence of other natural and anthropogenic factors on the NDVI. Furthermore, the traditional methods, such as correlation analysis, used in the existing studies are not suitable for studying the interaction between factors and to quantitatively analyze the factors affecting NDVI. Therefore, the aims of the present study are to analyze the spatial and temporal variability of the NDVI in the TRHR over a 34-year period from 1982 to 2015, using linear regression analysis, the Mann-Kendall statistical test, and the moving t-test, and quantitatively investigate the natural and anthropogenic driving factors of NDVI variability and their interactions using a geographical detector that capable of identifying spatial heterogeneity. The results of this study provide a scientific basis for ecological restoration and conservation in the TRHR.

2. Materials and Methods

2.1. Study Area

The TRHR (31°39′N–36°16′N, 89°24′E–102°23′E) (Figure ??) is located south of Qinghai Province, and it supplies the Yangtze, Yellow, and Lantsang Rivers. It includes 21 counties and Tanggula Township, covering a total area of 38.1×10^4 km^2. The topography is high in the West and low in the East (Figure ??c), with an average altitude of 3500–4800 m. It has a continental plateau climate, with temperature and precipitation decreasing from the southeast to the northwest (Figure ??g,h). The main vegetation types are alpine meadows and alpine grasslands. The Qinghai–Tibet Plateau is a vast semi-natural area with relatively little artificial influence [?]. The TRHR is located in the central part of the Qinghai–Tibet Plateau, with high altitude and sparse population. The TRHR is an important ecological barrier in China with a fragile ecological environment; therefore, its ecological conservation is crucial for the sustainable development of a vast area in China.

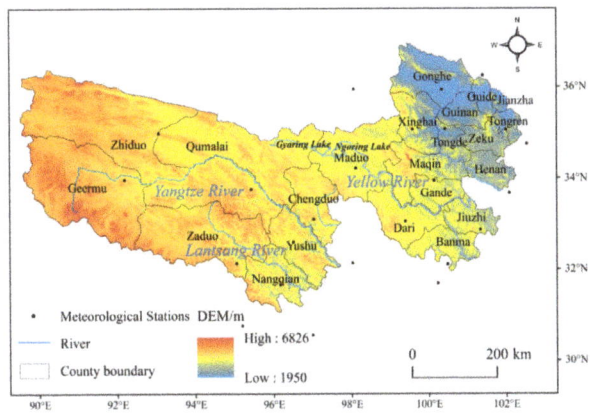

Figure 1. Sketch map of the three-river headwater region.

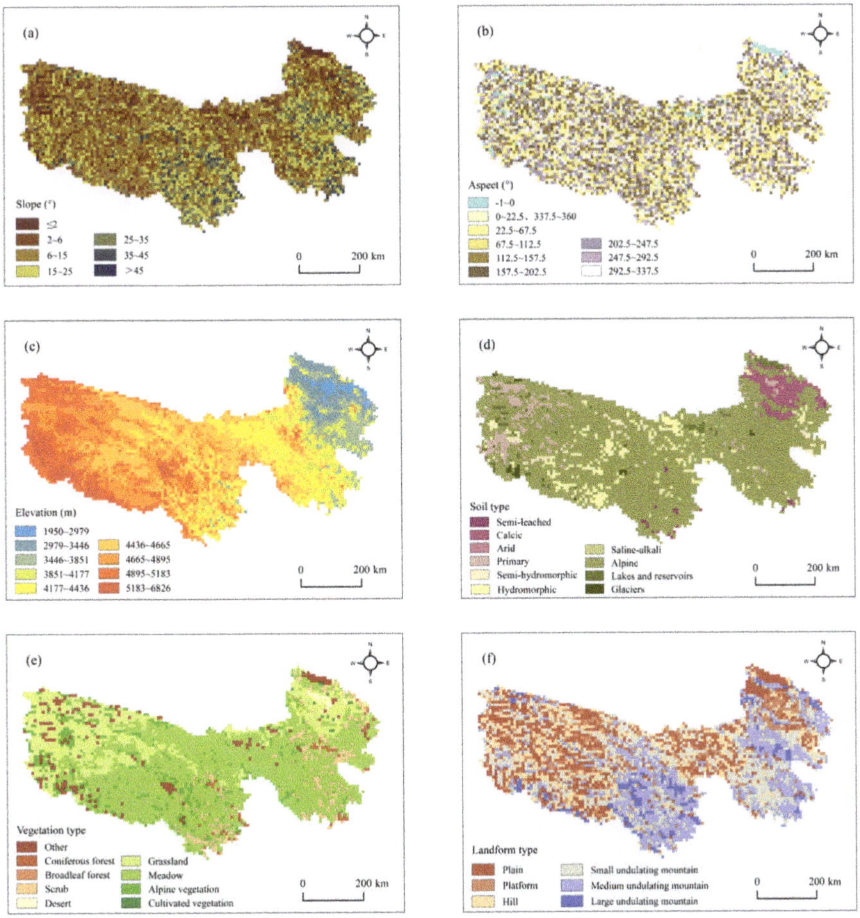

Figure 2. Cont.

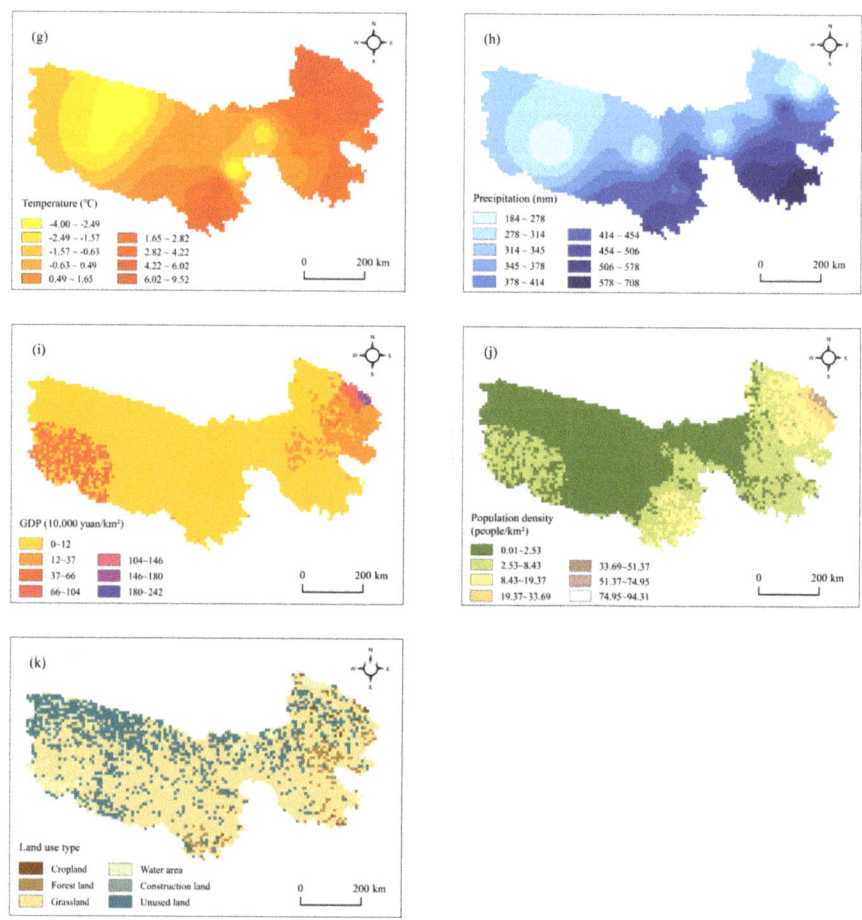

Figure 2. Classification of factors: (**a**) Slope; (**b**) Aspect; (**c**) Elevation; (**d**) Soil type; (**e**) Vegetation type; (**f**) Landform type; (**g**) Temperature; (**h**) Precipitation; (**i**) GDP; (**j**) Population density; (**k**) Land-use type.

2.2. Data and Processing

The data included in this study included natural factors, such as the NDVI, digital elevation model (DEM), climate data, landform type, soil type, and vegetation type, as well as anthropogenic factors, such as land use type, population density, and GDP in the TRHR. NDVI data were obtained from the Big Earth Data Platform for Three Poles, using GIMMS NDVI3g data with a spatial resolution of 8 km and a temporal resolution of 16 days [? ?], the NDVI images for each year from 1982 to 2015 were obtained by maximum value composite (MVC) [?]. Vegetation coverage was divided into five classes according to NDVI values: low coverage (≤0.2), medium–low coverage (0.2–0.4), medium coverage (0.4–0.6), medium–high coverage (0.6–0.8), and high coverage (>0.8). The annual mean temperature and annual precipitation data were obtained from daily standard meteorological data of 26 meteorological stations in and around the TRHR from 1982 to 2015 using the inverse distance weighting method. DEM data were GDEMV2 30 m resolution digital elevation data from the Geospatial Data Cloud of the Chinese Academy of Sciences (http://www.gscloud.cn/ (accessed on 3 August 2021)), and the elevation, slope, and aspect data were obtained from the DEM data. Other data were obtained from the Resource and Environmental Sciences Data Center of the Chinese Academy of Sciences (http://www.resdc.cn/ (accessed on 3 August 2021)). All the above data were extracted according to the vector boundary of the TRHR [?] and were resampled to make them consistent with the 8-km NDVI data image. Using ArcGIS to create a fishnet, 5853 random sampling points were generated based on 8 × 8 km grids according to the scope of the TRHR, and the spatial attribute values were extracted.

We selected NDVI as the dependent variable and six categories of topography (slope, aspect, elevation), soil (soil type), vegetation (vegetation type), landform (landform type), climate (annual mean temperature, annual precipitation), and human activity (GDP, population density, land use type), a total of 11 representative and easily quantifiable factors, as independent variables. Precipitation and temperature are important factors affecting vegetation changes [?], elevation, slope, aspect, and landform type affect vegetation distribution by changing water and heat conditions [?]. Soil and vegetation types are important environmental elements for vegetation growth [?]; economic development affects ecological environment, land use type, GDP, and population density are indicators that can quantify changes in socioeconomic development [?]. The independent variables in the geographical detector model must use discrete quantities, therefore we have to classify the factors. According to the actual situation of the TRHR, slope was divided into 7 categories according to the Technical Regulations for Land Use Status Survey; aspect was divided into 9 categories according to slope orientation; Soil type was divided into 10 categories according to the traditional "Soil Occurrence Classification" system; vegetation type was divided into 9 categories according to the 1:1,000,000 Chinese Vegetation Atlas; landform type was divided into 6 categories according to the 1:1,000,000 Landform Atlas of the People's Republic of China; land use type was divided into 6 categories according to the 1:1,000,000 Land Use Map of China; the elevation, annual mean temperature and annual precipitation were divided into 9 categories according to the natural breakpoint method [?], and the GDP and population density were divided into 7 categories according to the natural breakpoint method [?] (Figure ??).

2.3. Methods

2.3.1. Linear Regression Analysis

Linear regression analysis can analyze the trend of each raster in an image [?]. The raster calculator of ArcGIS was used to analyze the NDVI trend of each image element in the TRHR from 1982 to 2015, and categorized the NDVI change trend into seven classes according to the natural breakpoint method [?]: significant degradation, moderate degradation, slight degradation, basically unchanged, slight improvement, moderate improvement, and significant improvement. The slope can be calculated through Equation (1) [?]:

$$Slope = \frac{n \times \sum_{i=1}^{n}(i \times NDVI_i) - (\sum_{i=1}^{n} i)(\sum_{i=1}^{n} NDVI_i)}{n \times \sum_{i=1}^{n} i^2 - (\sum_{i=1}^{n} i)^2} \tag{1}$$

In Equation (1): n is the total number of the year series (n = 34 in this study), i ranges from 1 to n, $NDVI_i$ is the NDVI value of the ith year, and $Slope$ is the variation trend of the NDVI; if $Slope$ > 0, the vegetation coverage shows an increasing trend; if $Slope$ < 0, the vegetation coverage shows a decreasing trend; if $Slope$ = 0, there is no significant change in the vegetation coverage.

2.3.2. Mann-Kendall Test

The Mann-Kendall method is a nonparametric statistical test used to determine the significance of trends [?]. The change trend was significant when $|Z| > Z_{0.05}$. In this study, the Mann-Kendall statistical test was used to test the mutation points of the NDVI. The significance level was set at 0.05. The intersection of UF and UB is the mutation point; if there is more than one intersection, it is not certain whether it is the mutation point, and further testing is needed [?].

2.3.3. Moving t-Test

The moving t-test was used to test for mutations by examining whether the difference between the means of the two sample groups was significant [?]. If the difference between the mean values of the two subsequences exceeded the significance level of $p = 0.05$, the mutation was considered to be present; otherwise, no mutation was considered to be present.

2.3.4. Geographical Detector

The geographical detector is a statistical method used to detect spatial heterogeneity and its driving factors [?]. We used a geographical detector to compare the spatial distribution of NDVI vegetation with the spatial distribution characteristics of the detection factors; if a factor drives the NDVI variation, then the spatial distribution of the NDVI will be similar to the spatial distribution of that factor. This method has been successfully used to study the drivers of NDVI change [? ? ? ? ? ?].

(1) Factor detector. The factor detector q-statistic measures the degree of spatial stratified heterogeneity of a variable Y; and the determinant power of an explanatory variable X of Y. A factor detector is used to detect the spatial heterogeneity of the NDVI and the explanatory power of the independent variable X on the dependent variable Y, expressed by the q value [?]:

$$q = 1 - \frac{1}{N_{\sigma^2}} \sum_{h=1}^{L} N_h \sigma_h^2 = 1 - \frac{SSW}{SST} \tag{2}$$

$$SSW = \sum_{h=1}^{L} N_h \sigma_h^2, \quad SST = N_{\sigma^2} \tag{3}$$

In Equations (2) and (3): q is the explanatory power of the independent variable X on the dependent variable Y, with a value range of [0, 1]; the larger the q value, the more obvious the spatial heterogeneity and the stronger the explanatory power of X on Y. The study area is divided into $h = 1, 2 \ldots, L$ regions; N_h and N are the number of units in layer h and the whole region, respectively; σ_h^2 and σ^2 are the variances of the Y values of layer h and region, respectively; SSW and SST are the sum of variance within layer and total variance of region, respectively.

In this study, the independent variable X represents the detection factor Xs (s = 1, 2, 3, 4, 5, 6, 7, 8, 9, 10, and 11), as is shown in Table ??, and the dependent variable Y is the NDVI.

Table 1. Detection factors.

Type	Detection Factors	Index	Unit	Type	Detection Factors	Index	Unit
Topography	X1	Slope	°	Climate	X7	Annual mean temperature	°C
	X2	Aspect	°		X8	Annual precipitation	mm
	X3	Elevation	m	Human activity	X9	GDP	10,000 yuan/km²
Soil	X4	Soil type	-		X10	Population density	people/km²
Vegetation	X5	Vegetation type	-		X11	Land use type	-
Landform	X6	Landform type	-				

(2) Interaction detector. The interaction detector reveals whether the factors X1 and X2 (and more X) have an interactive influence on a response variable Y. Because the factors in nature do not exist independently, there are interactions between the factors, and the interactions between the factors need to be analyzed in the study. An interaction detector

was used to detect the interaction between NDVI detection factors. It can detect any relationship between factors as long as they exist. The assessment methods used are presented in Table ??.

Table 2. Types of interaction.

Foundation	Interaction
q (X1∩X2) < Min [q (X1), q (X2)]	Nonlinear weakening
Min [q (X1), q (X2) < q (X1∩X2) < Max (q (X1), q (X2)]	Univariate weakening
q (X1∩X2) > Max [q (X1), q (X2)]	Bivariate enhancement
q (X1∩X2) = q (X1) + q (X2)	Independent
q (X1∩X2) > q (X1) + q (X2)	Nonlinear enhancement

(3) Risk detector. A risk detector was used to compare whether there was a significant difference between the mean values of the dependent variables in the two regions. This study was used to detect the appropriate range or types of the driving NDVI factors. The t-statistic used was the following [?]:

$$t_{\bar{y}_{h=1} - \bar{y}_{h=2}} = \frac{\bar{Y}_{h=1} - \bar{Y}_{h=2}}{\left[\frac{Var(\bar{Y}_{h=1})}{n_{h=1}} + \frac{Var(\bar{Y}_{h=2})}{n_{h=2}}\right]^{1/2}} \quad (4)$$

In Equation (4): $\bar{Y}_h$ denotes the attribute mean within subregion h, n_h is the number of samples within subregion h, and Var denotes the variance. According to the null hypothesis H_0: $\bar{Y}_{h=1} = \bar{Y}_{h=2}$, if H_0 is rejected at confidence level α, it is considered that there is a significant difference in the attributed means between the two subregions.

(4) Ecological detector. An ecological detector was used to detect whether there was a significant difference in the influence of different factors on NDVI changes, as measured by the F-statistic [?]:

$$F = \frac{N_{X1}(N_{X2} - 1)SSW_{X1}}{N_{X2}(N_{X1} - 1)SSW_{X2}} \quad (5)$$

$$SSW_{X1} = \sum_{h=1}^{L1} N_h \sigma_h^2, \quad SSW_{X2} = \sum_{h=1}^{L2} N_h \sigma_h^2 \quad (6)$$

In Equations (5) and (6): N_{X1} and N_{X2} denote the sample sizes of the two factors X1 and X2, respectively, SSW_{X1} and SSW_{X2} denote the sum of within-layer variances of the strata formed by X1 and X2, respectively, and L1 and L2 denote the number of strata of the variables X1 and X2, respectively. According to the null hypothesis H_0: $SSW_{X1} = SSW_{X2}$, if H_0 is rejected at the significance level of α, this indicates that there is a significant difference in the effect of the two factors X1 and X2 on the spatial distribution of attribute Y.

3. Results

3.1. Spatial and Temporal Variation Characteristics of the NDVI in the TRHR

The NDVI values in the TRHR were high in the southeast and low in the northwest (Figure ??). Regions with low vegetation coverage were mainly distributed in the northwest, with most being low coverage grassland; regions with high vegetation coverage were mainly in the southeast, where the hydrothermal conditions were better, the elevation was relatively low, and the vegetation was mainly high coverage grassland and forest.

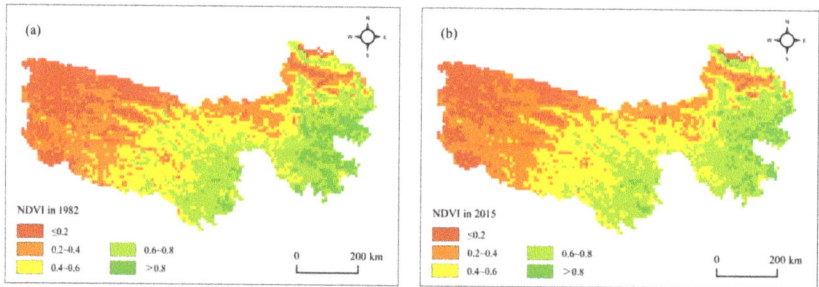

Figure 3. Spatial distributions of the NDVI: (**a**) NDVI in 1982; (**b**) NDVI in 2015.

In this study, the annual mean NDVI values from 1982 to 2015 were selected to represent the annual vegetation coverage status in the TRHR. The change observed exhibited an increasing trend, which is consistent with the findings of Zhai et al. [?]. With an increase rate of 0.002/10 a, the mean NDVI value increased from 0.454 in 1982 to 0.458 in 2015, and the maximum (0.493) and minimum (0.430) NDVI values occurred in 2010 and 1995, respectively (Figure ??). These results indicate that the vegetation coverage of the TRHR has been improving, but with small changes from 1982 to 2015. Due to overgrazing, the ecological degradation of the TRHR as serious and the vegetation coverage was low. After 2005, the vegetation coverage gradually increased due to the increase in artificial precipitation and the implementation of ecological projects, such as the return of grazing to grass.

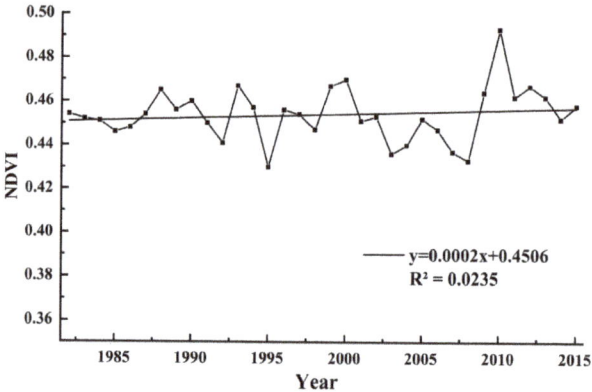

Figure 4. Change trend of the NDVI during 1982–2015 (The statistical significance level is 0.05).

3.2. Trend Analysis of NDVI Changes in the TRHR

In 1982 and 2015, high and medium-high vegetation coverage areas accounted for more than 33% and 31%, respectively, of the TRHR area, while low and medium–low areas accounted for approximately 42% and 41%, respectively, of the total TRHR area. From 1982 to 2015 the low and high vegetation coverage areas decreased, with the former type decreasing the most (by 3.08%). During the same period, the medium vegetation coverage area increased the most (by 2.78%; Table ??).

Table 3. The transfer matrix of NDVI changes during 1982–2015 (km^2).

NDVI Grade.	≤0.2	0.2–0.4	0.4–0.6	0.6–0.8	>0.8	Total in 2015	Shifted-In
≤0.2	52,380.30	3574.52	28.62	40.13	0.00	56,023.57	3643.27
0.2–0.4	15,250.20	79,604.60	7079.04	67.18	0.00	102,001.02	22,396.42
0.4–0.6	136.95	11,649.30	75,441.70	15,042.10	21.39	10,2291.44	26,849.74
0.6–0.8	0.00	234.22	9102.26	81,869.60	14,528.30	105,734.38	23,864.78
>0.8	0.00	0.00	40.50	4214.91	10,510.80	14,766.21	4255.41
Total in 1982	67,767.45	95,062.64	91,692.13	101,233.92	25,060.49	380,816.63	
Shifted-out	15,387.15	15,458.04	16,250.43	19,364.32	14,549.69		
Variation	−11,743.88	6938.38	10,599.31	4500.46	−10,294.28		
Percentage (%)	−3.08	1.82	2.78	1.18	−2.70		

The NDVI transfer matrix of the TRHR showed that there was a transformation in the NDVI at all levels from 1982 to 2015 (Table ??). The shifted-out areas were mainly medium–high vegetation coverage, which shifted mainly to medium vegetation coverage, and the shifted-in areas were mainly medium–low, medium, and medium–high vegetation coverage, with a significant increase in medium vegetation coverage and a substantial decrease in high vegetation coverage.

Although the trend of the NDVI value of the TRHR was increasing, it was still dominated by low, medium, and medium–high vegetation coverage, which all accounted for more than 25% of the area, while the high vegetation coverage area accounted for the smallest proportion and decreased significantly. Previously, the ecological environment was severely damaged, and the restoration was difficult and slow. Land use is still dominated by low-coverage grassland, thus, the status of the vegetation coverage of the TRHR was still not optimistic.

From the linear regression analysis, it was concluded that the vegetation coverage of the TRHR showed an increasing trend from 1982 to 2015 (Figure ??), thereby indicating that the vegetation coverage of the TRHR gradually recovered. The area with the largest increase in vegetation coverage was mainly distributed in the west and north, covering a total of 14.5×10^4 km^2 and accounting for 37.86% of the total area; this area was mainly dominated by grassland, meadow, and alpine vegetation. The area with the largest decrease in vegetation coverage was mainly concentrated in the center and the south, covering a total of 12.6×10^4 km^2 and accounting for 32.87% of the total area. Areas with unchanged vegetation were distributed throughout the region (Table ??). The NDVI change trend in the TRHR increased in the north and west and decreased in the south and center. The desert in the northeast of the TRHR has gradually transformed into grassland and meadow vegetation types [?]. The unused land in the Sanjiangyuan Ecological Protection Project area has been transformed into low-coverage grassland, and the area of high-coverage grassland has increased significantly (Table ??), therefore the implementation of ecological projects has significantly improved the vegetation coverage of Zhiduo, Qumalai and Mado counties in the north and northwest of the TRHR. The decrease in vegetation coverage in Yushu, Jiuzhi, and Banma counties in the south may be due to the decrease in precipitation.

The M-K test showed that none of the intersection points of the UF and UB exceeded the critical value. Significance test indicated that $|Z| = 0.048 < Z_{0.05} = 0.236$, thereby indicating that the trend of the NDVI change in the TRHR was not significant, but had multiple intersection points (Figure ??). Therefore, the mutation points needed to be further examined using the moving t-test.

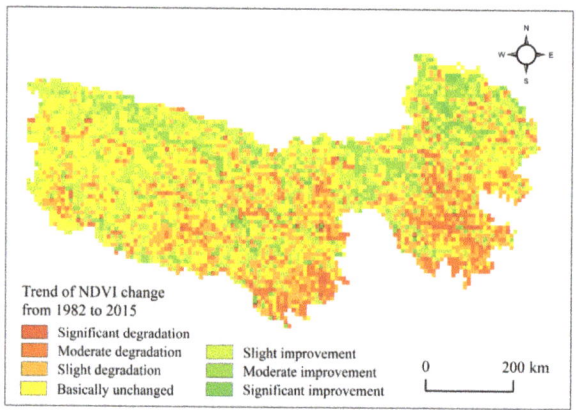

Figure 5. Distribution of the NDVI change trend during 1982–2015.

Table 4. Change trend of the NDVI during 1982–2015.

Change Trend	Gradient	Area/km^2	Percentage (%)
Significant degradation	−0.0107–−0.0024	5511.50	1.44
Moderate degradation	−0.0024–−0.0010	40,788.80	10.65
Slight degradation	−0.0010–−0.0002	79,577.90	20.78
Basically unchanged	−0.0002–0.0005	112,079.00	29.27
Slight improvement	0.0005–0.0013	94,305.40	24.63
Moderate improvement	0.0013–0.0026	44,664.90	11.66
Significant improvement	0.0026–0.0179	6025.25	1.57

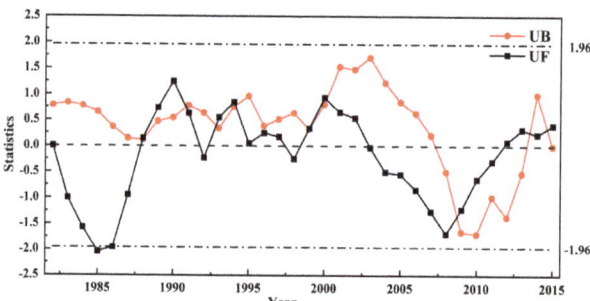

Figure 6. The results of the M-K test.

The moving t-test showed that 2008 was the mutation point of the NDVI (Figure ??), which experienced a decreasing trend before 2008 and a significantly increasing trend in 2008 according to the cumulative departure method (Figure ??). Therefore, the combination of the M-K test, moving t-test, and cumulative departure method led to the conclusion that the NDVI of the TRHR was mutated in 2008.

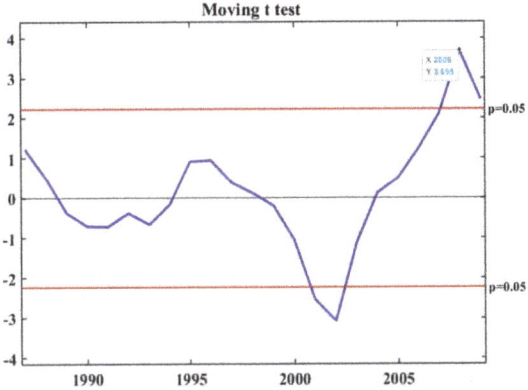

Figure 7. The results of the moving t-test.

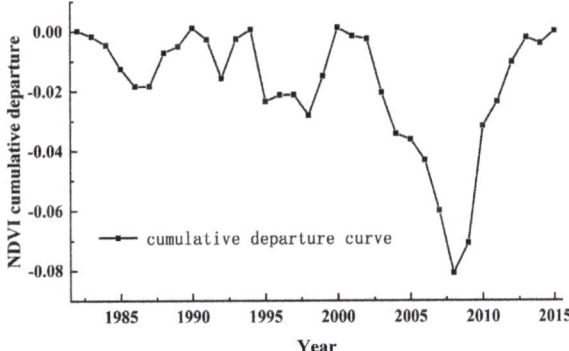

Figure 8. NDVI cumulative departure.

3.3. Factor Detection

According to the q values of each factor obtained from the factor detector (Table ??), the magnitude of the influence of each factor on the NDVI of the TRHR was as follows: annual precipitation (0.550) > annual mean temperature (0.463) > vegetation type (0.409) > elevation (0.350) > land use type (0.244) > landform type (0.216) > population density (0.204) > soil type (0.147) > slope (0.141) > GDP (0.088) > aspect (0.055).

Table 5. q values of factors.

Factors	X1	X2	X3	X4	X5	X6	X7	X8	X9	X10	X11
q value	0.141	0.055	0.350	0.147	0.409	0.216	0.463	0.550	0.088	0.204	0.244
p value	0.000	0.000	0.000	0.000	0.000	0.000	0.000	0.000	0.000	0.000	0.000

The q value of annual precipitation was the largest, with an explanatory power of 55%, which was much more influential than other factors; therefore, annual precipitation was the main driving factor of vegetation change in the TRHR, followed by annual mean temperature, vegetation type, and elevation, with an explanatory power of more than 30%; land use type, landform type, population density, soil type, and slope had an explanatory power of more than 10%; GDP and aspect had little explanatory power.

3.4. Ecological Detection

Ecological detection reflects whether there is a significant difference in the influence of each detection factor on the NDVI in the TRHR. The results showed that there were significant differences in the effects of all factors on NDVI, except for the effects of soil and slope, population, and landform on NDVI (Table ??). The effects of annual precipitation on the NDVI in the TRHR were significantly different from those of the other factors. The factor detection showed that annual precipitation was the dominant driver of NDVI changes in the TRHR, and the results of ecological detection further proved that the effects of annual precipitation were stronger than those of other factors. The non-significant differences between the effects of soil and slope, population density, and landform on the NDVI indicated that both had little influence on vegetation.

Table 6. Ecological detection of factors.

Factors	X1	X2	X3	X4	X5	X6	X7	X8	X9	X10	X11
X1											
X2	Y										
X3	Y	Y									
X4	N	Y	Y								
X5	Y	Y	Y	Y							
X6	Y	Y	Y	Y	Y						
X7	Y	Y	Y	Y	Y	Y					
X8	Y	Y	Y	Y	Y	Y	Y				
X9	Y	Y	Y	Y	Y	Y	Y	Y			
X10	Y	Y	Y	Y	Y	N	Y	Y	Y		
X11	Y	Y	Y	Y	Y	Y	Y	Y	Y	Y	

Note: Y indicates a significant difference in the influence of two factors on vegetation NDVI, while N indicates no significant difference (confidence level is 95%).

3.5. Interaction Detection

A single variable could not explain the spatial variation in the NDVI, and the synergistic effects of multiple natural and anthropogenic factors needed to be considered. The geographical detector can reveal the interactions among the factors and their effect on NDVI changes. The results showed that all factor interactions enhanced the influence of a single factor on the NDVI, showing a bivariate and non-linear enhancement effects. Among them, the interactions of aspect with elevation, annual mean temperature, GDP, and population density, the interactions of soil type with GDP and population density, and the interactions of GDP with population density and land use type showed a non-linear enhancement effect, and the interactions of other factors showed a bivariate enhancement effect. Among them, the q value of the interaction between annual precipitation and other factors was high, with the explanatory power reaching more than 58%. This was higher than the explanatory power of the single factor of annual precipitation on vegetation, whose q value of interacting with elevation, annual mean temperature, and vegetation type was the largest, reaching approximately 68% (Table ??). The annual precipitation was the dominant factor influencing the NDVI changes in the TRHR, and the interaction between the annual precipitation and other factors could further increase its influence on the NDVI in the TRHR. Among other factors, the q value of the interaction among vegetation type, elevation, and annual precipitation was larger, reaching approximately 60%; although the influence of GDP and aspect on the NDVI was small, their interaction with other factors greatly increased their explanatory power of the NDVI.

Table 7. Interaction detection of factors.

	X1	X2	X3	X4	X5	X6	X7	X8	X9	X10	X11
X1	0.141										
X2	0.176	0.055									
X3	0.473	0.415	0.350								
X4	0.258	0.164	0.461	0.147							
X5	0.468	0.431	0.599	0.468	0.409						
X6	0.265	0.255	0.548	0.356	0.507	0.216					
X7	0.522	0.532	0.611	0.597	0.660	0.535	0.463				
X8	0.583	0.585	0.680	0.610	0.677	0.605	0.679	0.550			
X9	0.227	0.149	0.376	0.247	0.487	0.301	0.487	0.586	0.088		
X10	0.304	0.266	0.485	0.371	0.558	0.352	0.527	0.607	0.352	0.204	
X11	0.324	0.263	0.501	0.324	0.492	0.369	0.573	0.628	0.334	0.360	0.244

3.6. Risk Detection

We used the risk detector to determine the range or types of factors suitable for vegetation growth (Table ??); the suitable range or types of factors is very important for vegetation growth, the larger the NDVI value, the better the vegetation growth. The results of the risk area detection can be applied to the ecological protection project of the TRHR. The suitable range or types of different factors can be combined with the spatial distribution of temperature, precipitation, and population density to increase the vegetation coverage.

Table 8. Suitable range or types of natural factors.

Factors	Suitable Range or Types	NDVI
Slope (°)	>25	0.610
Aspect	North, Northeast, East, West, Northwest	0.484
Elevation (m)	3446–3851	0.743
Soil type	Semi-leached	0.689
Vegetation type	Coniferous forest, broadleaf forest, scrub	0.712
Landform type	Medium undulating mountains	0.601
Annual mean temperature (°C)	1.65–3.82	0.681
Annual precipitation (mm)	578–708	0.770
GDP (10,000 yuan/km^2)	12–37, 104–242	0.609
Population density (people/km^2)	74.95–94.31	0.699
Land use type	Forest land, construction land	0.743

3.6.1. Annual Precipitation

The spatial distribution of vegetation coverage in the TRHR was consistent with the spatial distribution pattern of annual precipitation. The annual precipitation was divided into nine subzones. The mean NDVI value generally increased with the increase in annual precipitation and peaked in the 578 to 708 mm range, thereby indicating that this range promoted vegetation growth. The results showed that the annual precipitation subzone 9 was significantly different from the other subzones, so that the vegetation coverage was best in the 578 to 708 mm annual precipitation range in the TRHR (Table ??). The interaction detector showed that interaction with other factors can further enhance the influence of annual precipitation on the NDVI.

Table 9. Mean value of the NDVI and significant differences in annual precipitation between two regions.

Zones	1	2	3	4	5	6	7	8	9
1									
2	Y								
3	Y	Y							
4	Y	Y	Y						
5	Y	Y	Y	Y					
6	Y	Y	Y	Y	Y				
7	Y	Y	Y	Y	Y	Y			
8	Y	Y	Y	Y	Y	Y	N		
9	Y	Y	Y	Y	Y	Y	Y	Y	
NDVI	0.310	0.282	0.329	0.462	0.572	0.664	0.696	0.691	0.771

Note: Y indicates a significant difference in the influence of two regions on vegetation NDVI, while N indicates no significant difference (confidence level is 95%); numbers 1–9 indicate (unit: mm) 184–278, 278–314, 314–345, 345–378, 378–414, 414–454, 454–506, 506–578, and 578–708, respectively.

3.6.2. Annual Mean Temperature

The factor detector showed that the annual mean temperature also had an important influence on the NDVI in the TRHR. The annual mean temperature was divided into nine subzones. The mean NDVI value increased and then decreased with the increase in the annual mean temperature, and peaked in the 1.65 °C to 3.82 °C range. There were significant differences between the mean NDVI values in subzone 6 and other subzones (Table ??). The interaction of annual mean temperature with other factors enhanced the effect of the former on the NDVI. Temperature changes can cause changes in other factors in the region, and within the temperature range suitable for vegetation growth, the higher the temperature, the better the vegetation coverage, beyond which vegetation growth will be inhibited.

Table 10. Mean NDVI value and significant differences in the annual average temperature between two regions.

Zones	1	2	3	4	5	6	7	8	9
1									
2	N								
3	Y	Y							
4	Y	Y	Y						
5	Y	Y	Y	Y					
6	Y	Y	Y	Y	Y				
7	Y	Y	Y	Y	Y	Y			
8	Y	Y	Y	N	Y	Y	Y		
9	Y	Y	Y	Y	Y	Y	Y	Y	
NDVI	0.260	0.275	0.294	0.482	0.598	0.682	0.659	0.497	0.553

Note: Y and N same as Table ??; numbers 1–9 indicate (unit: °C) −4.00−−2.49, −2.49−−1.57, −1.57−−0.63, −0.63−0.49, 0.49−1.65, 1.65−2.82, 2.82−4.22, 4.22−6.02, and 6.02−9.52 respectively.

3.6.3. Vegetation Type

The vegetation type had an important influence on the NDVI of the TRHR, and the interaction with other factors further enhanced its influence on the NDVI. Vegetation types were divided into nine subzones. The mean NDVI values peaked in the coniferous forest vegetation type. There was no significant difference among the mean NDVI values in vegetation type subzones 2, 3, and 4. There were significant differences between the coniferous forest vegetation type and other vegetation type subzones; the coniferous forest, broadleaf forest, and scrub vegetation covers were better (Table ??).

Table 11. Mean NDVI value and significant differences in terms of vegetation types between two regions.

Zones	1	2	3	4	5	6	7	8	9
1									
2	Y								
3	Y	N							
4	Y	N	N						
5	Y	Y	Y	Y					
6	Y	Y	Y	Y	Y				
7	Y	Y	N	Y	Y	Y			
8	Y	Y	N	Y	Y	Y	Y		
9	Y	Y	N	Y	Y	Y	N	N	
NDVI	0.159	0.714	0.583	0.702	0.229	0.270	0.544	0.368	0.466

Note: Y and N same as Table ??; numbers 1–9 indicate other, coniferous forest, broadleaf forest, scrub, desert, grassland, meadow, alpine vegetation, and cultivated vegetation, respectively.

3.6.4. Elevation

Elevation affects the spatial distribution of natural elements and human activity. The elevation was divided into nine subzones. The mean NDVI value increased and then decreased with the elevation of the TRHR, and it was better in the 3446 to 3851 m range. There were significant differences between this elevation range and other elevation subzones (Table ??). At elevations higher than 3851 m, the NDVI decreased as the elevation increased.

Table 12. Mean NDVI values and significant differences between two regions in terms of elevation.

Zones	1	2	3	4	5	6	7	8	9
1									
2	Y								
3	Y	Y							
4	Y	Y	Y						
5	Y	Y	Y	Y					
6	Y	Y	Y	Y	Y				
7	Y	N	Y	Y	Y	Y			
8	N	Y	Y	Y	Y	Y	Y		
9	Y	Y	Y	Y	Y	Y	Y	Y	
NDVI	0.297	0.418	0.744	0.704	0.550	0.457	0.396	0.303	0.253

Note: Y and N same as Table ??; numbers 1–9 indicate (unit: m) 1950–2979, 2979–3446, 3446–3851, 3851–4177, 4177–4436, 4436–4665, 4665–4895, 4895–5183, and 5183–6826, respectively.

3.6.5. Land Use Type

The land use types were divided into six subzones. The NDVI value peaked in construction land, with no significant difference from the value obtained in forest land, and with significant differences from other land use types; therefore, construction land and forest land had the best vegetation coverage. The main land use type in the TRHR was grassland, accounting for 68%, of which low-coverage grassland accounts for 38%, followed by unused land, water area and forest land, which accounted for 23%, 5%, and 4%, respectively, of the total area. The cropland, forest land, middle-coverage grassland, and low-coverage grassland areas in the TRHR decreased from 1980 to 2015, while the high-coverage grassland, water area, construction land, and unused land areas increased, with the low-coverage grassland area decreasing the most and the high-coverage grassland and unused land area increasing the most (Table ??). Both the forest and construction lands were small, but both were distributed in the middle–high and high vegetation coverage areas east and south of the study area, with better hydrothermal conditions; the construction land was affected by human activities and had more green vegetation, thus the NDVI values were higher there.

3.6.6. Synergistic Effects of Other Factors

The factor detector demonstrated that the single factors of landform type, soil type, slope, aspect, GDP, and population density had small effects on NDVI changes in the TRHR, but the interactions of these factors with others could enhance the effects on NDVI changes.

The landform types of the TRHR were diverse and affected the distribution of vegetation. The landform types were divided into six subzones. The mean NDVI values peaked in the medium-undulating mountains; there were significant differences between this and other landform types, thereby indicating that the vegetation coverage in the medium-undulating mountains was the best. The soil types were divided into 10 subzones. The mean NDVI value peaked in semi-leached soil; there were significant differences between the mean NDVI value in this soil type and other soil types. Therefore, semi-leached soil had the best vegetation coverage. Different slopes and aspects led to differences in climatic elements, and suitable slopes and aspects were conducive to vegetation growth. The slope was divided into seven subzones. The mean NDVI value increased and then decreased with slope increases, and peaked in the 35° to 45° range. The vegetation of this slope consisted mainly of scrubs and alpine meadows. There were no significant differences between the mean NDVI value of this slope and those of slope subzones 6, 5, and 7, while there were significant differences with other subzones. Therefore, the vegetation growth conditions were better in the slope range of >25°. As shown by the q value (Table ??), aspect had a minimal effect on the NDVI. The aspect was divided into nine subzones. The mean NDVI value fluctuated little with aspect changes. The NDVI value of the eastern slope was the largest, with no significant NDVI differences between this and aspect subzones 2, 3, 8, and 9, and significant differences with the other aspect subzones. Therefore, the vegetation coverage of the northern, northeastern, eastern, western, and northwestern aspects was the best.

Among the anthropogenic factors, both GDP and population density had little influence. The GDP was divided into seven subzones. The NDVI value peaked at a GDP of 12×10^4–37×10^4 yuan/km^2 and was not significantly different from that of the area with a GDP of 1.04×10^4–2.42×10^4 yuan/km^2; therefore, the vegetation growth was good in both of these areas. The population density was divided into seven subzones. The largest NDVI value was observed in the area with a population density of 74.95–94.31 people/km^2, with no significant differences with the area with a population density of 8.43–19.37 people/km^2; therefore, the vegetation coverage was optimal in both areas. The area with a population density in the 74.95–94.31 people/km^2 range was very small, accounting for only 0.02% of the total area, which may have led to inaccurate results. If this area is not considered in the analysis, the NDVI will increase and then decrease with increasing population density, with larger values in the range of 8.43 to 19.37 people/km^2, this result would be more accurate.

4. Discussion

Global warming over the last few decades has led to changes in the regional environment. Under the influence of climate change and human activities, vegetation green has generally increased in China [?]; the NDVI has shown an increasing trend in northern China over the past 40 years [?]; the Qinghai–Tibet Plateau tends to become warm and wet, and the vegetation status has gradually improved [?]. This study showed that the NDVI of vegetation in the TRHR also showed an increasing trend from 1982 to 2015, which is consistent with the trend of the NDVI change in China and Qinghai–Tibet Plateau during this period.

In this study, four geographical detectors were used to quantify the main drivers of the NDVI in the TRHR and the interaction of the factors. In the following sections, we will discuss the effects of natural and anthropogenic factors separately.

4.1. Effects of Natural Factors

The Qinghai–Tibet Plateau is a sensitive area for climate change in China [?]. This study indicated that climate factors were the main drivers of the NDVI changes in the TRHR, which is consistent with the findings of Chen [?]. The factor detector showed that the q value of annual precipitation was the largest and was the dominant factor influencing NDVI changes in the TRHR, which is consistent with the findings of Zheng [?] and Xiong [?]. In contrast, Xu [?] and Zhu [?] considered temperature as the dominant factor influencing NDVI variation in the TRHR; the differences in the results may be attributed to the different time scales of the study or the different spatial resolutions of the NDVI used. The warming trend in the TRHR was greater than the Chinese, as well as global average during 1982–2015, and precipitation was lower compared to the global [?]. Extreme temperature increases, and extreme precipitation is relatively stable. The rapid increase in temperature and slow increase in precipitation in the TRHR has led to regional warming and drought [?], while studies have shown that precipitation is the main factor affecting changes in vegetation NDVI in arid and semi-arid alpine meadow and alpine grassland regions [?]. The M-K test showed that the annual precipitation in the TRHR changed abruptly in 2004 and 2006 (Figure ??), and extreme drought events occurred frequently. In 2006, the TRHR suffered an extreme drought, and the growth of forage grasses was disrupted and the grassland ecosystem was damaged [?], resulting in a decrease in NDVI. Precipitation increased abruptly around 2007 [?], since the NDVI has a lag effect on precipitation [?], the NDVI increased abruptly from 2008 onwards. The influence of extreme precipitation events on NDVI in the Qinghai–Tibet Plateau region is more pronounced than that of extreme temperature events, indicating that vegetation is more sensitive to changes in precipitation. Extreme wetness would offset the negative effects caused by extreme drought, and extreme high temperature events occurring in May would stimulate vegetation growth, while extreme low temperatures would inhibit vegetation growth [?]. The effects of extreme climatic events on the vegetation of the TRHR need further study. The influence of temperature on the NDVI gradually decreased, while precipitation occupied a more dominant position [?]. The annual precipitation and annual mean temperature of the TRHR decreased from southeast to northwest. The increasing trend of temperature in the TRHR was significantly greater than that of precipitation, can lead to the warming and drying of the TRHR, which will inhibit vegetation growth [?]. Seasonally, precipitation in the TRHR increases in spring and winter, and in summer when the temperature rises, precipitation also increases [?]. The growing season of vegetation in the TRHR is from May to September, and the climate is conducive for vegetation growth. Water resources are closely related to vegetation, and vegetation changes interact with hydrological processes [?]. Changes in temperature and precipitation lead to changes in vegetation patterns, which can alter surface hydrological characteristics, which, in turn, can affect changes in vegetation coverage [?]. The artificial rainfall implemented by the ecological project of the TRHR has restored the vegetation coverage and the increase in precipitation is beneficial to the growth of vegetation, but the excessive precipitation may cause soil erosion [?], which will instead damage vegetation, therefore the artificial rainfall project should be implemented scientifically and consistently to promote the growth of vegetation in the TRHR.

The vegetation types of the TRHR were mainly alpine meadows and alpine grasslands. During 1982–2015, part of the desert vegetation was converted to grassland and meadow vegetation types, increasing the vegetation coverage. Coniferous forests were mainly distributed in the elevation range of 3446 to 3851 m, and natural environmental conditions were more suitable for vegetation growth. The medium-undulating mountains were mainly dominated by meadow and scrub vegetation types, which were distributed in the southern part of the TRHR, with sufficient hydrothermal conditions and relatively suitable elevation, which are favorable for vegetation growth. The soil is the basis for vegetation growth. The fertilizer retention capacity of semi-leached soil is high, and the semi-leached soil of the TRHR is mainly distributed in mountainous areas, which are favorable for vegetation growth. In this study, the influence of soil type on vegetation change was small, but the interaction with other factors could enhance this influence; for example, the interaction of soil type with temperature and precipitation had a higher influence on the NDVI than did soil by itself. Soil temperature has an important effect on vegetation growth [?].

Topography affects vegetation distribution by changing water and heat conditions [?]. According to Chen [?], the 3500 to 3800 m elevation range is relatively low and precipitation and temperature conditions are good, thus the NDVI value is the largest in this elevation range. In elevations higher than 3800 m, the natural conditions become worse as the elevation increases, thus the NDVI value decreases as well. Slope affects vegetation growth by changing surface runoff, and vegetation coverage generally decreases with increasing slope. However, in this study, the gentle slope was more influenced by human activities; the vegetation coverage was low, while, with increasing slope, human influence decreased and vegetation coverage was relatively high. Aspect affects light intensity, which, in turn, changes the hydrothermal conditions for vegetation growth. The sunny slope has strong light, less soil water content, less nutrient accumulation, and lower vegetation coverage, while the shady, semi-shady, and semi-sunny slopes have sufficient soil water and high nutrient content [?], which are suitable for vegetation growth.

4.2. Effects of Anthropogenic Factors

According to the results of factor detection, anthropogenic factors had little influence on the NDVI. However, the combination of anthropogenic with natural factors can increase the impact. The population density in the TRHR was relatively small, and economic development was slow. Land use type had the greatest influence on the NDVI among anthropogenic factors. Low-coverage grassland is mainly located in the northwest, where water resources are scarce and the altitude is high, while high-coverage grassland is mainly located in the southeast where water and heat conditions are better. From 1980 to 2015, the conversion area between unused land and grasslands is large, and most unused land is converted into low-cover grasslands, but overall the increase in the area of unused land is greater than the decrease. Due to increase in population, land for construction has expanded. Ecological protection projects have increased the area of waters and lakes and improved the condition of wetlands. Before 2000, overgrazing led to the degradation of grassland; therefore, although the grassland area was large in the TRHR, the NDVI value was low. The implementation of the Sanjiangyuan Ecological Project in 2005 resulted in the slight recovery of the grassland, but the effects were short-term [?]. The areas with the highest NDVI values under the influence of GDP and population density were all located in the northeastern part of the TRHR, which is relatively densely populated, vegetation is affected by human activities, and the population is usually distributed in areas with better vegetation coverage [?] which have good survival conditions. Such natural conditions are also suitable for the growth of vegetation, but the increase in population will also cause some damage to vegetation, and the NDVI of vegetation will decrease beyond a certain range of population numbers.

The main conclusion of this study is that, compared with natural factors, anthropogenic factors had less influence on the NDVI of the TRHR. Natural factors, especially climatic factors, dominated the changes in the NDVI in the vegetation of the TRHR. In the context of global climate change, climatic factors have a strong association with vegetation change [?]. This is also verified by this study. The TRHR is at a high altitude, the population is sparse and the area of cultivated land is small. The impact of human farming activities is small, and although there is a certain degree of grazing, the impact is minimal relative to the climate, so the study area in this paper is basically equivalent to an undisturbed area. Therefore, the impact of human activities on the vegetation of the TRHR is very limited. As the impact of anthropogenic factors is short-lived, ecological engineering needs to be implemented continuously. Effective interventions for the restoration of vegetation in the TRHR can be based on the appropriate range or types of factors or a combination of factors. Separating natural factors from anthropogenic factors and quantitatively studying the influence of factors on vegetation is important for the ecological protection and sustainable development of the TRHR, as well as the middle and lower reaches of the region.

4.3. Effectiveness, Limitations, and Future Directions

To the best of our knowledge, this study is the first to use a geographical detector to quantify the effects of natural and anthropogenic factors on vegetation activity and effectively distinguish between the effects of natural and anthropogenic factors on the NDVI in the TRHR. The natural environment of the TRHR is complex, diverse, and spatially heterogeneous. Previous studies on vegetation drivers have used correlation analysis, which assumes a linear relationship between the NDVI and drivers, whereas correlation studies have shown a nonlinear relationship [?]; in contrast to traditional methods, the geographical detector can quantify the non-linear effects of factors and their interactions on vegetation change, making it well suited for this study. We also made the selection of factors with reference to existing studies, and the factors selected in this paper have been shown to be effective many times [? ? ? ? ?], so that the factors selected can be non-independent and the geographical detector method selected for this study allows the analysis of interactions between factors that have been neglected by traditional methods. However, the independent variable input to the geographical detector consists of type quantities, thus the numerical quantities must be classified. This study was based on the natural break method of classifying independent variables, which has been applied before and proven to be effective [?]; different methods of classification can affect the results. To ensure the length and completeness of the time series, NDVI data with a spatial resolution of 8 km were used in this study, which may have had some influence on the results owing to the low data resolution. Although NDVI is currently considered to be the most effective indicator for detecting vegetation change, it has shortcomings, such as the NDVI can reach saturation in dense vegetation canopies, which may lead to inaccurate trends in areas of dense biomass, and the effect on soil background in low vegetation coverage areas is not addressed [?], which were did not consider in this paper. Additionally, the different time ranges of the selected data may lead to some differences in the results, for example, if the growing season data are selected for analysis, the spatial and temporal distribution pattern of the NDVI in the growing season is basically the same as that of the whole year, the influence results are opposite to the annual data, and the influence of temperature (0.458) is slightly greater than the influence of precipitation (0.448). Although there are some differences in the results, the influence of climate factor is still the largest and is the dominant factor of vegetation coverage change in the TRHR, and this main result is unchanged. Therefore, for further research, data resolution should be further improved, while classification methods also need further improvement. To obtain a more accurate result, future studies could use the improved enhanced vegetation index (EVI) for comparison. The effect of growing season climate change on vegetation NDVI also needs further study.

5. Conclusions

In this study, which was based on GIMMS-NDVI data from 1982 to 2015 and 11 detection factors from the same period, we analyzed the spatial and temporal variation characteristics of the NDVI in the TRHR using linear regression analysis, the Mann-Kendall test, and the moving *t*-test. We also analyzed its spatial heterogeneity and driving factors using a geographical detector, and determined the appropriate range or types of factors suitable for vegetation growth. The main conclusions of the study are as follows:

(1) The NDVI distribution of the TRHR was high in the southeast and low in the northwest; the change had an increasing trend in the west and north and a decreasing trend in the center and south. The annual mean value of the NDVI from 1982 to 2015 generally followed a slow increasing trend with a growth rate of 0.002/10 a; regions with low and high vegetation coverage decreased, while other regions increased. The NDVI increased

abruptly in 2008. Overall condition of the TRHR has been improving, but vegetation coverage remains poor.

(2) The magnitude of the influence of each factor on the NDVI was as follows: annual precipitation > annual mean temperature > vegetation type > elevation > land use type > landform type > population density > soil type > GDP > aspect. Among them, annual precipitation had an explanatory power of more than 50% and was the dominant factor influencing NDVI changes in the TRHR. The annual mean temperature, vegetation type, and elevation had an explanatory power of more than 30% and also explained NDVI changes well. Land use type, landform type, and population density had an explanatory power of more than 20%, while other factors had less explanatory power. Compared with the natural factors, the influence of anthropogenic factors on the NDVI of vegetation in the TRHR was smaller. Climatic factors were the main drivers of NDVI changes in the TRHR.

(3) Interactions of bivariate and non-linear enhancements among the NDVI factors were observed, and there were no factors with weakening and independent effects. The interactions of annual precipitation, elevation, mean annual temperature, and vegetation type enhanced the influence of the factors to the greatest extent. Although factors such as the GDP and aspect had small influence on the NDVI, their interaction with other factors greatly increased their explanatory power on the NDVI.

(4) We analyzed the NDVI changes in the TRHR from 1982 to 2015, revealed the natural and anthropogenic factors driving NDVI changes, and determined the appropriate range or types of factors, which is important for ecological conservation and the sustainable development of the TRHR.

Author Contributions: Conceptualization, S.G.; Methodology, S.G.; software, S.G.; validation, T.N., formal analysis, T.N.; investigation, S.G.; resources, G.D.; data curation, S.G.; writing—original draft preparation, S.G.; writing—review and editing, G.D., H.Y., X.G.; visualization, G.D.; supervision, G.D.; project administration, G.D., X.J.; funding acquisition, G.D., X.J. All authors have read and agreed to the published version of the manuscript.

Funding: This research was supported by the National Natural Science Fund (51779099, 51779209, 51909099), and the National key research and development plan (2016YFC0402400).

Data Availability Statement: The data presented in this study are available on request from the corresponding author. The data are not publicly available due to privacy.

Acknowledgments: We sincerely thank the editor and anonymous reviewers for their valuable comments and suggestions to improve the quality of this paper.

Conflicts of Interest: The authors declare no conflict of interest.

Appendix A

Table A1. Land use transfer matrix (km^2).

Area/km^2	Cropland	Forest Land	High-Coverage Grassland	Middle-Coverage Grassland	Low-Coverage Grassland	Water Area	Construction Land	Unused Land	2015 Total
Cropland	2010.75	0.00	0.00	232.85	2.41	0.00	0.00	0.00	2246.01
Forest land	0.00	14,805.10	0.00	20.44	30.61	0.00	0.00	0.00	14,856.15
High-coverage grassland	78.42	0.00	20,236.10	254.87	156.84	0.00	0.00	80.83	20,807.07
Middle-coverage grassland	0.00	71.34	78.42	93,372.20	1250.00	0.28	0.00	160.96	94,933.20
Low-coverage grassland	78.42	80.83	19.61	663.11	141,923.00	225.49	0.00	182.09	143,172.54
Water area	313.68	0.00	0.00	50.22	8.78	16,702.20	0.00	289.61	17,364.50
Construction land	78.42	0.00	0.00	156.84	0.00	0.00	78.42	0.00	313.68
Unused land	0.00	0.00	30.61	346.55	738.81	414.77	0.00	85,102.60	86,633.34
1980 Total	2559.70	14,957.27	20,364.74	95,097.07	144,110.46	17,342.74	78.42	85,816.09	
Area change	−313.69	−101.12	442.33	−163.87	−937.92	21.76	235.26	817.25	

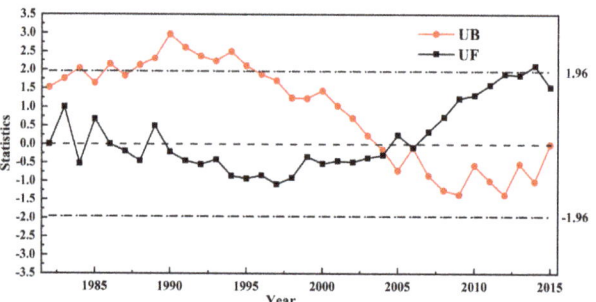

Figure 1. M-K test for annual precipitation.

References

1. Zhao, L.; Dai, A.G.; Dong, B. Changes in global vegetation activity and its driving factors during 1982–2013. *Agric. For. Meteorol.* **2018**, *249*, 198–209. [CrossRef]
2. Pettorelli, N.; Vik, J.O.; Mysterud, A.; Gaillard, J.M.; Tucker, C.J.; Stenseth, N.C. Using the satellite-derived NDVI to assess ecological responses to environmental change. *Trends Ecol. Evol.* **2005**, *20*, 503. [CrossRef] [PubMed]
3. Fensholt, R.; Langanke, T.; Rasmussen, K.; Reenberg, A.; Prince, S.D.; Tucker, C.; Scholes, R.J.; Le, Q.B.; Bondeau, A.; Eastman, R.; et al. Greenness in semi-arid areas across the globe 1981–2007—An Earth Observing Satellite based analysis of trends and drivers. *Remote Sens. Environ.* **2012**, *121*, 144–158. [CrossRef]
4. Liu, Y.; Li, Y.; Li, S.H.; Motesharrei, S. Spatial and Temporal Patterns of Global NDVI Trends: Correlations with Climate and Human Factors. *Remote Sens.* **2015**, *7*, 13233–13250. [CrossRef]
5. Jiang, L.L.; Jiapaer, G.; Bao, A.M.; Guo, H.; Ndayisaba, F. Vegetation dynamics and responses to climate change and human activities in Central Asia. *Sci. Total Environ.* **2017**, 599–600. [CrossRef]
6. Gong, Z.N.; Zhao, S.Y.; Gu, J.Z. Correlation analysis between vegetation coverage and climate drought conditions in North China during 2001–2013. *J. Geogr. Sci.* **2017**, *27*, 143–160. [CrossRef]
7. Hu, Y.F.; Dao, R.N.; Hu, Y. Vegetation Change and Driving Factors: Contribution Analysis in the Loess Plateau of China during 2000–2015. *Sustainability* **2019**, *11*, 1320. [CrossRef]
8. Huang, K.; Zhang, Y.J.; Zhu, J.T.; Liu, Y.J.; Zu, J.X.; Zhang, J. The Influences of Climate Change and Human Activities on Vegetation Dynamics in the Qinghai-Tibet Plateau. *Remote Sens.* **2016**, *8*, 876. [CrossRef]
9. Qu, S.; Wang, L.C.; Lin, A.W.; Zhu, H.J.; Yuan, M.X. What drives the vegetation restoration in Yangtze River basin, China: Climate change or anthropogenic factors. *Scientidicators* **2018**, *90*, 438–450. [CrossRef]
10. Chu, H.S.; Venevsky, S.; Wu, C.; Wang, M.H. NDVI-based vegetation dynamics and its response to climate changes at Amur-Heilongjiang River Basin from 1982 to 2015. *Sci. Total Environ.* **2018**, *650 Pt 2*, 2051–2062. [CrossRef]
11. Wang, J.F.; Xu, C.D. Geodetector: Principle and prospective. *Acta Geogr. Sin.* **2017**, *72*, 116–134. (In Chinese)
12. Zhao, W.; Hu, Z.M.; Guo, Q.; Wu, G.N.; Chen, R.R.; Li, S.G. Contributions of Climatic Factors to Interannual Variability of the Vegetation Index in Northern China Grasslands. *J. Clim.* **2020**, *33*, 175–183. [CrossRef]
13. Yuan, L.H.; Chen, X.Q.; Wang, X.Y.; Xiong, Z.; Song, C.Q. Spatial associations between NDVI and environmental factors in the Heihe River Basin. *J. Geogr. Sci.* **2019**, *29*, 1548–1564. [CrossRef]
14. Zhu, L.J.; Meng, J.J.; Zhu, L.K. Applying Geodetector to disentangle the contributions of natural and anthropogenic factors to NDVI variations in the middle reaches of the Heihe River Basin. *Ecol. Indic.* **2020**, *117*, 106545. [CrossRef]
15. Ran, Q.W.; Hao, Y.B.; Xia, A.Q.; Liu, W.J.; Hu, R.H.; Cui, X.Y.; Xue, K.; Song, X.N.; Xu, C.; Ding, B.Y.; et al. Quantitative Assessment of the Impact of Physical and Anthropogenic Factors on Vegetation Spatial-Temporal Variation in Northern Tibet. *Remote Sens.* **2019**, *11*, 1183. [CrossRef]
16. Zhang, Y.; Zhang, K.C.; An, Z.S.; Yu, Y.P. Quantification of driving factors on NDVI in oasis-desert ecotone using geographical detector method. *J. Mt. Sci.* **2019**, *16*, 2615–2624. [CrossRef]
17. Liu, Y.X.; Liu, S.L.; Sun, Y.X.; Li, M.Q.; An, Y.; Shi, F.N. Spatial differentiation of the NPP and NDVI and its influencing factors vary with grassland type on the Qinghai-Tibet Plateau. *Environ. Monit. Assess.* **2021**, *193*, 48. [CrossRef]
18. Liu, J.Y.; Xu, X.L.; Shao, Q.Q. Grassland degradation in the "Three-River Headwaters" region, Qinghai Province. *J. Geogr. Sci.* **2008**, *18*, 259–273. [CrossRef]
19. Wang, R.; Dong, Z.B.; Zhou, Z.C. Changes in the depths of seasonal freezing and thawing and their effects on vegetation in the Three-River Headwater Region of the Tibetan Plateau. *J. Mt. Sci.* **2019**, *16*, 2810–2827. [CrossRef]
20. Bai, Y.F.; Guo, C.C.; Degen, A.A.; Ahmad, A.A.; Wang, W.Y.; Zhang, T.; Li, W.Y.; Ma, L.; Huang, M.; Zeng, H.J.; et al. Climate warming benefits alpine vegetation growth in Three-River Headwater Region, China. *Sci. Total Environ.* **2020**, *742*, 140574. [CrossRef]

21. Pang, G.J.; Wang, X.J.; Yang, M.X. Using the NDVI to identify variations in, and responses of, vegetation to climate change on the Tibetan Plateau from 1982 to 2012. *Quat. Int.* **2016**, *444 Pt A*, 87–96. [CrossRef]
22. Jiang, C.; Zhang, L.B. Climate Change and Its Impact on the Eco-Environment of the Three-Rivers Headwater Region on the Tibetan Plateau, China. *Int. J. Environ. Res. Public Health* **2015**, *12*, 12057–12081. [CrossRef] [PubMed]
23. Li, F.X.; Li, X.D.; Zhou, B.R.; Qi, D.L.; Wang, L.; Fu, H. Effects of grazing intensity on biomass and soil physical and chemical characteristics in alpine meadow in the source of three rivers. *Pratacultural Sci.* **2015**, *32*, 11–18. (In Chinese)
24. Pinzon, J.E.; Tucker, C.J. A non-stationary 1981–2012 AVHRR NDVI3g time series. *Remote Sens.* **2014**, *6*, 6929–6960. [CrossRef]
25. National Oceanic and Atmospheric Administration. GIMMS NDVI3g dataset for Sanjiangyuan (1982–2015). *A Big Earth Data Platf. Three Poles* **2018**. CSTR: 18406.11.Ecolo.tpdc.271224. [CrossRef]
26. Holben, B.N. Characteristics of maximum-value composite images from temporal AVHRR data. *Int. J. Remote Sens.* **1986**, *7*, 1417–1434. [CrossRef]
27. Wei, Y.Q. Establishing Developing and Applying of the Space-Air-Field Integrated Eco-Monitoring and Data Infrastructure of the Three-River-Source National Park. The boundaries of the source regions in Sanjiangyuan region (2018). *A Big Earth Data Platf. Three Poles* **2018**. CSTR: 18406.11.Geogra.tpdc.270009. [CrossRef]
28. Feng, J.M.; Dong, B.Q.; Qin, T.L.; Liu, S.S.; Zhang, J.W.; Gong, X.F. Temporal and Spatial Variation Characteristics of NDVI and Its Relationship with Environmental Factors in Huangshui River Basin from 2000 to 2018. *Pol. J. Environ. Stud.* **2021**, *30*, 3043–3063. [CrossRef]
29. Peng, W.F.; Kuang, T.T.; Tao, S. Quantifying influences of natural factors on vegetation NDVI changes based on geographical detector in Sichuan, western China. *J. Clean. Prod.* **2019**, *233*, 353–367. [CrossRef]
30. Nie, T.; Dong, G.T.; Jiang, X.H.; Lei, Y.X. Spatio-Temporal Changes and Driving Forces of Vegetation Coverage on the Loess Plateau of Northern Shaanxi. *Remote Sens.* **2021**, *13*, 613. [CrossRef]
31. Liu, Y.S.; Li, J.T. Geographic detection and optimizing decision of the differentiation mechanism of rural poverty in China. *Acta Geogr. Sin.* **2017**, *72*, 161–173. (In Chinese)
32. Jiao, K.W.; Gao, J.B.; Wu, S.H. Climatic determinants impacting the distribution of greenness in China: Regional differentiation and spatial variability. *Int. J. Biometeorol.* **2019**, *63*, 523–533. [CrossRef]
33. Feng, X.Q.; Zhang, G.X.; Yin, X.R. Hydrological Responses to Climate Change in Nenjiang River Basin, Northeastern China. *Water Resour. Manag.* **2011**, *25*, 677–689. [CrossRef]
34. Liu, H.Y.; Jiao, F.S.; Yin, J.Q.; Li, T.Y.; Gong, H.B.; Wang, Z.Y.; Lin, Z.S. Nonlinear relationship of vegetation greening with nature and human factors and its forecast—A case study of Southwest China. *Ecol. Indic.* **2020**, *111*, 106009. [CrossRef]
35. Du, R.S.; Shang, F.H.; Ma, N. Automatic mutation feature identification from well logging curves based on sliding t test algorithm. *Clust. Comput.* **2019**, *22*, 14193–14200. [CrossRef]
36. Zhai, X.H.; Liang, X.L.; Yan, C.Z.; Xing, X.G.; Jia, H.W.; Wei, X.X.; Feng, K. Vegetation Dynamic Changes and Their Response to Ecological Engineering in the Sanjiangyuan Region of China. *Remote Sens.* **2020**, *12*, 4035. [CrossRef]
37. Rao, P.Z.; Wang, Y.C.; Wang, F. Analysis on the NDVI Change and Influence Factors of Vegetation Cover in the Three-River Headwaters Region. *Acta Agrestia Sin.* **2021**, *29*, 572–582. (In Chinese)
38. Wang, H.; Yan, S.J.; Liang, Z.; Jiao, K.W.; Li, D.L.; Wei, F.L.; Li, S.C. Strength of association between vegetation greenness and its drivers across China between 1982 and 2015: Regional differences and temporal variations. *Ecol. Indic.* **2021**, *128*, 107831. [CrossRef]
39. Lin, X.N.; Niu, J.Z.; Berndtsson, R.; Yu, X.X.; Zhang, L.; Chen, X.W. NDVI Dynamics and Its Response to Climate Change and Reforestation in Northern China. *Remote Sens.* **2020**, *12*, 4138. [CrossRef]
40. Piao, S.L.; Cui, M.D.; Chen, A.P.; Wang, X.H.; Ciais, P.; Liu, J.; Tang, Y.H. Altitude and temperature dependence of change in the spring vegetation green-up date from 1982 to 2006 in the Qinghai-Xizang Plateau. *Agric. For. Meteorol.* **2011**, *151*, 1599–1608. [CrossRef]
41. Li, W.H. An Overview of Ecological Research Conducted on the Qinghai-Tibetan Plateau. *J. Resour. Ecol.* **2017**, *8*, 1–4.
42. Chen, C.; Li, T.J.; Sivakumar, B.; Li, J.Y.; Wang, G.Q. Attribution of growing season vegetation activity to climate change and human activities in the Three-River Headwaters Region, China. *J. Hydroinform.* **2020**, *22*, 186–204. [CrossRef]
43. Zheng, Y.T.; Han, J.C.; Huang, Y.F.; Fassnacht, S.R.; Xie, S.; Lv, E.Z.; Chen, M. Vegetation response to climate conditions based on NDVI simulations using stepwise cluster analysis for the Three-River Headwaters region of China. *Ecol. Indic.* **2018**, *92*, 18–29. [CrossRef]
44. Xiong, Q.L.; Xiao, Y.; Halmy, M.W.; Dakhil, M.A.; Liang, P.H.; Liu, C.G.; Zhang, L.; Pandey, B.; Pan, K.W.; Kafraway, S.B.; et al. Monitoring the impact of climate change and human activities on grassland vegetation dynamics in the northeastern Qinghai-Tibet Plateau of China during 2000–2015. *J. Arid Land* **2019**, *11*, 637–651. [CrossRef]
45. Xu, W.X.; Gu, S.; Zhao, X.Q.; Xiao, J.S.; Tang, Y.H.; Fang, J.Y.; Zhang, J.; Jiang, S. High positive correlation between soil temperature and NDVI from 1982 to 2006 in alpine meadow of the Three-River Source Region on the Qinghai-Tibetan Plateau. *Int. J. Appl. Earth Obs. Geoinf.* **2011**, *13*, 528–535. [CrossRef]
46. Zhu, W.H.; Mao, F.; Xu, Y.; Zheng, J.; Song, L.X. Analysis on response of vegetation index to climate change and its prediction in the Three-Rivers-Source region. *Plateau Meteorol.* **2019**, *38*, 693–704. (In Chinese)
47. Bai, X.L.; Wei, J.H.; Xie, H.W. Characteristics of wetness/dryness variation and their influences in the Three-River Headwaters region. *Acta Ecol. Sin.* **2017**, *37*, 8397–8410. (In Chinese)

48. Sun, J.; Cheng, G.W.; Li, W.P.; Sha, Y.K.; Yang, Y.C. On the Variation of NDVI with the Principal Climatic Elements in the Tibetan Plateau. *Remote Sens.* **2013**, *5*, 1894–1911. [CrossRef]
49. Meng, X.H.; Chen, H.; Li, Z.G.; Zhao, L.R.; Lu, S.H.; Deng, M.S.; Liu, Y.M.; Li, G.W. Review of Climate Change and Its Environmental Influence on the Three-River Regions. *Plateau Meteorol.* **2020**, *39*, 1133–1143. (In Chinese)
50. Cao, L.; Pan, S. Changes in precipitation extremes over the "Three-River headwaters" region, hinterland of the Tibetan plateau, during 1960–2012. *Quat. Int.* **2014**, *321*, 105–115. [CrossRef]
51. Zhong, L.; Ma, Y.M.; Salama, M.S.; Su, Z.B. Assessment of vegetation dynamics and their response to variations in precipitation and temperature in the Tibetan Plateau. *Clim. Chang.* **2010**, *103*, 519–535. [CrossRef]
52. Liu, D.; Wang, T.; Yang, T.; Yan, Z.J.; Liu, Y.W.; Zhao, Y.T.; Piao, S.L. Deciphering impacts of climate extremes on Tibetan grasslands in the last fifteen years. *Sci. Bull.* **2019**, *64*, 446–454. [CrossRef]
53. Liu, X.F.; Zhang, J.S.; Zhu, X.F. Spatiotemporal changes in vegetation coverage and its driving factors in the Three-River Headwaters Region during 2000–2011. *J. Geogr. Sci.* **2014**, *24*, 288–302. [CrossRef]
54. Yi, X.S.; Li, G.S.; Yin, Y.Y. Spatio-temporal variation of precipitation in the Three-River Headwater Region from 1961 to 2010. *J. Geogr. Sci.* **2013**, *23*, 447–464. [CrossRef]
55. Wang, H.N.; Lv, X.Z.; Zhang, M.Y. Sensitivity and attribution analysis of vegetation changes on evapotranspiration with the Budyko framework in the Baiyangdian catchment, China. *Ecol. Indic.* **2021**, *120*, 106963. [CrossRef]
56. Kumari, N.; Srivastava, A.; Dumka, U.C. A Long-Term Spatiotemporal Analysis of Vegetation Greenness over the Himalayan Region Using Google Earth Engine. *Climate* **2021**, *9*, 109. [CrossRef]
57. Shao, Q.Q.; Cao, W.; Fan, J.W.; Huang, L.; Xu, X.L. Effects of an ecological conservation and restoration project in the Three-River Source Region, China. *J. Geogr. Sci.* **2017**, *27*, 183–204. [CrossRef]
58. Jiang, C.; Zhang, L.B. Ecosystem change assessment in the Three-river Headwater Region, China: Patterns, causes, and implications. *Ecol. Eng.* **2016**, *93*, 24–36. [CrossRef]
59. Chen, Q.; Zhou, Q.; Zhang, H.F.; Liu, F.G. Spatial disparity of NDVI response in vegetation growing season to climate change in the Three-River Headwaters region. *Ecol. Environ. Sci.* **2010**, *19*, 1284–1289. (In Chinese)
60. Liu, M.X.; Wang, G. Responses of plant community diversity and soil factors to slope aspect in alpine meadow. *Chin. J. Ecol.* **2013**, *32*, 259–265. (In Chinese)
61. Shen, X.J.; An, R.; Feng, L.; Ye, N.; Zhu, L.J.; Li, M.H. Vegetation changes in the Three-River Headwaters Region of the Tibetan Plateau of China. *Ecol. Indic.* **2018**, *93*, 804–812. [CrossRef]
62. Wang, S.H.; Jia, S.F.; Lv, A.F. The relationship be- tween NDVI and residential sites across Three-River-Source area. *Resour. Sci.* **2012**, *34*, 2045–2050. (In Chinese)
63. Hein, L.; de Ridder, N.; Hiernaux, P.; Leemans, R.; de Witd, A.; Schaepman, M. Desertification in the Sahel: Towards better accounting for ecosystem dynamics in the interpretation of remote sensing images. *J. Arid Environ.* **2011**, *75*, 1164–1172. [CrossRef]
64. Meng, X.Y.; Gao, X.; Li, S.Y.; Lei, J.Q. Spatial and Temporal Characteristics of Vegetation NDVI Changes and the Driving Forces in Mongolia during 1982–2015. *Remote Sens.* **2020**, *12*, 603. [CrossRef]
65. Yu, G.; Zhang, X.Y.; Wang, Q.B.; Chen, H.K.; Du, X.Y.; Ma, Y.P. Temporal changes in vegetation around a shale gas development area in a subtropical karst region in southwestern China. *Sci. Total Environ.* **2020**, *701*, 134769.
66. Huete, A.; Didan, K.; Miura, K.; Rodriguez, E.P.; Gao, X.; Ferreira, L.G. Overview of the radiometric and biophysical performance of the MODIS vegetation indices. *Remote Sens. Environ.* **2002**, *83*, 195–213. [CrossRef]

Article

Spatiotemporal Analysis of Sea Ice Leads in the Arctic Ocean Retrieved from IceBridge Laxon Line Data 2012–2018

Dexuan Sha [1], Younghyun Koo [2], Xin Miao [3], Anusha Srirenganathan [1], Hai Lan [1], Shorojit Biswas [3], Qian Liu [1], Alberto M. Mestas-Nuñez [2], Hongjie Xie [2] and Chaowei Yang [1],*

[1] NSF Spatiotemporal Innovation Center, Department of Geography and Geoinformation Science, George Mason University, Fairfax, VA 22030, USA; dsha@gmu.edu (D.S.); asrireng@gmu.edu (A.S.); hlan5@gmu.edu (H.L.); qliu6@gmu.edu (Q.L.)

[2] Center for Advanced Measurements in Extreme Environments, Department of Earth and Planetary Sciences, University of Texas at San Antonio, San Antonio, TX 78249, USA; younghyun.koo@utsa.edu (Y.K.); alberto.mestas@utsa.edu (A.M.M.-N.); hongjie.xie@utsa.edu (H.X.)

[3] Department of Geography, Geology and Planning, Missouri State University, Springfield, MO 65897, USA; xinmiao@missouristate.edu (X.M.); sb6835s@MissouriState.edu (S.B.)

* Correspondence: cyang3@gmu.edu

Citation: Sha, D.; Koo, Y.; Miao, X.; Srirenganathan, A.; Lan, H.; Biswas, S.; Liu, Q.; Mestas-Nuñez, A.M.; Xie, H.; Yang, C. Spatiotemporal Analysis of Sea Ice Leads in the Arctic Ocean Retrieved from IceBridge Laxon Line Data 2012–2018. *Remote Sens.* **2021**, *13*, 4177. https://doi.org/10.3390/rs13204177

Academic Editors: Baojie He, Ayyoob Sharifi, Chi Feng, Jun Yang and Yi Luo

Received: 7 August 2021
Accepted: 13 October 2021
Published: 19 October 2021

Publisher's Note: MDPI stays neutral with regard to jurisdictional claims in published maps and institutional affiliations.

Copyright: © 2021 by the authors. Licensee MDPI, Basel, Switzerland. This article is an open access article distributed under the terms and conditions of the Creative Commons Attribution (CC BY) license (https://creativecommons.org/licenses/by/4.0/).

Abstract: The ocean and atmosphere exert stresses on sea ice that create elongated cracks and leads which dominate the vertical exchange of energy, especially in cold seasons, despite covering only a small fraction of the surface. Motivated by the need of a spatiotemporal analysis of sea ice lead distribution, a practical workflow was developed to classify the high spatial resolution aerial images DMS (Digital Mapping System) along the Laxon Line in the NASA IceBridge Mission. Four sea ice types (thick ice, thin ice, open water, and shadow) were identified, and relevant sea ice lead parameters were derived for the period of 2012–2018. The spatiotemporal variations of lead fraction along the Laxon Line were verified by ATM (Airborne Topographic Mapper) surface height data and correlated with coarse spatial resolution sea ice motion, air temperature, and wind data through multiple regression models. We found that the freeboard data derived from sea ice leads were compatible with other products. The temperature and ice motion vorticity were the leading factors of the formation of sea ice leads, followed by wind vorticity and kinetic moments of ice motion.

Keywords: sea ice classification; ice motion vorticity; multiple linear regression; wind; temperature

1. Introduction

Arctic sea ice functions as a sensitive indicator of global warming because sea ice responds to even a small increase in temperature [1–3]. On the other hand, Arctic sea ice is also an important driver of climate change, and it plays an important role in the Earth's solar radiation budget. This is due to how sea ice has a significantly higher albedo compared to that of the water surface. Therefore, when the Arctic sea ice starts to melt, the oceans absorb more solar radiation and warm up, accelerating the melting of sea ice in a positive feedback [4].

Among all types of sea ice features, leads have unique characteristics. A lead is an elongated crack in the sea ice developed by the divergence or shear of floating ice floes when moving with currents and winds [5]. Leads vary in width from meters to hundreds of meters depending on their development and the directions of surrounding pressure and tension. Since a lead is physically an open water body, thin ice, or mixed open water and thin ice within (thicker) sea ice floe or between sea ice floes, it allows the direct interaction between the atmosphere and the ocean and is the only (or major) channel in the cold Arctic. Thus, leads play an important role in the local radiation energy budget, ship navigation, and the Arctic sea ice ecosystem [6]. In particular, they dominate the vertical exchange of energy during winter when turbulent heat fluxes over leads can be orders of magnitude larger than that over thick ice. The width of leads and their orientation markedly influence associated vertical sensible and latent heat fluxes and associated cloud formation [7,8]. Recent studies suggest that these fluxes could influence the atmospheric properties tens to hundreds of kilometers downstream [9–11]. Even a small fraction of thin ice and open water within the sea ice pack can significantly modify the total energy transfer between the ocean and the atmosphere [12]. Furthermore, leads are elusive and inconsistent features. If sea water temperature drops below around −1.8 °C, the open water within a lead quickly refreezes (in a few hours), and leads will be partly or entirely covered by a thin layer of new ice [13–15]. Therefore, leads are an important component of the Arctic surface energy budget, and more quantitative studies are needed to explore and model their impact on the Arctic climate system.

Arctic climate models require a detailed spatial distribution of leads to simulate interactions between the ocean and the atmosphere. Remote sensing techniques can be used to extract sea ice physical features and parameters and calibrate or validate climate models [16]. However, most of the sea ice leads studies focus on low-moderate resolution (~1 km) imagery such as Moderate Resolution Imaging Spectroradiometer (MODIS) or Advanced Very High-Resolution Radiometer (AVHRR) [17–20], which cannot detect small leads, such as those smaller than 100 m. On the other hand, high spatial resolution (HSR) images such as aerial photos are discrete and heterogeneous in space and time, i.e., images usually cover only a small and discontinuous area with time intervals between images varying from a few seconds to several months [21,22]. Therefore, it is difficult to weave these small pieces into a coherent large-scale picture, which is important for coupled sea ice and climate modeling and verification. Onana et al. used operational IceBridge airborne visible DMS (Digital Mapping System) imagery and laser altimetry measurements to detect sea ice leads and classify open water, thin ice (new ice, grease ice, frazil ice, and nilas), and gray ice [23]. Miao et al. utilized an object-based image classification scheme to classify water, ice/snow, melt ponds, and shadow [24]. However, the workflow used in Miao et al. was based on some independent proprietary software, which is not suitable for batch processing in an operational environment. In contrast, Wright and Polashenski developed an Open Source Sea Ice Processing (OSSP) package for detecting sea ice surface features in high-resolution optical imagery [25,26]. Based on the OSSP package, Wright et al. investigated the behavior of meltwater on first-year and multiyear ice during summer melting seasons [26]. Following this approach, Sha et al. further improved and integrated the OSSP modules into an on-demand service in cloud computing-based infrastructure for operational usage [22].

Following the previous studies, this paper focuses on the spatiotemporal analysis of sea ice lead distribution through NASA's Operation IceBridge images, which used a systematic sampling scheme to collect high spatial resolution DMS aerial photos along critical flight lines in the Arctic. A practical workflow was developed to classify the DMS images along the Laxon Line into four classes, i.e., thick ice, thin ice, water, and shadow, and to extract sea ice lead and thin ice during the missions 2012–2018. Finally, the spatiotemporal variations of lead fraction along the Laxon Line were verified by ATM surface height data (freeboard), and correlated with sea ice motion, air temperature, and wind data. The paper is organized as follows: Section 2 provides a background description of DMS imagery, the Laxon Line collection, and auxiliary sea ice data. Section 3 describes the methodology and workflow. Section 4 presents and discusses the spatiotemporal variations of leads. The summary and conclusions are provided in Section 5.

2. Dataset

2.1. IceBridge DMS Images and Study Area

This study uses IceBridge DMS images to detect Arctic sea ice leads along the Laxon Line one day over the course of 7 years in 2012–2018, since these are the longest continuous yearly data available in this Arctic region. The DMS images were collected during the IceBridge sea ice flights using an airborne digital camera. DMS has a high spatial resolution

0.1–2.5 m [27], depending on the aircraft flight height. It has three natural color (red, green, and blue) bands, and each image has a field of view of approximately 400 m by 600 m. The IceBridge campaigns had been designed to survey the Arctic region in March and April since 2009 to partially fill the temporal gap between the ICESat (2003–2009) and ICESat-2 (2018–present) missions.

DMS images are collected, processed, and maintained by the Airborne Sensor Facility located at the NASA AMES Research Center. We downloaded the Level 1B geolocated and orthorectified images for the Arctic Laxon Line in spring from 2012 to 2018 from the NASA National Snow and Ice Data Center Distributed Active Archive Center (NSIDC DAAC) (https://nsidc.org/data/iodms1b) (accessed on 6 August 2021). The Laxon Line starts from the Thule Air Base, Greenland to Fairbanks, AK, USA, transiting across the Arctic Ocean (Figure 1). It passes through both multiyear ice (MYI) regions in the north of the Canadian Archipelago and the first-year ice (FYI) regions in northern Alaska. Thus, sea ice data along this line provides useful insights on the transition of sea ice conditions over the Central Arctic in the spring. Furthermore, the IceBridge mission collected data along this track repeatedly every year from 2012 to 2018, which is appropriate for spatiotemporal analysis of sea ice leads. The overall DMS image collection along the Laxon Line is 106,674 aerial photos (1.54 TB) with an overlap of 60–90% along the track. The photo distribution from 2012–2018 is summarized in Table 1. The overall distance of the Laxon Line is around 3398 km, and the distance for the overlapped track through the years is around 2437 km.

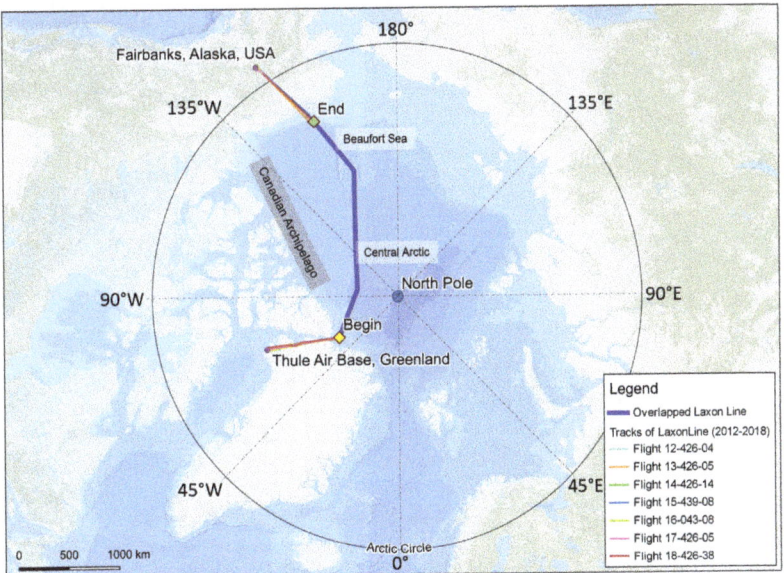

Figure 1. Spatial distribution of the seven tracks along the Laxon Line from 2012 to 2018. The tracks are highly overlapped.

Table 1. The DMS images selected for lead detection along the Laxon Line from 2012 to 2018.

Name	Date	Image #	# Image with Sea Ice Leads	Selected/Original Image Size (GB)	Lighting Condition
Flight 12-426-04	14 March 2012	16,544	1066	14.8/260	Cloudy
Flight 13-426-05	21 March 2013	18,480	993	13.8/290	Normal
Flight 14-426-14	14 March 2014	14,322	492	5.2/150	Cloudy
Flight 15-439-08	26 March 2015	20,038	816	9.3/250	Normal
Flight 16-043-08	20 April 2016	15,205	1069	18.4/270	Normal
Flight 17-426-05	10 March 2017	10,939	659	8.67/93	Cloudy
Flight 18-426-38	6 April 2018	11,146	1040	22.2/240	Normal

2.2. Auxiliary Sea Ice Data

2.2.1. AMSR Data

AMSR (Advanced Microwave Scanning Radiometer) is a passive microwave satellite sensor developed by the Japan Aerospace Exploration Agency. Due to its low spatial resolution, the AMSR data can only be used to examine sea ice concentrations at the regional scale. We collected the AMSR-E/AMSR-2 Unified Level 3 daily brightness temperature and sea ice concentration data which has a spatial resolution of 25 km through NSIDC (Table 2) [28]. The data contain vertically polarized and horizontally polarized brightness temperatures at four frequency channels: 18.7, 23.8, 36.5, and 89.0 GHz. The Arctic sea ice concentration (SIC) was calculated by the NASA Team 2 (NT2) algorithm, which provides <2% of error compared with the high-resolution optical data [29–31]. The collected AMSR data coincides with the days of the IceBridge mission from 2012 to 2018, so that the SIC can be compared with that retrieved from the DMS images. Furthermore, the passive microwave data can be used to calculate thin ice concentration (TIC). Röhrs and Kaleschke used brightness temperatures at the vertically polarized 18.7 and 89.0 GHz to identify water and thin ice (i.e., new ice, nilas, and pancake ice) from thick ice, and the sea ice leads and TIC showed a good agreement with the MODIS, Envisat ASAR, and CryoSat-2 data [14]. In this study, we calculated TIC following the Röhrs and Kaleschke's algorithm. The coarser spatial resolution of 25 km of TIC were compared with our lead and thin ice fractions retrieved from the DMS images.

Table 2. Auxiliary sea ice datasets.

Product Name	Type	Source	Spatial Resolution	Category
AMSR-E/AMSR2 Unified L3 Daily Brightness Temperatures & Sea Ice Concentration	Passive microwave	NSIDC	25 km	Sea Ice
IceBridge Airborne Topographic Mapper (ATM)	Laser altimeter	NSIDC	~1 m footprint (resampled to 2 m grid)	Sea Ice
Global sea ice type	Sea ice type	EUMETSAT OSI SAF	10 km	Sea Ice
Polar Pathfinder Daily EASE-Grid Sea Ice Motion Vectors	Sea ice motion	NSIDC	25 km	Dynamic
ERA5 (air temperature and wind velocity)	Climate reanalysis	European Centre for Medium-Range Weather Forecasts (ECMWF)	0.25°	Dynamic and thermodynamic

2.2.2. ATM Surface Height Data (DMS Level)

Our DMS-based lead detection results can be used to cross-validate sea ice freeboard products derived from IceBridge Airborne Topographic Mapper (ATM) Level 1B data [23]. The ATM is an airborne conically scanning laser altimeter with a wavelength of 532 nm. A laser pulse is emitted from the ATM at a rate of 5 kHz, and it has ~1 m of footprint at a typical 500 m altitude above the surface. Since ATM covers exactly the same location and time with the DMS images with a smaller cross-track width (~400 m), DMS images are usually used as good reference for extracting the ATM-based freeboard data [32,33]. In this study, the ATM data are resampled in a 2 m grid and projected to the same projection system as DMS (NSIDC sea ice polar stereographic North) to match the geographical location. After retrieving thin ice and leads through DMS images, we geographically linked the leads with the ATM data to extract freeboard variations along the Laxon Line, and compare with freeboard data derived from SILDAMS (Sea Ice Lead Detection Algorithm) [23,32].

2.3. Oceanic and Atmospheric Geophysical Parameters

NSIDC provides sea ice motion data (nsidc.org/data/NSIDC-0116) with a spatial resolution of 25 km on the Equal-Area Scalable Earth grid [34]. This sea ice motion vector is derived from multiple data sources, including AVHRR, AMSR-E, SMMR, SSMI, SSMI/S satellite sensors, International Arctic Buoy Program (IABP) buoys, and the National Center for Environmental Prediction/National Center for Atmospheric Research (NCEP/NCAR) reanalysis.

We also acquired a global sea ice type product provided by the European Organization for the Exploitation of Meteorological Satellites (EUMETSAT) Ocean and Sea Ice Satellite Application Facility (OSI SAF, www.osi-saf.org) (6 August 2021) [35]. This product assigns different sea ice types, such as multiyear ice (MYI), first-year ice (FYI), and open water, from various satellite data. This is a daily product and has 10 km of spatial resolution.

Other data we used included air temperature (2 m above sea level) and wind velocity data coincident with the DMS images acquired from the European Centre for Medium-Range Weather Forecasts ERA-5 reanalysis. The ERA-5 product has 0.25° spatial resolution and consists of hourly variables, and we integrated this hourly data into daily products and resampled them to 25 km resolution to match the ice motion data. This ERA-5 product was downloaded from the Climate Data Store (cds.climate.copernicus.eu) of the Copernicus Climate Change Service.

In this study, the high spatial resolution lead fractions derived from DMS along the Laxon Line were linearly regressed with the coarse spatial resolution sea ice motion, air temperature, and wind velocity products to identify potential significant drivers.

3. Methods

3.1. Batch Classification Processing Workflow

Since the IceBridge DMS images are highly overlapped along the track (60–90%), we selected one image from every three consecutive images along the Laxon Line to reduce the computation burden. All images in continental land masses and poor-quality images due to overwhelming cloud coverage and low lighting conditions were manually removed, finally generating a collection of sea ice lead images (Figure 2).

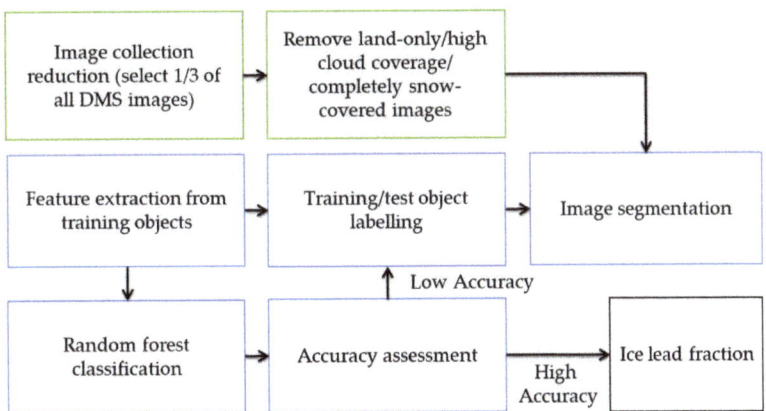

Figure 2. Sea ice lead detection workflow.

The object-based classification scheme was designed based on the color and texture of sea ice features on DMS images. Four sea ice classes were defined: (1) thick ice is usually thick ice or snow-covered ice with a high albedo; (2) thin ice is usually fresh and newly formed ice, which has a smooth surface with a low albedo, since solar radiation is partially absorbed by the water beneath it; (3) open water is dark and smooth due to its strong absorbance of solar radiation; and (4) shadow is within a thick-ice area and is a relative dark feature projecting on the ice surface by surrounding ridges or snow dunes. DMS images collected in different years have different lighting conditions, which affects the image quality (Table 1). Furthermore, even in the same year, the quality of images was quite distinctive due to the local cloud coverage and lighting conditions, as shown in Figure 3. For example, three subgroups were identified in 2012 DMS images: normal images contained regular sea ice scenes with an appropriate exposure and contrast, and all sea ice classes were recognizable by color and texture; gray images were partially cloudy images with a poor lighting condition, so they were relatively dark, and shadows were difficult to detect; and poor images were under extremely poor lighting conditions, and the boundaries between water, thick ice, and thin ice were blurred due to low contrast.

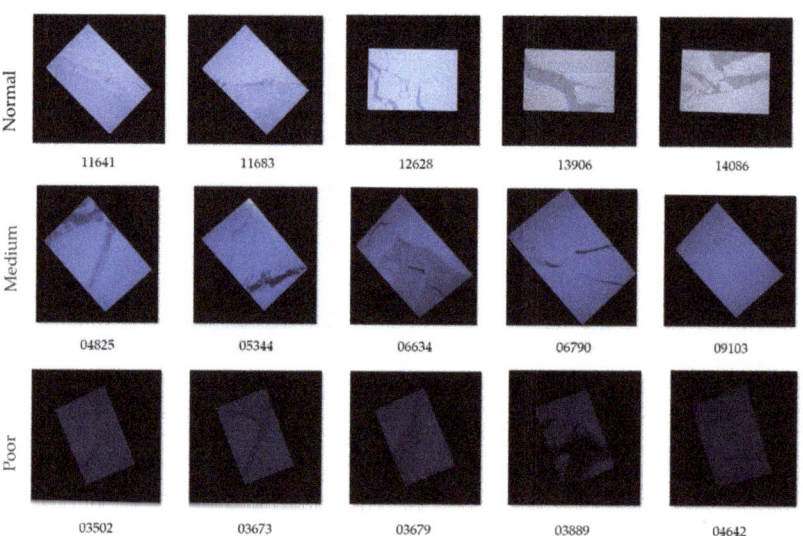

Figure 3. DMS sea ice sample images in 2012 were classified into three subgroups based on different lighting conditions.

Therefore, training samples were selected using a divide-and-conquer strategy based on image quality. All DMS images taken in 2013, 2015, 2016, and 2018 were under good lighting conditions, and training samples were selected for all four sea ice features. However, the images taken for the other three years were processed in different ways. The training samples for all images taken in 2012, 2014, and 2017 were only selected for thin ice, open water, and thick ice, without considering shadow due to low lighting conditions. Furthermore, the 2012 images were manually classified into three subgroups, i.e., normal, medium, and poor. The 2014 images were manually classified into two subgroups, i.e., normal and medium, and all poor images were abandoned due to serious vignetting, caused by light hitting the lens aperture at a large angle, and significantly reduced brightness values on the four corners of the image. The 2017 images were all classified into the medium subgroup only. In summary, the independent training samples were collected for each subgroup and year for supervised classification.

The OSSP package uses an object-based classification scheme. For each image, the watershed segmentation method is used to convert pixels into objects. Therefore, training samples are labelled at the object level. Only distinctive and typical sea ice objects are selected across the whole scene, and each sea ice class has around 120–250 objects. The attributes of objects such as color values, band ratios, textures, and shape indexes are calculated and served as supervised classification features. Based on these training datasets, the OSSP package uses the random forest classification method to label all unknown objects in DMS images [24,25].

To evaluate the accuracy of classification results, the independent test object samples were also collected. Table 3 lists the selected image and object numbers for the training and testing process of each classification group. Finally, the confusion matrix was generated at the pixel level and was used for calculating the overall accuracy, user's accuracy, producer's accuracy, and Kappa coefficient.

Table 3. The DMS images selected for lead detection in the Laxon Line from 2012 to 2018.

Testing Group	# Training Image	# Training Object	# Test Image	# Test object
DMS2012_normal	6	50	5	114
DMS2012_medium	7	90	5	94
DMS2012_poor	7	65	5	124
DMS2013	13	196	7	221
DMS2014_normal	8	106	6	178
DMS2014_medium	6	66	6	119
DMS2015	11	150	9	254
DMS2016	8	144	12	444
DMS2017	12	140	6	150
DMS2018	13	135	9	319

3.2. Sea Ice Leads Parameters Definitions

Based on the classified result in each surface type, we derived the sea ice leads by combining thin ice and open water. Then, the sea ice lead fraction, open water fraction, thin ice fraction, and sea ice concentration were calculated on a per-scene basis. The sea ice lead fraction for each DMS image can be calculated using the following equations:

Sea Ice Lead Fraction (SILF):

$$SILF = (ThinIce + OpenWater)/(ThickIce + ThinIce + OpenWater + Shadow) * 100, \quad (1)$$

where ThinIce, OpenWater, ThickIce, and Shadow are pixel numbers of classified thin ice area, open water, thick ice, and shadow for a DMS image, respectively.

3.3. Spatiotemporal Analysis with Auxiliary Sea Ice Data

The auxiliary sea ice datasets can be used to assess the DMS-based lead detection results to deepen the understanding of the formation mechanism of leads. In this research, first, our lead detection result was used to determine local sea reference height and calculate the sea ice freeboard. This retrieved freeboard was compared with the existing NSIDC freeboard data at the scale of 400 m [36]. Furthermore, the coincident AMSR thin ice concentration (TIC) data, and the geophysical atmosphere and ocean data, such as air temperature, wind velocity, and sea ice motion, were compared with the lead fraction results.

Based on our DMS lead detection algorithm, sea ice freeboards were retrieved from the ATM lidar data using the same method as in [32]. Specifically, we removed variations in the instantaneous sea surface height by subtracting geoid and ocean tide height. Then, we calculated the freeboard by subtracting locally determined leads surface height (z_{shh}) from the corrected height (H_{corr}).

$$Freeboard = H_{corr} - z_{shh}, \quad (2)$$

where z_{shh} is determined from the sets of individual lead elevation estimates through ordinary kriging. We calculated the mean freeboard for each DMS image (around 400 m by 600 m) and resampled the value to 400 m resolution. On the other hand, Kurtz et al. used an automated lead detection algorithm through the minimal signal transform [23,32] and then retrieved the freeboard at the resolution of 400 m. Therefore, the two products can be compared and cross-verified at this scale.

TIC could be calculated from the AMSR as described in Röhrs and Kaleschke [14] with a rather coarse spatial resolution of 25 km. This AMSR-based TIC represents the existence of open water and thin ice on sea ice leads. This TIC is conceptually equivalent to our SILF. Since the AMSR and DMS have different resolutions and geographical coverage, they cannot be compared directly. Therefore, we resampled and averaged the DMS-based ice lead fractions for every 25 km grid cell to match the spatial resolution of AMSR data, as shown in Figure 4. Then, the mean of sea ice lead fractions within the range of each 25 km block was calculated.

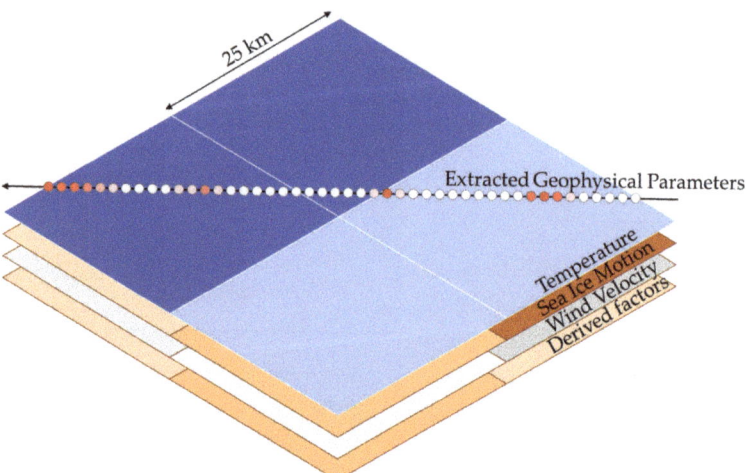

Figure 4. Data fusion diagram with derived geophysical parameters and DMS-based sea ice leads (each 25 km AMSR pixel covers around 5–70 point of HSR image locations).

Furthermore, the 25 km resampled lead fractions were also correlated with other 25 km resolution sea ice and atmospheric data including NSIDC sea ice motion, ERA5 air temperature, and wind velocity. Since kinetic moments of sea ice movement can play an important role in formations of leads, four kinetic moments or tensions were calculated from the NSIDC sea ice motion data by the following equations [37]:

$$divergence = \frac{\partial F_x}{\partial x} + \frac{\partial F_y}{\partial y} \quad (3)$$

$$vorticity = \frac{\partial F_y}{\partial x} - \frac{\partial F_x}{\partial y} \quad (4)$$

$$shearing\ deformation = \frac{\partial F_y}{\partial x} + \frac{\partial F_x}{\partial y} \quad (5)$$

$$stretching\ deformation = \frac{\partial F_x}{\partial x} - \frac{\partial F_y}{\partial y} \quad (6)$$

where F_x and F_y refer to the velocity of sea ice along the x and y axes, respectively. Divergence is a measure of parcel area change without the change of orientation or shape, and vorticity is a measure of orientation change without area or shape change. Shearing and stretching deformation are measures of shape change produced by differential motions parallel and normal to the boundary, respectively [37].

Finally, based on the assumption that these atmosphere and sea ice variables for a series of the previous days would contribute to the formation of sea ice leads, the average of these dynamic and thermodynamic variables up to 30 successive days before the DMS acquisition day were calculated (Table 4). By comparing these variables and the lead fractions, we hoped to identify the potential contribution of these explanatory variables to lead formation.

Multiple linear regression (MLR) was used for modelling the mean lead fractions in terms of large-scale sea ice dynamic–thermodynamic variables, including the NSIDC sea ice motion data with four kinetic moments, ERA-5 air temperature, and wind velocity data. The forward and backward stepwise regression methods were used to identify the most important explanatory variables. This strategy refers to the process of building a regression model by adding or removing explanatory variables in a stepwise manner until the predicted variable does not change significantly [38].

Table 4. Variables for the multiple linear regression models.

Department	Factors	Description
Sea Ice Leads	mean_leads	Mean lead fraction for 25 km segment
Temperature	tmpXX	Averaged air temperature for last XX days (e.g., tmp03 means average temperature of last 1, 2, 3 days)
Wind	U10_XX V10_XX wind_XX	Averaged u-component of wind velocity for last XX days Averaged v-component of wind velocity for last XX days Averaged wind velocity for last XX days (e.g., wind_10 means wind velocity for last 10 days)
Sea Ice Motion	u_ice_XX v_ice_XX vel_ice_XX divXX vorXX shrXX stcXX	Averaged u-component of ice velocity for last XX days (e.g., u_ice_10 means u-velocity for last 10 days) Averaged v-component of ice velocity for last XX days (e.g., v_ice_10 means v-velocity for last 10 days) Averaged ice velocity for last XX days (e.g., v_ice_10 means ice velocity for last 10 days) Averaged divergence of sea ice motion for last XX days (e.g., div10 means divergence for last 10 days) Averaged vorticity of sea ice motion for last XX days (e.g., vor10 means vorticity for last 10 days) Averaged shearing deformation of sea ice motion for last XX days (e.g., shr10 means shearing deformation for last 10 days) Averaged stretching deformation of sea ice motion for last XX days (e.g., stc10 means stretching deformation for last 10 days)

4. Result and Discussion

4.1. Classification Result

A total of 106,674 DSM images along the Laxon Line from 2012–2018 were processed, and a total of 6135 images with sea ice leads were visually selected (Table 1). All images were classified through the OSSP package integrated in the ArcCI online service [22].

Six classified images in 2012 are shown in Table 5. The first row shows the classification results for the subgroup of normal images, the second row for the medium images, and the third row for the poor images. All six images were selected to show a variety of sea ice features under different lighting conditions. The classified results illustrate four sea ice classes: open water, shadow, thin ice, and thick ice.

Table 5. Comparison of original 2012 DMS images and classified results for three subgroups. Two samples were selected for each subgroup.

	Sample Result 1		Sample Result 2	
	Raw Image	Classified Result	Raw Image	Classified Result
Normal				
Medium				
Poor				

LEGEND: Thick Ice, Thin Ice, Shadow, Open Water

The classification accuracies were evaluated at the pixel-level, and all calculated accuracies are summarized in Table 6. The overall accuracy across the 10 test samples selected by year and illumination conditions was 90.9 ± 3.5%, where the latter number is the standard deviation, and the Kappa coefficient was 0.85 ± 0.05. Since sea ice leads were defined as a combination of thin ice and open water, classification accuracy was determined by these two classes. The user's accuracy for thin ice and water were 90.7 ± 5.9% and 92.7 ± 11.0%, respectively. The low accuracy of 61.9% for open water in the 2012 poor subgroup was due to the confusion between water and thin ice under extremely poor lighting conditions.

Table 6. Pixel-level classification accuracy for each production group. All values except Kappa coefficient are in percentages.

Testing Group	Overall Accuracy	Kappa Coef.	UA_Thick **	UA_Thin	UA_Shadow	UA_Water	PA_Thick **	PA_Thin	PA_Shadow	PA_Water
DMS2012_normal	88.9	0.83	88.0	91.7	83.8	nan *	98.4	94.2	63.8	nan
DMS2012_medium	93.6	0.85	97.3	85.0	nan	95.5	93.8	93.1	nan	97.5
DMS2012_poor	93.8	0.86	95.0	96.0	nan	61.9	98.9	81.2	nan	94.9
DMS2013	96.4	0.95	92.2	100.0	99.4	95.5	99.7	96.5	88.3	99.9
DMS2014_normal	88.0	0.82	74.7	86.2	93.9	98.0	97.1	81.3	99.7	89.0
DMS2014_medium	93.7	0.89	91.7	96.3	nan	97.1	100.0	75.7	nan	97.1
DMS2015	86.4	0.78	86.6	83.5	98.6	93.4	99.8	80.9	82.2	57.9
DMS2016	87.9	0.83	82.1	89.3	95.0	95.7	99.4	68.8	89.7	90.2
DMS2017	86.7	0.75	87.4	82.8	nan	99.4	97.6	76.5	nan	60.7
DMS2018	93.5	0.88	91.9	96.5	95.2	97.9	98.5	79.1	89.4	98.4
Average Accuracy	90.9	0.84	88.7	90.7	94.3	92.7	98.3	82.7	85.5	87.3

*,** User's accuracy and producer's accuracy for each classified ice type represented as UA_XX, and PA_XX, and XX could be thick ice, thin ice, shadow, or open water.

4.2. Overall Integrated Statistical Analysis and Trend of Sea Ice Leads and Freeboard

4.2.1. Sea Ice Leads Fraction, Area, and Frequency

Figure 5a shows the averaged lead fraction for every 25 km along the Laxon Line. Relatively large lead fractions (>15%) were only observed near the Beaufort Sea area (track distance > 1200 km) in 2013, 2014, and 2016, where they were generally located in the FYI region or transition region between FYI and MYI. However, the smaller lead fraction region in the central Arctic (track distance < 1200 km) was primarily covered by MYI and thick ice. Although these observations of one day per year for seven years cannot represent the overall continuous spatiotemporal variations of lead fraction, this general spatial pattern agrees with that of previous lead studies [5,18,19,39]. Figure 5b portrays the averaged area of individual leads for the 25 km track segment, and Figure 5c portrays the ratio of the number of lead-included images to the total number of images for the 25 km segment. The lead fraction (Figure 5a) was determined by the individual lead area (Figure 5b) and the frequency of leads (Figure 5c). For example, although large leads were observed in 2013 for 0–500 km (Figure 5b), lead frequency for this part was low (Figure 5c) due to the small number of large leads. As a result, the averaged lead fraction for this segment was not high because of the low lead frequency. In addition, the lead frequency in 2018 for 1000–2500 km was relatively high, but the averaged lead fraction was not so high due to the large number of small leads.

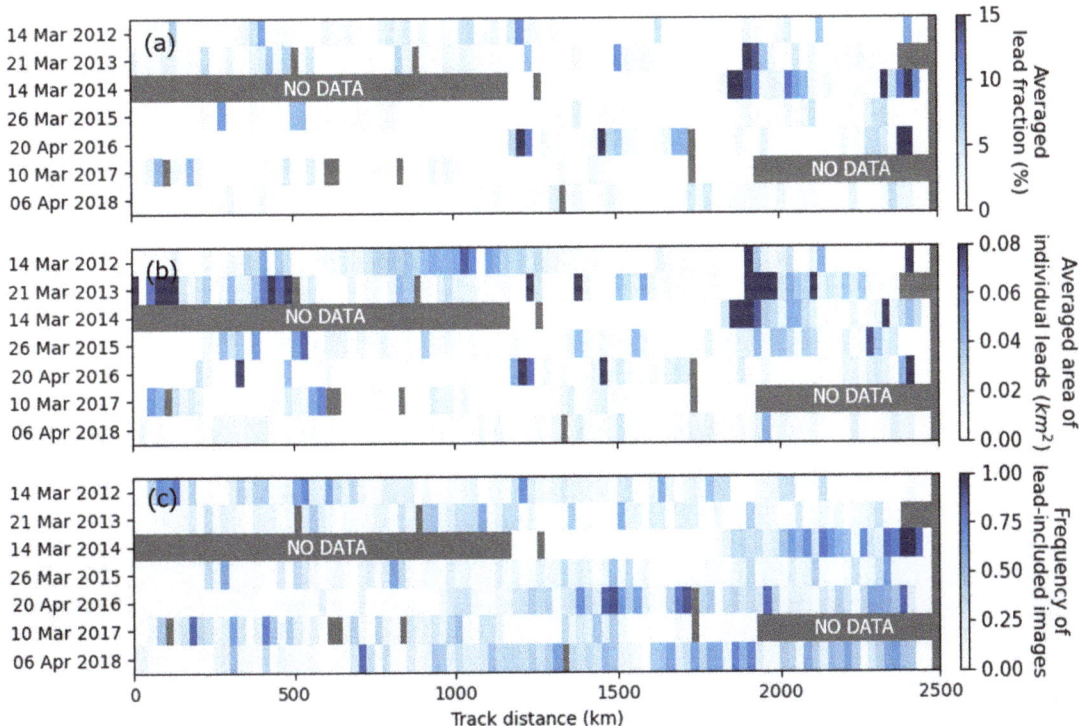

Figure 5. (**a**) Averaged lead fraction for every 25 km; (**b**) averaged area of individual leads for every 25 km; (**c**) frequency of lead-included images for every 25 km. Gray parts indicate missing/invalid data.

4.2.2. Retrieval of Freeboard

Based on the DMS lead detection result, we calculated the 400 m mean sea ice freeboard from the ATM surface height data (Figure 6). The MYI area (near central Arctic Ocean) at track distance <1200 km showed a higher freeboard (i.e., thicker ice) compared to that of the FYI area (near the Beaufort Sea with a track distance beyond 1200 km). As shown in Table 7, the FYI area always showed a lower freeboard than the MYI area. In addition, the freeboard retrieved from our lead detection shows a good correlation with the ATM freeboard product provided by NSIDC [32]—correlation coefficient (R) was 0.832, and root mean square difference (RMSD) was 0.105 m (Table 8). It is also noted that 2015, 2016, and 2017 showed relatively lower R and higher root mean square error (RMSE) than the other years (Table 8 and Figure 7), which might be due to the lower classification accuracy of these years (Table 6). Some misclassified leads can make substantial differences in estimation of sea surface height, eventually leading to the differences between our freeboard estimation and the NSIDC freeboard products. Nevertheless, the freeboard differences between MYI and FYI and the cross-validation with the NSIDC freeboard product showed that our lead detection result was reasonable and compatible with other lead detection products.

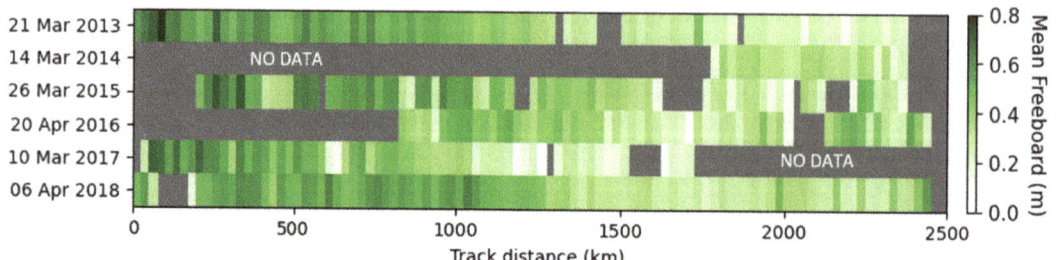

Figure 6. Averaged ATM freeboard for every 25 km for each year.

Table 7. ATM sea ice freeboard retrieved from the DMS lead detection.

Year	FYI	MYI	Total
2013	0.263	0.519	0.409
2014	0.277	0.339	0.320
2015	0.275	0.470	0.407
2016	0.335	0.398	0.354
2017	0.211	0.467	0.366
2018	0.320	0.505	0.414

Table 8. R and RMSE between our freeboard estimation and NSIDC freeboard estimation.

Year	R	RMSD (m)
2013	0.928	0.089
2014	0.907	0.063
2015	0.755	0.140
2016	0.784	0.114
2017	0.742	0.119
2018	0.869	0.082
Total	0.832	0.105

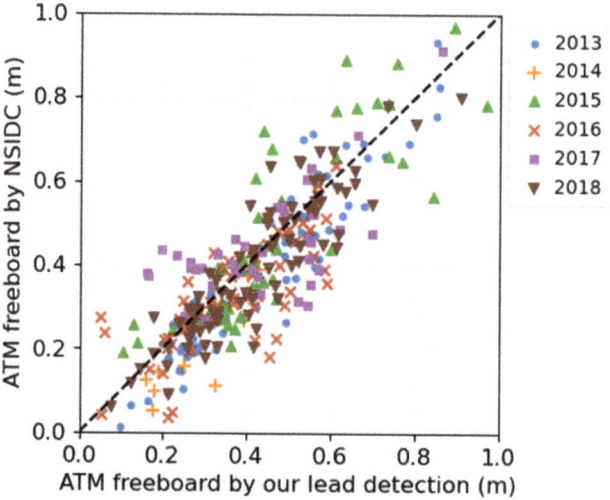

Figure 7. Scatter plot between ATM freeboard derived by our lead detection and NSIDC freeboard product for every 400 m (2% random selection of the total points).

4.3. Sea Ice Lead Fraction Modelling with Auxiliary Sea Ice Product

In general, March and April have the lowest lead fraction and lead frequency in a year because of the highly packed sea ice conditions [5,23]. Since the OIB missions were conducted during these months of packed sea ice, the widths of individual observed leads were usually less than 1 km. Indeed, as shown in Figure 5b, most leads had less than 0.1 km^2 of area, which accounts for a tiny portion of the entire 25 × 25 km grid cells. Hence, it is reasonable that the DMS-based lead detection and AMSR-based TIC were not highly correlated (R~0.21, Figure 8), because narrow leads are hardly detected by the coarse resolution satellite data [14,40]. For example, we found that most of AMSR-based TIC along the track was zero and AMSR-based SIC was 100% even though the DMS images clearly showed leads in that area.

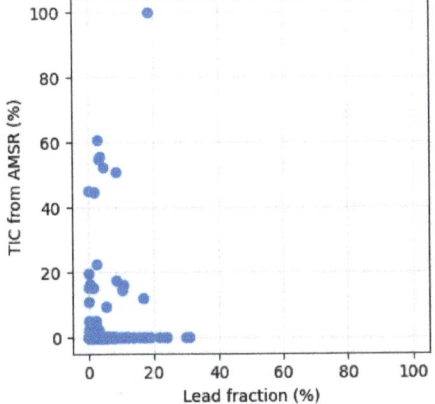

Figure 8. Scatter plot between DMS-based lead fraction (this study) and AMSR-based TIC.

Figure 9 shows the lead fractions and related dynamic and thermodynamic variables at the scale of 25 km on the same days that DMS images were taken from 2012 to 2018. In general, the lead fractions did not show significant correlation with any single auxiliary variable or kinetic property from sea ice motion data. This is reasonable because (1) these ancillary data have 25 km spatial resolution, which is much coarser than the spatial resolution of the DMS image; (2) the DMS images have only ~500 m of width, representing only a small portion along the Laxon Line; and (3) the formation of sea ice leads results from the accumulative and complex effects of multiple dynamic and thermodynamic variables, rather than just one variable.

Although the DMS images have different spatial scale with the ancillary datasets, we attempted to explore the potential relationship the DMS-based lead fractions and sea ice dynamic and thermodynamic variables from the ancillary datasets. Assuming that (1) these variables are the results of the large-scale atmosphere and ocean circulation and (2) the combination of these variables somehow affects the formation of leads, we normalized all explanatory variables and constructed a series of multiple-variables linear regression models, as shown in Equation (7).

$$\text{SILF} = \sum_{k=0}^{n} a_k x_k \qquad (7)$$

where x_k is one of the normalized dynamics-thermodynamic variables, and a_k are corresponding coefficients.

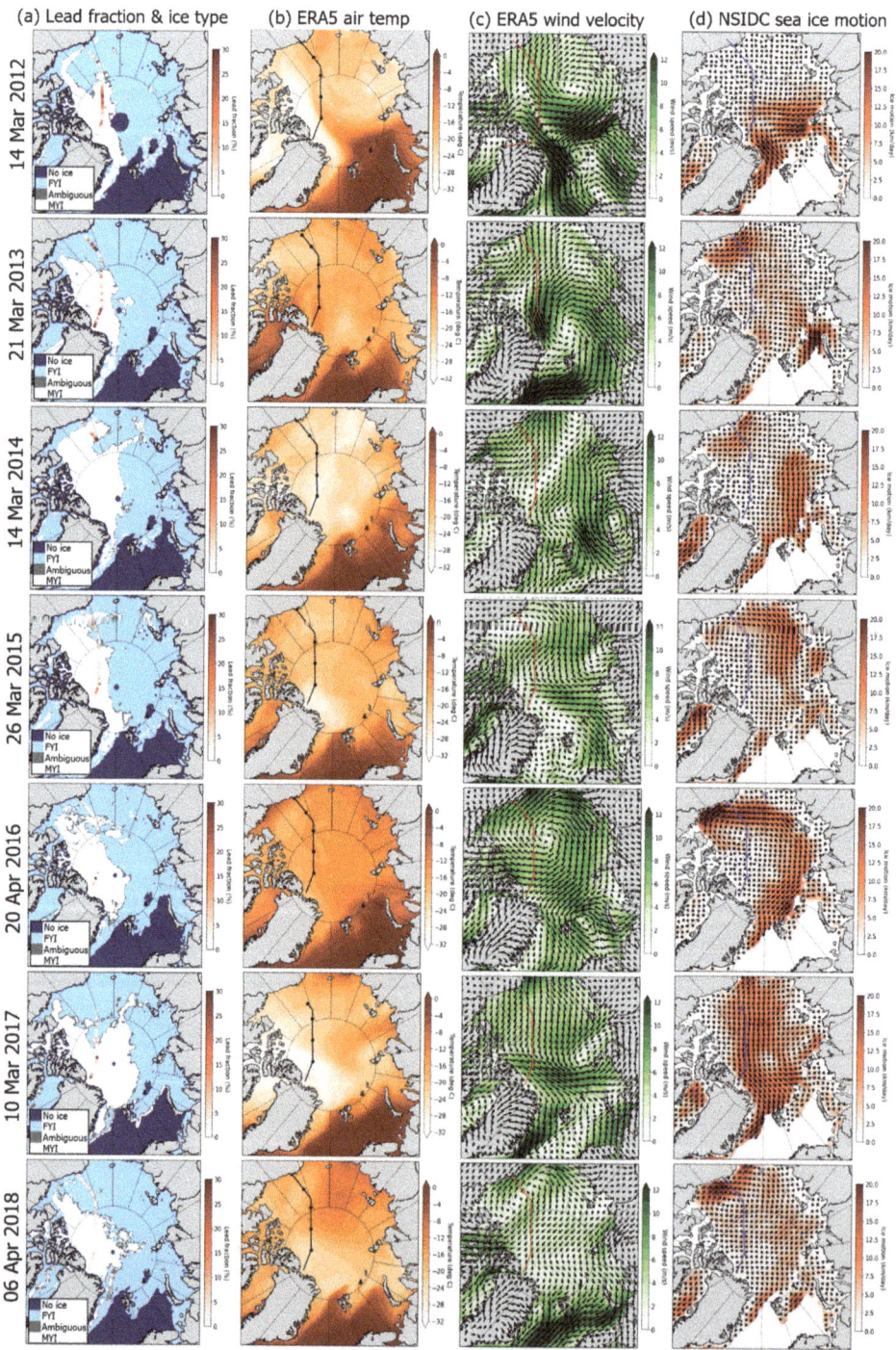

Figure 9. (**a**) DMS-based lead fraction and nearby ice types; (**b**) ERA5 air temperature; (**c**) ERA5 wind velocity; (**d**) sea ice motion for each year.

The lead fraction variable is the mean of all DMS image-based lead fractions within a 25 km block. On the other hand, all dynamic-thermodynamic variables, including four kinetic moments from the NSIDC sea ice motion data, ERA5 air temperature, and wind velocity data, were averaged by 1, 2, 5, 10, 20, and 30 days prior to the date when the DMS image was taken, considering the accumulative effects of these explanatory variables.

After exploring all possible multiple linear regression models, we found that dynamic-thermodynamic variables integrated by 10 days showed the highest correlation coefficient. Therefore, these explanatory variables were used to reconstruct the linear regression models using the forward and backward stepwise regression approach. The coefficients of all normalized explanatory variables for all models are illustrated in Table 9. There were 11 thermodynamic-dynamic variables, including one thermodynamic variable (temperature), six dynamic variables (velocity of wind and ice motion), and four kinetic moments caused by ice motion.

Table 9. Selected variables and coefficients in 14 stepwise linear regressions.

Year	Approach	R^2	Tmp10	U10_10	V10_10	Wind_10	U_Ice_10	V_Ice_10	Vel_Ice_10	Div10	Vor10	Shr10	Stc10	Constant
2012	Forward	0.26	/	/	/	/	−0.39	−0.38	0.16	−0.10	−0.08	/	/	0.41
	Backward	0.26	0.10	/	/	/	−0.34	−0.19	/	−0.12	/	/	/	0.31
2013	Forward	0.48	−1.19	/	/	0.35	−6.46	−2.78	9.51	/	−0.01	−0.14	/	0.60
	Backward	0.48	−1.18	/	/	0.35	−6.44	−2.75	9.45	/	/	−0.15	/	0.08
2014	Forward	0.87	4.61	−5.60	−0.97	1.09	1.24	15.31	−12.98	/	0.89	−0.55	/	−2.08
	Backward	0.87	4.64	−5.37	/	/	1.16	13.34	−11.25	−0.16	0.87	−0.59	/	−1.94
2015	Forward	0.34	/	/	−0.53	/	−1.35	/	1.19	0.15	0.14	0.28	−0.33	0.40
	Backward	0.34	/	/	−0.53	/	−1.35	/	1.19	0.15	0.14	0.28	−0.33	0.40
2016	Forward	0.29	/	−0.79	/	/	/	/	0.29	0.30	−0.39	0.57	0.15	0.21
	Backward	0.34	0.67	−4.62	−0.53	4.09	/	/	/	/	−0.36	0.46	/	0.22
2017	Forward	0.66	−1.17	−6.54	−3.08	6.77	2.98	−0.09	−2.01	−0.19	/	/	/	1.50
	Backward	0.66	−1.15	−6.57	−3.11	6.86	3.02	/	−2.09	−0.19	/	/	/	1.45
2018	Forward	0.30	0.34	−1.40	−1.40	1.83	/	/	/	/	−0.03	/	−0.31	0.45
	Backward	0.30	0.34	−1.31	−1.33	1.72	/	/	/	/	/	/	−0.32	0.42

The forward and backward stepwise regression models for each year identified different sets of explanatory variables. Both 2012 models identified ice motion velocity and divergence as the significant explanatory variables. The 2013 models mainly identified the ice motion velocity and temperature variables. Other than ice motion velocity and temperature, the 2014 models included wind velocity at u-direction, and the correlation coefficient was significantly higher than that of other models. The 2015 models emphasized the functions of wind and ice motion velocity. The 2016 forward model identified more kinetic moments, but the backward model emphasized wind velocity, which represents the possible correlation among these variables. Finally, the 2017 and 2018 models showed significant influence of wind velocity and temperature.

Except for that of 2014, all other models had only moderate correlation, and R^2 ranged from 0.26 to 0.66. This was because (1) the sea ice fractions were derived from high spatial resolution DMS images, and the dynamic-thermodynamic variables had a much coarser resolution of 25 km; (2) the atmospheric and oceanic dynamics that contribute to lead formation can occur in a much smaller scale (<25 km scale), which cannot be captured by coarse resolution products; and (3) the uncertainty of the DMS-based lead detection (accuracy of 90%) can be carried and exaggerated in the data fusion and resampling process.

Based on all stepwise regression results, the relative explanatory variable importance could be ranked based on their frequencies in a total of 14 regression models (Table 9), as summarized in Figure 10. It showed that temperature and ice motion vorticity were the leading factors of the formation of sea ice leads, followed by wind vorticity and kinetic moments or tensions of ice motion. However, since this result is only based on stepwise regression of several available variables, it cannot clearly explain the detailed mechanism of lead formation that is a complex combination of multiple ocean and atmospheric parameters. In addition, it is noted that the spatial resolution of the variables can be too coarse to represent the formation of leads in the DMS image scale. Therefore, more comprehensive studies are needed to clearly understand small-scale lead formations in the future.

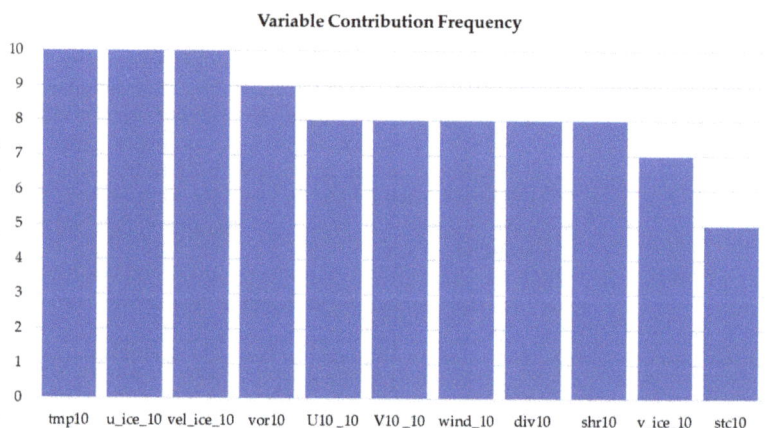

Figure 10. Relative importance of dynamic-thermodynamic explanatory variables.

5. Conclusions

This research demonstrates a scientific case study for sea ice lead detection during 2012–2018 along the IceBridge Laxon Line. To address the lack of standard image processing workflow for sea ice parameter extraction from massive and long-term HSR imagery, a practical object-based image classification workflow was implemented based on the OSSP package to extract multiscale multitype sea ice features and to calculate sea ice lead fractions and freeboard parameters. These sea ice products could be directly used to validate other coarse resolution remote sensing images/products. Furthermore, the high-spatial-resolution sea ice fractions were statistically modeled using large scale dynamic-thermodynamic models.

We found that thick ice, thin ice, water, and shadow could be successfully classified using an object-based classification algorithm or the OSSP package with reasonable overall accuracies of 86.4–96.4%. The sea ice lead fractions along the Laxon Line could be calculated for each DMS image accordingly. The temporal and spatial distribution of leads were verified by ATM surface height data and an independent freeboard product. Finally, the lead fractions were aggregated and modelled with 25 km resolution dynamic and thermodynamic variables including sea ice motion, air temperature, and wind data. All stepwise linear regression models had medium to high correlation coefficients. It seems that temperature and ice motion vorticity were the leading factors of the formation of sea ice leads, and each year could have different dominant factors. The results could provide insightful understanding of the mechanism of sea ice leads, which is useful for climate modelling.

In the future, novel image classification algorithms such as deep learning could be used to improve the traditional machine learning methods. The methods can be extended to other sea ice regions and data types. The results and parameters derived from this study can help the sea ice community to better understand the mechanisms driving sea ice variability so that they can be better represented in climate models.

Author Contributions: Conceptualization, D.S., X.M., H.X. and C.Y.; methodology, D.S., Y.K. and X.M.; software, D.S., A.S. and H.L.; investigation, D.S., Y.K. and X.M.; resources, Q.L. and S.B.; data curation, D.S. and Y.K.; writing—original draft preparation, D.S., Y.K. and X.M.; writing—review and editing, H.X., A.M.M.-N. and C.Y.; project administration, D.S. and X.M.; funding acquisition, C.Y. All authors have read and agreed to the published version of the manuscript.

Funding: This research was funded by NSF with grant numbers 1835507 and 1841520 (GMU), 1835784 (UTSA), 1835512 (MSU), and by NASA with grant numbers 80NSSC18K0843 and 80NSSC19 M0194 (UTSA).

Acknowledgments: The authors are thankful to Kevin Wang for providing technical support on testing the online web services and writing the user's manual. Jennifer Smith proofread the language.

Conflicts of Interest: The authors declare no conflict of interest.

References

1. Parkinson, C.L. A 40-y record reveals gradual Antarctic sea ice increases followed by decreases at rates far exceeding the rates seen in the Arctic. *Proc. Natl. Acad. Sci. USA* **2019**, *116*, 14414–14423. [CrossRef]
2. Peng, G.; Matthews, J.L.; Yu, J.T. Sensitivity Analysis of Arctic Sea Ice Extent Trends and Statistical Projections Using Satellite Data. *Remote Sens.* **2018**, *10*, 230. [CrossRef]
3. Marshall, M. Arctic ice low kicks off a cascade of tipping points. *New Sci.* **2013**, *217*, 6–7. [CrossRef]
4. Parkinson, C.L.; Comiso, J.C. On the 2012 record low Arctic sea ice cover: Combined impact of preconditioning and an August storm. *Geophys. Res. Lett.* **2013**, *40*, 1356–1361. [CrossRef]
5. Wang, Q.; Danilov, S.; Jung, T.; Kaleschke, L.; Wernecke, A. Sea ice leads in the Arctic Ocean: Model assessment, interannual variability and trends. *Geophys. Res. Lett.* **2016**, *43*, 7019–7027. [CrossRef]
6. Hui, F.; Zhao, T.; Li, X.; Shokr, M.; Heil, P.; Zhao, J.; Zhang, L.; Cheng, X. Satellite-Based Sea Ice Navigation for Prydz Bay, East Antarctica. *Remote Sens.* **2017**, *9*, 518. [CrossRef]
7. Hakkinen, S.; Proshutinsky, A.; Ashik, I. Sea ice drift in the Arctic since the 1950s. *Geophys. Res. Lett.* **2008**, *35*, 19704. [CrossRef]
8. Hirano, D.; Fukamachi, Y.; Watanabe, E.; Ohshima, K.I.; Iwamoto, K.; Mahoney, A.R.; Eicken, H.; Simizu, D.; Tamura, T. A wind-driven, hybrid latent and sensible heat coastal polynya off Barrow, Alaska. *J. Geophys. Res. Ocean.* **2016**, *121*, 980–997. [CrossRef]
9. Alam, A. Determination of surface turbulent fluxes over leads in Arctic sea ice. *J. Geophys. Res. C Ocean.* **1997**, *102*, 3331–3343. [CrossRef]
10. Andreas, E.L.; Murphy, B. Bulk Transfer Coefficients for Heat and Momentum over Leads and Polynyas. *J. Phys. Oceanogr.* **1986**, *16*, 1875–1883. [CrossRef]
11. Marcq, S.; Weiss, J. Influence of sea ice lead-width distribution on turbulent heat transfer between the ocean and the atmosphere. *Cryosphere* **2012**, *6*, 143–156. [CrossRef]
12. Worby, A.P.; Allison, I. Ocean-atmosphere energy exchange over thin, variable concentration Antarctic pack ice. *Ann. Glaciol.* **1991**, *15*, 184–190. [CrossRef]
13. Maykut, G.A. Large-scale heat exchange and ice production in the central Arctic. *J. Geophys. Res.* **1982**, *87*, 7971. [CrossRef]
14. Röhrs, J.; Kaleschke, L. An algorithm to detect sea ice leads by using AMSR-E passive microwave imagery. *Cryosphere* **2012**, *6*, 343–352. [CrossRef]
15. Heorton, H.D.B.S.; Radia, N.; Feltham, D.L. A Model of Sea Ice Formation in Leads and Polynyas. *J. Phys. Oceanogr.* **2017**, *47*, 1701–1718. [CrossRef]
16. National Research Council. *Earth Science and Applications from Space: National Imperatives for the Next Decade and Beyond*; National Academies Press: Washington, DC, USA, 2007.
17. Hall, D.K.; Riggs, G.A.; Salomonson, V.V. MODIS Snow and Sea Ice Products. In *Earth Science Satellite Remote Sensing*; Springer: Berlin/Heidelberg, Germany, 2006; pp. 154–181.
18. Willmes, S.; Heinemann, G. Sea-Ice Wintertime Lead Frequencies and Regional Characteristics in the Arctic, 2003–2015. *Remote Sens.* **2016**, *8*, 4. [CrossRef]
19. Hoffman, J.; Ackerman, S.; Liu, Y.; Key, J. The Detection and Characterization of Arctic Sea Ice Leads with Satellite Imagers. *Remote Sens.* **2019**, *11*, 521. [CrossRef]
20. Lindsay, R.W.; Rothrock, D.A. Arctic sea ice leads from advanced very high resolution radiometer images. *J. Geophys. Res.* **1995**, *100*, 4533. [CrossRef]
21. Li, L.; Changqing, K.E.; Xie, H.; Ruibo, L.; Anqi, T. Aerial observations of sea ice and melt ponds near the North Pole during CHINARE$_{2010}$. *Acta Oceanol. Sin.* **2017**, *36*, 64–72. [CrossRef]
22. Sha, D.; Miao, X.; Xu, M.; Yang, C.; Xie, H.; Mestas-Nuñez, A.M.; Li, Y.; Liu, Q.; Yang, J. An On—Demand Service for Managing and Analyzing Arctic Sea Ice High Spatial Resolution Imagery. *Data* **2020**, *5*, 39. [CrossRef]
23. Onana, V.-D.-P.; Kurtz, N.T.; Farrell, S.L.; Koenig, L.S.; Studinger, M.; Harbeck, J.P. A Sea-Ice Lead Detection Algorithm for Use With High-Resolution Airborne Visible Imagery. *IEEE Trans. Geosci. Remote Sens.* **2013**, *51*, 38–56. [CrossRef]
24. Miao, X.; Xie, H.; Ackley, S.F.; Perovich, D.K.; Ke, C. Object-based detection of Arctic sea ice and melt ponds using high spatial resolution aerial photographs. *Cold Reg. Sci. Technol.* **2015**, *119*, 211–222. [CrossRef]
25. Wright, N.C.; Polashenski, C.M. Open-source algorithm for detecting sea ice surface features in high-resolution optical imagery. *Cryosphere* **2018**, *12*, 1307–1329. [CrossRef]
26. Wright, N.C.; Polashenski, C.M.; Mcmichael, S.T.; Beyer, R.A. Observations of sea ice melt from Operation IceBridge imagery. *Cryosphere* **2020**, *14*, 3523–3536. [CrossRef]

27. Dominguez, R. *IceBridge DMS L1B Geolocated and Orthorectified Images*; National Snow and Ice Data Center: Boulder, CO, USA, 2010.
28. Markus, T.; Comiso, J.C.; Meier, W.N. *AMSR-E/AMSR2 Unified L3 Daily 25 km Brightness Temperatures & Sea Ice Concentration Polar Grids, Version 1*; NASA National Snow and Ice Data Center Distributed Active Archive Center: Boulder, CO, USA, 2018.
29. Meier, W.N.; Markus, T.; Comiso, J.; Ivanoff, A.; Miller, J. *Amsr2 Sea Ice Algorithm Theoretical Basis Document*; NASA Goddard Space Flight Center: Greenbelt, MD, USA, 2017.
30. Markus, T.; Cavalieri, D.J. An enhancement of the NASA Team sea ice algorithm. *IEEE Trans. Geosci. Remote Sens.* **2000**, *38*, 1387–1398. [CrossRef]
31. Markus, T.; Cavalieri, D. The AMSR-E NT2 Sea Ice Concentration Algorithm: Its Basis and Implementation. *J. Remote Sens. Soc. Jpn.* **2009**, *29*, 216–225. [CrossRef]
32. Kurtz, N.T.; Farrell, S.L.; Studinger, M.; Galin, N.; Harbeck, J.P.; Lindsay, R.; Onana, V.D.; Panzer, B.; Sonntag, J.G. Sea ice thickness, freeboard, and snow depth products from Operation IceBridge airborne data. *Cryosphere* **2013**, *7*, 1035–1056. [CrossRef]
33. Wang, X.; Guan, F.; Liu, J.; Xie, H.; Ackley, S. An improved approach of total freeboard retrieval with IceBridge Airborne Topographic Mapper (ATM) elevation and Digital Mapping System (DMS) images. *Remote Sens. Environ.* **2016**, *184*, 582–594. [CrossRef]
34. Tschudi, M.; Meier, W.N.; Stewart, J.S.; Fowler, C.; Maslanik, J. *Polar Pathfinder Daily 25 km EASE-Grid Sea Ice Motion Vectors, Version 4*; NASA National Snow and Ice Data Center Distributed Active Archive Center: Boulder, CO, USA, 2019. [CrossRef]
35. Aaboe, S.; Down, E.J.; Eastwood, S. *Product User Manual for the Global Sea-Ice Edge and Type Product*; Norwegian Meteorological Institute: Oslo, Norway, 2021.
36. NSIDC. IceBridge Sea Ice Freeboard, Snow Depth, and Thickness Quick Look, Version 1. Available online: https://nsidc.org/data/NSIDC-0708/versions/1 (accessed on 16 September 2021).
37. Molinari, R.; Kirwan, A.D. Calculations of Differential Kinematic Properties from Lagrangian Observations in the Western Caribbean Sea. *J. Phys. Oceanogr.* **1975**, *5*, 483–491. [CrossRef]
38. Wilkinson, L. Tests of significance in stepwise regression. *Psychol. Bull.* **1979**, *86*, 168–174. [CrossRef]
39. Qu, M.; Pang, X.; Zhao, X.; Lei, R.; Ji, Q.; Liu, Y.; Chen, Y. Spring leads in the Beaufort Sea and its interannual trend using Terra/MODIS thermal imagery. *Remote Sens. Environ.* **2021**, *256*, 112342. [CrossRef]
40. Lee, S.; Kim, H.; Im, J. Arctic lead detection using a waveform mixture algorithm from CryoSat-2 data. *Cryosphere* **2018**, *12*, 1665–1679. [CrossRef]

Article

Accuracy Evaluation on Geolocation of the Chinese First Polar Microsatellite (Ice Pathfinder) Imagery

Ying Zhang [1,2], Zhaohui Chi [3], Fengming Hui [2,4,5], Teng Li [2,4,5], Xuying Liu [1,2], Baogang Zhang [1,2], Xiao Cheng [2,4,5,*] and Zhuoqi Chen [2,4,5,†]

1. State Key Laboratory of Remote Sensing Science, College of Global Change and Earth System Science, Beijing Normal University, Beijing 100875, China; 201831490009@mail.bnu.edu.cn (Y.Z.); 201831490014@mail.bnu.edu.cn (X.L.); zhang_bob@bnu.edu.cn (B.Z.)
2. University Corporation for Polar Research, Beijing 100875, China; huifm@mail.sysu.edu.cn (F.H.); liteng28@mail.sysu.edu.cn (T.L.); chenzhq67@mail.sysu.edu.cn (Z.C.)
3. Department of Geography, Texas A&M University, College Station, TX 77843, USA; zchi@tamu.edu
4. School of Geospatial Engineering and Science, Sun Yat-sen University, Zhuhai 519000, China
5. Southern Marine Science and Engineering Guangdong Laboratory (Zhuhai), Zhuhai 519082, China
* Correspondence: chengxiao9@mail.sysu.edu.cn
† Joint first authors.

Abstract: Ice Pathfinder (Code: BNU-1), launched on 12 September 2019, is the first Chinese polar observation microsatellite. Its main payload is a wide-view camera with a ground resolution of 74 m at the subsatellite point and a scanning width of 744 km. BNU-1 takes into account the balance between spatial resolution and revisit frequency, providing observations with finer spatial resolution than Terra/Aqua MODIS data and more frequent revisits than Landsat-8 OLI and Sentinel-2 MSI. It is a valuable supplement for polar observations. Geolocation is an essential step in satellite image processing. This study aims to geolocate BNU-1 images; this includes two steps. For the first step, a geometric calibration model is applied to transform the image coordinates to geographic coordinates. The images calibrated by the geometric model are the Level1A (L1A) product. Due to the inaccuracy of satellite attitude and orbit parameters, the geometric calibration model also exhibits errors, resulting in geolocation errors in the BNU-1 L1A product. Then, a geometric correction method is applied as the second step to find the control points (CPs) extracted from the BNU-1 L1A product and the corresponding MODIS images. These CPs are used to estimate and correct geolocation errors. The BNU-1 L1A product corrected by the geometric correction method is processed to the Level1B (L1B) product. Although the geometric correction method based on CPs has been widely used to correct the geolocation errors of visible remote sensing images, it is difficult to extract enough CPs from polar images due to the high reflectance of snow and ice. In this study, the geometric correction employs an image division and an image enhancement method to extract more CPs from the BNU-1 L1A products. The results indicate that the number of CPs extracted by the division and image enhancements increases by about 30% to 182%. Twenty-eight images of Antarctica and fifteen images of Arctic regions were evaluated to assess the performance of the geometric correction. The average geolocation error was reduced from 10 km to ~300 m. In general, this study presents the geolocation method, which could serve as a reference for the geolocation of other visible remote sensing images for polar observations.

Keywords: geolocation; microsatellite; Ice Pathfinder; BNU-1; geometric correction; image division; image enhancement

1. Introduction

Visible remote sensing plays an important role in earth observations by providing super-width and high spatial resolution visual images. Along with its advantages, it has a wide range of applications in environmental surveying and mapping, disaster monitoring, resource investigation, vegetation monitoring, etc. [1–5]. In polar regions, visible remote sensing provides comprehensive observations of features on the earth's surface, and thus it is a supplement to limited field observations. With climate warming, dramatic changes have taken place in the polar regions where glaciers have retreated [6,7], ice flow has accelerated [8,9], and sea-ice has shrunk rapidly [10]. However, many of the rapid changes occurring in polar regions are difficult to monitor due to the trade-off between the temporal and spatial resolutions of existing satellite sensors (fine spatial resolution with a long revisit period; coarse resolution with a short revisit period) [4,11]. For example, the Moderate-Resolution Imaging Spectroradiometer (MODIS) sensors aboard the Terra/Aqua satellites can provide daily observations that facilitate the capture of rapid surface changes [4], but the coarse spatial resolution (250–1000 m) of MODIS sensors is often inadequate for monitoring the collapse of small glaciers or the disintegration of small icebergs. In contrast, the Landsat-8 OLI/Sentinel-2 MSI sensor has a higher spatial resolution (30 m/10 m) than MODIS, providing more details of the snow and ice surface changes. However, the 16-day/10-day revisit period of the Landsat-8 OLI/Sentinel-2 MSI sensor limits its application in the study of time-sensitive events, such as sea ice drift, which may evolve rapidly in a few days. Therefore, a sensor that provides high-resolution remote sensing data on a daily frequency or satellite constellations are needed for observing the rapid changes in polar regions.

Launched on 12 September 2019 and developed through the collaboration between Beijing Normal University, Sun Yat-sen University, led by Shenzhen Aerospace Dongfanghong HIT Satellite Ltd., Ice Pathfinder (Code: BNU-1) is the first Chinese polar-observing microsatellite. It is in a sun-synchronous orbit (SSO) with an altitude of 739 km above Earth's surface, a semi-major axis of 7,116,914.419 m, an inclination of 98.5238 Degrees, and an eccentricity of 0.000220908. Weighing only 16 kg, BNU-1 carries an optical payload with a panchromatic band and four multispectral bands. The spatial resolution at the sub-satellite point is approximately 74 m from the ground. The wide swath of BNU-1 (744 km) provides a 5-day revisit period of polar regions up to 85° latitude. BNU-1 takes into account the balance between spatial resolution and revisit frequency, providing observations with finer spatial resolution than Terra/Aqua MODIS data and more revisit frequency than Landsat-8 OLI and Sentinel-2 MSI, benefiting the environmental monitoring of the polar regions. Also, the low cost of BNU-1 makes it financially feasible to construct a constellation observation system [12]. A five-satellite constellation system provides the ability to observe polar environmental elements on a daily basis with a spatial resolution finer than 100 m.

Image geolocation is an essential process prior to the application of satellites. However, geolocation errors are commonly found in visible images. For example, the images from MODIS have a geolocation error of 1.3 km in the along-track direction and 1.0 km in the along-scan direction without correction [13]. Geolocation errors need to be corrected because they cause uncertainty in satellite data and have a serious impact on the applications of satellite data for environmental monitoring [14]. The geolocation errors are usually corrected by parametric and non-parametric correction models [13,15,16]. Both these models correct the errors of a satellite image by matching the CPs obtained from the target image (the image with geolocation errors) and the corresponding points from the reference image with high geolocation accuracy [17]. The parametric correction model corrects the errors by optimizing the inner and external orientation parameters in the geometric calibration model based on the differences between the CPs from the target and reference images [13,15,18], while the non-parametric model is performed by establishing the coordinates transformation model between coordinates of the target image and coordinates of the reference images based on the CPs [8,11,12].

Both the parametric and non-parametric correction models are highly dependent on the amount of CPs [17,19,20]. Various methods have been used to increase the number of CPs extracted from the images, such as image division [21,22] and histogram equalization [22], and GCP sampling optimization [17]. Other methods have been used to eliminate the mis-matched CPs such as random sample consensus (RANSAC) [17,19], etc. These methods are commonly used for images with rich textures at low- and mid-latitudes. However, due to the high reflectance of ice and snow surfaces at high latitudes, the texture of images in polar regions is rarely observed. It is necessary to explore methods for correcting the geolocation errors of the images of polar regions.

This study aims to develop a geolocation method for polar images from BNU-1. The BNU-1 images for several regions of Antarctica and Greenland were used to demonstrate the effectiveness of this proposed geolocation method to deliminate the geolocation errors. This paper is organized as follows. Section 2 describes the data and the study area. Section 3 describes the geolocation method of the BNU-1 images in detail. Section 4 shows the performance of the geolocation method. The discussion of the results is shown in Section 5. Conclusions are given in Section 6.

2. Data

BNU-1 Imagery. BNU-1 has obtained more than 6000 images covering Antarctica and Greenland since it was launched. It provides the observations in panchromatic, blue, green, red, and red-edge spectral bands. Twenty-eight images of Antarctica and fifteen images of Greenland in the panchromatic band were collected for geolocation and accuracy evaluation. As shown in Figure 1, the images of Antarctica are distributed over the Amery Ice Shelf, Victoria Land, Dronning Maud Land, and Pine Island Glacier. The images of Greenland cover the west and north of Greenland.

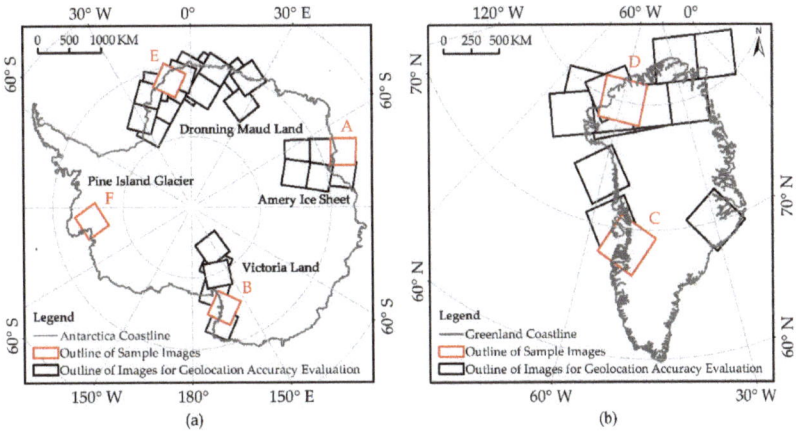

Figure 1. Distributions of the sample images of BNU-1 for (**a**) Antarctica and (**b**) Greenland. Black and red rectangles refer to the footprints of 43 sample BNU-1 image scenes, in which red rectangles show the 6 Sample Images (A–F) used for the analysis in this study. Sample Images A–F were acquired for the Amery Ice Shelf on 8 October 2019, Victoria Land on 11 October 2019, Greenland on 7 July 2020, Greenland on 18 July 2020, Dronning Maud Land on 18 December 2019, and Pine Island Glacier on 28 December 2019, respectively.

MODIS Imagery. MODIS is a key instrument onboard the Terra and Aqua satellites, which were launched on 18 December 1999, and 4 May 2002, respectively, providing global coverage every one to two days. Since the MODIS sensor has high geolocation accuracy (50 m for one standard deviation) [4] and a daily revisit capability, we used MOD02QKM and MYD02QKM products as the reference images for the geometric correction of the BNU-1 images in this study. The geolocation error of MOD02QKM (MYD02QKM) is 50 m or better, which is finer than the pixel size of the MODIS image [13,23,24]. It is reasonable to use MODIS images as the reference data in this study since the spatial resolution of BNU-1 images is 80 m.

Coastline dataset. We used the high-resolution vector polylines of the Antarctic coastline (7.4) [25] of 2021 from the British Antarctic Survey (BAS). We also used the

MEaSUREs MODIS Mosaic of Greenland (MOG) 2005, 2010, and 2015 Image Maps, Version 2 [26] from the NASA National Snow and Ice Data Center (NSIDC) to obtain the Greenland coastline. We evaluated the geolocation error of the BNU-1 image visually by comparing the coastline dataset with the geolocation of the BNU-1 image.

3. Methods

There are two steps for geolocating the BNU-1 images in this study. The first step is geometric calibration. In this step, a geometric calibration model is constructed to transform the image coordinates to geographic coordinates. The images with geographic coordinates are the BNU-1 Level 1A (L1A) product. The second step is the geometric correction. The geolocation errors of the BNU-1 L1A product are corrected by an automated geometric correction method in this step. This method is designed to correct the geolocation errors of the images of polar regions where surface textures rarely exist. The corrected BNU-1 L1A product, which has high geolocation accuracy, is named the BNU-1 Level 1B (L1B) product.

3.1. Geometric Calibration Model

3.1.1. Description of Geometric Calibration Model

A rigorous geometric calibration model was constructed for transforming the image coordinates to the geographic coordinates for the BNU-1 images. The timing, position, and altitude of satellites and camera position parameters are used as inputs of the model. The model is shown as follows [27]:

$$\begin{bmatrix} X \\ Y \\ Z \end{bmatrix}_{WGS84} = \begin{bmatrix} X_s(t) \\ Y_s(t) \\ Z_s(t) \end{bmatrix}_{WGS84} + m \left(R_{J2000}^{WGS84} R_{body}^{J2000} \right)_t \left[\begin{bmatrix} D_x \\ D_y \\ D_z \end{bmatrix} + \begin{bmatrix} d_x \\ d_y \\ d_z \end{bmatrix} + R_{camera}^{body} \begin{bmatrix} \tan \psi_x \\ \tan \psi_y \\ 1 \end{bmatrix} \right] \quad (1)$$

where t is the scanning time of an imaginary line. $\begin{bmatrix} X_s(t) & Y_s(t) & Z_s(t) \end{bmatrix}^T_{WGS84}$ indicates the coordinates of the Global Positioning System (GPS) antenna phase center, which are measured by a GPS receiver on the satellite in the WGS84 coordinate system (derived from ECEF) at t. m is the scale factor determined by the orbital altitude. $\left(R_{J2000}^{WGS84} \right)_t$, $\left(R_{body}^{J2000} \right)_t$, and R_{camera}^{body} are the rotation matrix of the coordinate system from J2000 to WGS84 at t, the rotation matrix from the satellite's body-fixed coordinate system to J2000 coordinate system at t, and the rotation matrix from the camera coordinate system to the satellite's body-fixed coordinate system, respectively. $\begin{bmatrix} D_x & D_y & D_z \end{bmatrix}^T$ is the coordinates of the GPS antenna phase center in the satellite's body-fixed coordinate system. $\begin{bmatrix} d_x & d_y & d_z \end{bmatrix}^T$ is the translations of the origin of the camera coordinate system relative to the satellite's body-fixed coordinate system. $\begin{bmatrix} \tan \psi_x & \tan \psi_y & 1 \end{bmatrix}^T$ represents the value of the coordinates of point (x, y) corresponding to the detector direction angle model composed of the camera's principal point, focal length, charge coupled device (CCD) installation position, and lens distortion. $\begin{bmatrix} X & Y & Z \end{bmatrix}^T_{WGS84}$ represents the ground coordinates of the point (x, y) in the World Geodetic System 1984 (WGS84) coordinate system.

$\tan \psi_x$ and $\tan \psi_y$ describe the directional angle of point (x, y) in the camera coordinate system [27–29], and this can be calculated by Equation (2), where f is the focal length of the camera.

$$\tan \psi_x = \frac{x}{f}$$
$$\tan \psi_y = \frac{y}{f} \quad (2)$$

This step is conducted on the Windows Server 2016 Standard operating system on the Intel(R) Xeon(R) Gold 5220R CPU @2.20 GHz, 256 GB RAM. It is a whole-day unattended automatic data production system.

3.1.2. Uncertainty Evaluation of Geometric Calibration Model

In addition to systematic errors, the geolocation of acquired images is also affected by random errors. The satellite imaging process is affected by various complex conditions such as attitude adjustment, attitude measurement accuracy, and imaging environment. In addition, as a microsatellite, BNU-1′s low cost and the imaging environment of polar regions limit, to a certain extent, the overall accuracy and stability of the measurement equipment of attitude and position. Moreover, due to the wide swath of BNU-1′s camera, the imaging time of its single-scene image is long. As the satellite attitude is adjusted along the imaging, random errors in attitude measurement cause random geolocation errors in single-scene images and multiple-scene images. Since the measurement error of GPS can be regarded as translation error (Equation (1)), which is equivalent to the satellite rotating at a small angle, we only designed and carried out an experiment to simulate the influence of the satellite's attitude angle change on the geolocation change through Equation 1. The satellite's attitude is determined by the roll angle, pitch angle, and yaw angle. We randomly selected an image and simulated the angles of roll, pitch, and yaw, which changed from 0° to 0.5° with a step size of 0.1°, to obtain 216 (6 * 6 * 6) groups of geolocations. The quintic polynomial method was used to fit the scatter plot.

3.2. Automated Geometric Correction Processing Method

Since the space environment is complex and variable during satellite launch and operation [15,28,30,31], the geographic coordinates calculated by geometric calibration models with the pre-launch laboratory measurement parameters usually have geolocation errors of about several hundred meters to several kilometers [32,33]. In addition, the random error of the attitude measurement cannot be eliminated due to the lack of ground control points in polar regions. An automated geometric correction method based on CPs matching was developed to improve the geolocation accuracy of the BNU-1 L1A product. There are three steps involved in the method. Firstly, we selected the reference image with a high geolocation accuracy for the BNU-1 images. Then, the Scale Invariant Feature Transform algorithm (SIFT) [33] was used to extract the CPs from both the BNU-1 image and the corresponding reference image. Finally, geometric correction was conducted on the BNU-1 image based on the CPs. The flow chart is shown in Figure 2. Our experiment was conducted on the operating system of Windows 10 on the Intel(R) Core (TM) i5-5200 CPU @2.20 GHz, 8 GB RAM. We used the programming language of python2.7 to implement the one-stop processing of the automatic geometric correction. In this process, the programming language of MATLAB was used to realize SIFT algorithm, and the software of ArcGIS 10.6 was used to realize data preprocessing and geometric correction.

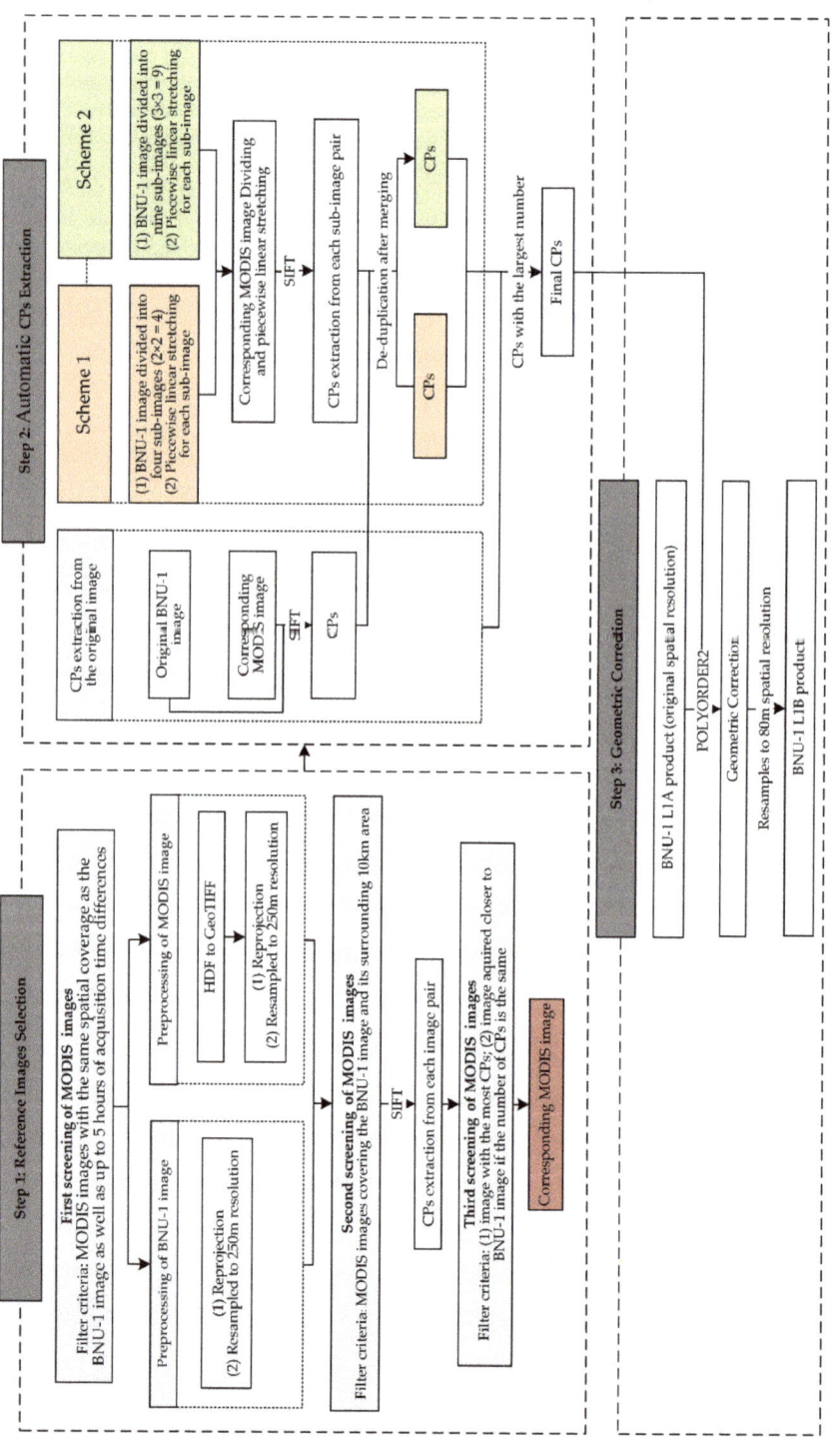

Figure 2. The flowchart for the automated geometric correction processing method of BNU-1 L1A product.

3.2.1. Step 1: Reference Images Selection

There were three criteria for selecting a reference image. Firstly, we selected the MODIS images with the same spatial coverage as the BNU-1 image as well as up to 5 h of different acquisition times. Secondly, the MODIS images covering the BNU-1 image and its surrounding 10 km area were chosen (one BNU-1 image corresponds to multiple MODIS images). Thirdly, the SIFT algorithm was adopted to extract the CPs from each image pair, where the BNU-1 image and the MODIS image are the target image and the reference image, respectively. The MODIS image with the most CPs was used as a reference image for the geometric correction. If more than one MODIS image has the highest number of CPs, the one whose acquisition time is closer to the BNU-1 image's acquisition time is preferred as the reference image. The reference image used for geometric correction of the BNU-1 image is referred to as the corresponding MODIS image hereinafter.

3.2.2. Step 2: Automatic CPs Extraction

The amount and spatial distribution of CPs are key factors for geometric correction because they have direct impacts on the geometric correction accuracy of the corrected images. In this study, we applied the SIFT algorithm based on MATLAB language to extract the CPs automatically from the BNU-1 L1A and the corresponding MODIS image. Due to the lack of texture features of snow and ice surfaces at high latitudes, CPs extracted from the original image pair are usually not sufficient for correcting the geolocation errors. When the number of CPs extracted from the image pair needs to be increased, image division and image enhancement methods are used to enhance the texture features of satellite images [1,21,22,34].

The combination of an image division method and an image enhancement method was applied to highlight surface features of the BNU-1 and MODIS images in this study. The extraction of CPs was carried out in three steps in Step 2 (Figure 2). Firstly, we extracted the CPs from the original BNU1-1 image and the corresponding MODIS image by using the SIFT algorithm. The Euclidean distances between each pair of CPs from the image pair were calculated. To avoid mismatches of the points, we eliminated the largest 10% points in the Euclidean distance. Secondly, we extracted the CPs from the image pair after processing by different image division schemes. The paired images were divided into $2 \times 2 = 4$ (Scheme 1) and $3 \times 3 = 9$ (Scheme 2) sub-images [22]. Then, an adaptive piecewise linear enhancement consisting of three rules for low, middle, and high reflectance ranges was used to enhance each sub-image [34]. We extracted the CPs from all the pairs of the sub-images again by the SIFT algorithm. The largest 30% of the extracted CPs in Euclidean distance were eliminated in this step. Finally, the CPs extracted in the above two steps were merged and de-duplicated to obtain the CPs with the largest number, which were taken as the final CPs for the geometric correction.

3.2.3. Step 3: Geometric Correction

This study performed the geometric correction of the BNU-1 L1A product with the original spatial resolution (74 m) by using a quadratic polynomial (POLYORDER2) model in ArcGIS 10.6 software. We obtained the BNU-1 L1B product by resampling the geolocation error-corrected image to 80 m spatial resolution using the nearest resampling method.

3.3. Geolocation Accuracy Evaluation

To evaluate the geolocation accuracy of the BNU-1 L1A/L1B product, we re-extracted the CPs from the BNU-1 L1A/L1B product by the SIFT algorithm as the verification points and calculated the root mean squared error (RMSE) of the verification points:

$$RMSE = \sqrt{\frac{\sum_{i=1}^{n}[R_{xi}^2 + R_{yi}^2]}{n}} \qquad (3)$$

where n is the number of points participating in the accuracy evaluation, R_{xi} and R_{yi} represent the residuals of the i-th extracted points from the BNU-1 and the reference MODIS image in the X and Y coordinates, respectively.

4. Results

4.1. Geolocation Accuracy of the BNU-1 L1A Product

The BNU-1 L1A images with 50% transparency are superimposed on the corresponding MODIS images in Figure 3. The sub-figures (a), (b), (c), and (d) correspond to the Sample Images A, B, C, and D shown in the red box in Figure 1. The prominent features in the images, such as coastlines, rocks, sea ice, etc., are blurred, indicating the mismatch of the geometric position between the BNU-1 images and the corresponding MODIS images. Obvious geolocation errors are observed in the BNU-1 L1A images. Table 1 shows the geolocation errors of the 42 scene BNU-1 L1A images. The errors of the BNU-1 L1A images range from 3 to 20 km, with an average of about 10 km. The geolocation errors of the sub-graphs in Figure 3a–d are 6544.83 m, 7919.60 m, 15,071.02 m, and 7778.63 m, respectively (Table 1).

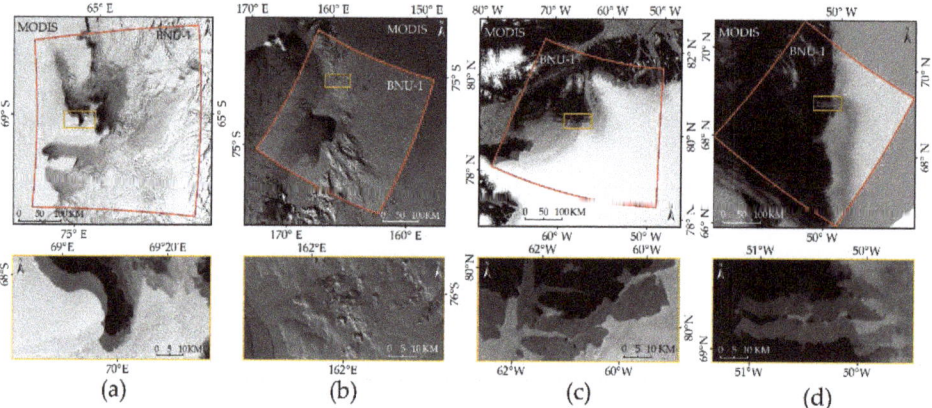

Figure 3. BNU-1 L1A images superimposed on the corresponding MODIS images. Red polygons refer to the outline of the Sample Image scenes shown in Figure 1. (**a**) Sample Image A; (**b**) Sample Image B; (**c**) Sample Image C; (**d**) Sample Image D. The images in the yellow boxes below are the enlarged versions of the part in the sample images.

Table 1. Geolocation errors of the BNU-1 L1A/L1B product (unit: meter).

Scene ID	RMSE		Scene ID	RMSE	
	L1A	L1B		L1A	L1B
	Amery Ice Sheet			Victoria Land	
1 (A) [1]	6544.83	270.19	1	7639.83	412.85
2	8613.73	180.87	2 (B)	7919.60	253.40
3	5061.16	234.17	3	4683.40	277.69
4	10,848.59	175.13	4	17,855.48	277.47
5	7079.49	245.66	5	8602.73	293.14
6	9572.81	237.79	6	17,380.35	339.21
	Dronning Maud Land			Greenland	
1	5642.45	362.99	1	16,236.81	283.33
2	6916.70	243.97	2	19,828.19	189.23
3	3625.91	189.88	3	15,870.7	229.51
4	5680.75	203.38	4	12,435.58	299.84
5	6761.34	258.69	5	9959.07	321.84
6	8142.78	220.03	6	10,836.43	258.28
7	8012.31	324.15	7	7854.03	339.87
8	8036.76	484.89	8	18,738.91	178.34
9	7759.36	279.76	9	13,244.36	216.14
10	7007.99	203.37	10 (C)	15,071.02	269.83
11	6786.16	242.65	11	13,489.89	265.7
12	5045.67	506.19	12	17,819.43	414.26
13	10,215.28	458.47	13	19,880.35	331.14
14	6880.00	575.88	14	16,219.35	292.06
15 (E)	7309.43	221.59	15 (D)	7778.63	302.29
	Pine Island Glacier				
1(F)	19,765.58	783.90			
Average			L1A: 10,480.31 L1B: 301.14		

[1] (A) indicates Sample Image A.

Figure 4 shows the distributions of the geolocation errors of the CPs in the X and Y directions for each image shown in Figure 3. The length and direction of the vectors in Figure 4 represent the magnitude and direction of the CPs' geolocation errors. The directions and the magnitude of geolocation errors for each image are not consistent (Figure 4). For example, the CPs' geolocation errors of Sample Image A, B, and D are less than 12 km, while most CPs' geolocation errors of Sample Image C are up to 19 km. And the geolocation errors in the middle part of Sample Image A are smaller than the errors in the edges of the image, while Sample Image B shows a quite different distribution of geolocation errors. In addition, the direction of the geolocation errors of the CPs shown in Sample Images B, C, and D are also different from the center to the periphery of the images. The CPs' geolocation errors within an image also vary significantly. For example, geolocation errors in Sample Image C are less than 2000 m in the center-west parts and more than 15,000 m in the east and southwest parts (Figure 4). The results illustrate that the distribution of the CPs' geolocation errors varies in each image and indicates that some local distortions exist in the BNU-1 L1A product.

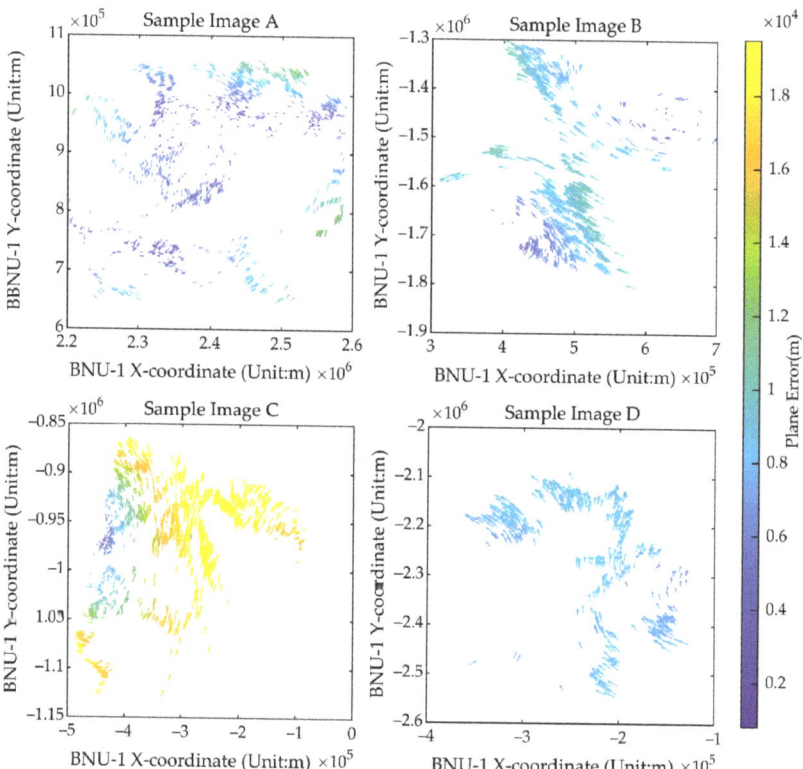

Figure 4. Displacement vectors of the CPs for the four sample BNU-1 L1A images and the corresponding MODIS images in Figure 3. The vectors start and end at the CPs' coordinates in polar stereographic projection on the BNU-1 L1A images and the corresponding MODIS images, respectively. The color of the vectors represents the error magnitude according to the legend.

4.2. Uncertainty Evaluation of Geolocation of BNU-1 L1A Product

Figure 5 shows the three-dimensional scatter diagram of the influence of the change in the satellite's attitude angle—roll, pitch, and yaw—on the geolocation change. The results show that when the angles of roll, pitch, and yaw change from 0.1° to 0.5°, the geolocation change in the upper left corner point of the image changes from −6256 m to 6594 m in the longitude and from −13,915 m to −112 m in the latitude. Similarly, the geolocation change in the center point of the image changes from −6431 m to 4998 m in the longitude and from −11,889 m to −251 m in the latitude. These geolocation changes are non-linear (Figure 5). Under imaging conditions in polar regions, the random error of attitude measurement cannot be eliminated due to the lack of ground control points in polar regions [28]. To obtain high-precision geolocation products, it is necessary to add CPs to the image for geometric correction. Since there are few textures observed on high-reflectance ice and snow surface in polar regions, the extraction of CPs is the key to geometric correction.

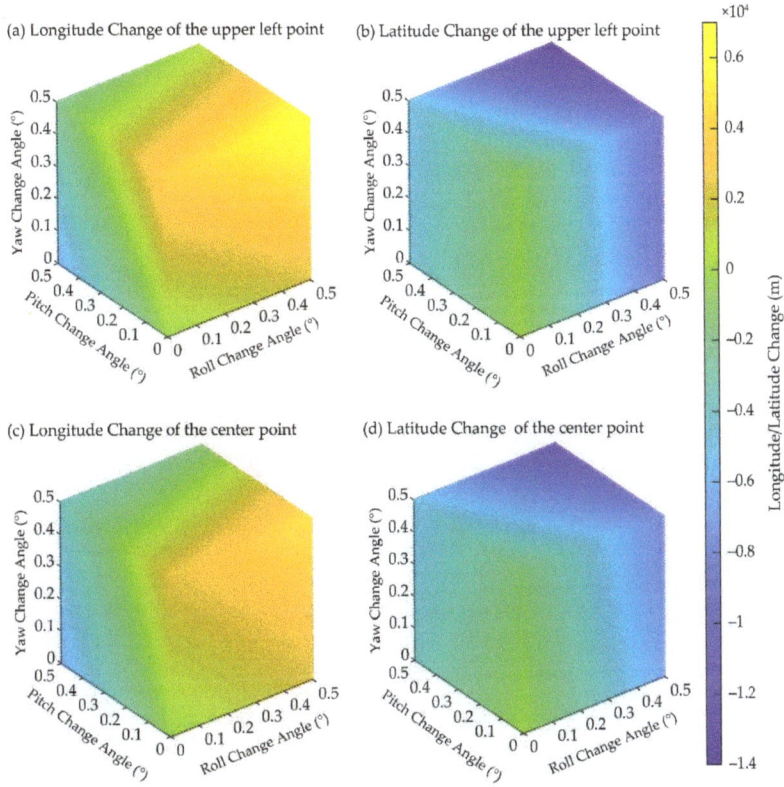

Figure 5. Three-dimensional scatter diagram of the influence of the satellite's attitude angle change on the geolocation change. (**a,b**) show the impact of the satellite's attitude changes on the longitude change and the latitude change of the upper left corner point of the image, respectively. (**c,d**) show the impact of the satellite's attitude changes on the longitude change and the latitude change of the center point of the image, respectively. The color of the scatters represents the longitude/latitude change according to the legend.

4.3. Influence of Image Division and Enhancement on the CPs Extraction

Sample Image E is a typical image for polar regions. Most of the features in the image are ice sheets and snow with limited boundary features, and only a few of them are sea ice with well-defined boundaries. However, due to the high reflectance of ice and snow surfaces in polar regions, the textures of the ice sheet and snow can rarely be observed in images. The ice sheet area of Sample Image E is a good case for evaluating the effectiveness of various image enhancement methods for increasing the control points on the ice sheet. Five image enhancement methods, which are linear enhancement, piecewise linear enhancement, Gaussian enhancement, equalization enhancement, and square root enhancement, were applied to enhance Sample Image E. Figure 6 shows the distributions of CPs extracted from the images enhanced by different image enhancement methods. The numbers of CPs extracted from the original image and the image stretched by the five enhancement methods were 31, 32, 76, 37, 27, and 22, respectively. By comparing these five enhancement methods, we found that the piecewise linear enhancement method makes the surface textures in the interior of the ice sheet more distinct, and as a result, the most CPs were extracted from the image. Therefore, the piecewise linear enhancement (Figure 6c) is considered to be more suitable for enhancing the images of polar regions.

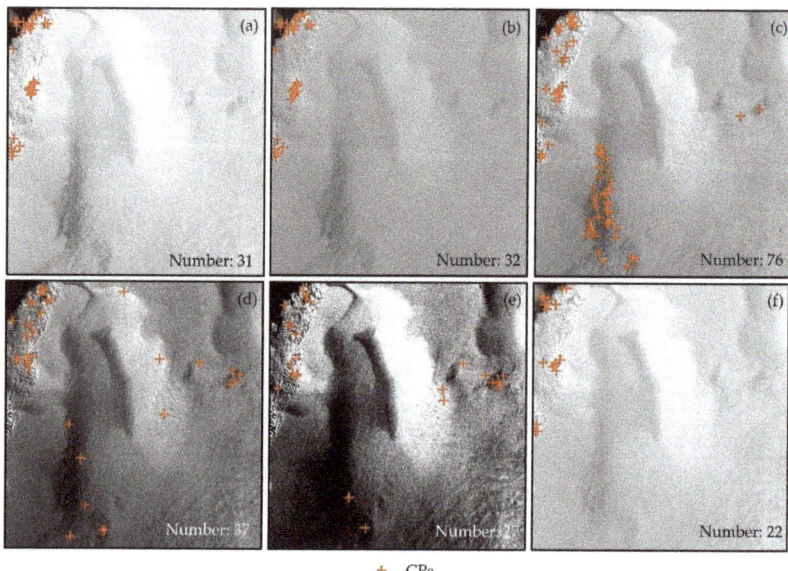

Figure 6. Comparisons of the number and distribution of CPs in the original image (**a**) and the image adopting different enhancement methods: (**b**) Linear; (**c**) Piecewise Liner; (**d**) Gaussian; (**e**) Equalization; (**f**) Square Root.

Sample Image E was also used to assess the influence of image division on CPs extraction. The image was divided into four sub-images (Scheme 1) and nine sub-images (Scheme 2) and then each sub-image was individually enhanced with the piecewise linear stretching method. Figure 7 shows the distribution of CPs extracted from the original image by Scheme 1, and by Scheme 2, respectively. The number of extracted points are 245, 435, and 596, respectively. More CPs are extracted in the center and the lower right corner of the image (the blue border area) divided by Scheme 2 (Figure 7c) compared to the original image (Figure 7a) and the image divided by Scheme 1 (Figure 7b). As shown in Table 2, the amount of CPs extracted from the image increases by 30% to 182% when the image division and piecewise linear enhancement were applied to the images. However, this does not mean we can get more CPs if the image is divided into more sub-images. The amount of CPs extracted from the Sample Image A, B, C, and D is less when Scheme 2 is applied to divide these images.

The CPs extracted from Sample Image E were used to correct the geolocation errors of the image. The BNU L1A/L1B image with 50% transparency is superimposed on the corresponding MODIS image in Figure 8. There is a distinct displacement between the BNU-1 L1A image and the MODIS image, while the displacement between the BNU-1 L1B image and the MODIS image can barely be discerned. This result indicates that the geolocation correction method improves the geolocation accuracy of the BNU-1 L1A product.

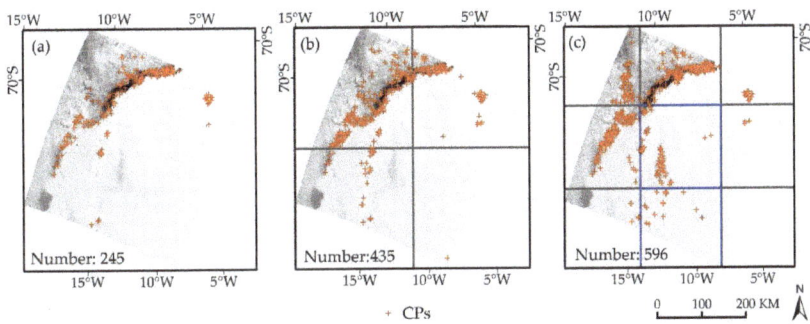

Figure 7. Schematic diagram of the number and distribution of the CPs extracted by different division strategies. (**a**) The original image; (**b**) Scheme 1; (**c**) Scheme 2.

Table 2. The number of the extracted CPs with different image division schemes.

Sample Image	Number of CPs Extracted from the Original Image	Number of CPs Extracted from Scheme 1	Number of CPs Extracted from Scheme 2	Optimal Increment of CPs (%) [2]
A	1071	2100 [1]	1840	96
B	935	2280 [1]	1624	144
C	1334	2236 [1]	1905	68
D	447	580 [1]	579	30
E	245	435	596 [1]	143
F	17	32	48 [1]	182

[1] represents the optimal number of the extracted CPs for geometric correction. [2] represents the ratio of the difference between the optimal number and the original image number to the original image number.

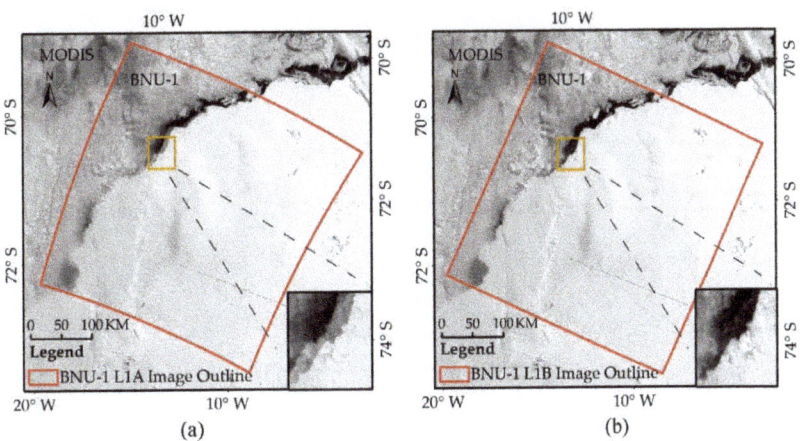

Figure 8. The BNU-1 images superimposed on the corresponding MODIS images. (**a**) BNU-1 L1A image; (**b**) BNU-1 L1B image. Red boxes refer to the extent of the BNU-1 L1A/L1B images.

In addition to Sample Image E, Sample Image F was also selected to evaluate the effectiveness of the CPs extraction scheme proposed. Most of the areas of Sample Image F are covered by the ice sheet, and only a few areas are fjords. Figure 9 shows the image after geometric correction of the BNU-1 L1A image using the CPs extracted from the original image and the optimal control point extraction scheme (Scheme 2). The CPs extracted from the original image are few and unevenly distributed. If these points are directly used for geometric correction of the BNU-1 L1A image, the corrected image will be severely distorted (Figure 9a). More, and more evenly distributed CPs are extracted of the image divided by Scheme 2 (Figure 9b) compared to the original image (Figure 9a). The corrected BNU-1 image overlaps well with the MODIS image (Figure 9b).

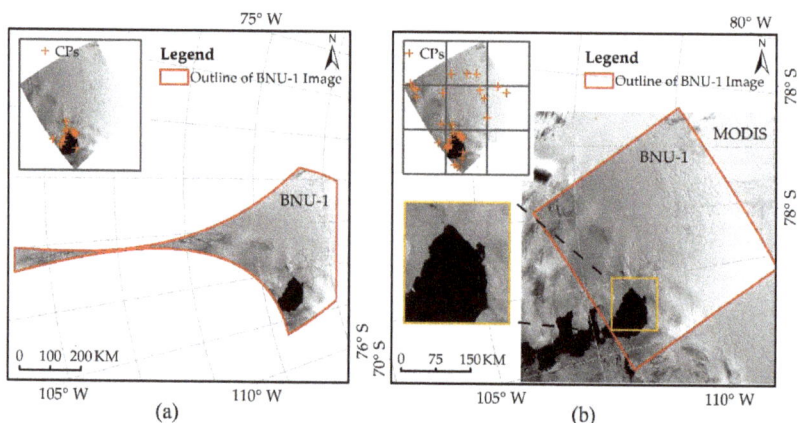

Figure 9. BNU-1 image after geometric correction of the BNU-1 L1A image using the CPs extracted from the original image (**a**) and Scheme 2 (**b**). The corrected image with 50% transparency is superimposed on the corresponding MODIS image in (**b**).

4.4. Geolocation Accuracy of the BNU-1 L1B Product

The verification points used to evaluate the geolocation errors of the BNU-1 L1B product for the Sample Images A, B, C, and D are shown in Figure 10a. We compared the coordinates of verification points in the BNU-1 L1A/L1B images with those in the corresponding MODIS images in Figure 10b, c. The verification points extracted from the BNU-1 L1A product (green dots) are distributed on one side of the 1:1 line (black diagonal line), which means that geolocation errors exist in the BNU-1 L1A product, while the verification points extracted from the BNU-1 L1B product (red dots) are almost scattered on the 1:1 line. We fitted the linear relationships between the coordinates of the verification points from BNU-1 L1A/BUN-1 L1B and the MODIS image. The regression coefficients, intercepts, and determined coefficients of the relationship fitted by the BNU-1 L1B product are significantly better than those fitted by BNU-1 L1A. The coordinates of the points from the BNU-1 L1B product show great consistency with the coordinates from the MODIS images. The geolocation accuracy of the BNU-1 L1B images was improved significantly. After geometric correction, the average geolocation error was reduced from 10,480.31 m to 301.14 m (Table 1).

We obtained the image mosaics of the Amery Ice Shelf and Victoria Land in Antarctica and northern Greenland in the panchromatic band of BNU-1 (Figure 11). Mismatches in the coastlines were found in the image mosaics from the BNU-1 L1A product. However, the coastlines in the image mosaics from the BNU-1 L1B product are consistent with the existing coastline dataset [25]. Even though the junction of adjacent images in the image mosaic from the BNU-1 L1B product has greatly improved coherence compared to the BNU-1 L1A product, the BNU-1 L1B product still has an average geolocation error of ~300 m. For example, the discontinuous waters at the junction are found in Victoria Land (obvious mismatches in the yellow circle in Figure 11a) from the BNU-1 L1A product, while the mosaic from the BNU-1 L1B product has consistent waters at the junction regions (slight mismatches in the green circle in Figure 11a).

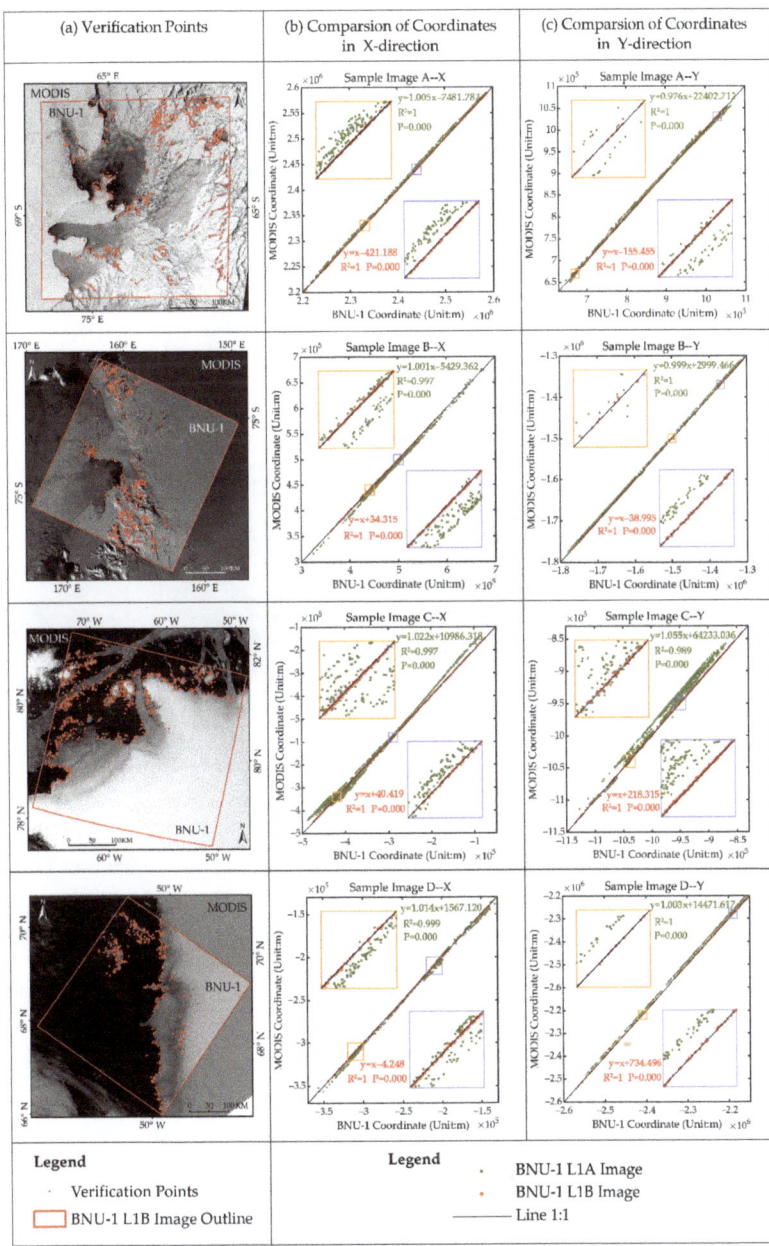

Figure 10. Spatial distribution of the verification points for accuracy evaluation of BNU-1 L1B Sample Images A, B, C, and D (**a**); and the coordinates comparison of the verification points of BNU-1 L1A/L1B Sample Image A, B, D, and E and their corresponding MODIS images, in X-direction (**b**) and Y-direction (**c**), respectively. Green and red dots represent the verification points on the BNU-1 L1A image and the BNU-1 L1B image, respectively. The black diagonal line in the sub-figures represents that the coordinates of points in the BNU-1 image are almost equal to those in the corresponding MODIS image. The two small graphs in each sub-graph are the enlarged version of the orange and blue rectangular areas on the black diagonal line.

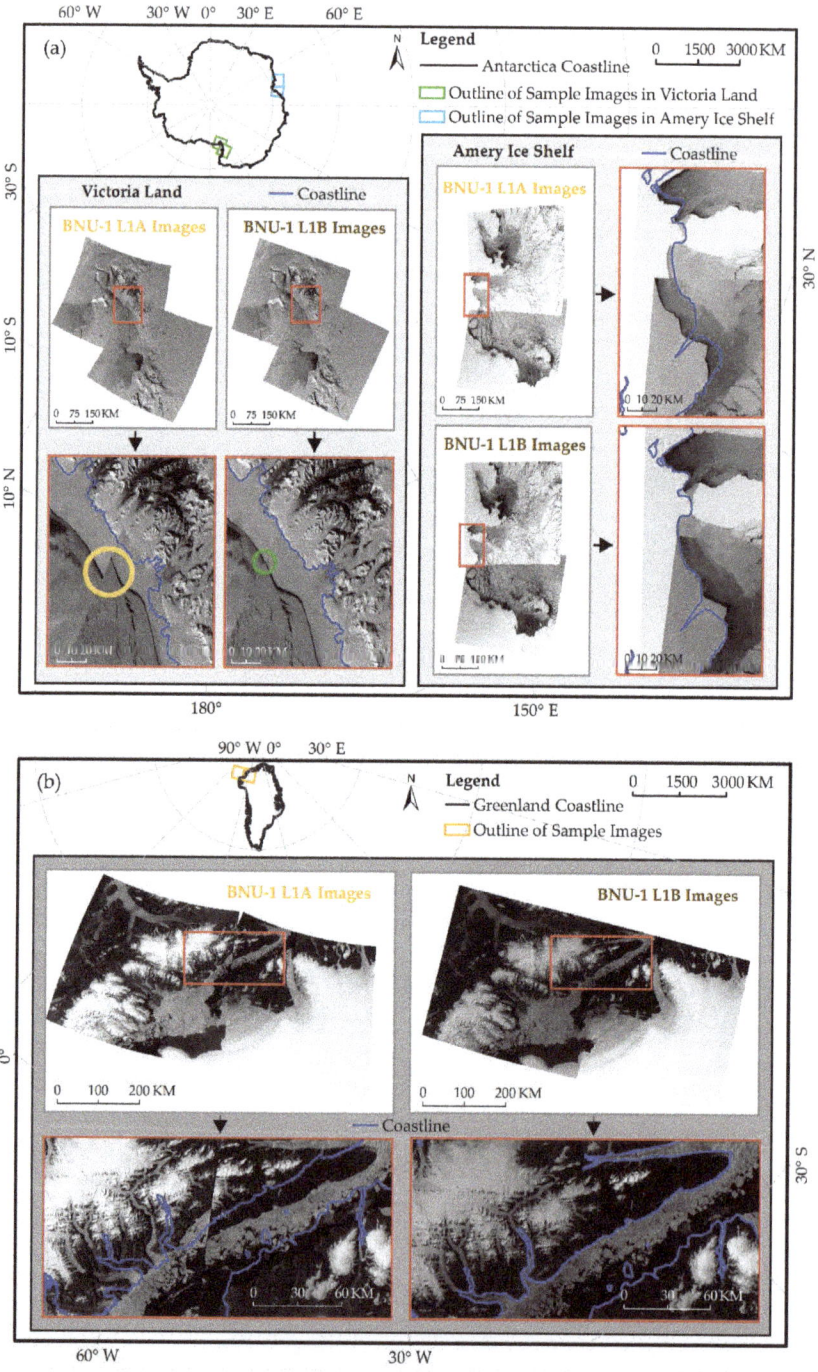

Figure 11. Mosaic of the panchromatic band images of some BNU-1 L1A/L1B Sample images: (**a**) Antarctica; (**b**) Greenland. The blue lines represent the existing coastline dataset.

5. Discussion

Although microsatellites have the advantages of compactness, low cost, and flexibility, their flight attitude may be unstable occasionally and the equipment measuring the attitude and position may be inaccurate or perform poorly, which leads to large geolocation errors in the images [35]. For example, the geolocation errors in UNIFORM-1's visible images are 50–100 km [35]. The BNU-1 L1A product has smaller geolocation errors, but the average geolocation error can still be up to 10 km, which is close to that of the Luojia 1-01 data [30].

The non-parametric geometric correction methods are widely applied to geolocation correction of images without distinguishing the error sources [16,31], such as HJ-1A/B CCD images [36] and Unmanned Aerial Vehicle (UAV) images [37]. The geolocation accuracy of these geometric correction methods relies mainly on adequate CPs. However, the limited texture features of the ice and snow surface in polar images make it difficult to extract the CPs. Some studies prove that image division and image enhancement have the ability to increase the amount of extracted CPs [1,21,22,34]. However, images used by these previous studies are from low- and mid- latitudes and contain rich land surface features. The correction for polar images with few texture features is rarely documented. This study proposes the geometric correction method to reduce the geolocation errors of the visible images for polar regions. The results indicate that piecewise linear enhancement highlights more surface features of ice and snow surfaces than other image enhancement methods. Some other studies have also proved that piecewise linear enhancement is effective in highlighting more texture features of the ice and snow surfaces in polar images [34]. More CPs can be observed after the image pair is processed by image division and piecewise linear enhancement. Different division schemes can be adopted to obtain more CPs for different image pairs.

In addition, we compared the geolocation accuracy of Sample Image A–F after the correction through the CPs extracted from the original image and the geolocation accuracy after correction through the optimal CPs extraction scheme (Table 3). It was found that the geolocation accuracy of Sample Image A–E was not significantly improved. The geolocation accuracy of Sample Image A–E after geometric correction based on the CPs extracted from the original image was close to the level of 250 m (the pixel size of MODIS image), and it was difficult to further improve by adding CPs on this basis. However, for some images where the ice sheet is widely distributed, such as Sample Image F, the proposed method effectively prevents the distortion of the corrected image caused by the lack of CPs on the ice sheet by adding CPs. The increase in CPs can remarkably improve the geolocation accuracy of such images. Therefore, the automatic geometric correction method proposed in this study is of great significance for the correction of images in polar regions with rare feature points.

Table 3. Comparison of the geolocation accuracy of the BNU-1 images corrected through different CPs extraction Scheme.

Sample Image	Geolocation Accuracy of the Corrected BNU-1 Image (m)		Improvement in Geolocation Accuracy (%)
	Corrected by the CPs Extracted from the Original Image	Corrected by the CPs Extracted from the Optimal Scheme	
A	279.42	270.19	3.30
B	260.80	253.40	2.84
C	275.79	269.83	2.16
D	305.28	302.29	0.98
E	242.43	221.59	8.60
F	/(Serious distortion)	783.90	/

Although the method presented in this study has some advantages in correcting the geolocation errors of polar images, it also has its limitations. For example, only two division schemes were applied in the BNU-1 images, and the division fractions for different images are still worth further study. Besides, the geolocation accuracy of some sub-images with relatively uniform surface features is difficult to improve by using the proposed method. This indicates that the SIFT algorithm has its limitation for finding more CPs. Therefore, it is necessary to explore some other CPs extraction methods for increasing the number of CPs. Since deep learning methods have been widely used in image registration [38,39], it is worth exploring the possible application of deep learning methods on CPs extraction from the images of polar regions.

6. Conclusions

In this study, we present the geolocation method for BNU-1 images including two steps. For the first step, a rigorous geometric calibration model was applied to transform the image coordinates to the geographic coordinates for the BNU-1 images. The images geolocated by the geometric calibration model are the BNU-1 L1A product. For the second step, an automated geometric correction method was used to reduce the geolocation errors of the BNU-1 L1A product. The images corrected by the geometric correction method are the BNU-1 L1B product.

The geometric correction method is commonly used for improving the geolocation errors of the visible image. However, the texture features of the ice and snow surfaces are rarely seen in polar images, which makes it difficult to find the CPs. The combination of the image division method and piecewise linear image enhancement method was applied to the BNU-1 L1A product and the corresponding MODIS images, and the results indicate that the CPs extracted increased by 30% to 182%, which can effectively improve the geometric accuracy of the BNU-1 images.

The geolocation method was applied to 28 images of Antarctica and 15 images of Arctic regions. The average geolocation error was reduced from 10 km to ~300 m. The coastlines in the image mosaics from the BNU-1 L1B product were consistent with the coastline dataset. These results suggest that the geolocation method has the ability to improve the geolocation errors of BNU-1 images and other satellite images in polar regions.

Author Contributions: Conceptualization, X.C.; methodology, Y.Z. and Z.C. (Zhuoqi Chen); software, Y.Z.; validation, Y.Z.; formal analysis, Y.Z.; data curation, Y.Z.; writing—original draft preparation, Y.Z. and Z.C. (Zhuoqi Chen); writing—review and editing, Y.Z., Z.C. (Zhaohui Chi), Z.C. (Zhuoqi Chen), F.H., T.L., B.Z. and X.L.; visualization, Y.Z.; supervision, Z.C. (Zhuoqi Chen), X.C.; project administration, X.C.; funding acquisition, X.C. All authors have read and agreed to the published version of the manuscript.

Funding: This research was funded by National Natural Science Foundation of China (Grant No. 41925027), National Key Research and Development Program of China (Grant No. 2019YFC1509104 and 2018YFC1406101), and Innovation Group Project of Southern Marine Science and Engineering Guangdong Laboratory (Zhuhai) (No. 311021008).

Institutional Review Board Statement: Not applicable.

Informed Consent Statement: Not applicable.

Data Availability Statement: The data presented in this study are available on request from the corresponding author.

Acknowledgments: We greatly thank NASA for providing the MODIS Level 1B calibrated radiances data (MOD02QKM and MYD02QKM) (https://ladsweb.modaps.eosdis.nasa.gov/search/, accessed on 11 September 2020), BAS for the high resolution vector polylines of the Antarctic coastline (7.4) (https://data.bas.ac.uk/items/e46be5bc-ef8e-4fd5-967b-92863fbe2835/#item-details-data, accessed on 1 August 2021), and NSIDC for the MEaSUREs MODIS Mosaic of Greenland (MOG) 2005, 2010, and 2015 Image Maps, Version 2 (https://nsidc.org/data/nsidc-0547, accessed on 20 October 2020). We greatly thank Zhuoyu Zhang for her help with the preprocessing of MODIS images.

Conflicts of Interest: The authors declare no conflict of interest.

References

1. Chen, Z.; Chi, Z.; Zinglersen, K.B.; Tian, Y.; Wang, K.; Hui, F.; Cheng, X. A new image mosaic of greenland using Landsat-8 OLI images. *Sci. Bull.* **2020**, *65*, 522–524. [CrossRef]
2. Ban, H.-J.; Kwon, Y.-J.; Shin, H.; Ryu, H.-S.; Hong, S. Flood monitoring using satellite-based RGB composite imagery and refractive index retrieval in visible and near-infrared bands. *Remote Sens.* **2017**, *9*, 313. [CrossRef]
3. Schroeder, W.; Oliva, P.; Giglio, L.; Quayle, B.; Lorenz, E.; Morelli, F. Active fire detection using Landsat-8/OLI data. *Remote Sens. Environ.* **2016**, *185*, 210–220. [CrossRef]
4. Gao, F.; Hilker, T.; Zhu, X.; Anderson, M.; Masek, J.; Wang, P.; Yang, Y. Fusing landsat and MODIS data for vegetation monitoring. *IEEE Geosci. Remote Sens. Mag.* **2015**, *3*, 47–60. [CrossRef]
5. Miller, R.L.; McKee, B.A. Using MODIS Terra 250 m imagery to map concentrations of total suspended matter in coastal waters. *Remote Sens. Environ.* **2004**, *93*, 259–266. [CrossRef]
6. Benn, D.I.; Cowton, T.; Todd, J.; Luckman, A. Glacier calving in greenland. *Curr. Clim. Chang. Rep.* **2017**, *3*, 282–290. [CrossRef]
7. Yu, H.; Rignot, E.; Morlighem, M.; Seroussi, H. Iceberg calving of Thwaites Glacier, West Antarctica: Full-Stokes modeling combined with linear elastic fracture mechanics. *Cryosphere* **2017**, *11*, 1283–1296. [CrossRef]
8. Liang, Q.I.; Zhou, C.; Howat, I.M.; Jeong, S.; Liu, R.; Chen, Y. Ice flow variations at Polar Record Glacier, East Antarctica. *J. Glaciol.* **2019**, *65*, 279–287. [CrossRef]
9. Shen, Q.; Wang, H.; Shum, C.K.; Jiang, L.; Hsu, H.T.; Dong, J. Recent high-resolution Antarctic ice velocity maps reveal increased mass loss in Wilkes Land, East Antarctica. *Sci Rep* **2018**, *8*, 4477. [CrossRef] [PubMed]
10. Onarheim, I.H.; Eldevik, T.; Smedsrud, L.H.; Stroeve, J.C. Seasonal and regional manifestation of arctic sea ice loss. *J. Clim.* **2018**, *31*, 4917–4932. [CrossRef]
11. Luo, Y.; Guan, K.; Peng, J.; Wang, S.; Huang, Y. STAIR 2.0: A generic and automatic algorithm to fuse modis, landsat, and Sentinel-2 to generate 10 m, daily, and cloud-/gap-free surface reflectance product. *Remote Sens.* **2020**, *12*, 3209. [CrossRef]
12. Singh, L.A.; Whittecar, W.R.; DiPrinzio, M.D.; Herman, J.D.; Ferringer, M.P.; Reed, P.M. Low cost satellite constellations for nearly continuous global coverage. *Nat. Commun.* **2020**, *11*, 200. [CrossRef]
13. Wolfe, R.E.; Nishihama, M.; Fleig, A.J.; Kuyper, J.A.; Roy, D.P.; Storey, J.C.; Patt, F.S. Achieving sub-pixel geolocation accuracy in support of MODIS land science. *Remote Sens. Environ.* **2002**, *83*, 31–49. [CrossRef]
14. Moradi, I.; Meng, H.; Ferraro, R.R.; Bilanow, S. Correcting geolocation errors for microwave instruments aboard NOAA satellites. *IEEE Trans. Geosci. Remote Sens.* **2013**, *51*, 3625–3637. [CrossRef]
15. Wang, M.; Cheng, Y.; Chang, X.; Jin, S.; Zhu, Y. On-orbit geometric calibration and geometric quality assessment for the high-resolution geostationary optical satellite GaoFen4. *ISPRS-J. Photogramm. Remote Sens.* **2017**, *125*, 63–77. [CrossRef]
16. Toutin, T. Review article: Geometric processing of remote sensing images: Models, algorithms and methods. *Int. J. Remote Sens.* **2010**, *25*, 1893–1924. [CrossRef]
17. Wang, J.; Ge, Y.; Heuvelink, G.B.M.; Zhou, C.; Brus, D. Effect of the sampling design of ground control points on the geometric correction of remotely sensed imagery. *Int. J. Appl. Earth Obs. Geoinf.* **2012**, *18*, 91–100. [CrossRef]
18. Jiang, Y.; Zhang, G.; Wang, T.; Li, D.; Zhao, Y. In-orbit geometric calibration without accurate ground control data. *Photogramm. Eng. Remote Sens.* **2018**, *84*, 485–493. [CrossRef]
19. Wang, X.; Li, Y.; Wei, H.; Liu, F. An ASIFT-based local registration method for satellite imagery. *Remote Sens.* **2015**, *7*, 7044–7061. [CrossRef]
20. Feng, R.; Du, Q.; Shen, H.; Li, X. Region-by-region registration combining feature-based and optical flow methods for remote sensing images. *Remote Sens.* **2021**, *13*, 1475. [CrossRef]
21. Goncalves, H.; Corte-Real, L.; Goncalves, J.A. Automatic image registration through image segmentation and SIFT. *IEEE Trans. Geosci. Remote Sens.* **2011**, *49*, 2589–2600. [CrossRef]
22. Wang, L.; Niu, Z.; Wu, C.; Xie, R.; Huang, H. A robust multisource image automatic registration system based on the SIFT descriptor. *Int. J. Remote Sens.* **2011**, *33*, 3850–3869. [CrossRef]
23. Khlopenkov, K.V.; Trishchenko, A.P. Implementation and evaluation of concurrent gradient search method for reprojection of MODIS level 1B imagery. *IEEE Trans. Geosci. Remote Sens.* **2008**, *46*, 2016–2027. [CrossRef]
24. Xiong, X.; Che, N.; Barnes, W. Terra MODIS on-orbit spatial characterization and performance. *IEEE Trans. Geosci. Remote Sens.* **2005**, *43*, 355–365. [CrossRef]
25. Gerrish, L.; Fretwell, P.; Cooper, P. High Resolution Vector Polylines of the Antarctic Coastline (7.4). 2021. Available online: 10.5285/e46be5bc-ef8e-4fd5-967b-92863fbe2835 (accessed on 20 October 2021).
26. Haran, T.; Bohlander, J.; Scambos, T.; Painter, T.; Fahnestock, M. MEaSUREs MODIS Mosaic of Greenland (MOG) 2005, 2010, and 2015 Image Maps, Version 2. 2018. Available online: https://nsidc.org/data/nsidc-0547/versions/2 (accessed on 20 October 2021).
27. Tang, X.; Zhang, G.; Zhu, X.; Pan, H.; Jiang, Y.; Zhou, P.; Wang, X. Triple linear-array image geometry model of ZiYuan-3 surveying satellite and its validation. *Int. J. Image Data Fusion* **2013**, *4*, 33–51. [CrossRef]
28. Wang, M.; Yang, B.; Hu, F.; Zang, X. On-orbit geometric calibration model and its applications for high-resolution optical satellite imagery. *Remote Sens.* **2014**, *6*, 4391–4408. [CrossRef]
29. Cao, J.; Yuan, X.; Gong, J.; Duan, M. The look-angle calibration method for on-orbit geometric calibration of ZY-3 satellite imaging sensors. *Acta Geod. Cartogr. Sin.* **2014**, *43*, 1039–1045.

30. Guan, Z.; Jiang, Y.; Wang, J.; Zhang, G. Star-based calibration of the installation between the camera and star sensor of the Luojia 1-01 satellite. *Remote Sens.* **2019**, *11*, 2081. [CrossRef]
31. Dave, C.P.; Joshi, R.; Srivastava, S.S. A survey on geometric correction of satellite imagery. *Int. J. Comput. Appl. Technol.* **2015**, *116*, 24–27.
32. Zhang, G.; Xu, K.; Zhang, Q.; Li, D. Correction of pushbroom satellite imagery interior distortions independent of ground control points. *Remote Sens.* **2018**, *10*, 98. [CrossRef]
33. Lowe, D.G. Distinctive image features from scale-invariant keypoints. *Int. J. Comput. Vis.* **2004**, *60*, 91–110. [CrossRef]
34. Bindschadler, R.; Vornberger, P.; Fleming, A.; Fox, A.; Mullins, J.; Binnie, D.; Paulsen, S.; Granneman, B.; Gorodetzky, D. The landsat image mosaic of antarctica. *Remote Sens. Environ.* **2008**, *112*, 4214–4226. [CrossRef]
35. Kouyama, T.; Kanemura, A.; Kato, S.; Imamoglu, N.; Fukuhara, T.; Nakamura, R. Satellite attitude determination and map projection based on robust image matching. *Remote Sens.* **2017**, *9*, 90. [CrossRef]
36. Hu, C.M.; Tang, P. HJ-1A/B CCD IMAGERY Geometric distortions and precise geometric correction accuracy analysis. In Proceedings of the International Geoscience and Remote Sensing Symposium, Vancouver, BC, Canada, 24–29 July 2011; pp. 4050–4053.
37. Li, Y.; He, L.; Ye, X.; Guo, D. Geometric correction algorithm of UAV remote sensing image for the emergency disaster. In Proceedings of the 2016 IEEE International Geoscience and Remote Sensing Symposium (IGARSS), Beijing, China, 10–15 July 2016; pp. 6691–6694.
38. Wang, S.; Quan, D.; Liang, X.; Ning, M.; Guo, Y.; Jiao, L. A deep learning framework for remote sensing image registration. *ISPRS-J. Photogramm. Remote Sens.* **2018**, *145*, 148–164. [CrossRef]
39. Ma, L.; Liu, Y.; Zhang, X.; Ye, Y.; Yin, G.; Johnson, B.A. Deep learning in remote sensing applications: A meta-analysis and review. *ISPRS-J. Photogramm. Remote Sens.* **2019**, *152*, 166–177. [CrossRef]

Article

Spatial Variability and Temporal Heterogeneity of Surface Urban Heat Island Patterns and the Suitability of Local Climate Zones for Land Surface Temperature Characterization

Ziqi Zhao [1,2,3], Ayyoob Sharifi [4], Xin Dong [1,2], Lidu Shen [5] and Bao-Jie He [1,2,*]

- [1] Centre for Climate-Resilient and Low-Carbon Cities, School of Architecture and Urban Planning, Chongqing University, Chongqing 400045, China; ziqizhao@iaesy.cn (Z.Z.); xin.dong@cqu.edu.cn (X.D.)
- [2] Key Laboratory of New Technology for Construction of Cities in Mountain Area, Ministry of Education, Chongqing University, Chongqing 400045, China
- [3] Institute of Atmospheric Environment, China Meteorological Administration, Shenyang 110166, China
- [4] Graduate School of Humanities and Social Science, Graduate School of Advances Science and Engineering, Network for Education and Research on Peace and Sustainability (NERPS), Hiroshima University, Higashi Hiroshima 739-8530, Japan; sharifi@hiroshima-u.ac.jp
- [5] Institute of Applied Ecology, Chinese Academy Sciences, Shenyang 110016, China; shenlidu@iae.edu.cn
- * Correspondence: baojie.he@cqu.edu.cn

Citation: Zhao, Z.; Sharifi, A.; Dong, X.; Shen, L.; He, B.-J. Spatial Variability and Temporal Heterogeneity of Surface Urban Heat Island Patterns and the Suitability of Local Climate Zones for Land Surface Temperature Characterization. *Remote Sens.* **2021**, *13*, 4338. https://doi.org/10.3390/rs13214338

Academic Editor: Yuji Murayama

Received: 26 September 2021
Accepted: 20 October 2021
Published: 28 October 2021

Publisher's Note: MDPI stays neutral with regard to jurisdictional claims in published maps and institutional affiliations.

Copyright: © 2021 by the authors. Licensee MDPI, Basel, Switzerland. This article is an open access article distributed under the terms and conditions of the Creative Commons Attribution (CC BY) license (https://creativecommons.org/licenses/by/4.0/).

Abstract: This study investigated monthly variations of surface urban heat island intensity (SUHII) and the applicability of the local climate zones (LCZ) scheme for land surface temperature (LST) differentiation within three spatial contexts, including urban, rural and their combination, in Shenyang, China, a city with a monsoon-influenced humid continental climate. The monthly SUHII and LST of Shenyang were obtained through 12 LST images, with one in each month (within the period between 2018 and 2020), retrieved from the Thermal InfraRed Sensor (TIRS) 10 in Landsat 8 based on a split window algorithm. Non-parametric analysis of Kruskal-Wallis H test and a multiple pairwise comparison were adopted to investigate the monthly LST differentiations with LCZs. Overall, the SUHII and the applicability of the LCZ scheme exhibited spatiotemporal variations. July and August were the two months when Shenyang underwent strong heat island effects. Shenyang underwent a longer period of cool than heat island effects, occurring from November to May. June and October were the transition months of cool–heat and heat–cool island phenomena, respectively. The SUHII analysis was dependent on the definition of urban and rural boundaries, where a smaller rural buffering zone resulted in a weaker SUHI or surface urban cool island (SUCI) phenomenon and a larger urban area corresponded to a weaker SUHI or SUCI phenomenon as well. The LST of LCZs did not follow a fixed order, where in July and August, the LCZ-10 (Heavy industry) had the highest mean LST, followed by LCZ-2 (Compact midrise) and then LCZ-7 (Lightweight low-rise). In comparison, LCZ-7, LCZ-8 (Large low-rise) and LCZ-9 (Sparsely built) had the highest LST from October to May. The LST of LCZs varied with urban and rural contexts, where LCZ-7, LCZ-8 and LCZ -10 were the three built LCZs that had the highest LST within urban context, while LCZ-2, LCZ-3 (Compact low-rise), LCZ-8, LCZ-9 and LCZ-10 were the five built LCZs that had the highest LST within rural context. The suitability of the LCZ scheme for temperature differentiation varied with the month, where from July to October, the LCZ scheme had the strongest capability and in May, it had the weakest capability. Urban context also made a difference to the suitability, where compared with the whole study area (the combination of urban and rural areas), the suitability of built LCZs in either urban or rural contexts weakened. Moreover, the built LCZs had a higher level of suitability in an urban context compared with a rural context, while the land-cover LCZs within rural had a higher level of suitability.

Keywords: land surface temperature; local climate zones; spatial variability; temporal heterogeneity; urban heat island intensity

1. Introduction

Cities are already the main human settlements since global urban population exceeded rural population in 2007, and about 68% of the world population is projected to live in cities by 2050 [1]. The increasing urbanization trend has brought a variety of challenges to cities, such as urban climate change, limited access to open and public spaces, low-quality housing conditions and constrained access to public transport [2]. One of the most acknowledged urban climate change phenomena is the urban heat island (UHI) effect, referring to the fact that cities are warmer than their surrounding suburban or rural areas [3,4]. Addressing the UHI effect is critical considering its significant impacts on urban systems, citizens' living and ecosystems [5]. Many studies have revealed that urban warming leads to an increase in energy and water consumption for cooling [6,7], deteriorates outdoor thermal comfort and air quality [8], and thereby results in an increase in mortality and morbidity [9]. What is worse, the heat-induced impacts can be more severe because of the interaction of UHI and heat waves that are getting more frequent, longer and more severe along with global warming [10,11].

An accurate measurement, assessment and identification of UHI effects is essential for the better communication of urban climate knowledge to urban planners and decision makers [12]. Understanding surface UHI (SUHI) effects based on land surface temperature (LST) and its associated drivers has been an important theme to achieve this, apart from studies on the canopy UHI effects [13]. Many studies have adopted SUHI intensity (SUHII), an important indicator that has significant implications for land use and land cover change changes [14,15], energy demand [16] and urban living suitability [17], to measure SUHI effects, and examined the drivers of SUHII variations. For instance, Peng, Piao [18] examined the UHII of 419 cities around the world, pointing out that average annual daytime SUHII was higher than the nighttime one and the driving factors for the daytime and nighttime ones were different. Li et al. [19] analyzed the SUHI effects of Berlin based on several hypothesized scenarios and pointed out that SUHII was affected by city size, urban density and compactness. Liu et al. [15] further analyzed the SUHI of 1288 urban clusters in China, concluding that the daytime SUHII was also more prominent than nighttime one and the SUHII was a function of urban size, shape, centrality and background conditions.

UHII calculation is a relative value of urban–rural/suburban temperature differences according to its original definition, implying that the UHII can be sensitive to the selection of urban and rural/suburban sites [20]. On the one hand, the LST in cities is strongly associated with land use/land cover (LULC), where LST of buildings and roads could be up to 10 °C higher than that of water bodies and grass land and the locations of different LULCs in both radial and circumferential directions influence the LST [21–23]. Moreover, urban artificial landscapes exhibit a high degree of heterogeneity, making it difficult to accurately choose urban sites for SUHI assessment. On the other hand, the determination of representative rural sites is also important considering the different ecosystems of the rural surface properties [24]. For instance, Peng et al. [18] assessed the sensitivity of UHII to the suburban areas in Beijing, suggesting that SUHII could have a similar magnitude when the suburban areas were 50%, 100% and 150% of the urban areas. These results may indicate that the suburban/rural sites should be at least at the suburban ring–buffer zone border of the 50% of the urban areas. Liu et al. [15] found that the sensitivity of SUHII to the rural sites decreased with the increase of ex-urban distance or area, implying that an arbitrary selection of rural sites could under-estimate SUHII magnitude [15,18].

To overcome such challenges in urban climate studies, the identification of the homogeneity and heterogeneity of local morphology has been prioritized as an important approach [25–27], among which, the local climate zones (LCZ) scheme was proposed by Stewart and Oke [20] to standardize surface structure and cover description and thereby standardize urban and suburban/rural sites for temperature comparisons. The LCZ scheme has been widely adopted as an objective framework to analyze intra-urban temperature differences in numerous cities such as Berlin [28], Dublin [29], Hong Kong [30], Nagpur [31], Nairobi [32], Olomouc [33], Shanghai [34], Vancouver [35] across various continents [36–39].

Overall, the LCZ scheme that is comprised of 10 built types and seven land cover types indicates that both air and surface temperatures reduce from compact to sparse built form and at the same time, from high-rise to low-rise built form [20]. Nevertheless, the responses of LST to the LCZ types can vary significantly because of the combination of different built-land cover types, geographical conditions, background climates and landscape effect (e.g., topography, distance from the sea) [31,37], making it significant to examine the suitability of the application of the LCZ scheme for SUHI studies. The work conducted by Eldesoky, Gil [40] confirmed that the LCZ scheme is applicable for tropical, temperate and cold climates, but not arid climates, with different levels of uncertainty.

Pending queries relevant to the LST responses to LCZ types include the seasonality or thermal anisotropy [41]. For instance, Du, Chen [42] analyzed LST variability of LCZ types in different seasons in Nanjing, China, concluding that the warmest or coolest zones varied with seasons, while the LST of built-up types increased with the reduction of building height. Geletič et al. [41] analyzed the seasonal LST variations of different LCZ types in Prague, Brno and Novi Sad, reporting that SUHII of a specific LCZ type exhibited the largest difference in summer and spring and the lowest in winter. Meanwhile, the dense built-up type and industrial type had the largest SUHII, and the sparse built-up types had the smallest. Some other studies have also examined the seasonal responses of LST or SUHII to LCZ types [43,44]. Nevertheless, the seasonal variation of SUHI to LCZ types has not been well understood, especially in different geographic and macroclimatic contexts. Given the seasonal variability, the applicability of the LCZ scheme to assess SUHI should be examined. Moreover, the impact of landscape effect on the SUHI responses to LCZ types should be further analyzed.

To address the above-mentioned challenges and expand the application of the LCZ scheme in urban planning and design, this study aims to detect monthly variations of LST responses to different LCZ types in the city of Shenyang, China. This study will analyze the impact of landscape effects on such variations through analyzing the variations of SUHII and the LST responses to LCZ types in urban (not fully urbanized) and rural landscapes. Built upon this, the applicability of the LCZ scheme to assess SUHI will be examined. This paper is structured as follows. Section 2 introduces the case study area and its urban heat challenges. Section 3 presents data sources and research methods. Section 4 analyzes the results in aspects of overall LST and SUHII variations, seasonal variations of SUHI responses to LCZ types and the applicability of LCZ schemes for SUHI studies. Following this, Section 5 discusses the results and Section 6 concludes this paper.

2. Study Area

This study is conducted in Shenyang (41°11′51″N–43°02′13″N, 122°25′09″E–123°48′24″E), the capital city of Liaoning Province in the southern part of the Northeast China (Figure 1). The terrain of Shenyang is flat, gradually extending eastward into hilly areas, with an altitude of 5–441 m. Shenyang has a monsoon-influenced humid continental climate (Dwa), with four distinctive seasons. Summers are hot and humid, where July is the hottest month, with an average temperature of 24.6 °C. Winters are dry and cold, where January is the coldest month with the average temperature of −11.2 °C. The annual average temperature is 8.5 °C. Recently, subject to climate change, Shenyang is also undergoing extreme climatic conditions like many other cities, with the extreme temperature of 39.3 °C on 2 August 2018. South-dominant wind prevails in Spring, Summer and Autumn, while in winter the prevailing wind is north-dominant. The annual average wind speed is about 2.6 m/s, where the wind is the strongest in April and the weakest in August. The rainfall of Shenyang ranges between 600 and 800 mm, while the average annual rainfall of the central city (Figure 1) is about 716.2 mm. In addition, the annual average relative humidity of Shenyang is about 55.7% [45].

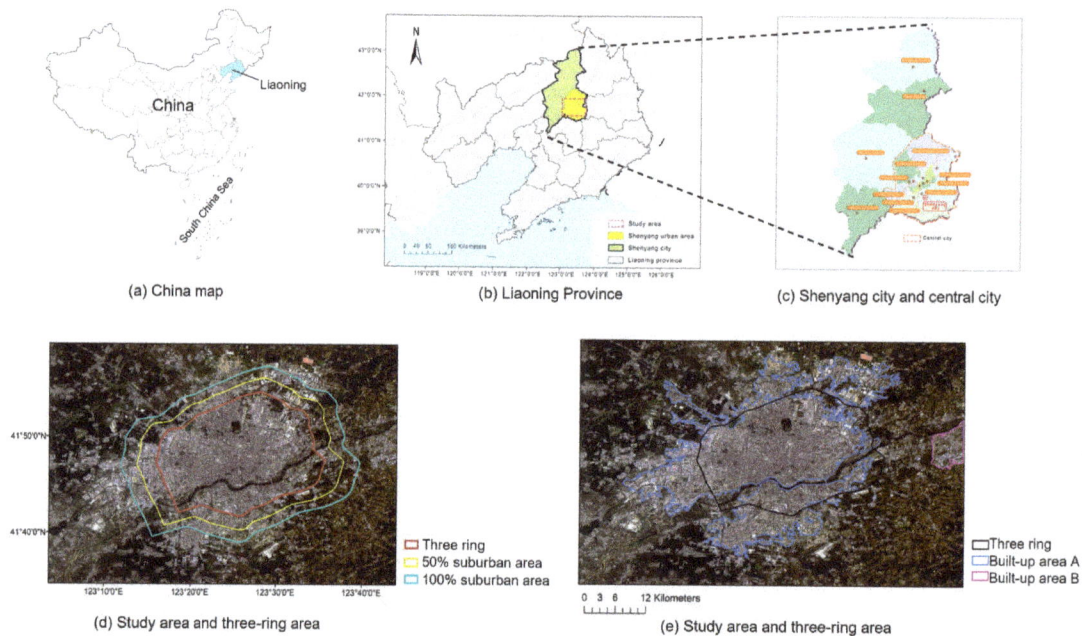

Figure 1. Location of study area and its composition according to China's land use status remote sensing monitoring database (http://www.resdc.cn, accessed on 19 October 2021).

Shenyang is the geographic center of Northeast Asia. Economically, it is in the center of the Northeast Asia Economic Circle and the Bohai Rim Economic Circle. Being a major city of the Greater Shenyang Metropolitan Area, Shenyang covers an area of 12,948 km^2, where the central city is about 3495 km^2 [45]. The city has seen rapid population increase in the past 20 years from 7.20 million in 2000 to 9.07 million in 2020 [46]. Accordingly, Shenyang is now a megacity, the only one in Northeast China. Meanwhile, Shenyang is urbanizing rapidly with the urbanization ratio increasing from 70.33% in 2000 to 84.52% in 2020 [47]. More than 50% of the population and infrastructure are within the three-ring area (the core city, Figure 1), consisting of Heping, Shenhe, Dadong, Huanggu, Tiexi and some parts of other districts such as Dongling, Yuhong and Hunnan New District, with the area of about 455 km^2. Within the three-ring area, the SUHII could reach 4–5 °C [23].

Along with the upward trend of urbanization, Shenyang is undergoing three trends including urban densification (inner city), urban sprawl (outer ring) and industrial structure change (China's land use status remote sensing monitoring database, http://www.resdc.cn, accessed on 19 October 2021). First, the population within the three-ring area saw an upward trend from 2.998 million in 1985 to 3.772 million in 2015 [23]. Second, upon the three-ring area, the urbanized area of central city has been expanding towards different directions, forming a new urban pattern (built-up area A, 635.36 km^2 in area in Figure 1). Third, Shenyang has traditionally been an old industrial base for heavy industry and manufacture, while such factories are relocating from the middle of Shenyang to the surrounding cities, counties and suburban areas to ensure environmental and living quality during urbanization. Fourth, to create the Greater Shenyang Metropolitan Area, some factories have been relocated to the middle of Shenyang and surrounding cities (e.g., Fushun), along which a new built-up area B (29.03 km^2 in area) forms (Figure 1).

3. Data and Methodology

3.1. Data Pre-Processing and Date Selection

The LST information, as well as the information of SUHII of the study area was obtained through Landsat 8 images (at the path/row of 119/31). To investigate the monthly variation of LST and SUHII, 12 remotely sensed thermal-infrared images (resolution: 30 m) (Table 1) collected by the United States Geological Survey were downloaded from http://earthexplorer.usgs.gov, accessed on 25 September 2021. These images present the thermal information of 10:27 a.m. local Shenyang time with a cloud coverage below 0.1%. Other meteorological conditions at 10:00–11:00 a.m. (UTC +8) were obtained from the local Bureau of Meteorology (Table 1).

Table 1. Date list and weather conditions in different months corresponding to 12 thermal-infrared imageries.

Date	Air Temperature/°C	Relative Humidity/%	Wind Speed/m·s^{-1}	Surface Temperature/°C
24 December 2018	−7.1	59	0.9	−9.5
25 January 2019	−6.9	35	2.6	−3.5
26 February 2019	1.5	36	1.9	6.9
14 March 2019	4.8	37	2.1	13.9
15 April 2019	18.9	21	4.0	30.4
1 May 2019	16.7	24	4.2	34.2
20 June 2020	29.4	44	5.4	49.9
4 July 2019	28.5	47	4.6	48.5
2 August 2018	33.2	38	4.0	40.5
22 September 2019	20.4	46	5.2	33.7
8 October 2019	11.4	30	8.0	24.6
9 November 2019	7.5	47	1.3	18.7

3.2. Retrieval of LST

The split window algorithm proposed by Qin, Karnieli [48] was used to retrieve LST from the only spectral band of Thermal InfraRed Sensor (TIRS) 10 in Landsat 8. In particular, the LST was obtained after the atmospheric correction of reflective and thermal bands. According to Equation (1), the digital number was converted to the spectral radiance L_λ at the top of the atmosphere.

$$L_\lambda = M_L \cdot DN + A_L \tag{1}$$

where L_λ is spectral radiance, W/(m^2 sr μm), M_L is the re-scaled gain corresponding to a specific band, W/(m^2 sr μm) and A_L is the re-scaled bias corresponding to a specific band, W/(m^2 sr μm).

At-sensor brightness temperature T_b (Unit: K) was calculated, based on Equation (2), from TIRS 10, corresponding to the OLI sensor.

$$T_b = k_2/(\ln(k_1/L_\lambda) + 1) \tag{2}$$

where k_1 and k_2 are constants, with values of 774.89 (Unit: W/(m^2 sr μm)) and 1321.08 (Unit: W), respectively. Next, the LST was subsequently obtained after emissivity correction of ground radiance B(LST) (Unit: K) via mono window algorithm Equations (3)–(6).

$$B(LST) = \{a(1 - C - D) + [(b - 1)(1 - C - D) + 1]T_b - DT_a\}/C \tag{3}$$

$$C = \varepsilon\tau \tag{4}$$

$$D = (1 - \varepsilon)[1 + (1 - \varepsilon)\tau] \tag{5}$$

$$LST = B(LST)/(\ln \varepsilon \ (\lambda \cdot B(LST)/\rho + 1) \tag{6}$$

where a and b are constants, ε and τ are the land surface emissivity and atmospheric transmittance of band i, respectively. T_a is effective mean atmospheric temperature (Unit: K).

λ is the wavelength of emitted radiance (11.5 μm). ρ is a constant (1.438×10^{-2} m·K), calculated by Planck's constant, light velocity and Boltzmann's constant.

The effective mean atmospheric temperature T_a (Unit: K) was calculated based on the following empirical formula [48].

For mid-latitude summer :
$$T_a = 16.0110 + 0.92621\, T_0$$
For mid-latitude winter : \hfill (7)
$$T_a = 19.2704 + 0.91118\, T_0$$

where T_0 actual air temperature at the time when Landsat images are captured (Unit: K).

3.3. Local Climate Zone Classification

This study follows the LCZ classification scheme developed by Stewart and Oke [20] that consists of 10 types of built-up zones and seven types of land-cover areas to characterize the land surface properties of the study area. First, the training LCZ samples were selected based on the Google Earth image of 2 August 2018, without clouds and with 30 m resolution through visual interpretation. In general, 5–28 training samples were selected for the 17 types of LCZ types, respectively. Each training area should have an area of at least 1 km² and the length/width should be at least 200 m. In addition, urban morphological characteristics should be homogeneous, so that small areas that may be heterogeneous or irregular could be excluded. A buffer zone between different LCZ training zones should have a width of at least 100 m to avoid fuzzy recognition of boundaries. Through the random forest algorithm on the SAGA GIS platform, the LCZ training samples were classified, during which the Landsat TM image (on 2 August 2018) was also used to generate the LCZ map. Through several rounds of iteration and verification, the LCZ map of the study area was generated, as shown in Figure 2. More statistics of the area of different LCZ types of the study area, built-up area and rural areas are presented in Table 2.

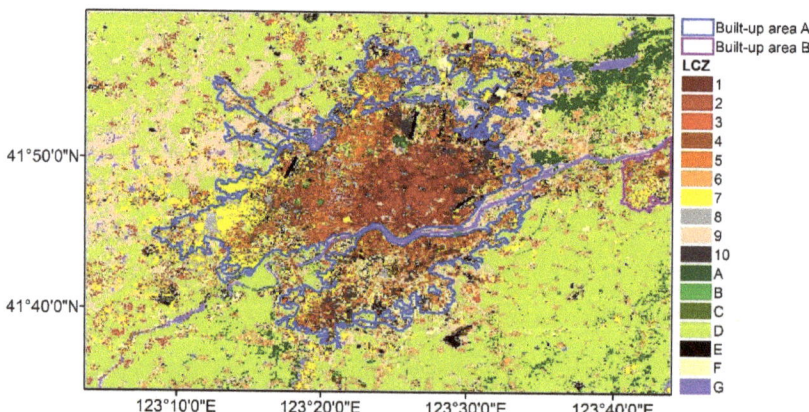

Figure 2. Distribution of local climate zones in the study area.

Table 2. The area of different LCZ types in the study area, built-up area and rural areas.

LCZ Types	Study Area /km²	Built-Up Area (A and B) /km² (%)	Rural Area/km² (%)
LCZ 1 (Compact high-rise)	22.02	21.34 (96.9%)	0.67 (3.1%)
LCZ 2 (Compact midrise)	63.19	61.87 (97.9%)	1.37 (2.1%)
LCZ 3 (Compact low-rise)	77.84	30.6 (39.3%)	47.24 (60.7%)
LCZ 4 (Open high-rise)	73.25	53.69 (73.3%)	19.79 (26.7%)
LCZ 5 (Open midrise)	93.89	75.79 (80.7%)	17.97 (19.3%)
LCZ 6 (Open low-rise)	103.49	24.54 (23.7%)	78.93 (76.3%)
LCZ 7 (Lightweight low-rise)	160.93	90.64 (56.3%)	70.25 (43.7%)
LCZ 8 (Large low-rise)	77.53	50.77 (65.5%)	26.94 (34.5%)
LCZ 9 (Sparsely built)	359.88	65.26 (18.1%)	294.71 (81.9%)
LCZ 10 (Heavy industry)	77.98	68.55 (87.9%)	8.98 (12.1%)
LCZ A (Dense trees)	75.29	2.13 (2.8%)	73.17 (97.2%)
LCZ B (Scattered trees)	16.21	3.58 (22.1%)	12.59 (77.9%)
LCZ C (Bush, scrub)	4.53	0.77 (17.0%)	3.76 (83.0%)
LCZ D (Low plants)	1132.68	30.28 (2.7%)	1102.25 (97.3%)
LCZ E (Bare rock or paved)	77.01	42.55 (55.3%)	34.4 (44.7%)
LCZ F (Bare soil or sand)	125.38	40.16 (32.0%)	85.28 (68.0%)
LCZ G (Water)	39.09	1.32 (3.4%)	37.74 (96.6%)

3.4. Data Analysis

This study presents the LST distribution and the SUHII of the study area, with the division of two pairs of 'urban' and 'suburban/rural' areas including (i) three-ring and suburban areas and (ii) built-up areas and rural areas (Figure 1d,e). The SUHII is the LST difference between 'urban' and 'suburban/rural' areas (Equation (8)), in which the definition of suburban/rural areas is critical. First, consistent with existing studies [15,18,49], rural/suburban areas are buffer zones which have an area of 50% or 100% of the core area (three-ring or built-up area), excluding water pixels. Moreover, the literal suburban/rural area was also defined by the whole area excluding the core area. Accordingly, there are three types of 'suburban/rural' areas in two pairs of 'urban' and 'suburban/rural' areas.

$$SUHII = LST_{urban} - LST_{suburban/rural} \qquad (8)$$

where LST_{urban} is average land surface temperature in urban area and $LST_{suburban/rural}$ is average land surface temperature of suburban/rural areas.

Furthermore, the monthly LST of different LCZs was analyzed within three contexts including the whole study area, the built-up areas and the rural areas (Figure 1e). The analysis was conducted in aspects of LST range and mean value, and departure of the average LST of a specific LCZ from the average LST of all LCZs was analyzed to examine their positive or negative contributions to the urban temperature. Moreover, the monthly suitability of LCZ scheme to indicate LST differentiation of the study area in Shenyang was also examined. The suitability was assessed by the significant difference between the LSTs of a pair of LCZs. Non-parametric analysis of Kruskal–Wallis H test was performed to determine the significance of differences between LSTs, since the LST dataset did not follow a normal distribution, after which a multiple pairwise-comparison between groups was conducted to examine which pairs of groups were different [5].

4. Results and Analysis

4.1. Monthly LST Variation

Figure 3 presents the monthly LST variation in the study area from 2018 to 2020, in which the three-ring area (inner circle) and 100% suburban area (outer circle) were marked to compare the urban–suburban–rural temperature for heat/cool island phenomenon. The results indicate an upward trend in the monthly maximum temperature from 4.35 °C in January to 45.56 °C in August and then a downward trend to 5.53 °C in December. The monthly minimum temperature increased from −32.33 °C in December to 22.52 °C in August and then decreased to −15.91 °C in November. December and January were the two coldest months of the year, with the lowest minimum and maximum temperatures, consistent with the lowest air and surface temperatures in Table 1.

August was the hottest month with the highest minimum and maximum temperatures, consistent with the data given in Table 1 as well. However, it should be noted that July has been statistically the hottest month of the year (Section 2). The disparity might be caused by air temperature differences in this study for 2019 compared to the climatological values, that August was hotter than July (Table 1). The similar scenario was applicable for the minimum temperatures in April and May.

Moreover, Shenyang is enduring a longer period of cool island phenomenon than heat island. The heat island phenomenon was obvious in July, August and September where urban–suburban temperature was generally higher than rural one. Along the urban–suburban–rural profile, there was an obvious gradient of LST reduction. In comparison, the cool island phenomenon was observed in November, December, January, February, March, April and May when rural temperature was generally higher. Along the urban–suburban–rural profile, there was a LST increase gradient, particularly in January, February, March, April and May. In both October and June, the LST distributed evenly within the study area and a clear pattern of cool/heat island phenomenon was not recognizable. Therefore, October was a month corresponding to the transition of heat island to cold island, while June was the transition month towards heat island from cold island.

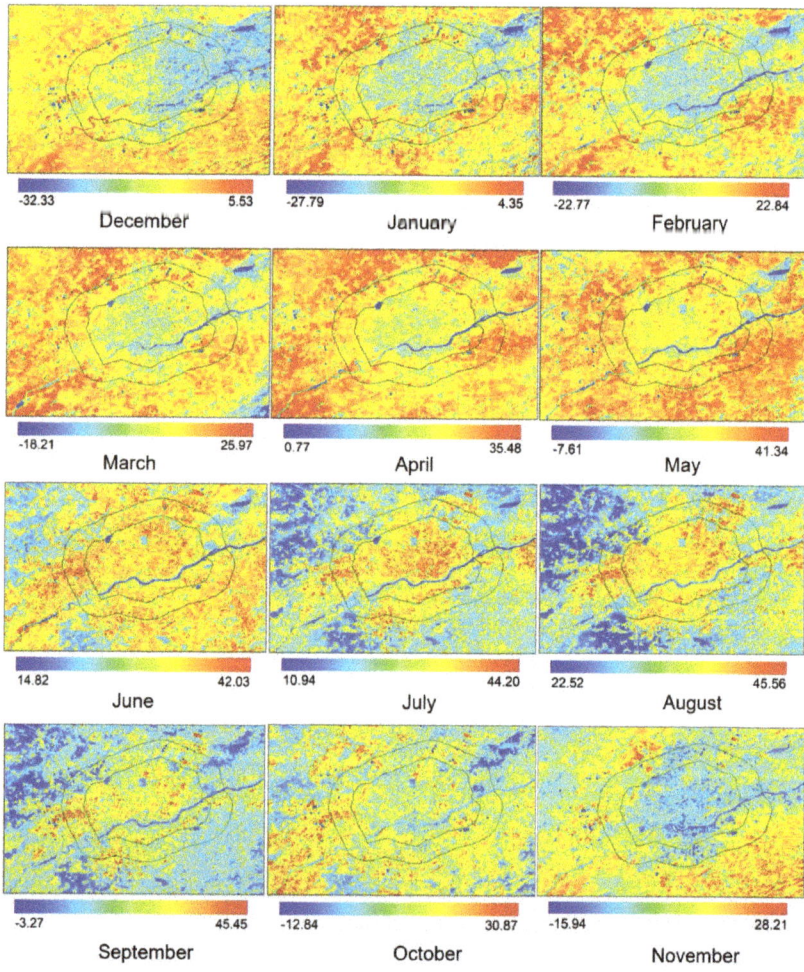

Figure 3. Monthly (instantaneous) land surface temperature variations and the identification of heat/cool island phenomenon from 2018 to 2020.

4.2. Monthly Variation of Urban Heat/Cool Island Intensity

This section quantifies SUHII and analyses the monthly variation of SUHII of the study area. Figure 4 presents the monthly LST and SUHII variations of the three-ring, the Central city, in buffer 50%, buffer 100% and buffer 0% scenarios. Overall, the maximum, mean and minimum LSTs in the central city, suburban-50%, suburban-100% and rural scenarios exhibited similar patterns. Nevertheless, some differences were observed, and the extent of such differences was dependent on the month. For instance, mean LSTs of the Central city were lower than those of rural/suburban areas, indicating cool island phenomenon, from November to May. In comparison, mean LSTs of the Central city were higher, indicating heat island phenomenon, from July to September. There were small differences among rural/suburban LSTs and central city LST in June and October. Such results were consistent with the recognition of cool/heat island phenomenon and the transition months in Section 4.1.

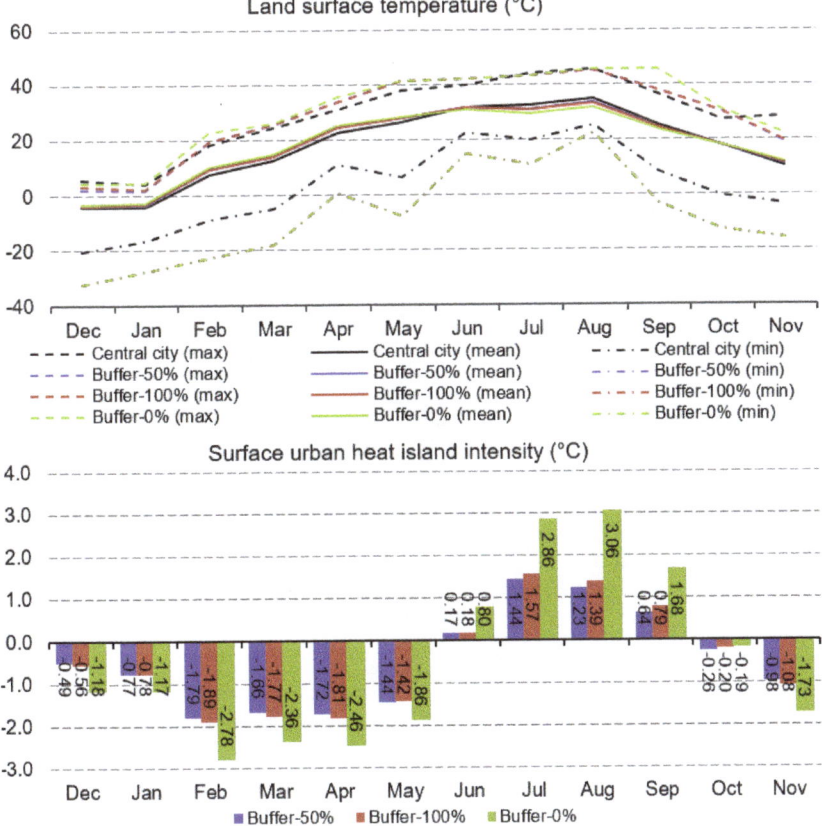

Figure 4. Monthly LST and SUHII variations of the three-ring area in buffer 50%, buffer 100% and buffer 0% scenarios from 2018 to 2020.

Both maximum and minimum LSTs exhibited different patterns, in aspect of the temperature order of Central city, buffer 50%, buffer 100% and buffer 0% scenarios. Regarding the maximum LST, the Central city had the lowest value from February to October and the rural (buffer 0%) area had the highest value from January to October, compared with those of buffer 50% and buffer 100%. Nevertheless, the central city had the highest minimum LST throughout the year and the buffer 50%, buffer 100% and buffer 0% areas had the same

LST. The minimum LST of Central city was about 2.86–13.72 °C higher than those of buffer 50%, buffer 100% and buffer 0% scenarios.

The SUHII varied seasonally and the SUHII was different depending on the definition of rural/suburban areas. The months of June, July, August and September underwent heat island phenomenon, in which the heat island phenomenon in July and August was the most intense with the highest SUHII of 2.86 and 3.06 °C according to the Buffer 0% scenario. The study area underwent cool island phenomenon from October to December, in which the cool island phenomenon was the most intense in February, March and April with an intensity of 2.78, 2.36 and 2.46 °C (Buffer 0% scenario), respectively. It should be noted that whilst June and October experienced heat and cool island phenomena, respectively, their intensities were very weak, further indicating that these two months were transition months. Furthermore, in the buffer 50% scenario, both SUHII and surface urban cool island intensities (SUCII) were the weakest, compared with those in both buffer 100% and buffer 0% scenarios. Such results indicate that a larger rural area corresponded to stronger SUHII or SUCII on the one hand, and it is important to define appropriate rural/suburban areas.

Figure 4 presents the monthly variation of temperature with respect to the three-ring area, while Figure 5 presents monthly variation of LST of the built-up areas and those of SUHII, in buffer 50%, buffer 100% and buffer 0% scenarios. Overall, the monthly variation of maximum, mean and minimum LSTs in the case of built-up area (Figure 5) were similar to those in the case of three-ring area (Figure 5). For instance, the urban area (built-up area) had the highest minimum LST and buffer 50%, buffer 100% and buffer 0% areas had the same LST. From July to September, the mean LSTs of the built-up area was much higher, while the ones of the built-up area were lower from October to June. SUHII based on the built-up area also exhibited a similar pattern to that based on a three-ring area. July and August underwent the highest SUHI effects, and February, March and April had the most intense cool island impacts. However, the cool island phenomenon in May was also obvious, with a SUHII of 2.91 °C (Buffer 0% scenario). June underwent cool island effects according to the built-up area (Figure 5), different from heat island effects according to the three-ring area (Figure 4). Under such conditions, June, September and October were the transition months with very weak heat island or cool island phenomenon. In addition, the SUHII in the buffer 0% scenario was the strongest, followed by the buffer 100% scenario and then buffer 50% scenario.

Moreover, the heat island phenomenon according to built-up area was less intense than that according to three-ring area, and the cool island phenomenon according to the built-up area was more intense. Overall, according to the results in Figures 4 and 5, the definition of urban area also influences the SUHII, apart from the definition of rural/suburban area. A different definition of urban area could lead to diverse results in SUHII magnitude, cool/heat island phenomenon and transition months.

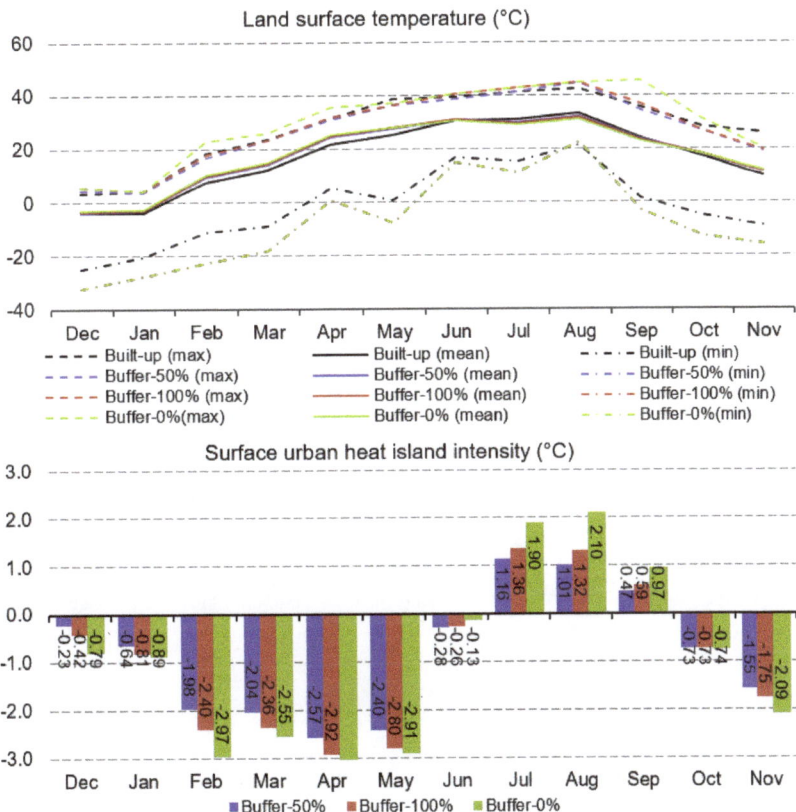

Figure 5. Monthly LST and SUHII variations of the built-up area in buffer 50%, buffer 100% and buffer 0% scenarios from 2018 to 2020.

4.3. Temporal Variation of Land Surface Temperature with Local Climate Zones

Table 3 presents the LST of different types of LCZs in 12 months. The results indicate that in July and August, when the SUHII was the strongest, Heavy industry areas (LCZ-10) had the highest mean LST of 34.58 and 36.87 °C, followed by Compact midrise (LCZ-2, 34.34 and 36.10 °C) and then Lightweight low-rise (LCZ-7). In comparison, the land cover of water was the strongest heat sink with the mean LST of 24.91 and 28.91 °C, respectively, in July and August. Both dense trees and low plants were also heat sinks with the second and third lowest LST.

Table 3. Land surface temperature of different local climate zones in different months from 2018 to 2020 (°C).

LCZ	December		January		February		March	
	Range	Mean	Range	Mean	Range	Mean	Range	Mean
LCZ-1 (Compact high-rise)	−20.28~1.63	−4.65	−16.26~−0.05	−4.80	−9.04~13.13	6.21	−4.89~18.78	10.73
LCZ-2 (Compact midrise)	−18.43~2.81	−4.15	−14.45~1.46	−4.21	−6.30~16.34	7.27	−1.08~21.18	12.04
LCZ-3 (Compact low-rise)	−26.15~3.74	−3.91	−22.25~4.35	−3.48	−12.47~15.19	9.11	−9.83~23.37	14.21
LCZ-4 (Open high-rise)	−20.80~4.43	−4.54	−19.65~3.86	−4.52	−6.54~15.52	6.71	−3.09~21.65	11.42
LCZ-5 (Open midrise)	−23.72~4.03	−4.52	−18.91~2.98	−4.27	−8.72~15.36	7.42	−3.01~23.30	12.47
LCZ-6 (Open low-rise)	−13.75~2.17	−4.01	−12.38~1.82	−3.58	−0.97~16.05	8.89	−1.19~21.61	13.67
LCZ-7 (Lightweight low-rise)	−25.58~3.32	−3.87	−24.72~3.36	−3.66	−12.77~18.41	9.10	−7.00~24.26	14.31
LCZ-8 (Large low-rise)	−27.04~3.35	−3.73	−22.01~2.50	−3.60	−11.93~18.60	8.90	−9.09~25.71	14.25
LCZ-9 (Sparsely built)	−26.81~4.36	−3.70	−21.74~3.88	−3.26	−12.69~18.13	9.46	−9.58~24.55	13.91
LCZ-10 (Heavy industry)	−17.34~4.66	−4.03	−13.74~3.35	−3.69	−6.36~19.79	8.64	−2.35~26.64	13.79
LCZ-A (Dense trees)	−9.26~4.50	−4.52	−9.22~3.26	−4.76	0.35~15.75	7.21	−4.75~20.80	10.98
LCZ-B (Scattered trees)	−9.62~2.35	−4.24	−9.17~1.59	−3.83	0.77~15.20	8.44	3.79~20.41	13.37

Table 3. Cont.

LCZ								
LCZ-C (Bush, scrub)	−7.58~3.10	−3.20	−6.76~3.78	−2.80	3.95~15.56	10.26	7.33~20.66	15.50
LCZ-D (Low plants)	−27.50~2.93	−2.66	−21.34~3.95	−2.39	−12.18~22.84	11.59	−9.23~25.97	15.82
LCZ-E (Bare rock or paved)	−28.64~4.05	−4.12	−22.09~3.98	−3.64	−16.53~16.04	8.85	−13.12~25.15	13.91
LCZ-F (Bare soil or sand)	−32.33~2.76	−3.69	−27.79~2.42	−3.19	−22.77~16.62	9.56	−18.21~23.09	14.48
LCZ-G (Water)	−8.97~5.53	−4.77	−9.30~4.16	−5.04	−0.89~14.78	3.77	−4.82~19.09	6.90

LCZ	April		May		June		July	
	Range	Mean	Range	Mean	Range	Mean	Range	Mean
LCZ-1 (Compact high-rise)	11.87~29.62	21.33	6.29~34.15	25.36	23.14~38.19	31.80	19.99~40.09	32.93
LCZ-2 (Compact midrise)	13.23~30.09	22.44	11.79~35.20	26.77	23.58~38.04	32.98	23.23~40.59	34.34
LCZ-3 (Compact low-rise)	7.36~31.86	24.42	1.84~35.45	27.58	19.21~41.46	32.53	17.36~40.92	32.16
LCZ-4 (Open high-rise)	10.62~30.18	22.02	6.88~35.50	25.44	20.72~39.05	31.37	19.37~38.31	31.66
LCZ-5 (Open midrise)	12.31~31.11	22.99	8.20~38.25	26.69	21.84~41.46	32.46	20.92~40.06	32.87
LCZ-6 (Open low-rise)	13.34~30.61	24.23	15.06~35.36	27.74	24.23~37.17	32.23	21.68~39.98	31.57
LCZ-7 (Lightweight low-rise)	6.56~33.08	24.27	3.36~39.63	27.64	19.28~41.82	32.96	16.28~42.95	32.95
LCZ-8 (Large low-rise)	7.24~33.81	24.24	1.91~41.34	27.79	18.91~42.03	33.05	17.24~43.34	33.02
LCZ-9 (Sparsely built)	6.20~32.18	24.24	1.87~36.46	27.03	19.18~38.55	29.85	16.78~39.83	29.05
LCZ-10 (Heavy industry)	13.34~32.57	23.81	9.81~39.29	27.92	23.43~42.60	33.59	22.68~44.20	34.58
LCZ-A (Dense trees)	13.35~31.64	22.93	15.07~37.05	24.55	21.25~38.30	28.28	22.22~36.89	27.87
LCZ-B (Scattered trees)	15.22~32.92	24.56	16.35~35.03	27.37	23.56~37.73	30.91	23.88~38.61	30.27
LCZ-C (Bush, scrub)	18.62~31.22	25.93	20.52~35.33	29.13	26.59~38.84	32.24	25.62~37.58	30.87
LCZ-D (Low plants)	5.15~35.48	26.18	0.28~37.45	29.22	15.70~39.86	30.97	15.89~39.72	28.55
LCZ-E (Bare rock or paved)	5.00~31.36	24.05	−2.75~38.29	27.52	17.85~41.97	32.61	13.85~40.77	32.43
LCZ-F (Bare soil or sand)	0.77~31.19	24.65	−7.61~35.43	28.00	14.82~42.15	32.48	10.94~41.52	31.69
LCZ-G (Water)	10.36~29.27	16.29	12.34~34.90	18.14	20.71~36.06	24.83	20.76~36.39	24.91

LCZ	August		September		October		November	
	Range	Mean	Range	Mean	Range	Mean	Range	Mean
LCZ-1 (Compact high-rise)	27.82~41.69	34.79	8.41~32.27	25.02	−0.27~23.89	17.08	−3.36~17.41	9.04
LCZ-2 (Compact midrise)	27.15~41.55	36.10	12.78~35.71	26.47	2.93~26.55	18.34	−1.34~20.44	9.88
LCZ-3 (Compact low-rise)	23.64~42.65	34.50	4.33~34.04	25.37	−4.09~26.42	18.87	−8.29~18.13	11.18
LCZ-4 (Open high-rise)	24.31~42.21	34.18	5.07~33.06	24.45	1.26~25.94	17.16	−3.45~16.78	9.67
LCZ-5 (Open midrise)	26.77~42.99	35.38	9.04~37.76	25.50	1.97~28.35	17.98	−6.92~18.49	10.10
LCZ-6 (Open low-rise)	26.30~41.79	34.15	13.76~32.89	24.87	6.58~24.96	18.41	1.61~17.74	11.22
LCZ-7 (Lightweight low-rise)	22.62~45.48	35.54	2.24~37.64	26.06	−5.15~30.87	19.04	−7.07~19.85	11.48
LCZ-8 (Large low-rise)	23.93~45.19	35.54	4.09~38.02	26.14	−3.99~30.36	19.18	−9.97~20.55	11.34
LCZ-9 (Sparsely built)	22.64~42.29	31.24	1.49~33.66	23.03	−5.79~26.26	17.15	−8.97~18.09	11.10
LCZ-10 (Heavy industry)	28.86~45.56	36.87	12.26~37.85	27.19	3.03~30.28	19.34	−1.50~22.39	10.98
LCZ-A (Dense trees)	25.09~39.13	30.70	18.93~28.96	21.59	11.05~25.11	14.91	5.92~18.12	11.20
LCZ-B (Scattered trees)	26.89~41.82	33.45	19.24~32.62	−1.00	11.48~24.65	17.27	4.99~17.28	11.50
LCZ-C (Bush, scrub)	28.21~41.89	33.82	21.18~30.51	24.36	14.55~23.59	19.05	6.90~20.28	12.45
LCZ-D (Low plants)	22.52~41.71	30.52	1.04~45.45	22.92	−6.11~27.09	18.13	−10.03~20.06	12.54
LCZ-E (Bare rock or paved)	24.32~42.93	35.23	0.05~38.23	25.55	−5.98~29.51	18.78	−11.45~28.21	−1.00
LCZ-F (Bare soil or sand)	22.59~43.61	34.24	−3.27~35.96	25.08	−12.84~29.03	18.86	−15.94~19.63	11.46
LCZ-G (Water)	25.72~40.09	28.19	18.57~28.97	20.78	11.95~22.71	15.40	5.38~17.93	8.72

In February and March, when the cool island phenomenon was the most obvious, the strongest heat sinks and sources were found in the land cover types of the LCZ scheme. The land cover of Low plants (LCZ-D) had the highest average LST of 11.59 and 15.82 °C, followed by Bush, scrub (LCZ-C) with the average LST of 10.26 and 15.50 °C, implying the strongest heat sources. Water (LCZ-G) exhibited the lowest average temperature of −5.04 and 6.90 °C, indicating the strongest heat sink. In addition, that built from Compact high-rise (LCZ-1) exhibited the lowest LST of 6.21 and 10.73 °C, followed by the Open high-rise (LCZ-4) with the average LST of 6.71 and 11.42 °C.

In June, a transition month from cool to heat island, the land cover of Water (LCZ-G) was the strong heat sink (24.83 °C), followed by Dense trees (LCZ-A, 28.28 °C), while Bare rock or paved was a strong heat source (LCZ-E, 32.61 °C), followed by Bush, scrub (LCZ-C, 32.24 °C) and Bare soil or sand (LCZ-F, 32.48 °C). For the built LCZs, the Sparsely built (LCZ-9) exhibited the lowest LST (29.85 °C), while the Heavy industry (LCZ-10) indicated the highest LST (33.59 °C). In October, a transition month from heat to cool island, the land cover of Bush, scrub (LCZ-C) had the highest average LST of 19.05 °C, followed by Bare soil or sand (LCZ-F, 18.86 °C), while dense trees had the lowest average LST of 14.91 °C and then the second lowest LST of Water (LCZ-G, 15.40 °C). The built LCZ of Heavy industry (LCZ-10) had the highest average LST (19.34 °C) while the Compact high-rise (LCZ-1) had the lowest LST of 17.08 °C.

Overall, LST of different types of LCZs varied temporally. In hot seasons (e.g., July, August), Heavy industry (LCZ-10) had the highest LST, forming the strongest heat sources among the built LCZs. Such results were relevant to the waste heat emissions during factory operation in hot seasons [50] and the strong solar radiation incidence due to large sky view factor [20]. Following this, the Compact midrise (LCZ-2), Lightweight low-rise (LCZ-7) and Large low-rise (LCZ-8) were also strong heat sources (Figure 6), which may be because of strong solar radiation incidence. In comparison, in cold seasons (from December to March), Sparsely built (LCZ-9) and Compact low-rise (LCZ-3) had the highest LST among all built LCZs, whilst the Compact high-rise (LCZ-1) had the lowest LST (Figure 6).

Temporal variation of LST to LCZs was also observed in land-cover LCZs. Whilst the Water had the lowest LST in almost all months, the Scattered trees (LCZ-B), Dense trees (LCZ-A) and Low plants (LCZ-D) exhibited the lowest average LST in September, October and November, respectively (Figure 6). Both Bush, scrub (LCZ-C) and Low plants (LCZ-D) had the highest average LST from December to May, which may be because of the dual impacts of the acceptance of solar radiation and vegetation. In comparison, the Bare-rock or paved (LCZ-E) and Bare soil or sand (LCZ-F) had the highest average LST from June to September, and the Bush, scrub (LCZ-C) had the highest LST in October and November (Figure 6).

4.4. Spatial Variation of the Responses of Land Surface Temperature to Local Climate Zones

Responses of LST to different types of LCZs within urban and rural areas is examined to analyze the impact of spatial context. The LSTs of different LCZs in urban and rural contexts are given in Appendices A and B. Table 4 presents the two maximum LSTs and two minimum LSTs within urban and rural areas in 12 months. Overall, there was an obvious difference between the LST patterns of urban and rural built LCZs. In urban area, LCZ-7 (Lightweight low-rise), LCZ-8 (Large low-rise) and LCZ-10 (Heavy industry) had the highest LSTs among ten built LCZs depending on months, while the LCZ-2 (Compact midrise), LCZ-3 (Compact low-rise), LCZ-8 (Large low-rise), LCZ-9 (Sparsely built) and LCZ-10 (Heavy industry) underwent the highest LSTs within rural context. The LCZ-10 (Heavy industry) within an urban context had the highest temperatures from May to October with the average temperature ranging between 19.27 and 36.88 °C, while the LCZ-10 (Heavy industry) within a rural context showed the highest temperature throughout the year, excluding two months of January and April, with average temperature ranging from −3.43 to 36.83 °C.

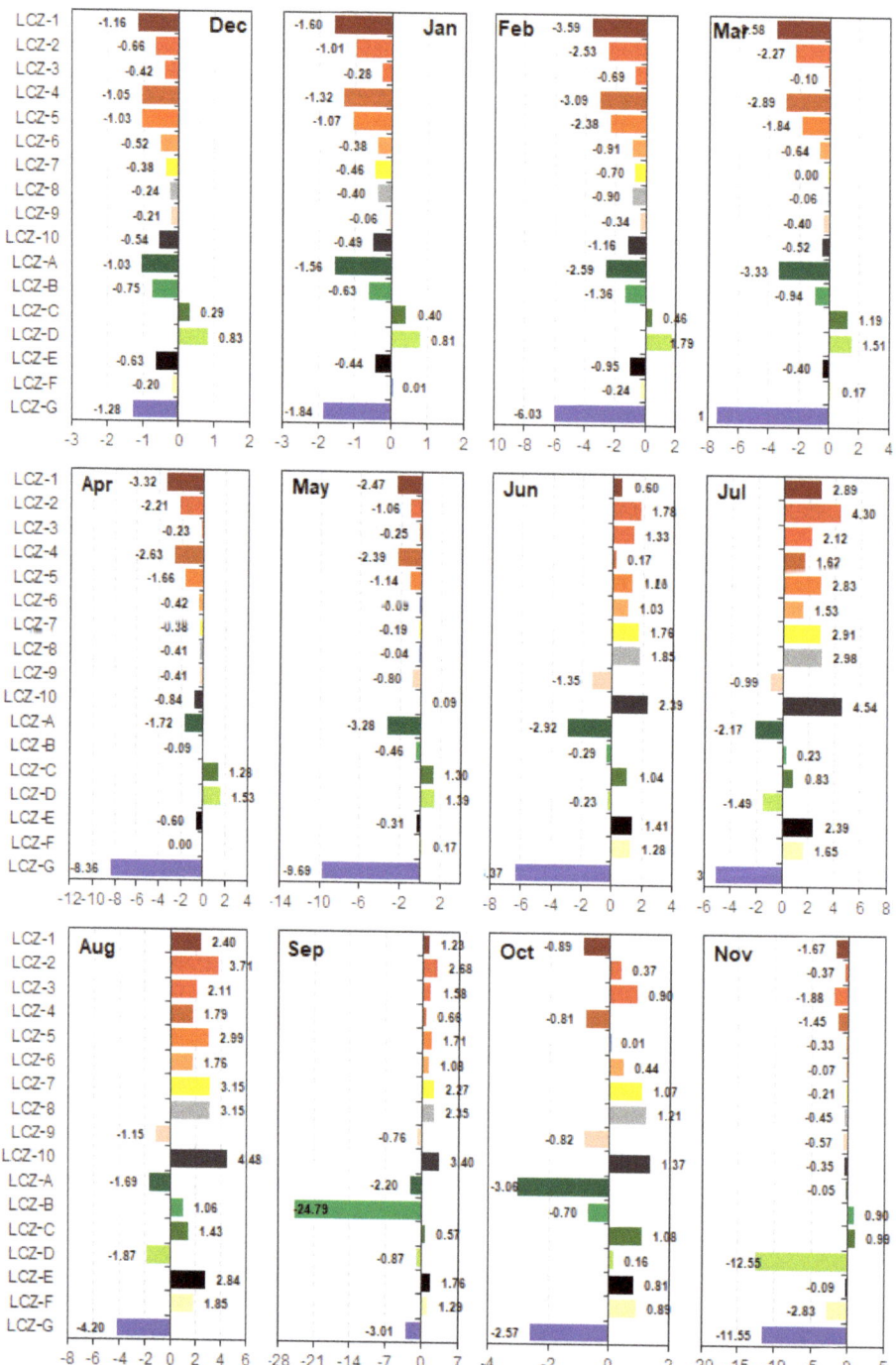

Figure 6. Deviation of the LST of different types of LCZs from the average LST of whole study area from 2018 to 2020 (°C).

Table 4. The maximum and minimum temperatures in different months and corresponding built local climate zones within urban and rural contexts (°C).

	Urban		Rural		Urban		Rural	
	Max-1	Max-2	Max-1	Max-2	Min-1	Min-2	Min-1	Min-2
December	LCZ-8 (−3.86)	LCZ-7 (−3.92)	LCZ-10 (−3.43)	LCZ-8 (−3.47)	LCZ-4 (−4.74)	LCZ-1 (−4.69)	LCZ-1 (−4.35)	LCZ-5 (−4.29)
January	LCZ-7 (−3.70)	LCZ-10 (−3.75)	LCZ-3 (−3.14)	LCZ-9 (−3.14)	LCZ-1 (−4.87)	LCZ-4 (−4.82)	LCZ-1 (−3.81)	LCZ-4 (−3.80)
February	LCZ-7 (9.00)	LCZ-8 (8.60)	LCZ-10 (10.03)	LCZ-9 (9.76)	LCZ-1 (6.12)	LCZ-4 (6.22)	LCZ-1 (7.79)	LCZ-4 (7.93)
March	LCZ-7 (14.28)	LCZ-8 (13.91)	LCZ-10 (15.33)	LCZ-8 (14.89)	LCZ-1 (10.65)	LCZ-4 (10.97)	LCZ-1 (12.36)	LCZ-4 (12.47)
April	LCZ-7 (24.19)	LCZ-8 (23.85)	LCZ-8 (24.98)	LCZ-3 (24.97)	LCZ-1 (21.25)	LCZ-4 (21.64)	LCZ-4 (22.87)	LCZ-1 (22.97)
May	LCZ-10 (27.82)	LCZ-7 (27.80)	LCZ-10 (28.73)	LCZ-8 (28.08)	LCZ-4 (25.24)	LCZ-1 (25.31)	LCZ-4 (25.86)	LCZ-1 (25.97)
June	LCZ-10 (33.59)	LCZ-7 (33.33)	LCZ-10 (33.58)	LCZ-2 (32.81)	LCZ-9 (31.44)	LCZ-4 (31.56)	LCZ-9 (29.49)	LCZ-4 (30.78)
July	LCZ-10 (34.63)	LCZ-2 (34.37)	LCZ-10 (34.26)	LCZ-2 (33.36)	LCZ-9 (31.44)	LCZ-4 (32.06)	LCZ-9 (28.52)	LCZ-1 (30.49)
August	LCZ-10 (36.88)	LCZ-7 (36.42)	LCZ-10 (36.83)	LCZ-2 (36.03)	LCZ-9 (34.09)	LCZ-4 (34.54)	LCZ-9 (30.62)	LCZ-1 (32.99)
September	LCZ-10 (27.18)	LCZ-7 (26.65)	LCZ-10 (27.32)	LCZ-2 (26.40)	LCZ-4 (24.54)	LCZ-9 (24.57)	LCZ-9 (22.69)	LCZ-1 (23.94)
October	LCZ-10 (19.27)	LCZ-7 (19.25)	LCZ-10 (19.92)	LCZ-8 (19.35)	LCZ-4 (16.93)	LCZ-1 (17.04)	LCZ-9 (17.03)	LCZ-1 (17.58)
November	LCZ-7 (11.43)	LCZ-8 (11.11)	LCZ-10 (12.37)	LCZ-8 (11.80)	LCZ-1 (8.96)	LCZ-4 (9.26)	LCZ-1 (10.62)	LCZ-4 (10.71)

Note: Max-1 means the highest temperature and Max-2 means the second highest temperature. Min-1 means the lowest temperature and Min-2 means the second lowest temperature. LCZ-1, Compact high-rise; LCZ-2, Compact midrise; LCZ-3, Compact low-rise; LCZ-4, Open high-rise; LCZ-5, Open midrise; LCZ-7, Lightweight low-rise; LCZ-8, Large low-rise; LCZ-9, Sparsely built; LCZ-10, Heavy industry.

Apart from the LCZ-10 (Heavy industry), within the urban area, the LCZ-7 (Lightweight low-rise) and LCZ-8 experienced the top two highest temperature throughout the year excluding July (11 times), and LCZ-8 (Large low-rise) was prominent for five times. In comparison, within a rural context, apart from LCZ-10 (Heavy industry), the top two highest temperatures were observed in different built LCZs, including LCZ-2 (Compact midrise, 4 times), LCZ-3 (Compact low-rise, 2 times), LCZ-8 (Large low-rise, 6 times) and LCZ-9 (Sparsely built, 2 times). For the minimum temperature, LCZ-1 (Compact high-rise), LCZ-4 (Open high-rise) and LCZ-9 (Sparsely built) were the built LCZs undergoing the lowest temperature within an urban context. Likewise, it was such three built LCZs that experienced the lowest LST within a rural context. It should be noted that the built LCZs of Compact high-rise, Compact midrise and Compact low-rise were not the case with the highest temperatures, different from the situation of the whole study area (Figure 6), while the Compact high-rise underwent the lowest temperature.

Table 5 compares the LSTs of land-cover LCZs within urban and rural contexts. In both urban and rural contexts, LCZ-C (Bush, scrub), LCZ-D (Low plants) and LCZ-E (Bare rock or paved) had the highest temperature among seven land-cover LCZs. Moreover, the LCZs with peaked temperatures with urban and rural contexts were generally the same throughout a year excluding December, May and September. From January to April, LCZ-D (Low plants) had the highest temperature within both urban and rural contexts, LCZ-E (Bare rock or paved) had the highest temperature from July to September, and LCZ-C (Bush, scrub) had the highest temperature from October to November. Within urban area, the LCZ-G (Water) was a strong heat sink, where it had the lowest temperature from December to August, and it had the second lowest temperature from September to November. A similar case was observed in the rural context; LCZ-G (Water) had the lowest temperature from November to August, and it had the second lowest temperature in September and October. Likewise, the LCZ-A (Dense trees) had the minimum temperature within both urban and rural contexts.

Table 5. The maximum and minimum temperatures in different months and corresponding land-cover local climate zones within urban and rural contexts (°C).

	Urban		Rural		Urban		Rural	
	Max-1	Max-2	Max-1	Max-2	Min-1	Min-2	Min-1	Min-2
December	LCZ-C (−3.65)	LCZ-B (−3.80)	LCZ-D (−2.63)	LCZ-C (−3.10)	LCZ-G (−4.84)	LCZ-E (−4.41)	LCZ-G (−4.77)	LCZ-A (−4.54)
January	LCZ-D (−3.08)	LCZ-C (−3.34)	LCZ-D (−2.37)	LCZ-C (−2.69)	LCZ-G (−5.03)	LCZ-A (−4.51)	LCZ-G (−5.04)	LCZ-A (−4.77)
February	LCZ-D (9.47)	LCZ-C (9.16)	LCZ-D (11.65)	LCZ-C (10.48)	LCZ-G (4.40)	LCZ-A (6.69)	LCZ-G (3.75)	LCZ-A (7.22)
March	LCZ-D (14.52)	LCZ-C (14.41)	LCZ-D (15.83)	LCZ-C (15.14)	LCZ-G (7.96)	LCZ-A (11.63)	LCZ-G (6.83)	LCZ-A (11.02)
April	LCZ-D (24.89)	LCZ-C (24.84)	LCZ-D (26.13)	LCZ-C (25.67)	LCZ-G (18.53)	LCZ-A (21.87)	LCZ-G (16.16)	LCZ-A (22.90)
May	LCZ-C (28.38)	LCZ-D (28.36)	LCZ-D (29.29)	LCZ-C (29.21)	LCZ-G (20.56)	LCZ-A (21.66)	LCZ-G (18.00)	LCZ-A (24.65)
June	LCZ-C (32.84)	LCZ-E (32.80)	LCZ-E (32.43)	LCZ-C (32.42)	LCZ-G (27.02)	LCZ-A (27.42)	LCZ-G (24.73)	LCZ-A (28.21)
July	LCZ-E (32.93)	LCZ-F (32.45)	LCZ-E (31.92)	LCZ-F (31.33)	LCZ-G (27.40)	LCZ-A (27.57)	LCZ-G (24.80)	LCZ-A (27.81)
August	LCZ-E (35.77)	LCZ-F (35.31)	LCZ-E (34.71)	LCZ-C (34.00)	LCZ-G (30.29)	LCZ-A (30.34)	LCZ-G (28.09)	LCZ-D (30.49)
September	LCZ-E (25.90)	LCZ-F (25.54)	LCZ-E (25.27)	LCZ-F (24.88)	LCZ-A (21.62)	LCZ-G (22.02)	LCZ-B (-1.00)	LCZ-G (20.71)
October	LCZ-C (18.84)	LCZ-E (18.71)	LCZ-C (19.04)	LCZ-F (18.97)	LCZ-A (14.71)	LCZ-G (15.61)	LCZ-A (14.88)	LCZ-G (15.37)
November	LCZ-C (11.63)	LCZ-D (11.28)	LCZ-C (12.63)	LCZ-D (12.57)	LCZ-E (−1.00)	LCZ-G (8.87)	LCZ-A (8.71)	LCZ-A (11.22)

Note: Max-1 means the highest temperature and Max-2 means the second highest temperature. Min-1 means the lowest temperature and Min-2 means the second lowest temperature. LCZ-A, Dense trees; LCZ-B, Scattered trees; LCZ-C, Bush, scrub; LCZ-D, Low plants; LCZ-E, Bare rock or paved; LCZ-F, Bare soil or sand; LCZ-G, Water.

Overall, the results indicate the urban context affected the responses of LST to LCZs. The highest temperature within built LCZs within an urban context exhibited a higher divergence compared with that within a rural context. Compared with built LCZs, the temperature of land-cover LCZs was more convergent. Moreover, there were limited differences between the land-cover LCZs with maximum (LCZ-C, -D, and -E) and minimum (LCZ-A, -G) temperatures within urban and rural contexts. Furthermore, the overall LST patterns in urban and rural contexts were different, as presented in Appendices C and D.

4.5. Land Surface Temperature Difference among Different Local Climate Zones

Given the spatiotemporal variations of the responses of LST to LCZs, the suitability of the LCZ scheme for LST differentiation was further examined, as shown in Figure 7. The results indicate that LSTs of different LCZ types were generally different throughout the year, while some types of LCZs failed to differentiate LST. For instance, there was no significant difference between the LST of LCZ 1 (Compact high-rise) and LCZ-2 (Compact midrise) in December. The same results were found among other pairs including LCZ-2 (Compact midrise) and LCZ-5 (Open midrise), LCZ-4 (Open high-rise) and LCZ-5 (Open midrise), LCZ 3 (Compact low-rise) and LCZ-6 (Open low-rise), LCZ-3 (Compact low-rise) and LCZ-7 (Lightweight low-rise), LCZ-6 (Open low-rise) and LCZ-7 (Lightweight low-rise), and LCZ-8 (Large low-rise) and LCZ-9 (Sparsely built) in December. Such results indicate the suitability of the LCZ scheme was compromised in differentiating urban temperatures. The insignificant difference occurred among the pairs with the same characteristics in terms of compactness (e.g., compact, open) and building height (e.g., low-rise).

The suitability of the LCZ scheme for surface temperature differentiation varied temporally. For instance, in July, there was no significant difference between the LST of LCZ-1 (Compact high-rise) and that of LCZ-5 (Open midrise). The same case was also found in the pairs of LCZ-1 (Compact high-rise) and LCZ-8 (Large low-rise), LCZ-5 (Open midrise) and LCZ-8 (Large low-rise), and LCZ-7 (Lightweight low-rise) and LCZ-8 (Large low-rise). In August, there was no significant difference in the three pairs including LCZ-6 (Open low-rise) and LCZ-4 (Open high-rise), LCZ-5 (Open midrise) and LCZ-8 (Large low-rise), and LCZ-7 (Lightweight low-rise) and LCZ-8 (Large low-rise). In July and August, the insignificant pairs including LCZ-5 and LCZ-8, and LCZ-7 and LCZ-8 were observed. However, two pairs of LCZ-1 and LCZ-5, and LCZ-1 and LCZ-8 were found in July, but not in August.

Suitability of LCZ scheme for LST differentiation was also compromised among land-cover LCZs and it exhibited temporal variations. In December, there was no significant difference between five pairs, including LCZ-A (Dense trees) and LCZ-C (Bush, scrub), LCZ-B (Scattered trees) and LCZ E (Bare rock or paved), LCZ-D (Low plants) and LCZ-F (Bare soil or sand), LCZ-D (Low plants) and LCZ G (Water), and LCZ-F (Bare soil or sand)

and LCZ-G (Water). Only temperatures of LCZ-A (Dense trees) and LCZ-G (Water) did not have significant differences in July. The pair of LCZ-C (Bush, scrub) and LCZ-F (Bare soil or sand) did not have significant differences in their surface temperatures.

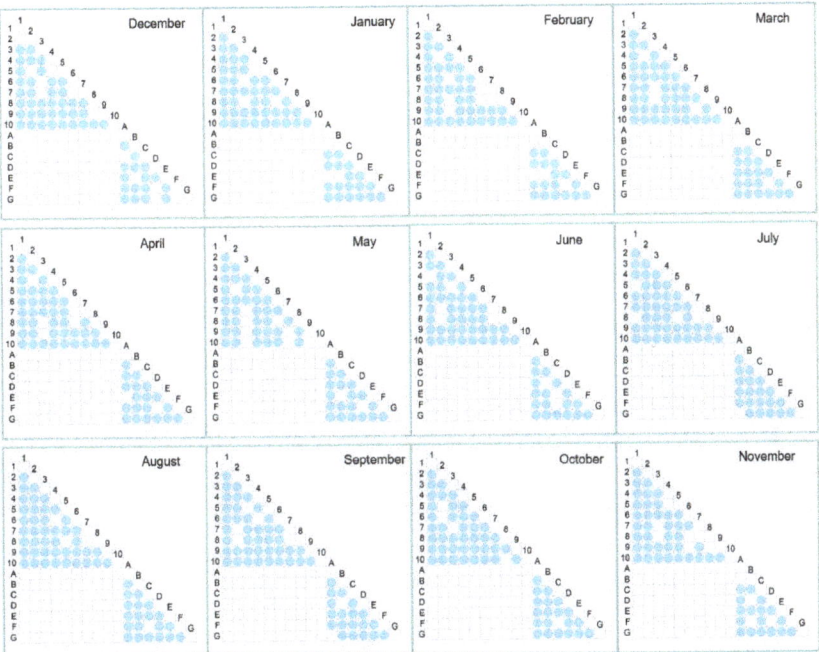

Figure 7. Difference of land surface temperatures of local climate zone types within the whole study area (Circle denotes significant difference at $p < 0.05$ level, blank demotes no significant difference).

Table 6 assesses the capability of the LCZ scheme in differentiating LST in different months. It shows that for the built LCZs within the whole study area, the LCZ scheme could differentiate 93.9% of the LST in August, indicating the strongest capability. This was followed by that in January, March and October, with a proportion of 91.1%. In comparison, the LCZ scheme could only differentiate 80.0% of the LST in May, indicating the weakest capability. For land-cover LCZs, the LCZ scheme could differentiate 95.2% of the temperature in July, August and October. However, the capability was lowest in December with a proportion of 76.2%.

Table 6. An assessment of the capability of local climate zone scheme in differentiating land surface temperatures.

	Whole Study Area		Urban (Built-Up) Area		Rural Area	
	Built LCZs	Land-Cover LCZs	Built LCZs	Land-Cover LCZs	Built LCZs	Land-Cover LCZs
December	86.7%	76.2%	71.1%	47.6%	48.9%	76.2%
January	91.1%	85.7%	77.8%	71.4%	57.8%	85.7%
February	84.4%	81.0%	91.1%	81.0%	62.2%	85.7%
March	91.1%	90.5%	95.6%	81.0%	77.8%	85.7%
April	88.9%	90.5%	88.9%	81.0%	75.6%	90.5%
May	80.0%	90.5%	84.4%	85.7%	62.2%	85.7%
June	88.9%	85.7%	88.9%	76.2%	75.6%	90.5%
July	91.1%	95.2%	86.7%	71.4%	88.9%	95.2%
August	93.3%	95.2%	86.7%	81.0%	86.7%	95.2%
September	88.9%	90.5%	88.9%	81.0%	33.3%	57.1%
October	91.1%	95.2%	82.2%	81.0%	75.6%	81.0%
November	84.4%	81.0%	91.1%	61.9%	62.2%	71.4%

Figures 8 and 9 reveal the suitability of LCZ scheme in differentiating LST in urban and rural area, respectively. Compared with the whole study area, the suitability weakens within the purely urban area. For instance, in December, there were 13 pairs of LCZ types, indicating insignificant LST difference within the urban area, while there were only six pairs when analyzing the whole study area. Compared with the urban area, there were 23 pairs of LCZs indicating insignificant LST difference in December. Such results indicate the spatial variation of the suitability of LCZ scheme in dividing LST. In hot seasons when urban thermal environments were a critical problem, there were six pairs of LCZs, exhibiting insignificant LST difference in both July and August within the urban area.

In comparison, there were five and six pairs in such two months within the rural area. Nevertheless, the pairs were different. For instance, in July, the six pairs in the urban area were LCZ-1 (Compact high-rise) and LCZ-3 (Compact low-rise), LCZ-1 (Compact high-rise) and LCZ-5 (Open midrise), LCZ-1 (Compact high-rise) and LCZ-7 (Lightweight low-rise), LCZ-3 (Compact low-rise) and LCZ-5 (Open midrise), LCZ-4 (Open high-rise) and LCZ-6 (Open low-rise), LCZ-7 (Lightweight low-rise) and LCZ-8 (Large low-rise). In August, such six pairs in urban area were LCZ-1 (Compact high-rise) and LCZ-4 (Open high-rise), LCZ-1 (Compact high-rise) and LCZ-6 (Open low-rise), LCZ-2 (Compact midrise) and LCZ-7 (Lightweight low-rise), LCZ-2 (Compact midrise) and LCZ-8 (Large low-rise), LCZ-4 (Open high-rise) and LCZ-6 (Open low-rise), and LCZ-7 (Lightweight low-rise) and LCZ-8 (Large low-rise). In addition, the results indicate that many pairs of land-cover LCZs did not have significant differences in their average LSTs.

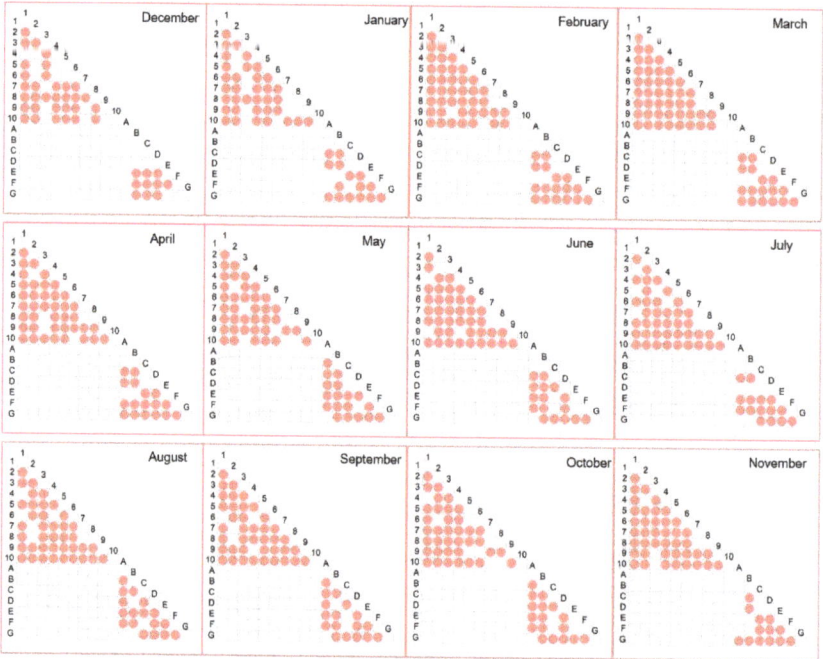

Figure 8. Difference of land surface temperatures of local climate zone types within an urban context (Circle denotes significant difference at $p < 0.05$ level, blank demotes no significant difference).

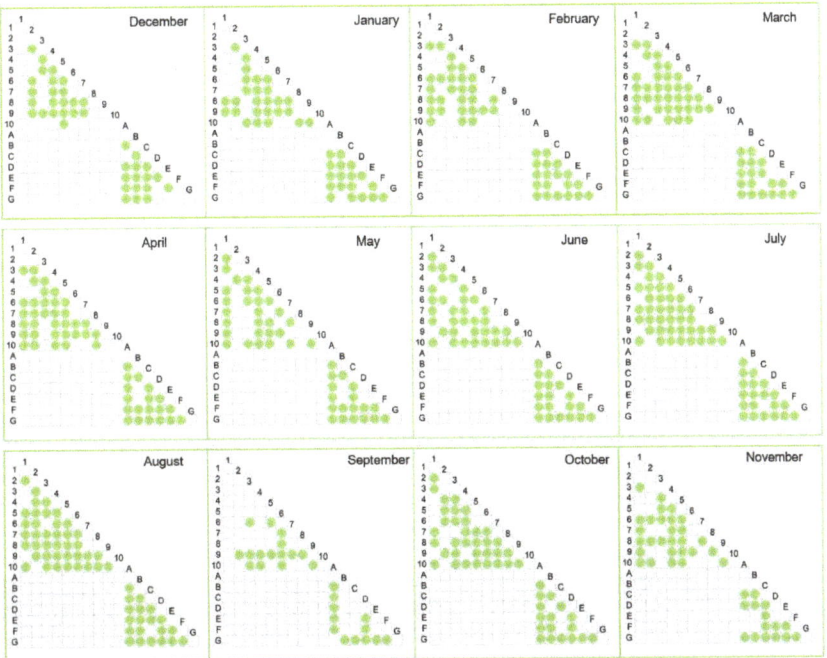

Figure 9. Difference of land surface temperatures of local climate zone types within a rural context (Circle denotes significant difference at $p < 0.05$ level, blank demotes no significant difference).

Table 6 also presents the capability of the LCZ scheme in differentiating urban temperatures of urban and rural areas. It is observed that the built LCZs had the strongest capability in March by differentiating 95.6% of the urban temperatures. Following this, in both February and November, the built LCZs could differentiate 91.1% of the urban temperatures. However, the weakest capability was observed in December where only 71.1% of the urban temperatures was distinct significantly. The capability of built LCZs was far weaker when differentiating rural temperatures, compared with urban temperature. The strongest capability of built LCZs in differentiating rural LSTs was found in July, with 89.9% of the temperature distinguishable. However, only 48.9% of the temperature was differentiated by built LCZs in December, indicating the weakest capability. Moreover, the land-cover LCZs, compared with built LCZs, showed a much weaker capability in differentiating LSTs. The strongest capability of land-cover LSTs was recognized in May, with 85.7% of (urban) LSTs differentiable, while the weakest, with only 47.6% of the (urban) LSTs distinguishable, was observed in December. In July and August, the rural land-cover LCZs exhibited the strongest capability (95.2%), while the lowest (57.1%) was found in September.

5. Discussion and Implications

5.1. Spatiotemporal Variations of Urban Thermal Environments and the Proper Month and Method Selection for Accurate Assessment

This study revealed that in July and August, Shenyang underwent the worst thermal environments with the highest LST, and both December and January were the coldest months throughout the year. Shenyang underwent an obvious heat island phenomenon in summertime, while it had a longer period of cool island phenomenon from November to May. In comparison, both June and October were transition months, implying cool–heat and heat–cool island alterations, respectively. Such results, on the one hand, indicate the dependency of LST and SUHII on month variations because of solar incidence, cooling performance and thermal inertia of vegetation and water bodies, aerodynamic properties and their interactions with urban form (e.g., materials, typology). On the other hand, LST and SUHII have been important indicators for the quantification and assessment of urban thermal environment and heat stress so that an appropriate selection of the month for study is essential, in order to avoid an underestimation of heat-related impacts and hazards. For instance, whilst June and September were also typical months in summer and autumn, the LST and SUHII were much lower than those in July and August.

Apart from temporal variations, the SUHII experienced spatial variations which is relevant to the definition of urban and suburban/rural boundaries. The adoption of economic (three-ring, Figure 4) and administrative (built-up area, Figure 5) border to screen urban and suburban/rural areas resulted in the upper limits of SUHII or SUCII. The definition of 50% and 100% buffering zones, adjacent to urban areas, as the suburban/rural areas, resulted in a lower SUHII or SUCII, and a smaller buffering zone corresponded to a weaker SUHI or SUCI phenomenon. This result was different from the findings, reported in a study by Peng et al. [18] which originally introduced the buffering method for quantifying the SUHI phenomenon, that the definition of 50%, 100% and 150% buffering zones as suburban/rural areas did not make differences to SUHII in Beijing. Our results indicate that such a hypothesis was not applicable to the case study city of Shenyang, where the SUHII based on a 100% buffering zone was 0.13–0.16 °C higher than that based on a 50% buffering zone in July and August. With a larger buffering zone, the SUHII went higher until the upper SUHII limits, about two times that based on the 50% buffering zone.

The definition of the urban area also made a difference to the SUHII or SUCII because of the variation of land use/land cover included in the study area. Our results indicate that the SUHI for the built-up (urban) area was weaker than that calculated for the three-ring (urban) area, while the SUCI for the built-up area was stronger. With the change of urban area, the months that underwent cool island or heat island phenomenon varied as well so that the transition months migrated. Such results further exhibited the significance of definition of urban/suburban/rural area and the importance of determining proper months for investigating urban thermal environment and assessing heat-induced impacts.

Overall, the same definition of urban/suburban/rural area resulting in distinct SUHII within different cities may be relevant to the fact that the SUHII formation is associated with urban form (e.g., urban size, shape, density, centrality) and urban macro climatic background [15,19,51,52]. Therefore, there is a need to seek for a flexible method for suburban/rural definition that could result in stable SUHII [15,24,53].

5.2. Spatiotemporal Variations of the Responses of Land Surface Temperature to Local Climate Zone

The LCZ scheme was developed to differentiate temperature across different urban zones and existing studies have indicated that Compact high-rise (LCZ-1), Compact midrise (LCZ-2), Large low-rise (LCZ-8) and Heavy industry (LCZ-10) could generally have the highest LST [20,54]. Nevertheless, our results revealed that the heat sinks/sources varied significantly with the month. In hot months (e.g., June, July, August), Compact midrise (LCZ-2), Heavy industry (LCZ-10), Open low-rise (LCZ-6) and Lightweight low-rise (LCZ-7) were the built LCZs with the highest LST. Such results were partially similar to the existing results [20,54]. However, this was not a fixed pattern. Lightweight low-rise (LCZ-7), Large low-rise (LCZ-8) and Sparsely built (LCZ-9) had the highest LST, while Compact high-rise (LCZ-1) and Open high-rise (LCZ-4) had the lowest LST among ten built LCZs from October to May. Urban greenery and water bodies and their relationship with built LCZs could show different cooling/heating influences due to thermal inertia, resulting in the change of temperature pattern of LCZs [45,55]. The inconsistent pattern of LST temperature of different built LCZs also indicate the variations of different combinations of heat sources and sinks, such as deciduous trees for shading and evapotranspiration (a lower urban greenery ratio in winter compared that in summer), the intermittent operation of factories, the use of heating/cooling system in cold/hot seasons and heat linkages, and the alternation of macro climate in hot–cold seasons.

For land-cover LCZs, the Water (LCZ-G) and Dense trees (LCZ-A) were the strong heat sinks throughout the year, while in other studies, the Water (LCZ-G) could be heat sources in winter because of its high inertia [42]. Such results were consistent with previous studies that water bodies and forestry land were conducive to lowering urban temperature [23]. Moreover, icing and snow, in combination with Water and Dense trees, fostered its lowest LST in winter. Nevertheless, the contribution of other land-cover LCZs varied temporally. The Bare rock and paved (LCZ-E) was a heat source in warm months (e.g., from June to October), while it was heat sinks from December to May (Figure 6). Such results may be relevant to the capability of heat absorption and storage of rock and paved materials (e.g., stone, concrete, asphalt) in summer, while in winter the higher radiative capacity, compared with other land-cover LCZS, made it have a lower temperature. Likewise, the Low plants (LCZ-D) made different contributions depending on cold or warm seasons, where it was a heat source from December to May and it was a heat sink from June to November. This result might be relevant to the growth cycle of low vegetation (e.g., rice with irrigation, grass), where in warm months vegetation grows and generates cooling performance, while it changes to the land with dry grasses (e.g., straw, detritus) contributing to heat sources in cold months. Such a result may imply a change of LCZ types in different months.

The response of LST to LCZs diversified with urban and rural contexts. For instance, Lightweight low-rise (LCZ-7), Large low-rise (LCZ-8) and Heavy industry (LCZ -10) were the three built LCZs that had the highest LST throughout the year within an urban context. Different from this, the Compact midrise (LCZ-2), Compact low-rise (LCZ-3), Large low-rise (LCZ-8), Sparsely built (LCZ-9) and Heavy industry (LCZ-10) were the five built LCZs that had the highest LST within rural context. Both the hottest built LCZs within urban and urban contexts were different from those within the whole study area. Nevertheless, the Compact high-rise (LCZ-1), Open high-rise (LCZ-4) and Sparsely built (LCZ-9) were the three built LCZs that had the lowest LST within both urban and rural contexts. Moreover, Bush, scrub (LCZ-C), Low plants (LCZ-D), Bare rock or paved (LCZ-E) and Bare soil or sand (LCZ-F) had the highest LST within urban contexts, which was consistent with that within rural context. Both Dense trees (LCZ-A) and Water (LCZ-G) had the lowest LST within both urban and rural areas. Overall, such results indicate that the responses of LST to LCZs was a function of the scope of area of interest, particularly the highest urban temperature. LCZ scheme have been thought of as an effective tool [12], to support climate-sensitive urban planning and design (e.g., outdoor thermal comfort, heat exposure, heat stress) [56–58]. The spatiotemporal change of the hottest LCZs in our study implies the consideration of only a month or an improper selection of study area for identifying heat stress may lead to inaccurate results and mislead actions of urban heat mitigation and adaptation.

5.3. Suitability of Local Climate Zone Scheme for Urban Temperature Differentiation

Existing studies have found that the suitability of the LCZ scheme for LST differentiation can be affected by macroclimate (e.g., tropical, arid, temperate and cold) [40] and seasons (e.g., spring, summer, autumn and winter) [42]. It has been indicated that the LCZ scheme had the weakest capability of differentiating LST within arid climates, and it had a moderate level of capability within cold climates [40]. The study area of Shenyang is in cold regions and the LCZ scheme could differentiate LST, with 80.0–93.3% and 76.2–95.2% of the urban and rural temperature being differentiated, respectively. Nevertheless, the capability was dependent on different months and urban context (e.g., urban and rural areas). The case study in Nanjing indicated the capability was about 86.1%, 93.1%, 77.88% and 82.0% in spring, summer, autumn and winter, respectively [42]. In comparison, our study indicated in hot months (from July to October), the LCZ scheme had the strongest capability (89.9-93.3%) and in May it had the weakest capability (80.0%). For land-cover

LCZs, our study indicated that the LCZ scheme had the strongest capability from July to October (90.5–95.2%) and the lowest from November to February (76.2–85.7%).

Our study further added a new finding that the capability of the LCZ scheme in differentiating temperature is dependent on the urban context (e.g., urban and rural), apart from macroclimate, seasons and months. Compared with the whole study area (a combination of urban and rural areas), the capability of built LCZs in either urban or rural contexts weakened. From July to October, the built LCZs could differentiate less than 90% of the urban temperature and only 71.1% and 77.8% of the temperature were differentiated in December and January, respectively. Furthermore, in a rural context, apart from 86.7% and 88.9% in August and July, less than 80% of the urban temperature was differentiated in all other months and even peaked at 48.9% in December. Such results can indicate that built LCZs had a higher level of applicability in urban context (highly urbanized area) compared with rural context (barely urbanized area). However, land-cover LCZs indicated a different result from built LCZs that within rural context the land-cover LCZs had a higher level of capability compared with that within urban context. Overall, spatiotemporal suitability of LCZ scheme for differentiating LST implies that the adoption of LCZ in urban planning and design should be pre-examined to avoid misleading results. One-month (e.g., in summer) LCZ data within a specific urban/rural context cannot fully represent and identify heat-induced impacts such as outdoor thermal comfort, heat stress and heat exposure of a city. It is essential to document several-month data based on specific contexts to overall support urban planning and design.

6. Conclusions

An accurate quantification of urban thermal environments is the premise of mitigating and avoiding the several consequences of urban heat challenges which are experienced by many cities. LST and SUHII are two important indicators, and the LCZ scheme is an important tool to differentiate surface temperatures and intra-urban temperatures. However, the landscape effects on monthly variation of SUHII and the applicability of LCZ scheme are not well understood. This study investigated the variation of SUHI effects and the suitability of the LCZ scheme for LST differentiation in Shenyang, China. The findings indicated that both the SUHII and the suitability of the LCZ scheme exhibited spatiotemporal variations. An accurate analysis of SUHII should, therefore, properly define urban and rural contexts and specify the month, and one-month analysis cannot fully represent urban thermal environments of a season. The LST of both built and land-cover LCZs could not follow a fixed order, particularly for the highest temperature that varied significantly with both month and spatial boundary. Moreover, the suitability of LCZ scheme for LST differentiation depended on both urban context and the month. In hot months such as July and August, the LCZ scheme in aspects of both built types and land-cover types had a high level of suitability while in cold months such as November, December and January the suitability weakened. The built LCZs within urban areas could have a higher level of suitability than that within rural areas, while the land-cover LCZs exhibited a reverse pattern. Overall, this paper added new findings on spatial variability and temporal heterogeneity of urban temperature and the applicability of LCZ scheme for LST differentiation. It can also provide important implications for the assessment of heat-induced impacts and supports climate-sensitive planning and design.

This paper has some possible limitations and future studies are needed to reveal the spatiotemporal variations of LST distribution, responses of LST to LCZs and the suitability of the LCZ scheme. First, built upon the landscape in August, this study only applied one suite of the LCZ scheme, which as a result cannot respect the possible LCZ variation (e.g., changes of both built and land-cover LCZs with deciduous trees and vegetation death in cold seasons) throughout the year. Therefore, it is essential to reproduce seasonal and monthly LCZ images to further verify the results and conclusions. Second, the urban and rural contexts considered in this study were differentiated within a same city, which might be not capable of representing highly urbanized and barely urbanized cities. Therefore, future studies are needed to verify the results and conclusions relevant to the urban context with the selection of metropolitans and their surrounding satellite cities. Third, this study analyzed the monthly variation of LST distribution, responses of LST to LCZs and the suitability of LCZ scheme within only one city with Dwa in cold regions and it is wise to conduct further investigations in other cities with diverse climates to gain a better understanding of the dynamics of the LCZ–LST relationships. Fourth, this study analyzed the single-year LST responses to the LCZ scheme, while the multi-year analysis could be of interest to indicate the LCZ–LST relationships in a dynamic context along with urbanization and climate change.

Author Contributions: Z.Z.: formal analysis, investigation, methodology, software, validation, visualization, writing—review & editing; A.S.: conceptualization, funding acquisition, project administration, resources, writing—review & editing; X.D.: investigation, resources, software, visualization, writing—review & editing; L.S.: formal analysis, visualization, writing—review & editing; B.-J.H.: conceptualization, data curation, formal analysis, funding acquisition, investigation, methodology, project administration, resources, visualization, roles/writing—original draft, writing—review & editing. All authors have read and agreed to the published version of the manuscript.

Funding: This project is partially supported by JSPS KAKENHI Grant Number 19K20497. Many thanks go to the projects NO. 2021CDJQY-004 and NO. 2021CDJQY-023, supported by the Fundamental Research Funds for the Central Universities.

Data Availability Statement: The data presented in this study are available on request from the corresponding author.

Conflicts of Interest: The authors declare no conflict of interest.

Appendix A

Table A1. Land Surface Temperature of Different Local Climate Zones in Different Months within Urban Context.

LCZ	December		January		February		March	
	Range	Mean	Range	Mean	Range	Mean	Range	Mean
LCZ-1 (Compact high-rise)	−20.28~1.63	−4.69	−16.26~−0.13	−4.87	−9.04~12.74	6.12	−4.89~17.95	10.65
LCZ-2 (Compact midrise)	−18.43~2.81	−4.15	−14.45~1.46	−4.22	−6.30~16.34	7.23	−1.08~21.18	12.00
LCZ-3 (Compact low-rise)	−24.17~3.74	−4.36	−20.47~3.72	−4.01	−11.48~15.19	8.22	−9.37~21.75	13.29
LCZ-4 (Open high-rise)	−9.54~1.62	−4.74	−8.45~1.97	−4.82	1.02~14.49	6.22	2.96~19.05	10.97
LCZ-5 (Open midrise)	−19.14~0.70	−4.57	−17.30~0.49	−4.40	−6.30~13.82	7.16	−3.01~20.11	12.18
LCZ-6 (Open low-rise)	−9.28~1.02	−4.55	−8.38~0.54	−4.07	1.21~14.21	7.69	5.30~19.39	12.68
LCZ-7 (Lightweight low-rise)	−21.00~3.56	−3.92	−18.69~4.35	−3.70	−8.51~18.41	9.00	−3.57~25.09	14.28
LCZ-8 (Large low-rise)	−25.42~3.35	−3.86	−22.25~2.02	−3.80	−12.47~18.88	8.60	−9.83~25.71	13.91
LCZ-9 (Sparsely built)	−26.81~1.58	−4.32	−21.74~2.07	−3.83	−12.69~16.74	8.11	−9.58~20.13	13.13
LCZ-10 (Heavy industry)	−17.34~4.66	−4.11	−13.72~3.21	−3.75	−6.36~19.79	8.47	−2.35~24.74	13.60
LCZ-A (Dense trees)	−7.35~−0.48	−3.89	−6.63~−1.41	−4.51	2.41~12.19	6.69	5.06~17.59	11.63
LCZ-B (Scattered trees)	−9.43~0.08	−3.80	−7.99~0.26	−3.99	2.32~14.36	7.80	6.77~19.91	12.98
LCZ-C (Bush, scrub)	−7.58~0.08	−3.65	−6.68~−0.74	−3.34	4.41~13.72	9.16	8.64~18.53	14.41
LCZ-D (Low plants)	−27.50~0.45	−3.83	−21.84~0.84	−3.08	−12.18~15.67	9.47	−9.23~21.13	14.52
LCZ-E (Bare rock or paved)	−18.09~3.12	−4.41	−14.37~3.36	−3.85	−3.10~15.18	8.42	−0.09~20.33	13.45
LCZ-F (Bare soil or sand)	−19.68~1.34	−4.15	−17.41~1.03	−3.56	−8.43~15.17	8.65	−3.24~20.42	13.67
LCZ-G (Water)	−8.57~−0.09	−4.84	−7.88~0.50	−5.03	1.23~12.42	4.40	2.23~16.99	7.96

Table A1. Cont.

LCZ	April		May		June		July	
	Range	Mean	Range	Mean	Range	Mean	Range	Mean
LCZ-1 (Compact high-rise)	11.87~27.52	21.25	6.29~33.23	25.31	23.14~38.19	31.82	19.99~40.13	33.00
LCZ-2 (Compact midrise)	13.42~29.88	22.40	11.79~35.20	26.74	24.99~38.04	32.99	23.63~40.59	34.37
LCZ-3 (Compact low-rise)	7.93~28.92	23.53	3.66~34.73	27.20	20.70~37.93	32.73	18.60~40.39	32.99
LCZ-4 (Open high-rise)	15.00~29.17	21.64	16.11~33.46	25.24	24.02~37.08	31.56	24.46~38.31	32.06
LCZ-5 (Open midrise)	12.31~29.58	22.70	8.20~38.25	26.48	21.84~38.99	32.45	20.92~40.06	32.95
LCZ-6 (Open low-rise)	17.68~28.89	23.21	18.59~35.18	26.80	25.71~37.17	32.15	26.08~39.98	32.17
LCZ-7 (Lightweight low-rise)	7.83~33.64	24.19	7.01~40.99	27.80	20.31~42.03	33.33	19.29~43.34	33.65
LCZ-8 (Large low-rise)	7.36~33.81	23.85	1.84~41.34	27.63	18.91~41.77	33.19	17.36~43.11	33.47
LCZ-9 (Sparsely built)	6.20~29.52	23.54	1.87~34.83	26.77	19.18~38.55	31.44	16.78~39.83	31.44
LCZ-10 (Heavy industry)	13.34~32.96	23.66	9.81~40.41	27.82	24.54~41.63	33.59	22.68~44.20	34.63
LCZ-A (Dense trees)	16.01~29.27	21.87	16.93~33.90	21.66	24.71~34.97	27.42	25.15~35.36	27.57
LCZ-B (Scattered trees)	18.40~28.98	23.27	19.96~33.98	25.90	26.49~36.75	30.73	26.56~37.20	30.47
LCZ-C (Bush, scrub)	19.04~28.83	24.84	22.38~35.27	28.38	28.42~36.07	32.84	27.70~37.58	31.67
LCZ-D (Low plants)	5.15~29.77	24.89	0.28~35.04	28.36	17.69~37.58	31.75	15.89~39.15	30.96
LCZ-E (Bare rock or paved)	12.91~29.37	23.56	9.37~34.96	27.24	24.45~38.46	32.80	22.14~40.76	32.93
LCZ-F (Bare soil or sand)	12.54~29.51	23.92	10.62~34.19	27.66	22.82~38.33	32.70	20.79~39.12	32.45
LCZ-G (Water)	13.33~26.61	18.53	15.63~30.26	20.56	21.41~35.41	27.02	22.19~34.78	27.40

LCZ	August		September		October		November	
	Range	Mean	Range	Mean	Range	Mean	Range	Mean
LCZ-1 (Compact high-rise)	27.82~41.69	34.84	8.41~32.27	25.03	−0.27~24.25	17.04	−3.36~14.72	8.96
LCZ-2 (Compact midrise)	28.98~41.55	36.10	12.78~35.71	26.48	2.93~26.11	18.33	−1.34~20.44	9.85
LCZ-3 (Compact low-rise)	24.30~42.52	35.63	6.48~33.72	25.87	−1.84~25.29	18.53	−5.47~18.13	10.59
LCZ-4 (Open high-rise)	27.80~41.30	34.54	15.51~31.60	24.54	11.36~23.63	16.93	2.33~15.53	9.26
LCZ-5 (Open midrise)	26.77~41.78	35.44	9.04~32.95	25.50	1.97~24.53	17.84	−2.71~16.55	9.90
LCZ-6 (Open low-rise)	27.92~41.79	34.85	19.75~32.89	25.01	12.14~24.12	17.86	5.50~15.71	10.15
LCZ-7 (Lightweight low-rise)	24.00~45.48	36.42	4.48~37.95	26.65	−0.49~30.16	19.25	−6.05~20.62	11.13
LCZ-8 (Large low-rise)	23.64~45.19	36.06	4.12~37.86	26.44	−4.09~30.36	19.09	−7.54~20.55	11.11
LCZ-9 (Sparsely built)	22.64~42.12	34.09	1.49~33.66	24.57	−5.79~26.17	17.67	−8.97~16.44	10.73
LCZ-10 (Heavy industry)	28.86~45.56	36.88	12.26~37.43	27.18	3.03~29.81	19.27	−1.50~22.39	10.81
LCZ-A (Dense trees)	26.98~36.70	30.34	20.13~27.15	21.62	13.00~20.80	14.71	6.84~14.41	10.61
LCZ-B (Scattered trees)	29.66~38.81	33.52	20.38~29.61	23.79	12.89~23.51	16.88	5.92~15.91	10.82
LCZ-C (Bush, scrub)	30.65~40.36	34.83	22.39~29.27	25.20	14.55~22.89	18.84	7.11~17.64	11.63
LCZ-D (Low plants)	22.52~41.15	33.35	1.04~31.69	24.34	−6.11~24.67	18.25	−10.03~16.80	11.28
LCZ-E (Bare rock or paved)	27.49~42.40	35.77	10.50~33.25	25.90	2.70~25.43	18.71	−0.90~28.21	−1.00
LCZ-F (Bare soil or sand)	25.76~41.44	35.31	9.67~32.60	25.54	1.02~25.31	18.64	−2.33~16.73	10.80
LCZ-G (Water)	26.93~37.58	30.29	19.64~28.27	22.02	13.09~20.90	15.61	6.17~13.77	8.87

Appendix B

Table A2. Land Surface Temperature of Different Local Climate Zones in Different Months within Rural Context.

LCZ	December Range	Mean	January Range	Mean	February Range	Mean	March Range	Mean
LCZ-1 (Compact high-rise)	−9.25~1.00	−4.35	−8.58~−0.05	−3.81	0.45~13.13	7.79	5.10~18.78	12.36
LCZ-2 (Compact midrise)	−8.62~1.39	−4.05	−8.10~0.99	−3.48	1.23~14.72	8.97	2.53~21.02	14.02
LCZ-3 (Compact low-rise)	−20.21~2.08	−3.61	−16.59~2.62	−3.14	−5.83~15.00	9.69	−1.77~22.67	14.80
LCZ-4 (Open high-rise)	−20.80~4.43	−4.07	−19.65~3.86	−3.80	−6.54~15.52	7.93	−3.09~21.65	12.47
LCZ-5 (Open midrise)	−17.00~4.03	−4.29	−14.46~2.98	−3.76	−1.76~15.36	8.50	2.00~20.92	13.55
LCZ-6 (Open low-rise)	−13.75~2.17	−3.84	−12.38~1.82	−3.43	−0.97~16.05	9.26	−1.48~21.61	13.91
LCZ-7 (Lightweight low-rise)	−25.58~2.88	−3.81	−24.72~2.83	−3.60	−12.77~16.78	9.24	−7.00~23.26	14.38
LCZ-8 (Large low-rise)	−16.87~1.96	−3.47	−16.28~2.50	−3.19	−5.57~16.55	9.50	1.56~22.89	14.89
LCZ-9 (Sparsely built)	−17.26~4.36	−3.56	−15.85~3.88	−3.14	−4.97~18.13	9.76	−3.19~24.55	14.07
LCZ-10 (Heavy industry)	−15.81~2.86	−3.43	−13.74~3.35	−3.16	−2.03~16.77	10.03	2.86~23.11	15.33
LCZ-A (Dense trees)	−9.26~4.50	−4.54	−9.22~3.26	−4.77	0.35~15.75	7.22	−1.90~20.80	11.02
LCZ-B (Scattered trees)	−9.62~2.35	−4.36	−9.17~1.59	−3.79	0.77~15.20	8.63	3.79~20.07	13.24
LCZ-C (Bush, scrub)	−7.03~3.10	−3.10	−5.71~3.78	−2.69	3.95~15.56	10.48	6.71~20.66	15.14
LCZ-D (Low plants)	−21.72~2.93	−2.63	−19.35~3.95	−2.37	−9.85~22.84	11.65	−9.00~25.97	15.83
LCZ-E (Bare rock or paved)	−28.64~4.05	−3.78	−22.09~3.98	−3.37	−16.53~16.04	9.37	−13.12~21.71	14.25
LCZ-F (Bare soil or sand)	−32.33~2.76	−3.47	−27.79~2.42	−3.02	−22.77~16.62	9.98	−18.21~23.09	14.87
LCZ-G (Water)	−8.88~5.53	−4.77	−9.30~4.16	−5.04	−0.89~14.78	3.75	−2.30~19.05	6.83

LCZ	April Range	Mean	May Range	Mean	June Range	Mean	July Range	Mean
LCZ-1 (Compact high-rise)	15.15~29.62	22.97	17.42~34.15	25.97	24.96~36.63	30.91	25.77~37.50	30.49
LCZ-2 (Compact midrise)	13.23~30.09	24.04	14.95~33.94	27.91	23.58~36.47	32.81	23.23~38.27	33.36
LCZ-3 (Compact low-rise)	9.70~30.32	24.97	5.45~35.45	27.85	21.00~37.79	32.41	18.51~37.90	31.65
LCZ-4 (Open high-rise)	10.62~29.54	22.87	6.88~35.50	25.86	20.72~37.78	30.78	19.37~37.96	30.52
LCZ-5 (Open midrise)	13.71~29.78	24.05	10.71~34.16	27.49	22.12~40.02	32.45	21.18~39.95	32.47
LCZ-6 (Open low-rise)	13.34~30.21	24.45	15.06~35.36	28.00	24.23~37.16	32.25	21.68~38.43	31.39
LCZ-7 (Lightweight low-rise)	6.56~30.60	24.37	3.36~35.53	27.45	19.28~40.51	32.49	16.28~40.62	32.07
LCZ-8 (Large low-rise)	12.73~30.76	24.98	9.56~34.86	28.08	21.88~38.29	32.80	21.64~40.32	32.17
LCZ-9 (Sparsely built)	11.72~31.23	24.35	9.75~36.46	27.10	21.15~37.78	29.49	21.55~39.49	28.52
LCZ-10 (Heavy industry)	14.29~30.81	24.94	13.89~35.29	28.73	23.43~40.80	33.58	24.18~42.88	34.26
LCZ-A (Dense trees)	13.35~31.64	22.90	15.07~37.05	24.65	21.25~36.92	28.21	22.22~36.89	27.81
LCZ-B (Scattered trees)	15.44~30.28	24.60	16.35~35.03	27.65	23.98~37.73	30.84	23.88~38.61	30.17
LCZ-C (Bush, scrub)	18.40~29.39	25.67	20.36~35.33	29.21	27.29~38.84	32.42	25.62~37.09	31.03
LCZ-D (Low plants)	7.48~35.48	26.13	1.81~37.00	29.29	15.70~39.66	30.96	16.66~38.73	28.55
LCZ-E (Bare rock or paved)	5.00~31.19	24.42	−2.81~35.08	27.78	17.85~39.04	32.43	13.85~38.80	31.92
LCZ-F (Bare soil or sand)	0.77~31.04	24.97	−7.61~35.43	28.21	14.82~39.08	32.38	10.94~40.33	31.33
LCZ-G (Water)	10.36~29.27	16.16	12.34~34.90	18.00	20.71~36.06	24.73	20.76~35.05	24.80

LCZ	August Range	Mean	September Range	Mean	October Range	Mean	November Range	Mean
LCZ-1 (Compact high-rise)	28.10~39.92	32.99	19.38~30.54	23.94	11.86~23.13	17.58	4.62~17.41	10.62
LCZ-2 (Compact midrise)	27.15~40.44	36.03	19.90~33.69	26.40	14.43~26.55	19.19	7.31~17.78	11.39
LCZ-3 (Compact low-rise)	25.10~40.60	33.79	6.42~32.82	25.06	2.21~26.10	19.08	−4.64~17.54	11.58
LCZ-4 (Open high-rise)	24.31~42.21	33.17	5.07~33.06	24.13	1.26~26.19	17.67	−3.45~16.78	10.71
LCZ-5 (Open midrise)	26.81~42.52	35.08	9.18~34.49	25.45	3.10~28.35	18.47	−0.74~18.49	10.89
LCZ-6 (Open low-rise)	26.30~41.25	33.97	13.76~32.16	24.86	6.58~24.96	18.57	1.61~17.74	11.55
LCZ-7 (Lightweight low-rise)	22.62~44.59	34.44	2.24~37.64	25.33	−5.15~30.87	18.80	−7.07~19.85	11.55
LCZ-8 (Large low-rise)	26.15~44.23	34.56	10.33~34.04	25.62	4.29~26.28	19.35	−2.23~19.05	11.80
LCZ-9 (Sparsely built)	25.22~42.23	30.62	9.11~33.25	22.69	2.55~24.89	17.03	−2.73~18.09	11.19
LCZ-10 (Heavy industry)	28.90~45.01	36.83	13.17~36.66	27.32	6.06~30.28	19.92	1.21~19.47	12.37
LCZ-A (Dense trees)	25.09~39.13	30.67	18.93~29.11	21.56	11.17~23.92	14.88	5.92~18.12	11.22
LCZ-B (Scattered trees)	27.09~41.82	33.36	19.24~32.62	−1.00	11.34~24.65	17.32	4.99~17.28	11.70
LCZ-C (Bush, scrub)	28.23~41.89	34.00	21.18~30.51	24.53	14.78~23.46	19.04	6.90~20.28	12.63
LCZ-D (Low plants)	23.88~41.71	30.49	2.05~45.45	22.91	−3.64~27.09	18.19	−8.46~20.06	12.57
LCZ-E (Bare rock or paved)	24.10~42.93	34.71	0.05~33.23	25.27	−6.37~25.64	18.86	−11.45~20.26	11.62
LCZ-F (Bare soil or sand)	22.59~43.61	33.76	−3.27~35.96	24.88	−12.84~29.03	18.97	−15.94~19.63	11.77
LCZ-G (Water)	25.72~40.09	28.09	18.57~28.97	20.71	12.31~22.71	15.37	5.38~17.93	8.71

Appendix C

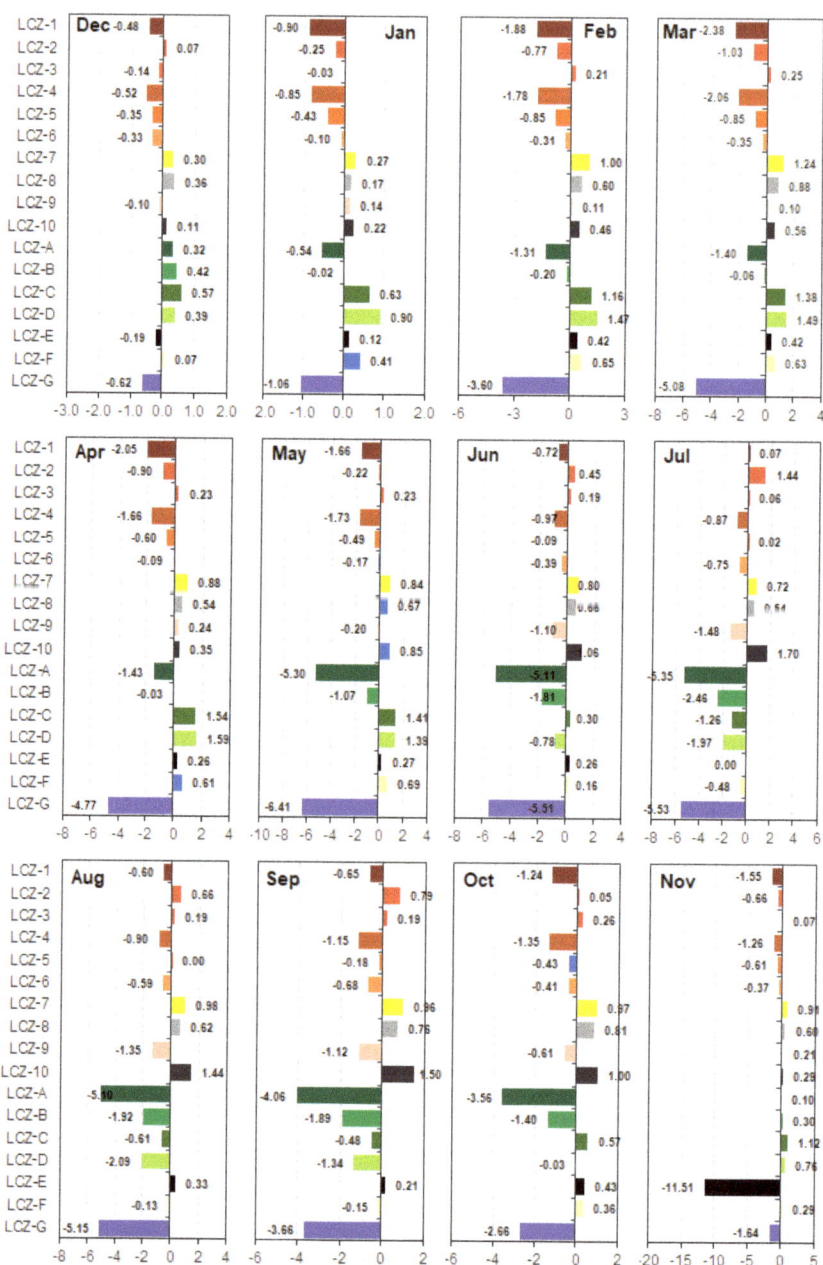

Figure A1. Deviation of the LST of Different Types of LCZs from the Average LST within an Urban Context (°C).

Appendix D

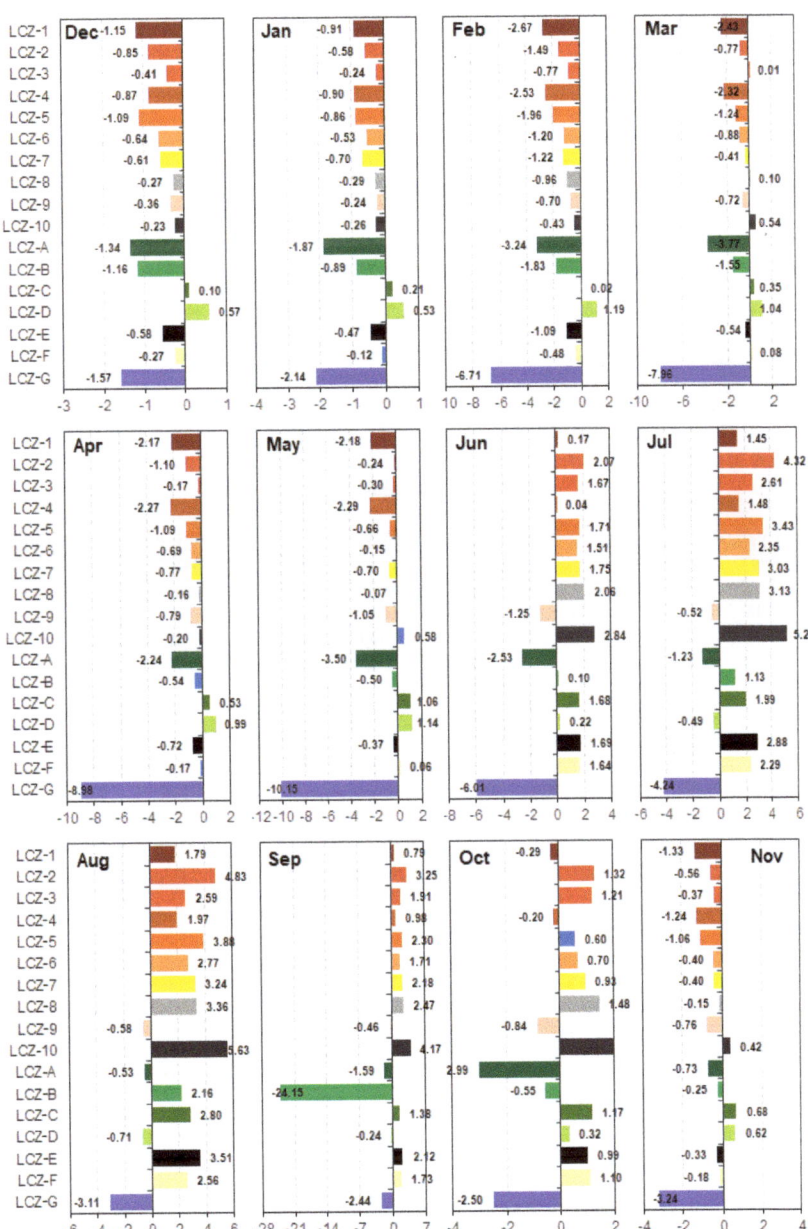

Figure A2. Deviation of the LST of Different Types of LCZs from the Average LST within a Rural Context (°C).

References

1. United Nations. 68% of the World Population Projected to Live in Urban Areas by 2050. In *2018 Revision of World Urbanization Prospects*; United Nations: New York, NY, USA, 2018.
2. Department of Economic and Social Affairs, Sustainable Development, United Nations. Goal 11: Make Cities and Human Settlements Inclusive, Safe, Resilient and Sustainable. 2021. Available online: https://sdgs.un.org/goals/goal11 (accessed on 19 October 2021).
3. Howard, L. *The Climate of London: Deduced from Meteorological Observations Made in the Metropolis and at Various Places Around It*; Darton, J.H., Longman, A.A., Highley, S.H., Hunter, R., Eds.; Joseph Rickerby: London, UK, 1833; Volume 3. Available online: https://books.google.co.jp/books?id=-yllMDVOz1IC&printsec=frontcover&hl=zh-CN&source=gbs_ge_summary_r&cad=0#v=onepage&q&f=false (accessed on 19 October 2021).
4. Oke, T.R. The energetic basis of the urban heat island. *Q. J. R. Meteorol. Soc.* **1982**, *108*, 1–24. [CrossRef]
5. He, B.-J.; Zhao, D.; Xiong, K.; Qi, J.; Ulpiani, G.; Pignatta, G.; Prasad, D.; Jones, P. A framework for addressing urban heat challenges and associated adaptive behavior by the public and the issue of willingness to pay for heat resilient infrastructure in Chongqing, China. *Sustain. Cities Soc.* **2021**, *75*, 103361. [CrossRef]
6. Santamouris, M. On the energy impact of urban heat island and global warming on buildings. *Energy Build.* **2014**, *82*, 100–113. [CrossRef]
7. Guhathakurta, S.; Gober, P. The Impact of the Phoenix Urban Heat Island on Residential Water Use. *J. Am. Plan. Assoc.* **2007**, *73*, 317–329. [CrossRef]
8. Santamouris, M.; Kolokotsa, D. *Urban Climate Mitigation Techniques*; Routledge: Oxfordshire, UK, 2016.
9. Lowe, S.A. An energy and mortality impact assessment of the urban heat island in the US. *Environ. Impact Assess. Rev.* **2016**, *56*, 139–144. [CrossRef]
10. Meehl, G.A.; Tebaldi, C. More Intense, More Frequent, and Longer Lasting Heat Waves in the 21st Century. *Science* **2004**, *305*, 994. [CrossRef]
11. He, B.-J.; Wang, J.; Liu, H.; Ulpiani, G. Localized synergies between heat waves and urban heat islands: Implications on human thermal comfort and urban heat management. *Environ. Res.* **2021**, *193*, 110584. [CrossRef]
12. Perera, N.G.R.; Emmanuel, R. A "Local Climate Zone" based approach to urban planning in Colombo, Sri Lanka. *Urban Clim.* **2018**, *23*, 188–203. [CrossRef]
13. Oke, T.R.; Mills, G.; Christen, A.; Voogt, J.A. *Urban Climates*; Cambridge University Press: Cambridge, UK, 2017.
14. Derdouri, A.; Wang, R.; Murayama, Y.; Osaragi, T. Understanding the Links between LULC Changes and SUHI in Cities: Insights from Two-Decadal Studies (2001–2020). *Remote Sens.* **2021**, *13*, 3654. [CrossRef]
15. Liu, H.; Huang, B.; Zhan, Q.; Gao, S.; Li, R.; Fan, Z. The influence of urban form on surface urban heat island and its planning implications: Evidence from 1288 urban clusters in China. *Sustain. Cities Soc.* **2021**, *71*, 102987. [CrossRef]
16. Yang, J.; Wang, Y.; Xue, B.; Li, Y.; Xiao, X.; Xia, J.; He, B. Contribution of urban ventilation to the thermal environment and urban energy demand: Different climate background perspectives. *Sci. Total. Environ.* **2021**, *795*, 148791. [CrossRef] [PubMed]
17. Luo, X.; Yang, J.; Sun, W.; He, B. Suitability of human settlements in mountainous areas from the perspective of ventilation: A case study of the main urban area of Chongqing. *J. Clean. Prod.* **2021**, *310*, 127467. [CrossRef]
18. Peng, S.; Piao, S.; Ciais, P.; Friedlingstein, P.; Ottle, C.; Bréon, F.-M.; Nan, H.; Zhou, L.; Myneni, R.B. Surface Urban Heat Island Across 419 Global Big Cities. *Environ. Sci. Technol.* **2012**, *46*, 696–703. [CrossRef] [PubMed]
19. Li, Y.; Schubert, S.; Kropp, J.P.; Rybski, D. On the influence of density and morphology on the Urban Heat Island intensity. *Nat. Commun.* **2020**, *11*, 2647. [CrossRef]
20. Stewart, I.D.; Oke, T.R. Local Climate Zones for Urban Temperature Studies. *Bull. Am. Meteorol. Soc.* **2012**, *93*, 1879–1900. [CrossRef]
21. Pal, S.; Ziaul, S. Detection of land use and land cover change and land surface temperature in English Bazar urban centre. *Egypt. J. Remote. Sens. Space Sci.* **2017**, *20*. [CrossRef]
22. Amiri, R.; Weng, Q.; Alimohammadi, A.; Alavipanah, S.K. Spatial–temporal dynamics of land surface temperature in relation to fractional vegetation cover and land use/cover in the Tabriz urban area, Iran. *Remote Sens. Environ.* **2009**, *113*, 2606–2617. [CrossRef]
23. Zhao, Z.-Q.; He, B.-J.; Li, L.-G.; Wang, H.-B.; Darko, A. Profile and concentric zonal analysis of relationships between land use/land cover and land surface temperature: Case study of Shenyang, China. *Energy Build.* **2017**, *155*, 282–295. [CrossRef]
24. Li, H.; Zhou, Y.; Li, X.; Meng, L.; Wang, X.; Wu, S.; Sodoudi, S. A new method to quantify surface urban heat island intensity. *Sci. Total Environ.* **2018**, *624*, 262–272. [CrossRef] [PubMed]
25. Adolphe, L. A Simplified Model of Urban Morphology: Application to an Analysis of the Environmental Performance of Cities. *Environ. Plan. B Plan. Des.* **2001**, *28*, 183–200. [CrossRef]
26. Osmond, P. The urban structural unit: Towards a descriptive framework to support urban analysis and planning. *Urban Morphol.* **2010**, *14*, 5–20.
27. He, B.-J.; Ding, L.; Prasad, D. Enhancing urban ventilation performance through the development of precinct ventilation zones: A case study based on the Greater Sydney, Australia. *Sustain. Cities Soc.* **2019**, *47*, 101472. [CrossRef]
28. Fenner, D.; Meier, F.; Bechtel, B.; Otto, M.; Scherer, D. Intra and inter 'local climate zone' variability of air temperature as observed by crowdsourced citizen weather stations in Berlin, Germany. *Meteorol. Z.* **2017**, *26*, 525–547. [CrossRef]

29. Alexander, P.J.; Mills, G. Local Climate Classification and Dublin's Urban Heat Island. *Atmosphere* **2014**, *5*, 755. [CrossRef]
30. Lau, K.K.-L.; Chung, S.C.; Ren, C. Outdoor thermal comfort in different urban settings of sub-tropical high-density cities: An approach of adopting local climate zone (LCZ) classification. *Build. Environ.* **2019**, *154*, 227–238. [CrossRef]
31. Kotharkar, R.; Bagade, A. Evaluating urban heat island in the critical local climate zones of an Indian city. *Landsc. Urban Plan.* **2018**, *169*, 92–104. [CrossRef]
32. Ochola, E.M.; Fakharizadehshirazi, E.; Adimo, A.O.; Mukundi, J.B.; Wesonga, J.M.; Sodoudi, S. Inter-local climate zone differentiation of land surface temperatures for Management of Urban Heat in Nairobi City, Kenya. *Urban Clim.* **2020**, *31*, 100540. [CrossRef]
33. Lehnert, M.; Geletič, J.; Husák, J.; Vysoudil, M. Urban field classification by "local climate zones" in a medium-sized Central European city: The case of Olomouc (Czech Republic). *Theor. Appl. Climatol.* **2015**, *122*, 531–541. [CrossRef]
34. Yang, J.; Jin, S.; Xiao, X.; Jin, C.; Xia, J.; Li, X.; Wang, S. Local climate zone ventilation and urban land surface temperatures: Towards a performance-based and wind-sensitive planning proposal in megacities. *Sustain. Cities Soc.* **2019**, *47*, 101487. [CrossRef]
35. Stewart, I.D.; Oke, T.R.; Krayenhoff, E.S. Evaluation of the 'local climate zone' scheme using temperature observations and model simulations. *Int. J. Climatol.* **2014**, *34*, 1062–1080. [CrossRef]
36. Bechtel, B.; Demuzere, M.; Mills, G.; Zhan, W.; Sismanidis, P.; Small, C.; Voogt, J. SUHI analysis using Local Climate Zones—A comparison of 50 cities. *Urban Clim.* **2019**, *28*, 100451. [CrossRef]
37. Chen, X.; Yang, J.; Ren, C.; Jeong, S.; Shi, Y. Standardizing thermal contrast among local climate zones at a continental scale: Implications for cool neighborhoods. *Build. Environ.* **2021**, *197*, 107878. [CrossRef]
38. Brousse, O.; Georganos, S.; Demuzere, M.; Vanhuysse, S.; Wouters, H.; Wolff, E.; Linard, C.; van Lipzig, N.P.M.; Dujardin, S. Using Local Climate Zones in Sub-Saharan Africa to tackle urban health issues. *Urban Clim.* **2019**, *27*, 227–242. [CrossRef]
39. Demuzere, M.; Bechtel, B.; Middel, A.; Mills, G. Mapping Europe into local climate zones. *PLoS ONE* **2019**, *14*, e0214474. [CrossRef] [PubMed]
40. Eldesoky, A.H.M.; Gil, J.; Pont, M.B. The suitability of the urban local climate zone classification scheme for surface temperature studies in distinct macroclimate regions. *Urban Clim.* **2021**, *37*, 100823. [CrossRef]
41. Geletič, J.; Lehnert, M.; Savić, S.; Milošević, D. Inter-/intra-zonal seasonal variability of the surface urban heat island based on local climate zones in three central European cities. *Build. Environ.* **2019**, *156*, 21–32. [CrossRef]
42. Du, P.; Chen, J.; Bai, X.; Han, W. Understanding the seasonal variations of land surface temperature in Nanjing urban area based on local climate zone. *Urban Clim.* **2020**, *33*, 100657. [CrossRef]
43. Gémes, O.; Tobak, Z.; Leeuwen, B.V. Satellite Based Analysis of Surface Urban Heat Island Intensity. *J. Environ. Geogr.* **2016**, *9*, 23–30. [CrossRef]
44. Ziaul, S.; Pal, S. Analyzing control of respiratory particulate matter on Land Surface Temperature in local climatic zones of English Bazar Municipality and Surroundings. *Urban Clim.* **2018**, *24*, 34–50. [CrossRef]
45. He, B.-J.; Zhao, Z.-Q.; Shen, L.-D.; Wang, H.-B.; Li, L.-G. An approach to examining performances of cool/hot sources in mitigating/enhancing land surface temperature under different temperature backgrounds based on landsat 8 image. *Sustain. Cities Soc.* **2019**, *44*, 416–427. [CrossRef]
46. Shenyang Statistics Bureau. *Bulletin No.1, Shenyang Seventh National Census. 2021-06-01*; Shenyang Statistics Bureau: Shenyang, China, 2021.
47. Shenyang Statistics Bureau. *Bulletin No.6, Shenyang Seventh National Census. 2021-06-01*; Shenyang Statistics Bureau: Shenyang, China, 2021.
48. Qin, Z.; Karnieli, A.; Berliner, P. A mono-window algorithm for retrieving land surface temperature from Landsat TM data and its application to the Israel-Egypt border region. *Int. J. Remote. Sens.* **2001**, *22*, 3719–3746. [CrossRef]
49. Zhou, D.; Zhao, S.; Liu, S.; Zhang, L.; Zhu, C. Surface urban heat island in China's 32 major cities: Spatial patterns and drivers. *Remote Sens. Environ.* **2014**, *152*, 51–61. [CrossRef]
50. Zhang, L.; Meng, Q.; Sun, Z.; Sun, Y. Spatial and Temporal Analysis of the Mitigating Effects of Industrial Relocation on the Surface Urban Heat Island over China. *ISPRS Int. J. Geo-Inf.* **2017**, *6*, 121. [CrossRef]
51. Debbage, N.; Shepherd, J.M. The urban heat island effect and city contiguity. *Comput. Environ. Urban Syst.* **2015**, *54*, 181–194. [CrossRef]
52. Yue, W.; Liu, X.; Zhou, Y.; Liu, Y. Impacts of urban configuration on urban heat island: An empirical study in China mega-cities. *Sci. Total Environ.* **2019**, *671*, 1036–1046. [CrossRef]
53. Imhoff, M.L.; Zhang, P.; Wolfe, R.E.; Bounoua, L. Remote sensing of the urban heat island effect across biomes in the continental USA. *Remote Sens. Environ.* **2010**, *114*, 504–513. [CrossRef]
54. Wang, R.; Cai, M.; Ren, C.; Bechtel, B.; Xu, Y.; Ng, E. Detecting multi-temporal land cover change and land surface temperature in Pearl River Delta by adopting local climate zone. *Urban Clim.* **2019**, *28*, 100455. [CrossRef]
55. Wang, Y.; Ouyang, W. Investigating the heterogeneity of water cooling effect for cooler cities. *Sustain. Cities Soc.* **2021**, *75*, 103281. [CrossRef]
56. Geletič, J.; Lehnert, M.; Savić, S.; Milošević, D. Modelled spatiotemporal variability of outdoor thermal comfort in local climate zones of the city of Brno, Czech Republic. *Sci. Total Environ.* **2018**, *624*, 385–395. [CrossRef] [PubMed]

57. Kotharkar, R.; Ghosh, A.; Kotharkar, V. Estimating summertime heat stress in a tropical Indian city using Local Climate Zone (LCZ) framework. *Urban Clim.* **2021**, *36*, 100784. [CrossRef]
58. Gilabert, J.; Deluca, A.; Lauwaet, D.; Ballester, J.; Corbera, J.; Llasat, M.C. Assessing heat exposure to extreme temperatures in urban areas using the Local Climate Zone classification. *Nat. Hazards Earth Syst. Sci.* **2021**, *21*, 375–391. [CrossRef]

Article

Predicting the Impact of Future Land Use and Climate Change on Potential Soil Erosion Risk in an Urban District of the Harare Metropolitan Province, Zimbabwe

Andrew K. Marondedze * and Brigitta Schütt

Physical Geography, Institute of Geographical Sciences, Freie Universität Berlin, 12449 Berlin, Germany; brigitta.schuett@fu-berlin.de
* Correspondence: ak.marondedze@fu-berlin.de; Tel.: +49-30-838-70239

Abstract: Monitoring urban area expansion through multispectral remotely sensed data and other geomatics techniques is fundamental for sustainable urban planning. Forecasting of future land use land cover (LULC) change for the years 2034 and 2050 was performed using the Cellular Automata Markov model for the current fast-growing Epworth district of the Harare Metropolitan Province, Zimbabwe. The stochastic CA–Markov modelling procedure validation yielded kappa statistics above 80%, ascertaining good agreement. The spatial distribution of the LULC classes CBD/Industrial area, water and irrigated croplands as projected for 2034 and 2050 show slight notable changes. For projected scenarios in 2034 and 2050, low–medium-density residential areas are predicted to increase from 11.1 km^2 to 12.3 km^2 between 2018 and 2050. Similarly, high-density residential areas are predicted to increase from 18.6 km^2 to 22.4 km^2 between 2018 and 2050. Assessment of the effects of future climate change on potential soil erosion risk for Epworth district were undertaken by applying the representative concentration pathways (RCP4.5 and RCP8.5) climate scenarios, and model ensemble averages from multiple general circulation models (GCMs) were used to derive the rainfall erosivity factor for the RUSLE model. Average soil loss rates for both climate scenarios, RCP4.5 and RCP8.5, were predicted to be high in 2034 due to the large spatial area extent of croplands and disturbed green spaces exposed to soil erosion processes, therefore increasing potential soil erosion risk, with RCP4.5 having more impact than RCP8.5 due to a higher applied rainfall erosivity. For 2050, the predicted wide area average soil loss rates declined for both climate scenarios RCP4.5 and RCP8.5, following the predicted decline in rainfall erosivity and vulnerable areas that are erodible. Overall, high potential soil erosion risk was predicted along the flanks of the drainage network for both RCP4.5 and RCP8.5 climate scenarios in 2050.

Keywords: land use land cover (LULC); Cellular Automata Markov model; representative concentration pathways; climate scenarios

Citation: Marondedze, A.K.; Schütt, B. Predicting the Impact of Future Land Use and Climate Change on Potential Soil Erosion Risk in an Urban District of the Harare Metropolitan Province, Zimbabwe. *Remote Sens.* **2021**, *13*, 4360. https://doi.org/10.3390/rs13214360

Academic Editor: Nicola Montaldo

Received: 17 September 2021
Accepted: 26 October 2021
Published: 29 October 2021

Publisher's Note: MDPI stays neutral with regard to jurisdictional claims in published maps and institutional affiliations.

Copyright: © 2021 by the authors. Licensee MDPI, Basel, Switzerland. This article is an open access article distributed under the terms and conditions of the Creative Commons Attribution (CC BY) license (https://creativecommons.org/licenses/by/4.0/).

1. Introduction

Soil erosion by water has become a global threat undermining environmental sustainability [1]. This is attributed to various controlling factors related to Land Use and Land Cover (LULC) changes influenced by population growth, rising economic activities, unsustainable agricultural practices and climate change [2,3]. LULC change has been reviewed as one of the main driving forces of global environmental change, making it an important factor to assess at different spatio-temporal levels [4,5]. The LULC changes at both local and global levels are dynamic processes [2] and their drivers correspond to complex systems with dependent characteristics and interactions having a wide array of implications for the future ecological balance and environmental sustainability. Urbanization, as one among the major drivers of LULC change, depends on population growth, migration and desires to change the current state of the Earth. These actions could be for the betterment of livelihoods and in turn could be detrimental to the environment and humankind [6,7]. The resulting ramifications include the modification of the landscape due to the sprawling of unplanned urban built-up areas, development of urban heat islands and over-exploitation of natural resources as direct impacts, and collateral land degradation, climate change, soil erosion and siltation [7–9].

The United Nation's World Urbanization Prospects reveal that the global urban population increased from about 30% in 1950 to approximately 54% in 2014, with almost 2.5 billion urban dwellers expected by 2050 [10]. For India, approximately 50% of the population have been projected to be living in cities by 2050 as a result of rural–urban migration due to increased economic activities in the urban areas, which have become a strong pulling factor [11]. Rapid urbanization in Africa has been reported due to population growth and it has been projected to almost triple by 2030 [11]. However, according to information from the World Economic Forum, in 2020 56.2% of the global population already lived in cities [12], with highly variable rates between regions, ranging from 81.2% urban dwellers in Latin America and the Caribbean to 43.5% in Africa [13]. Breaking these data down to Zimbabwe, about a quarter of the country's population lives in urban areas. Focusing on the case study of Epworth district, being part of the Harare Metropolitan Province, approximately 47% of the population increase was registered between 1992 and 2012 [14], with a triplication of built-up areas from about 19.5% in 1984 to 61.3% in 2018 [15]. Such trends in urban population growth directly impact the ecosystem of the urbanizing area, including the peri-urban area. This earmarks a gap which requires monitoring of the impacts driven by rampant LULC changes through urban expansion on the ecosystem as a basis to implement a proper spatial policy to enable effective decision-making processes [16,17]. This implies a rich understanding of the trends of urban expansion and development, and it requires the integration of spatially differentiated data, applying geomatics to quantify and predict future spatial distributions [18,19]. By the case study of the Epworth district in the Harare Metropolitan Province, it will be demonstrated that future land use models provide a valuable basis for foresight spatial planning to ensure environmental sustainability.

The LULC changes occurring at unprecedented levels threaten multiple ecological processes such as surface runoff, soil erosion, siltation and agricultural non-point source pollution, resulting in landscape degradation, habitat loss and inaccessibility to properties [20,21]. Focusing on sub-Saharan metropolitan areas, the example of the Harare Metropolitan Province documents a rapid transformation of urban agricultural land and shrub lands to built-up areas and other sealed settlement areas over the past decades [9,22]. For example, Epworth district, as part of the Harare Metropolitan Province, has witnessed an increase in built-up areas linked with high soil erosion risk due to increased impervious surfaces and construction activities which facilitate surface runoff [23]. This results in accelerated soil loss in sensitive areas mostly within active built-up areas. The radical LULC changes in this area also include the loss of water bodies due to siltation resulting from sand mining and brick moulding along the river banks; encroachment of wetlands by construction activities; and grading of unpaved roads which later facilitate accelerated surface runoff due to compaction [24].

Furthermore, climate change is reiterated to be heavily associated with locally increasing rainfall intensity, frequency and extent, resulting in increasing rainfall erosivity [25]. The Fifth Assessment Report (AR5) of the IPCC (Intergovernmental Panel on Climate Change) highlights that global mean precipitation and surface temperatures have significantly changed with reference to observed changes between 1850 and 1900, and these changes are likely to continue to be experienced in the 21st century [26]. Several studies point out that accelerated soil erosion by water due to climate change accentuates processes that alter soil physiochemical and biological properties [27–29]. This entails the need to curb soil erosion through minimizing the removal of vegetation cover, improving surface roughness to facilitate infiltration capacity and reducing rainfall-runoff processes [30]. Climate change also inevitably triggers a shift in land use, forcing the adoption of new management practices and planting new crops in order to mitigate detrimental impacts [31,32].

The sketched interrelations between LULC change and climate change and its possible environmental impacts emphasize the need to investigate future potential impacts of LULC change and climate change on potential soil erosion risks caused by water. For the coming decades, for wider areas, the increasing intensity of the hydrological cycle is projected by multiple global circulation models (GCMs), pronouncing more intense rainfall events that directly influence rainfall erosivity [26]. We want to investigate these interrelations using the example of Epworth district, a fast-growing urban area of the Harare Metropolitan Province. Soil erosion by water has been repeatedly investigated in different regions of Zimbabwe, focusing on either catchments or arable areas [33–35]. There is limited knowledge regarding estimated future soil loss rates and potential soil erosion risk in Zimbabwe as impacted by future climate change and land use changes, knowledge indispensable for future policy decision-making processes. The current study examines the potential future effects of land use change as well as of climate change on soil erosion risk. Overall, climate change scenarios as provided by the IPCC [26] and forecasts of LULC change were applied for the assessment of future potential soil erosion risk for the years 2034 and 2050.

1.1. Modelling Land Use Changes in Urban Areas

Multiple studies on future soil erosion focus mainly on the dynamics of climate variables such as temperature and rainfall [32,36], while land use changes are rarely considered regardless of the high awareness of processes such as population growth, immigration and urbanization occurring at alarming rates. There is a wide range of spatial models able to simulate and predict land use changes based on the application of remote sensing techniques [37,38]. The spatial transition model and statistical description model are the two major models widely used for the assessment and monitoring of land use changes [8,37,39]. Furthermore, the Markov chain model is widely applied to simulate urban growth due to its capability of quantifying land use changes, their trends and their dimensions [9,22,40–43]. Markov chain models correspond to stochastic processes [44] that summarize changes by developing a transition probability matrix of land use change, indicating that the probability of a system being in one state at a given time can be determined if the state at an earlier period is known [40,45]. The Cellular Automata (CA) are simple and flexible dynamic spatial systems able to integrate complex urban systems in order to simulate future urban growth patterns [46–48]. The CA are based on the supposition that land use change for any given location (grid cell) can be explained by its present state and the transformations in its neighbouring cells [49]. Therefore, the inability of the Markov chain model to simulate spatial changes over time is superseded by integrating it into the CA to enhance the spatial predictive accuracy of the urban land use dynamics [47,50–53].

Previous studies have adopted simulation models that apply GIS and remote sensing techniques for land use change modelling and monitoring of dynamic urban growth patterns [40,46,50]. In the case of Harare Metropolitan Province, due to the dynamic nature of urban growth, some parts of its districts were simulated using the CA–Markov model in order to predict the impact of urban land use change on future microclimate [9], while Sibanda and Ahmed [52] predicted the future LULC and their impacts on wetland areas in the Shashe sub-catchment of Zimbabwe. According to Mushore et al. [9], accelerated urban growth without the conservation of green spaces and adherence to mitigation policies contribute to locally increasing microclimate temperatures, causing thermal discomfort in urban areas. The CA–Markov model was also applied to project future LULC scenarios for Arasbaran biosphere reserve in Iran [54]. Future LULC distribution patterns were also simulated with high accuracy using the CA–Markov model for Jordan's Irbid governorate, with built-up areas predicted to increase from about 19.5% to approximately 64.6% between 2015 and 2050 [55]. Due to the plausible outcomes, recommendations indicate that the CA–Markov model is an effective tool in monitoring and assessing future land use patterns for policy and decision-making processes [40,51–53].

1.2. Climate Change Emission Scenarios

The establishment of the Representative Concentration Pathways (RCPs) as future climate change mitigation scenarios followed a response call on the effectiveness of climate policy inclusion in future climate change modelling and research [26,56,57]. The RCPs illustrate how the future climate may evolve, considering a range of variables which encompass socio-economic changes, technological advancement, energy, greenhouse gas emissions and land use changes [26]. Most precipitation projections from GCMs have been widely used on land surface processes for the assessment of climate change impacts and adaptation [1,58,59]. However, uncertainties in GCMs primarily exist on biases of raw outputs, resulting in either over or underestimation of climate variables due to erroneous assumptions in the model's development [60,61]. As such, many studies have embarked on the use of multi-modelling techniques to minimize the uncertainty of future predictions in order to obtain plausible future projections [62–66].

The climate change emission scenarios approximate radiative forcing levels of greenhouse gas concentrations, aerosols, and tropospheric ozone precursors by 2100 [57]. The RCP8.5 scenario is characterized by increasing levels of greenhouse gas concentrations [67]. Further, the RCP8.5 is a highly energy-intensive scenario attributed to high population growth and a lower rate of technology development; this is a scenario with little to no climate policy, making it possible to represent all future climatic possibilities [26,57]. For the RCP4.5 scenario, historical emissions and land cover information are integrated in order to follow a cost-effective pathway through stabilization of anthropogenic components to reach the target radiative forcing [56,68]. The RCP4.5 considers technological advances such as combining bioenergy production with CO_2 capture and geologic storage to enhance more energy production with negative carbon emissions [68,69].

2. Materials and Methods

2.1. Study Area

The Harare Metropolitan Province is the capital city of Zimbabwe, with Epworth district 17°40′–18°00′S, 30°55′–31°15′E located approximately 12 km southeast of the Central Business District (CBD) (Figure 1). Epworth district is a high-density residential suburb of Harare Metropolitan Province and the smallest in terms of area-wide coverage among the four districts which comprise the Harare Metropolitan Province, occupying an estimated area of 35 km^2; the area is characterized by the densification of built-up structures and overcrowdings [70] and an above-average increase in informal urban development in comparison to other urban districts in Zimbabwe [71,72]. There has been rampant population growth and mushrooming urban built-up structures due to rural–urban migration which dates back to the pre-and post-independence phase (1980) in search of better livelihoods, employment and a hive of economic activities in the capital city [14,72,73]. Since then, Epworth district has grown from about 500 families recorded in 1950, to a total population of approximately 114,047 in 2002, to a total population of 167,462 in 2012 [14,73,74].

The Harare Metropolitan area is located on the Highveld at an elevation between 1455 m and 1556 m a.s.l., with a general topography characterized by undulating to slightly rolling terrain in the plateau areas. Annual precipitation in Harare Metropolitan Province varies between 470 mm and 1350 mm, falling mainly during the four months of the rainy season between mid-November to mid-March. Daily temperature ranges between 13 °C and 28 °C during the hot-dry season (September to mid-November) and low temperature averages between 7 °C and 20 °C are experienced during the cool-dry season (mid-May to August) [22]. Dominating soil types in Epworth district are the widely spread Paraferrallitic soils (coarse grained) covering the high-altitude areas and clayey Fersiallistic soils developed predominantly from dolerite in the central plateau [75]. Both soil types are largely influenced by nutrient loss through moderately to strongly occurring leaching processes [75,76].

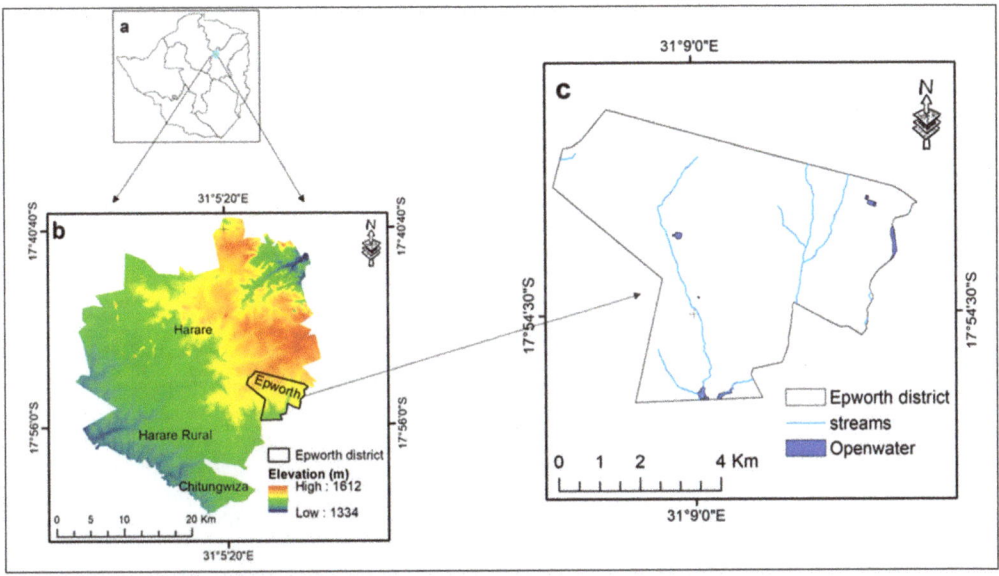

Figure 1. Study site—Epworth district of the Harare Metropolitan Province. (**a**) Zimbabwe provincial boundaries including the Harare Metropolitan Province. (**b**) Elevation and district boundaries of the Harare Metropolitan Province. (**c**) Epworth district with hydrological network, retrieved from OSM data (OSM-Geofabric).

2.2. Urban Land Use Change Modelling Using CA–Markov

The CA–Markov analysis was adopted to predict land use future scenarios. The CA–Markov model is embedded into the IDRISI software (Clarks Lab), an image processing software useful for improved digital image display and spatial analysis [77]. The Markov chain analysis describes the probability of LULC changes from one state to another at given times t_1 and t_2 by developing a transition probability [49,78,79]. The Markov chain model simulates land use changes and generates a transition probability matrix, which indicates the probability of each LULC to change from one state to another, and this is obtained by cross tabulation of the earlier and later LULC maps. The proportional changes become the transition probability, indicating that each land use class will change to other categories using Equation (2). The conditional suitability maps are produced and display the probability that each land use category might be found at each pixel, with values standardized between 0 and 255 [9,42,80–82]. The transition probability of converting the current state of a system to another state in the next time step is determined using the mathematical expression Equation (1) [80,83]:

$$P = (P_{ij}) = \begin{vmatrix} P_{11} & P_{12} & \dots & P_{1n} \\ P_{21} & P_{22} & \dots & P_{2n} \\ \dots & \dots & & \dots \\ P_{n1} & P_{n2} & & P_{nm} \end{vmatrix} \qquad (1)$$

where P_{ij} is the probability from state i to state j and P_n is the state probability of any time. Equation (1) must satisfy the following conditions:

$$\sum_{j=1}^{n} P_{ij} = 1 \ (i, j = 1, 2, 3 \dots \dots , n) \qquad (2)$$

$$0 \leq P_{ij} \leq 1 \ (i, j = 1, 2, 3 \dots \dots \dots , n) \qquad (3)$$

These steps are performed to obtain the Markov chain model's primary matrix and the matrix of the transition probability (P_{ij}). The Markov prediction model is expressed as:

$$P_{(n)} = P_{(n-1)} \;\; P_{ij} = P_{(0)} P_{ij}^n \tag{4}$$

where P_n refers to the state probability of any time and $P_{(0)}$ stands for the primary matrix. High transitions have probabilities near 1, while low transitions attract probabilities near 0 [80,84].

The Markov chain probabilities of change represent all multi-directional LULC changes between land use classes [82]. The Markov chains were selected as a result of their simplicity, robustness and capability in mapping LULC transitions in complex urban areas [9,81]. Despite forecasting transition probabilities per land use category and their growth trends, the major limitation of the Markov chain model is its inability to simulate the spatial distribution of each land use category's occurrence [42,79,82]. Due to the heterogeneity of urban systems and structures, historical information is essential for a better understanding and interpretation of simulated future spatial trends [19]. The subsequent limitations of the Markov chains can be addressed by combining their outputs with other models that have open structures, including the Cellular Automata (CA), Multi-Layer Perceptron (MLP) and the Stochastic Choice [77,84]. In the present study, we integrated the Cellular Automata (CA) into the Markov chain approach to address the spatiality limitations of the Markov chain model and the probable spatial transitions occurring in the study area over the given time [40,47,54,81].

The CA have high spatial resolution and computational efficiency, enabling the prediction of future urban growth trends based on the supposition that the state of each cell at the present time depends on the previous state of cells within the neighbourhood [16,85,86]. Thus, the CA models are based on four major attributes, which include the cell, the state, the neighbourhood, and the transition rule [47,87]. The cell element of the CA signifies spatial shapes and sizes on the ground, while real characteristics of the area (land use) at a discrete time, represented as grid cells, show the state [47,48,87]. The neighbourhood cells are the immediate adjacent cells that form the kernel, and the transition rules theoretically code for the transformation from one cell state to another state resulting from the changes in neighbouring cells at a discrete time and state [39,47]. Despite being a powerful and simple tool in modelling urban growth patterns, the CA models have a limited capability for quantifying aspects, and the simulation processes do not include urban growth driving forces [50,51].

The CA–Markov modules embedded in the IDRISI GIS software were used to simulate LULC distribution patterns for the year 2018 and to predict future LULC for the years 2034 and 2050. Primarily, the simulation phase of the 2018 LULC scenarios applied the Markov chain to generate a transition probability matrix, and transition suitability images between 1990 and 2008 using the LULC maps of the same period were generated using support vector machines (SVMs) by Marondedze and Schütt [15]. A proportional error of 15% was set during the modelling of the transition probability matrix [77]. The Markov chain analysis outputs from 1990 and 2008 formed the basis of input parameters for the probable simulation of LULC spatial characteristics and their occurrence in the CA for the prediction of LULC patterns for 2018 (Figure 2). The contiguity filter specified the spatial characteristics applied by the CA modelling approach [40,77]. For this study, a contiguity filter of 5*5 pixels was applied to define the kernel due to higher spatial characterization when applied to determine the occurrence or position of the simulated LULC category compared to 3*3 or 7*7 [88,89].

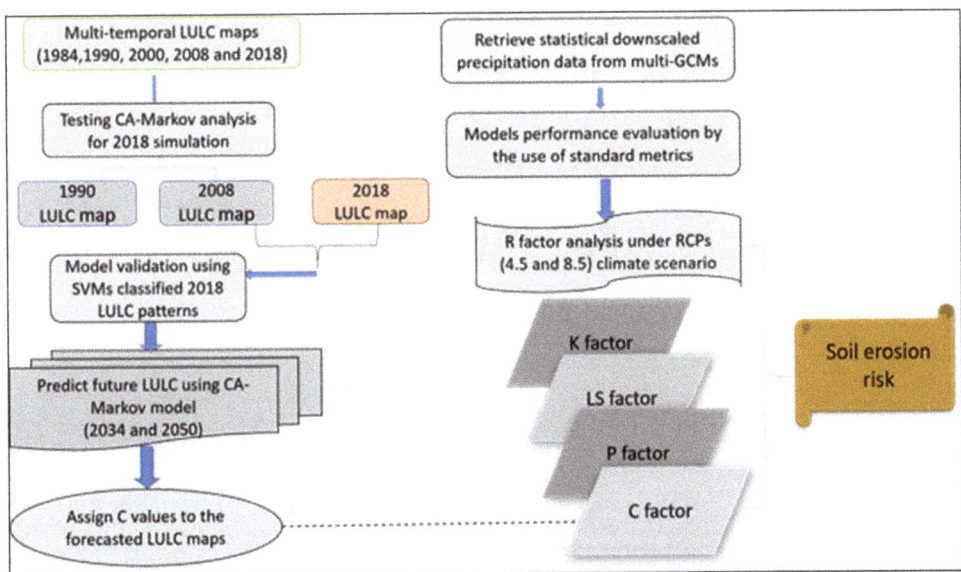

Figure 2. Conceptual framework for the prediction of future LULC and soil erosion risk for Epworth district. LULC: land use and land cover. RCPs: representative concentration pathways, GCMs: global circulation models.

The spatiality characteristics in the CA approach were developed in a spatially explicit weighting that enabled the transformation of single and random grid cells in areas closer to the existing and widely spread land use [54,90]. This is further simplified by assuming that a pixel that is near one specific land cover class is more likely to be transformed to that category than pixels farther apart [78]. This assumption was used to initially test the predictive capability of the CA–Markov model set of the LULC distribution patterns for 2018. The cross validation of the 2018 simulated LULC patterns was performed applying the LULC patterns as provided by a support vector machines (SVMs) supervised classification map [15]. Finally, the CA–Markov techniques were applied between the LULC patterns of 2000 and 2018 for the prediction of future LULC distribution patterns for 2034, whilst the LULC distribution patterns of 1984 and 2018 were applied for the future prediction of 2050 LULC patterns. A 5*5 contiguity filter was applied for the prediction of future LULC patterns for the years 2034 and 2050.

2.3. Cellular Automata–Markov Chain Validation

The simulated LULC distribution patterns for 2018 were compared with the SVMs classified map for the same year to test the level of agreement. A two-phase validation approach was performed, which includes visual inspection and quantitative evaluation [9,91]. Visual inspection allowed close comparison and the agreement assessment between the simulated 2018 LULC map and the SVMs supervised classification LULC map. The kappa index of agreement (KIA) was used to assess the prediction accuracy for the 2018 actual map and the simulated LULC maps [54,91,92]. In general, kappa is referred to as a member of a family of indices with the properties (a) kappa = 1, when the level of agreement is perfect, and (b) kappa = 0, when the observed agreement is equal to the expected proportion due to chance [18]. Considering the model validity and performance in predicting LULC patterns for 2018, the LULC patterns for 2000–2018 and 1984–2018 were used in the prediction of 2034 and 2050 LULC spatial trends in the CA–Markov model. This introduces kappa indices to assess the performance and agreement of the model: the traditional kappa, which measures a simulation's ability to attain perfect classification, that is, the closer to 1 the values are, the higher the level of agreement ($K_{standard}$); the improved general kappa statistic, which is described as kappa for no ability (K_{no}); followed by the sophisticated kappa statistics ($K_{quantity}$ and $K_{location}$) used for distinguishing placement accuracies in both the quantity and location [54,92]. The K_{no} denotes the proportion classified correctly relative to the expected proportion classified correctly by a simulation without the ability to accurately specify quantity or location [18,92].

2.4. Predicting Future Soil Erosion Risk

The empirical RUSLE model was used to predict the spatially differentiated risk of long-term average annual soil loss. The selection of the empirical RUSLE model to assess future potential soil erosion risk considered the availability of data, robustness, complexity of the landscape and calibration [93,94]. The RUSLE model is widely used and a powerful tool to quantitatively assess spatial interactions of land use, topographic characteristics, climate, and soil characters in order to predict the spatial distribution of soil erosion [31,34,95–97]. The wide use of the empirical RUSLE model is based on its simplicity and easy accessibility of data compared to complex physical models [1,98]. Unlike other physical and process-based soil erosion models, the stochastic RUSLE model does not address soil deposition but mainly displays areas of sheet and rill erosion processes [98], allowing land managers to direct limited resources for landscape management [99]. The estimation of spatial soil erosion risk by the RUSLE model makes use of the factors soil erodibility (K), rainfall erosivity (R), slope length and steepness (LS), land cover and management (C) and the support practices (P) [97]. The RUSLE model calculates the risk of long-term average annual soil loss rates by multiplying the different factors:

$$A = K * R * C * LS * P \qquad (5)$$

where A: annual average soil loss (t ha^{-1} yr^{-1}), R: rainfall erosivity factor (MJ mm ha^{-1} h^{-1} yr^{-1}), K: soil erodibility factor (t ha h ha^{-1} MJ^{-1} mm^{-1}), C: cover-management factor (dimensionless), LS: slope length and slope steepness factor (dimensionless) and P: support practices factor (dimensionless).

The RUSLE factors harmonized at 30 × 30 m spatial resolution for the compatibility of data from different sources [100] are multiplied to predict the soil erosion risk for the district using raster calculator in ArcGIS® 10.2. The computation of the RUSLE model integrates remote sensing and GIS techniques to analyse factors and geostatistics for the graphical interpretation [97,101].

Soil erodibility factor (K). The soil erodibility factor (K) represents the susceptibility of the soil to detachment due to rainfall erosivity (R) [97]. The soil erodibility factor varies corresponding to soil properties such as soil texture, type and size of aggregates, shear strength, soil structure, infiltration capacity, bulk density, soil depth, organic matter and other chemical constituents [97,102]. Based on the RUSLE model, the estimated K-factor values range between 0 and 1, indicating the degree of soils' susceptibility to erosion [97]. Thus, soils being highly susceptible to erosion have soil erodibility values near 1, whereas the corresponding values close to 0 designate the resistive ability of a particular soil to erosion processes [103]. For this research, available data for the computation of the K factor were retrieved from ISRIC (International Soil Reference Information Centre), available at 250 m spatial resolution [104]. The estimation of the K factor was performed using the equation by Sharpley and Williams [105], which excludes soil structure and profile permeability due to the unavailability of experimental based information.

Slope length and slope steepness factor (LS). The RUSLE model considers the effects of topography on soil erosion, including slope length (L) and slope steepness (S). The Shuttle Radar Topography Mission (SRTM) digital elevation model (DEM) with a spatial resolution of 30 × 30 m (https://earthexplorer.usgs.gov/SRTM1Arc; accessed on 19 September 2020) was used for the computation of the LS factor using the Hydrology module (field-based), embedded in SAGA 2.3 software [106,107].

Land cover and management factor (C) and support practice factor (P). The land cover and management factor of the RUSLE model represents the effects of vegetation cover on soil erosion rates [97]. The C factor ranges from 0 for high-density vegetation to 1 for barren land; bare land is frequently used as the reference land use for C factor calibration [108,109]. The vegetation cover plays a vital role in dissipating raindrop energy before reaching the surface, thereby reducing the harsh effects posed by raindrop impact on the soil surface [101,108]. The C factor values in Table 1 result from the weighted field-based observations, and additional biophysical characterizations were adopted [23]. The support practice factor (P) was assigned to be 1, corresponding to the lack of support practice all over the study area [23].

Table 1. The weighted C factor values.

Land Use Class	Weighted C Factors
CBD/industrial areas	0.017
LMD (less concentrated residential area)	0.066
HD (concentrated residential area)	0.083
Irrigated cropland	0.166
Rainfed cropland	0.239
Green spaces	0.03
Water	0

Rainfall erosivity factor (R) estimation: The R factor describes the soil loss potential triggered by rainfall [97,102,110]. As such, the analysis of the spatial distribution of rainfall erosivity was computed following the empirical relations developed by El-Swaify et al., [111] Equation (6), as cited [34,112],

$$R = 38.5 + 0.35 \times M \qquad (6)$$

where R = rainfall erosivity factor (MJ mm ha^{-1} h^{-1} yr^{-1}), and M = mean annual rainfall.

The further analysis highlights the likely potential effects of climate change on the R factor. The representative concentration pathway (RCP) 4.5 and 8.5 climate scenarios projected by multiple general circulation models (GCMs) were selected for the assessment of future climate change, primarily variations in precipitation magnitudes on soil erosion risk (downloaded from https://earthobservatory.nasa.gov/images/86027/; accessed on 2 October 2020). These climate change scenarios constitute a set of greenhouse gas concentration and emission pathways to facilitate decision and policy makers in the crafting of sustainable climate policies due to their plausibility [57,68]. To predict future rainfall erosivity, future RCP 4.5 and 8.5 climate scenarios proposed by the Intergovernmental Panel on Climate Change [26] were applied (Table 2). Annual rainfall, as required for Equation (6), was the sum of mean monthly rainfall data retrieved from the NASA Exchange Global Daily Downscaled Projections (NEX-GDDP), as listed in Table 2, which was statistically downscaled to a 0.25° by 0.25° spatial resolution [62,113]. The NEX-GDDP general circulation models grid point data locations do not match with the Harare Meteorological gauging points, as the spatial coverage of station data is not uniform; to cope with the varying spatial resolutions, annual averages were interpolated using the inverse distance-weighted methods [2].

Table 2. Global circulation models (GCMs) used for data retrieval.

Global Circulation Model	Source	Original Resolution
		(Lat × Lon) °
ACCESS1-0	Commonwealth Scientific and Industrial Research Organization/Bureau of Meteorology, Australia	1.875 × 1.25
BNU-ESM	College of Global Change and Earth System Science, Beijing Normal University, China	2.8 × 2.8
CanESM2	Canadian Centre for Climate Modeling and Analysis, Canada	2.8 × 2.8
CCSM4	National Centre for Atmospheric Research, United States	1.25 × 0.94
CNRM-BGC	National Centre for Meteorological Research, France	1.4 × 1.4
GFDL-ESM2G	NOAA/Geophysical Fluid Dynamics Laboratory, United States	2.5 × 2.0
GFDL-ESM2M	NOAA/Geophysical Fluid Dynamics Laboratory, United States	2.5 × 2.0
IPSL-CM5A-MR	L'Institut Pierre-Simon Laplace, France	2.5 × 1.25
MIROC-ESM	Japan Agency for Marine-Earth Science and Technology, Atmosphere and Ocean Research Institute (The University of Tokyo), and National Institute for Environmental Studies	2.8 × 2.8
MIROC-ESM-CHEM	Japan Agency for Marine-Earth Science and Technology, Atmosphere and Ocean Research Institute (The University of Tokyo), and National Institute for Environmental Studies	2.8 × 2.8
MIROC5	Atmosphere and Ocean Research Institute (The University of Tokyo), National Institute for Environmental Studies, and Japan Agency for Marine-Earth Science and Technology, Japan	1.4 × 1.4
MPI-ESM-LR	Max Planck Institute for Meteorology, Germany	1.9 × 1.9
MPI-ESM-MR	Max Planck Institute for Meteorology, Germany	1.9 × 1.9
MRI-CGCM3	Meteorological Research Institute, Japan	1.1 × 1.1
NorESMI-M	Norwegian Climate Center, Norway	2.5 × 1.9

General circulation models' performance was assessed, comparing their average annual rainfall data as provided per grid cell between 1980 and 2005 with the observed data from Harare gauging stations. This evaluation was processed by applying the interpolated GCMs average rainfall data from six available grid points within the Harare Metropolitan Province in parallel with observed average precipitation from the Harare Meteorological stations (Table 3) using the standard statistical metrics [114]. The evaluation of the GCMs performance was assessed using the standard metrics to outweigh GCMs that are not representative: the coefficient of determination (R^2), relative root mean square error (rRMSE) (%), correlation coefficient (r), and index of agreement (d) [63,115,116]. With values ranging between 0 and 1, the lower the values of the rRMSE, the better the model's performance, while the higher the value for R^2, the better the goodness of fit of the model [115–117]. For the index of agreement (d), the closer values are to 1, the better they document the increasing goodness of the fit of the model, ascertaining that there is good agreement between the simulated and observed annual precipitation [63,118,119].

Table 3. Location of Harare Metropolitan Province gauging stations.

Rain Stations	Coordinates	Altitude (m.a.s.l)	Mean Annual Precipitation (mm)
			1980–2005
Kutsaga	17°55′S, 31°08′E	1488	825.3
Belvedere	17°50′S, 31°01′E	1474	862.6
Airport	17°55′S, 31°06′E	1502	798.2

Separate runs of the GCMs ensemble averages from 2019 to 2034 and 2035 to 2050 were used for the assessment of climate variability and its impact on future soil erosion risk under RCP4.5 and RCP8.5 climate scenarios. Estimations of future climate change scenarios from single GCMs relay limited information required for the direct calculation of the R factor [120,121]. Therefore, the application of multi-GCM ensemble averages decreases individual model errors and provides more robust predictions for future climate change [61,63,64,122,123]. Accordingly, empirical relations were used between monthly and annual precipitation in order to analyse GCM outputs relative to R factor changes [120]. Thus, long-term model ensemble averages were analysed for trends in rainfall erosivity factor (R) using suitable empirical relations [97,121,124].

3. Results
3.1. Land Use Land Cover Changes

The LULC maps (1990–2008, 2000–2018 and 1984–2018) generated by supervised classification applying SVMs [15] were used to simulate LULC distribution patterns for 2018; simultaneously, they were used as the reference for the simulation accuracy and to forecast future land use for 2034 and 2050 (Figure 3). The adopted supervised classification maps of the years 1984–2018 [15] show seven distinct classes (Table 4). The overall classification of each LULC map for 1984, 1990, 2000, 2008 and 2018 was estimated to be 90.1, 85.1, 88.9, 87.6 and 89.7%, respectively. The overall Kappa coefficient values produced were 0.87, 0.82, 0.86, 0.85 and 0.87 [15]. The data reveal that spatial LULC patterns will significantly change during the forecasted periods, indicating that the expansion of the built-up areas will be at the expense of green spaces and croplands (Figure 3). The built-up areas will continue to grow towards the peripheries and into the southward direction of the Epworth district (Figure 3).

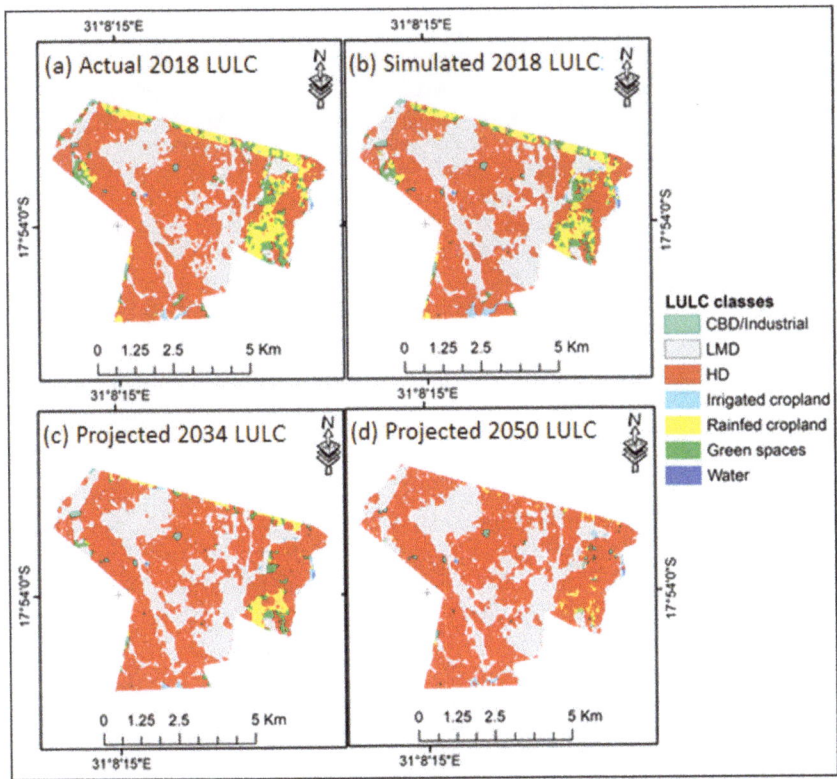

Figure 3. Land use and land cover maps for Epworth district from the support vector machines supervised classification, simulated and predicted using the CA–Markov chain model: (**a**) actual 2018 supervised classification [15], (**b**) simulated 2018, (**c**) projected 2034 and (**d**) projected 2050. CBD: central business department, LMD: low–medium density, HD: high density.

Table 4. Description of LULC classes for the study (source: [15]).

LULC Class	Description
CBD/industries	Industries and central business district defined with high fraction of impervious surfaces, mainly buildings, and a low proportion of vegetation
LMD residential	Leafy and well-established low- and medium-density suburbs surrounded by high vegetation
HD residential	High-density residential areas with low vegetation cover or clustered settlements with areas undergoing developments and bare exposed land
Irrigated cropland	Cultivated land under irrigation schemes
Rainfed cropland	Cultivated land or land with crop residues after harvesting
Vegetation	All wooded areas, shrubs and bushes, riverine vegetation and grass-covered areas
Water	Areas occupied by water, rivers, wetlands, reservoirs and dams

Comparison of LULC areas for 2018, resulting from the supervised classification applying SVMs, with 2018 simulated LULC classes shows that the land use land cover classes CBD/industrial, croplands, green spaces and water (Figure 3a,b) fit reasonably when comparing each class category, while slight differences between mapped and simulated distribution patterns occur for low–medium density and high-density residential areas (Figure 4). To summarize, for the period 2018 to 2050, the LULC class of CBD/industrial areas are estimated to remain stable, with an area expansion of +/−0.5–0.6% (Table 5). The spatial distribution of the LULC classes CBD/industrial area, water and irrigated croplands as projected for 2034 and 2050 widely correspond to those as mapped for 2018 (Figure 4). For both projected scenarios 2034 and 2050, the low–medium residential areas are predicted to increase slightly from 11.1 km^2 to approximately 11.9 km^2 between 2018 and 2034 and up to 12.3 km^2 in scenario 2050. Similarly, high-density residential areas are predicted to increase from 18.6 km^2 to 20.3 km^2 between 2018 and 2034, and to reach 22.4 km^2 in 2050 (Figure 4).

Low–medium-density residential areas (LMD) are predicted to increase in coverage from 31.5% to 34.8% between 2018 and 2050, while high-density (HD) residential areas are predicted to increase in coverage from 52.6% to 63.3% between 2018 and 2050 (Table 5). During the period 2018–2050, the spatial distribution of croplands is predicted to decrease from 9.5% to 1.1% of the total Epworth district area, while green spaces will shrink from 5.8% to 0.1%, largely due to the spatial expansion of built-up areas.

Table 5. Relative proportions of LULC classes by area extent (km^2) and percentage (%) for the adapted 2018 and the projected 2034 and 2050.

LULC Class	2018		2034		2050	
	Km^2	%	Km^2	%	Km^2	%
CBD/industrial	0.2	0.5	0.2	0.6	0.2	0.6
LMD residential	11.1	31.5	11.9	33.7	12.3	34.8
HD residential	18.6	52.6	20.3	57.3	22.4	63.3
Irrigated cropland	0.1	0.4	0.1	0.2	0.1	0.1
Rainfed cropland	3.2	9.1	1.6	4.6	0.3	1.0
Green spaces	2.1	5.8	1.2	3.5	0.1	0.1
Water	0.01	0.04	0.01	0.03	0.01	0.03

CBD: central business department, LMD: low–medium density, HD: high density.

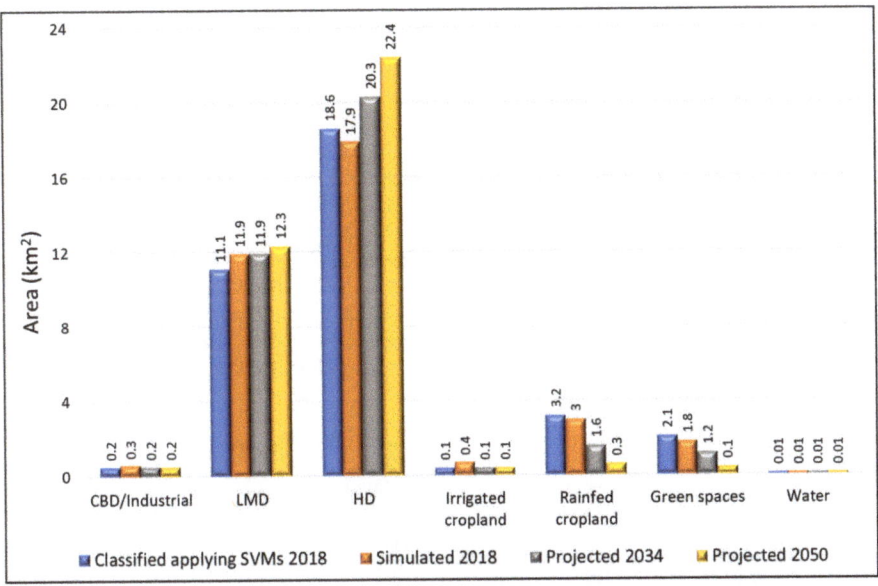

Figure 4. The spatial area extent of different land use land cover classes for Epworth district: the depiction shows the variation between actual 2018 support vector machines (SVMs) supervised classification and the simulated 2018 LULC classes, including the predicted area LULC class extent for 2034 and 2050. CBD: central business department, LMD: low–medium density, HD: high density.

The summary of the probability matrix for major LULC conversions that occurred in Epworth district between 1990 and 2008 is documented in Table 6. The probability of change for CBD/industrial areas to remain CBD/industrial areas between 1990 and 2008 was 96.5%, displaying that built-up areas widely remained stable and will remain stable (Table 6). In contrast, irrigated croplands had a probability of change of 19.1%, that is, to remain irrigated cropland between 1990 and 2008, while the probability of change of irrigated cropland to rainfed cropland was 7.3% and to high-density residential areas was 47.2%. For green spaces, the probability to remain as green spaces between 1990 and 2008 was as low as 18.3%, while the probability of the change of green spaces to low–medium-density residential areas was 18.5%, to high-density residential areas was 40.8% and to croplands was 13.9% (Table 6).

Table 6. Markov chain transition probability matrix from LULC maps between 1990 and 2008.

Changing from:	Probability of Changing to Another Land Use Class by 2008:							Total
1990	CBD/Industrial	LMD	HD	Irrigated Cropland	Rainfed Cropland	Green Spaces	Water	
CBD/industrial	0.9650	0.0183	0.0129	0.0033	0.0000	0.0005	0.0000	1.000
LMD residential	0.0062	0.9716	0.0150	0.0051	0.0000	0.0005	0.0016	1.000
HD residential	0.0071	0.0138	0.9712	0.0027	0.0052	0.0000	0.0000	1.000
Irrigated cropland	0.0708	0.1630	0.4721	0.1910	0.0725	0.0273	0.0033	1.000
Rainfed cropland	0.0416	0.2110	0.4357	0.0574	0.2041	0.0502	0.0000	1.000
Green spaces	0.0850	0.1848	0.4080	0.0412	0.0976	0.1834	0.0000	1.000
Water	0.0295	0.0838	0.0521	0.1121	0.0000	0.0213	0.7012	1.000

CBD: central business department, LMD: low–medium density, HD: high density.

The Markov chain transition probability matrix computed LULC maps between 2000 and 2018 for the prediction of 2034 future LULC distribution patterns (Table 7), which indicates that in 2018 the built-up area classes have a probability of more than 95% to remain as built-up areas in the future, documenting a stable distribution at least until 2034. For the irrigated croplands, a probability of 10.1% is indicated to remain as irrigated croplands until 2034, while at the same time 24.1% of the irrigated croplands have a probability to be converted into low–medium-density residential areas, and even 40.4% of the irrigated croplands underly a probability to be converted into high-density residential areas until 2034. For rainfed cropland, a probability of 33% is indicated to remain as rainfed cropland until 2034, while there is a 42.8% probability that rainfed cropland will be converted into high-density residential areas. There is a probability of 14.1% that rainfed cropland will be converted into low–medium-density residential areas by 2034, while at the same time there is an 8.3% probability that the rainfed croplands will be converted into green spaces. Similarly, green spaces have a probability of 24.7% to remain as green spaces until 2034, while for the same period, green spaces have a 30.6% probability to be converted into high-density residential areas, and a 16.1% probability to be converted into low–medium-density residential areas.

Table 7. Markov chain transition probability matrix from LULC maps between 2000 and 2018.

Changing from:	Probability of Changing to Another Land Use Class by 2018:							Total
2000	CBD/Industrial	LMD	HD	Irrigated Cropland	Rainfed Cropland	Green Spaces	Water	
CBD/industrial	0.9523	0.0109	0.0186	0.0081	0.0043	0.0058	0.0000	1.000
LMD residential	0.0000	0.9507	0.0212	0.0164	0.0000	0.0102	0.0015	1.000
HD residential	0.0064	0.0185	0.9694	0.0000	0.0057	0.0000	0.0041	1.000
Irrigated cropland	0.0500	0.2405	0.4036	0.1011	0.1310	0.0097	0.0033	1.000
Rainfed cropland	0.0000	0.1405	0.4282	0.0183	0.3297	0.0833	0.0000	1.000
Green spaces	0.0370	0.1606	0.3062	0.0641	0.1852	0.2469	0.0000	1.000
Water	0.0026	0.1332	0.1071	0.1290	0.0000	0.0000	0.6281	1.000

CBD: central business department, LMD: low–medium density, HD: high density.

Based on the period 1984–2018, the transition probability matrix for the prediction of 2050 LULC distribution patterns was calculated (Table 8). The results indicate that built-up areas have probabilities higher than 90% to remain as built-up areas until 2050. In contrast, irrigated croplands have only a probability of 15% to remain as irrigated croplands until 2050, while they simultaneously have a probability of 41% to be transformed into high-density residential areas and a 21.4% probability to be transformed into low–medium-density residential areas. The rainfed croplands have a probability of 22.3% to remain as rainfed cropland until 2050; simultaneously, a 5.1% probability occurs that rainfed cropland will be transformed into irrigated croplands, a 5.3% probability occurs that rainfed cropland will be transformed into green spaces and a 42.5% probability occurs that rainfed cropland will be transformed into high-density residential areas.

Table 8. Markov chain transition probability matrix from LULC maps between 1984 and 2018.

Changing from:	Probability of Changing to Another Land Use Class by 2018:							Total
1984	CBD/Industrial	LMD	HD	Irrigated Cropland	Rainfed Cropland	Green Spaces	Water	
CBD/industrial	0.9240	0.0308	0.0404	0.0000	0.0017	0.0031	0.0000	1.000
LMD residential	0.0000	0.9467	0.0251	0.0162	0.0000	0.0103	0.0017	1.000
HD residential	0.0064	0.0191	0.9621	0.0041	0.0060	0.0000	0.0023	1.000
Irrigated cropland	0.0612	0.2140	0.4104	0.1501	0.1268	0.0363	0.0012	1.000
Rainfed cropland	0.0640	0.1813	0.4251	0.0534	0.2229	0.0513	0.0002	1.000
Green spaces	0.0454	0.2102	0.4305	0.0313	0.0904	0.1922	0.0000	1.000
Water	0.0142	0.0965	0.1013	0.1199	0.0500	0.0000	0.6181	1.000

CBD: central business department, LMD: low–medium density, HD: high density.

3.2. Validation of CA–Markov Model

A two-stage model validation approach was performed, including the visual inspection and quantitative assessment. The visual inspection shows that there is close agreement between the 2018 LULC distribution patterns derived from the support vector machines supervised classification (actual) and the 2018 LULC patterns simulated using the CA–Markov model (Figure 3). The computed kappa statistics recorded a kappa for a no ability K_{no} of 0.8893, a kappa for quantity accuracy $K_{locationStrata}$ of 0.8943, a traditional kappa $K_{standard}$ of 0.9044 and a kappa for location accuracy $K_{location}$ of 0.925. To summarize, the kappa index of agreement values indicates that there is good agreement between the actual and simulated 2018 LULC maps. Therefore, the model can be applied with a high confidence in its reliability to forecast LULC maps for 2034 and 2050 (Table 9).

Table 9. Kappa indices computed between the actual and simulated 2018 LULC maps.

K Indices	2018
K_{no}	0.8893
$K_{location}$	0.9251
$K_{standard}$	0.9044
$K_{location}Strata$	0.8943

3.3. Future Climate Data Analysis

The predicted meteorological data, as provided by the global circulation model ensemble, show slightly diverging data in terms of precipitation regimes by the different climate scenarios for the observation period 2019–2050. Comparing annual rainfall predictions as provided by the RCP8.5 climate scenario and RCP4.5 climate scenario (Figure 5) indicates similar trends with varying magnitude. In climate scenario RCP4.5, the predicted annual rainfall oscillates with an overall decrease until 2050; the maximum predicted annual precipitation reaches around 950 mm in the years 2022, 2025, 2029 and 2031 and then decreases, reaching 855 mm in 2041 and around 785 mm in 2045 and 2050 (Figure 5). Underlying the same overall decline in precipitation, the minimum annual precipitation as predicted by climate scenario RCP4.5 varies between 814 mm in 2027 and 770–780 mm in 2034 and 2046. In climate scenario RCP8.5, the predicted annual rainfall also oscillates but does not show a distinct decrease during the forecasted period until 2050, as shown by the outcomes of RCP8.5. Maximum predicted annual precipitation varies between 800 and 900 mm and minimum predicted annual precipitation varies between 705 and 740 mm. The years of maximum predicted annual precipitation in RCP4.5 and RCP8.5 widely concur, but offsets can also be repeatedly observed (Figure 5).

3.4. Model Performance Evaluation

The performance evaluation carried out for each of the 15 statistically downscaled global circulation models' outcomes with in situ historical observations from the Harare gauging stations varied, as displayed in Table 10. The global circulation model performance evaluations show that fourteen GCMs (ACCESS1-0, BNU-ESM, CanESM2, CNRM-BGC, GFDL-ESM2G, GFDL-ESM2M, MIROC-ESM, MIROC5, MPI-ESM-LR, MPI-ESM-MR and NorESM1-M, CCSM4, IPSL-CM5A-LR, MIROC-ESM-CHEM) have sufficient performance when evaluated against observations ($d > 0.7$, $r > 0.7$ and $R^2 > 0.5$). The least successful performance in terms of accuracy when evaluating historical observations and global circulation models' average precipitation data was observed for MRI-CGCM3 ($R^2 < 0.5$), but the results show that the model has a strong positive correlation ($r > 0.7$) with a high index of agreement ($d > 0.7$), and an rRMSE below 20% (Table 10). As such, there is confidence to apply the GCM data for future soil erosion risk estimation for Epworth district (Table 10).

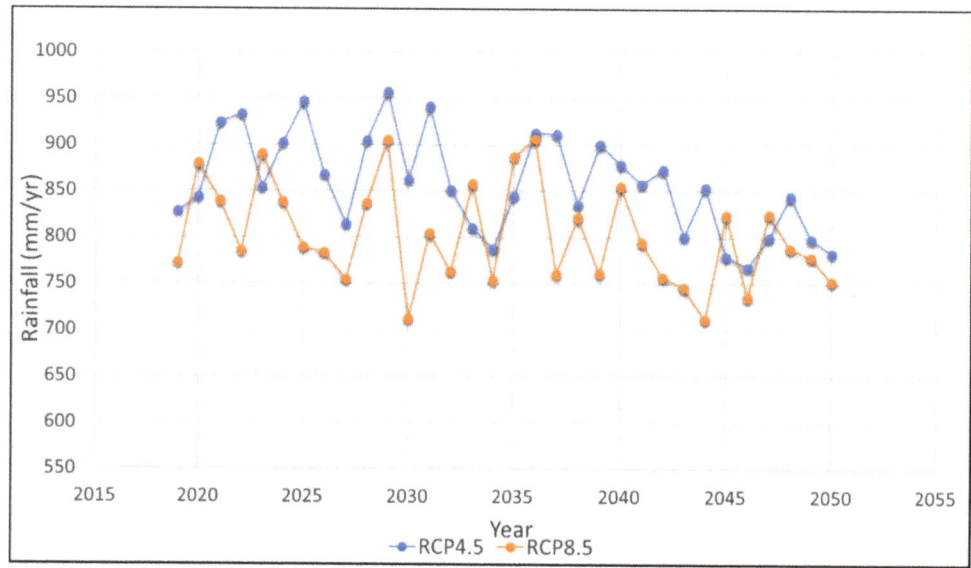

Figure 5. Annual rainfall variations for Epworth district, 2019–2050, based on global circulation model ensemble climate scenarios RCP4.5 and RCP8.5.

Table 10. The GCMs' performance evaluation against the observed precipitation dataset from 1980 to 2005.

GCM	RRMSE (%)	d	r	R^2
ACCESS1-0	15.64	0.84	0.77	0.60
BNU-ESM	16.36	0.79	0.78	0.62
CanESM2	16.49	0.80	0.78	0.61
CCSM4	18.70	0.75	0.71	0.51
CNRM-BGC	16.34	0.85	0.78	0.61
GFDL-ESM2G	17.65	0.82	0.79	0.61
GFDL-ESM2M	17.73	0.80	0.79	0.62
IPSL-CM5A-LR	17.43	0.77	0.71	0.51
MIROC-ESM	18.69	0.78	0.78	0.61
MIROC-ESM-CHEM	17.82	0.77	0.71	0.55
MIROC5	15.38	0.80	0.79	0.64
MPI-ESM-LR	16.55	0.80	0.77	0.60
MPI-ESM-MR	16.43	0.81	0.78	0.61
MRI-CGCM3	18.10	0.75	0.69	0.47
NorESMI-M	14.45	0.87	0.79	0.63

3.5. RUSLE Model Factor Maps

To be able to later assess the impact of future climate change on the future long-term potential soil erosion risk for Epworth district, the analysis of future predicted precipitation was split into two time intervals, 2019–2034 and 2035–2050; applying the RCP4.5 climate scenario between 2019 and 2034, annual rainfall averages 886 mm, and for the time interval 2035–2050, annual rainfall averages 839 mm; applying the RCP8.5 climate scenario between 2019 and 2034, annual rainfall averages 827 mm, and for the time interval 2035–2050 annual rainfall averages 799 mm. For the time period 2019–2034, rainfall erosivity factor (R) values, as derived from RCP4.5 model ensemble, are on average between 333 and 338 MJ mm ha^{-1} h^{-1} yr^{-1} and significantly exceed the values of the R factor based on the RCP8.5 model ensemble of 318–324 MJ mm ha^{-1} h^{-1} yr^{-1} (Figure 6). For the period

2035–2050, the R factor calculated on the basis of the RCP4.5 climate scenario varies between 321 and 328 MJ mm ha^{-1} h^{-1} yr^{-1}, again exceeding the R factor derived from the RCP8.5 model ensemble, which varies between 313 and 318 MJ mm ha^{-1} h^{-1} yr^{-1}. The variation in the R factor values dictates the temporal variation in annual rainfall for different climate scenarios. High R factor values were recorded from the RCP4.5 model ensemble averages for both future periods considered, the highest R factor being predicted for the period 2019–2034.

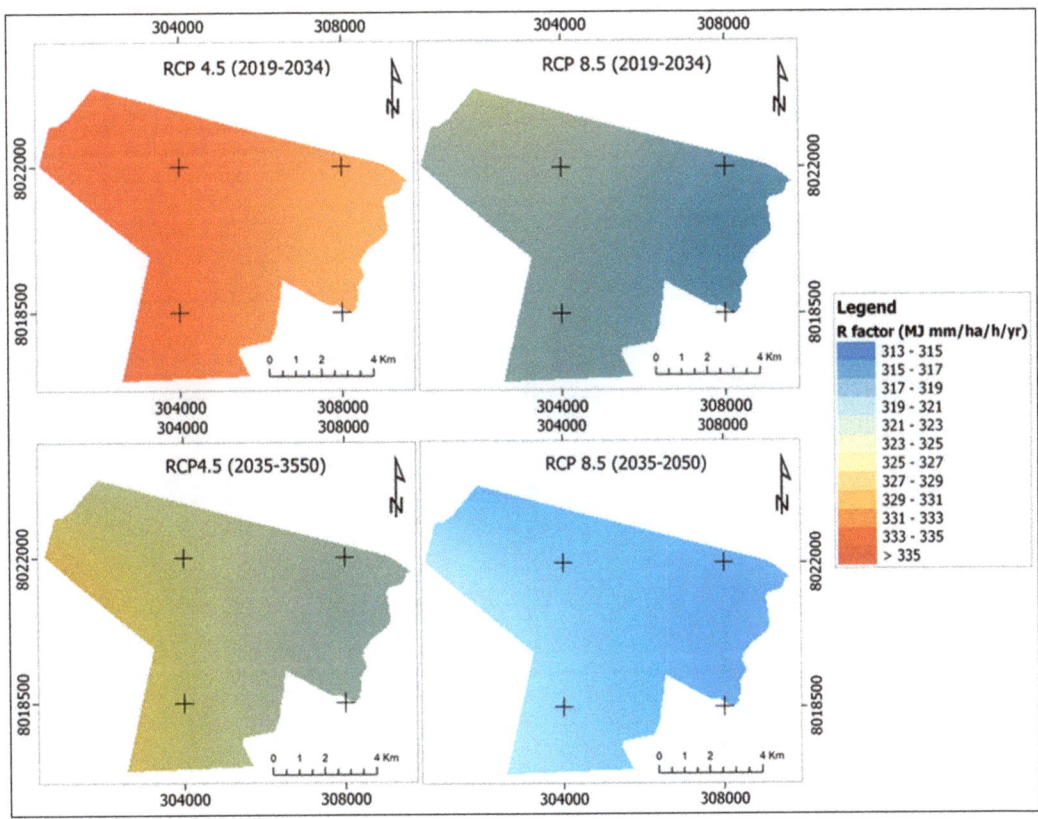

Figure 6. The rainfall erosivity factors (R) for Epworth district for the time periods: (**Top right**) 2019–2034 (RCP4.5); (**Top left**) 2019–2034 (RCP8.5); (**Bottom right**) 2035–2050 (RCP4.5); and (**Bottom left**) 2035–2050 (RCP8.5).

The soil texture in Epworth district corresponds largely to sand, sandy loam and clayey loam; only along the alluvial plains do predominantly sandy loams occur. Correspondingly, soil erodibility factor values (K) range between 0.06 and 0.09 (Figure 7b). The topography of Epworth district is undulating to gently rolling, with steep sloping areas along the river banks and at the intersections of tributary channels into the major receiving streams. Related topographic factor values (LS) range from 0 in the plateau areas up to approximately 22 on the steep sloping areas (Figure 7a). The width of the weighted land cover and management factor values (C) range between 0 and 0.239, with different distribution patterns in 2034 and 2050 (Figure 7c,d). Major differences in land cover and management relate to shifts in land use over time, as predicted by the CA–Markov model (Figure 3). Due to the lack of support practices in the study area, the support practice factor values (P) are set as 1.

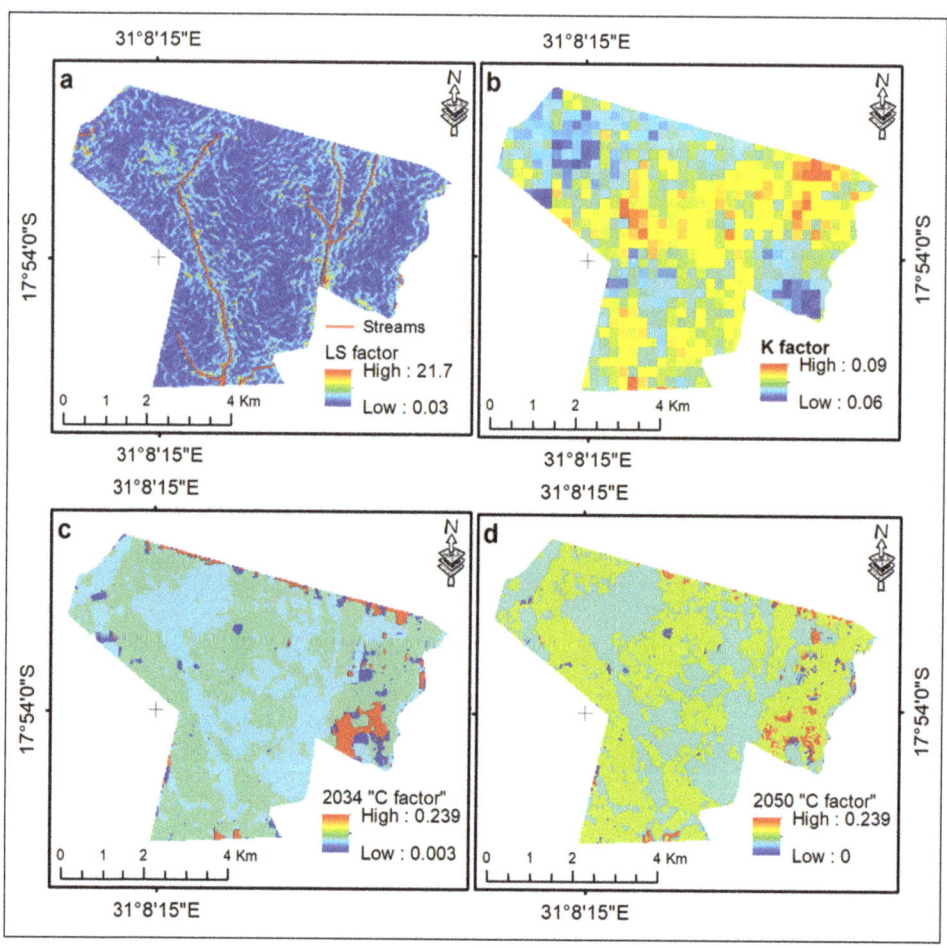

Figure 7. RUSLE input factors for modelling potential soil erosion risk for Epworth district. (**a**) Topographic factor (LS); (**b**) soil erodibility factor (K); (**c**) the crop cover and management factor (C) for 2034; (**d**) the crop cover and management factor (C) for 2050.

3.6. Potential Soil Erosion Risk

Potential soil erosion risk mapping was performed independently for the years 2034 and 2050 as selected time slices, considering the two climate scenarios RCP4.5 and RCP8.5. The predicted average soil erosion risk, applying precipitation data as provided by RCP4.5 for the period 2019–2034, totals 1.2 t ha^{-1} yr^{-1} for 2034 and 1.1 t ha^{-1} yr^{-1} for the period 2035–2050. Applying the R factor based on the annual precipitation data, as provided by climate scenario RCP8.5, the predicted average potential soil erosion risk amounts to 1.1 t ha^{-1} yr^{-1} in 2034 and 1.0 t ha^{-1} yr^{-1} in 2050. The estimated soil loss rate for the climate scenario RCP4.5 in 2034 varies between 0 and 69.3 t ha^{-1} yr^{-1} and 0 and 48.9 t ha^{-1} yr^{-1} in 2050. Applying the R factor based on the annual precipitation data, as provided by climate scenario RCP8.5, soil loss rates ranged between 0 and 62.4 t ha^{-1} yr^{-1} in 2034 and 0 and 42.3 t ha^{-1} yr^{-1} in 2050. Future potential soil erosion risk predictions for climate scenarios RCP4.5 and RCP8.5 were significantly different ($p < 0.05$) for each time interval, 2034 and 2050, highlighting that the presented changes can be attributed to various predicted factors, including land use and rainfall erosivity changes.

High potential soil erosion risk areas are predicted for the south-eastern periphery of Epworth district and along the tributaries, as well moving downwards in the south direction along the stream, as depicted in Figure 8. The predicted spatial patterns of potential soil erosion risk applying annual precipitation data, as provided by the RCP4.5 and RCP8.5 climate scenarios for the time slices 2034 and 2050, reveal in all cases high potential soil erosion risk along the flanks of the major rivers and along the flanks of steep tributaries (Figure 8). The displayed potential soil erosion risk maps in Figure 8 reveal that the predicted decrease in the R factor in the long term, corresponding to decrease in annual rainfall averages, reduces soil erosion processes, which simultaneously is on the rise in some localized parts of the district, and this is purportedly triggered by land use changes. Environmental characters, predominantly topography and soil properties (Figure 7), control the overall vulnerability of the area to soil erosion, finally displayed as potential soil erosion risk, including rainfall and land use.

The area-wide potential soil erosion risk predicted for the year 2034, applying R factors derived from the RCP4.5 climate scenario, indicates that 62.0% of the Epworth district will be exposed to low potential soil erosion risk and 27.9% to moderate potential soil erosion risk, while 8.1% will be exposed to high potential soil erosion risk and 2.0% to very high and extreme potential soil erosion risk (Table 11). The predicted results evidently show that there is an extensive distribution of areas of low potential soil erosion risk across the district, while high potential soil erosion risk is predicted predominantly along the channel networks. Applying R factors from the same climate scenario, RCP4.5, for the year 2050, approximately 74.3% of the entire district will be exposed to low potential soil erosion risk, 14.7% will be exposed to moderate potential soil erosion risk, 5.6% will be exposed to high potential soil erosion risk and 5.4% to very high and extreme potential soil erosion risk. The area-wide proportion of low potential soil erosion risk extended extensively across the entire district in 2050, attributed to the decline in the average rainfall erosivity in climate scenario RCP4.5. Applying R factors based on the RCP8.5 climate scenario in the year 2034, about 66.7% of the Epworth district is predicted to be exposed to low potential soil erosion risk, 24.6% to moderate potential soil erosion risk, 7.4% to high potential soil erosion risk and 1.3% to very high and extreme potential soil erosion risk (Table 11). Furthermore, for the year 2050, based on RCP8.5 climate scenario, the predicted area of Epworth district exposed to low potential soil erosion risk will be 77.7%, 14.1% will be exposed to moderate potential soil erosion risk, 4.6% to high potential soil erosion risk and 3.6% of the entire district will be exposed to very high and extreme potential soil erosion risk (Table 11). Applying climate scenario RCP8.5, similar to the application of climate scenario RCP4.5, high-intensity potential soil erosion is predicted predominantly along channel networks and predominates in the southern area of Epworth district (Figure 8).

Table 11. Predicted proportion of the spatial area of Epworth district exposed to potential soil erosion risk.

Soil Loss (t ha^{-1} yr^{-1})	Soil Erosion Risk	Area (%) in 2018	Area (%) in 2034		Area (%) in 2050	
			RCP4.5	RCP8.5	RCP4.5	RCP8.5
0–1	Low	59.5	62.0	66.7	74.3	77.7
1–2	Moderate	29.3	27.9	24.6	14.7	14.1
2–5	High	10.0	8.1	7.4	5.6	4.6
5–10	Very high	1.1	1.6	1.1	3.5	2.3
>10	Extreme	0.1	0.4	0.2	1.9	1.3

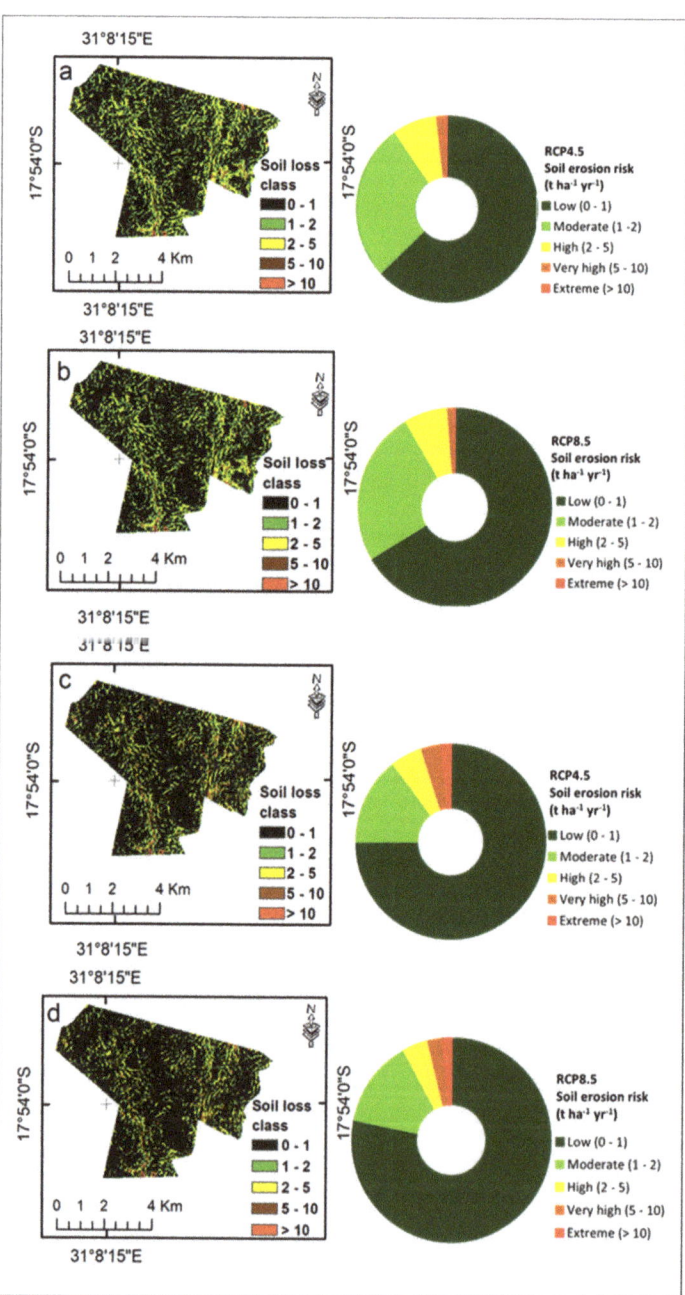

Figure 8. Predicted spatio-temporal potential soil erosion risk for Epworth district. (**a**) Potential soil erosion risk for 2034 applying R factors based on RCP4.5; (**b**) potential soil erosion risk for 2034 applying R factors based on RCP8.5; (**c**) potential soil erosion risk for 2050 applying R factors based on RCP4.5; and (**d**) potential soil erosion risk for 2050 applying R factors based on RCP8.5.

The average area-wide potential soil erosion risk in Epworth district predicted for the time slices 2034 and 2050 shows extended areas exposed to low potential soil erosion rates between 0 and 1 t ha^{-1} yr^{-1}, considering annual precipitation as provided by the RCP8.5 climate scenario. In contrast, the average area-wide potential soil erosion risk predicted for the time slices 2034 and 2050, considering annual precipitation as provided by RCP4.5 climate scenario, distinctively exposes a smaller area to low soil loss rates between 0 and 1 t ha^{-1} yr^{-1} compared to the respective predictions applying climate scenario RCP8.5. In relation to the study on the present-day soil erosion risk in Epworth district [23], the current area exposed to low soil erosion risk amounts to 59.5%; thus, it is predicted to distinctly increase in the future (Table 11). In contrast, currently 10% of the Epworth district is exposed to high soil erosion risk and up to 1.2% is exposed to very high and extreme soil erosion risk (Table 11). Correspondingly, it is expected that in the future, the areas in Epworth district exposed to high potential soil erosion risk with soil loss rates between 2 and 5 t ha^{-1} yr^{-1} will markedly decrease, and most likely will even halve by 2050. Furthermore, areas exposed to very high to extreme potential soil erosion risk with soil loss rates of more than 5 t ha^{-1} yr^{-1} will massively increase under future changes in land use and climate, while in 2034, under the RCP8.5 climate scenario, areas exposed to very high potential soil erosion risk will be widely stable compared to 2018 area coverage. By 2050, the spread of this category will double and might even triple when applying R-factors from the RCP4.5 climate scenario (Table 11). This development is even more distinctive when focusing on areas exposed to extreme potential soil erosion risk compared to the present-day situation until 2034, where areas exposed to extreme potential soil erosion risk will steadily increase by doubling the area extent when applying R-factors resulting from the RCP8.5 climate scenario, and up to 4 times the area extent when applying R-factors resulting from the RCP4.5 climate scenario. In 2050, areas exposed to extreme potential soil erosion risk will have increased by more than tenfold, independent of whether applying R-factors resulting from the RCP8.5 or RCP4.5 climate scenario. However, the total area exposed to extreme potential soil erosion risk remains small and predominantly will occur along the river banks (Table 11).

4. Discussion

The predicted CA–Markov model results reveal an increase in the spatio-temporal pattern of built-up area, with built-up area expected to cover over 95% in 2050 from an approximated total of 84.5% in 2018 (Figure 3, Table 5). The forecasted results indicate that green spaces and croplands will continue to decline at the expense of built-up area (Table 5). Thus, the transition probability matrices for different periods reveal the probability of each class (n) in the LULC maps changing in the next distinct period (t_{n+1}) in respect of the surrounding cells [22,81]. These predictions of built-up area growth at the expense of green spaces and croplands in the Harare Metropolitan Province concur with the conversion rates predicted by Mushore et al. [9] using CA–Markov model analysis. The same analysis agrees with the predicted urban growth and the development of Irbid's governorate of Jordan, with projected built-up area growth amounting to almost 65% in total area from an estimated 14.5% between 2015 and 2050, at the expense of vegetation and farmlands [55]. Therefore, such developments indicate the core principle of the CA models, which stipulates that the present state of development is a continuation of historical changes induced by the neighbourhood interactions [40,49,81]. This predicted expansion pattern is a result of the neighbourhood effect, which exhibits that the converted land use is next or close to the existing dominant land use, and predominantly built-up area exists for this scenario [39,47,49].

The predicted loss of green spaces and croplands may result in the detrimental loss of urban agricultural land and areas of aesthetic value to the ecosystem, which provide environmental protection. With the escalating socioeconomic woes and poverty in the city [70], the loss of urban agricultural land to urban development will leave many poorly resourced Epworth residents with detrimental food insecurities, threatening their livelihoods since many survive on market gardening and other urban farming activities [125,126]. The loss of green spaces also results in the reduction in vegetation cover and biomass which dissipates rainfall, reducing its direct impacts on the soil surface and facilitating percolation [101]. Further, with the current economic meltdown and population growth, the surge of urban built-up area predicted by the CA–Markov model can be justified; the Epworth district will be no exception in terms of absorbing more inhabitants from other spheres of the Harare Metropolitan Province. This push could be exacerbated by unaffordable rental charges and cost of living in other affluent suburbs of the Harare Metropolitan Province, resulting in further densification and overcrowding in Epworth district. However, due to excessive demand for shelter and anticipated population growth, the conversion of croplands and green spaces to a built-up area will intensify impervious surfaces across the district [15,127].

The GCM ensembles were used to quantify the hydrological impacts of climate change under different climate scenarios, RCP4.5 and RCP8.5, to obtain reliable projections [61,65,122,128]. Based on statistical metrics, the evaluation of the performance showed that fourteen GCMs (Table 10) have sufficient performance when evaluated with observations from Harare Metropolitan gauging stations (d > 0.7, r > 0.7 and R^2 > 0.5), with the exception of MRI-CGCM3, observed to have the lowest determination coefficient of 0.47. This may suggest that the general circulation model could have other specific years that were not properly simulated [63]; however, the analysis shows that most GCMs displayed good simulation. Above all, the GCMs have an rRMSE below 20%, which is reasonably acceptable [116,117,129]. Further, coarse grid resolutions from GCMs make it difficult to match, with few in situ observations which are not uniformly distributed attributed to increases in spatial variation and uncertainty to clearly define local precipitation characteristics, therefore increasing the simulation bias [130–132].

For the RUSLE model, potential soil erosion risk maps were produced using the geostatistical ArcGIS package (raster calculator) to multiply the RUSLE factor maps (Figures 6 and 7). The predicted potential soil erosion risk averaged at 1.2 t ha^{-1} yr^{-1} in 2034 and 1.1 t ha^{-1} yr^{-1} in 2050 for the RCP4.5 climate scenario, while 1.1 t ha^{-1} yr^{-1} and 1.0 t ha^{-1} yr^{-1} were the predicted averages for 2034 and 2050 for the RCP8.5 climate scenario. Meanwhile, studies on the influence of land use change or the impact of soil erosion risk on crop productivity indicated that a tolerable soil loss rate at 1 t ha^{-1} yr^{-1} was sustainable for the tropics [95,133–135]. Based on the slow rate of soil formation across the tropics, including Europe and America (<1 t ha^{-1} yr^{-1}) [95,133,136,137], the sustainable soil loss tolerance at 1 t ha^{-1} yr^{-1} was considered across the entire Epworth district. The resulting arguments around the proposed 10 t ha^{-1} yr^{-1} as the estimated soil erosion tolerance threshold for tropical ecosystems showed that it was highly overestimated, considering threats to the landscape and impacts on crop productivity likely to occur at such a high risk threshold [138]. Furthermore, other studies indicated that average soil loss rates of 5 t ha^{-1} yr^{-1} may be sustainable soil loss rates in the tropics [139,140]. Nevertheless, an estimated 1 t ha^{-1} yr^{-1} soil loss threshold subsisted for the current study and the predicted area-wide averages were unsustainable in that they slightly surpassed the recommended soil loss threshold, except for the RCP8.5 climate scenario in 2050. However, the slight notable deviation from the 1 t ha^{-1} yr^{-1} sustainable threshold can be justified as the averages fall within the applicable tolerable range of c.a 1.4 t ha^{-1} yr^{-1} proposed for some parts of the tropics, including America and Europe [137]. Thus, the estimated soil loss tolerance threshold was used to describe a sustainable soil loss rate [141].

The integrated average annual precipitation between 2019 and 2034, based on the climate scenario RCP4.5 results, shows high average annual soil loss rates ranging between 0 and 69.3 t ha^{-1} yr^{-1} and 0 and 62.4 t ha^{-1} yr^{-1} for the RCP8.5 climate scenario in 2034. In contrast, applying average annual precipitation between 2035 and 2050, the R factor-based values show a decline in soil loss rates for the year 2050 in both climate scenarios ranging between 0 and 48.9 t ha^{-1} yr^{-1} for RCP4.5 and 0 and 42.3 t ha^{-1} yr^{-1} for RCP8.5. However, these results show a continuous declining trend of soil loss rates when compared with the baseline period that applied the R factor based on the average annual precipitation data derived from in situ observations between 1984 and 2000 for Epworth district, estimating high soil erosion risk with average annual soil loss rates between 0 and 92.8 t ha^{-1} yr^{-1} in 2000 [23]. In summary, the soil loss rates for both the RCP4.5 and RCP8.5 climate scenarios are observed to be decreasing in spatial coverage over the years 2034 and 2050. Regardless of the high rainfall erosivity predicted between 2019 and 2034 in comparison with soil loss rates estimated for the year 2000 [23], it is revealed that land use changes, including the shrinking of croplands and disturbed shrublands, predominantly reduce the soil loss impact due to increases in impervious surfaces across the Epworth district.

The increasing potential soil erosion risk predicted for Epworth district along the channel networks has been attributed to the steep slopes along the streams in combination with massive impervious surfaces, resulting in the accumulation of overland flow [142]. Correspondingly, high topographic factor values appear on valley flanks (Figure 7), exposing surfaces to severe runoff and flooding resulting from the increased slope inclination and reduced infiltration capacity [143,144]. Displayed soil loss rates exceeding 1 t ha^{-1} yr^{-1} for Epworth district will be considered unsustainable [95,137], and therefore, the need for sound policy implementation to avoid detrimental environmental damage. Such estimates, as indicated in Table 11, reveal that a larger proportion of the study area will be exposed to tolerable soil loss rates [95,133,134]. Nevertheless, there is a predicted increase in soil erosion risk in vulnerable areas, mainly downslope and low-lying areas along the flanks of the channel networks [23,142,145].

The study results predict that soil loss rates vary with precipitation and land use changes for all the climate scenarios. The results suggest that the soil erosion response with regard to climate change could be complex, as it varies with time and on a climate scenario basis [25]. Consequently, the proportion of area exposed to high potential soil erosion risk with average soil loss rates between 2 and 5 t ha^{-1} yr^{-1} will markedly decline and most likely will even halve by 2050, as opposed to the doubling and triplicating proportional areas exposed to very high and extreme potential soil erosion risk for both climate scenarios in 2050. This is linked with the increasing vulnerability to smaller proportional area occupied by sparse green spaces and bare areas along channel networks. Such increasing trends in potential soil erosion risks are primarily accelerated by concentrated overland flow resulting from reduced infiltration processes across the Epworth district [99,143,146]. This vulnerability and response to rainfall impact and runoff processes with regard to reduced spatial area exposed to direct soil displacement in 2050 underpins the effects of land use changes and sloping topography along the channel network [114,147].

The decreasing rainfall erosivity for both scenarios over time concurs with the future analysis that incorporated regional climate models (RCMs) by Hudson and Jones [130], in which they highlighted the likelihood of increasing consecutive dry days in southern Africa; however, with some increases in other parts of the region [148]. Additionally, interannual high rainfall intensity impact is relatively expressed as this would be masked in annual rainfall averages due to low rain-day frequency [148]. The contraction of the rainfall season was projected following the observed late onset and early rainfall cessation in sub-Saharan Africa, mostly in central Mozambique, large parts of Botswana and the northern and southern parts of Zimbabwe [148]. Such responses to climate change tally with the predicted decline in overall soil erosion risk in 2050, which, however, still require more robust regional analysis on precipitation uncertainties to global climate change [130,148,149]. Nevertheless, the use of model ensemble averages could have limited the impact of other predicted extreme rainfall events [61,65,122]. Such changes and manipulations of rainfall intensities could negatively impact the final soil erosion prediction outcome [150,151]. Furthermore, the use of coarse grid resolutions and numerical methods reduces models' data independency, and therefore increases the bias and uncertainty range of the outcomes [61,122,123]. The empirical RUSLE model is also limited only to the predictive capacity of sheet, inter-rill and rill soil erosion processes spanning over long periods, as it is not an event-based model, which also does not consider gullying erosion processes [1,93,97,152,153]. Other data-driven processes integrated in the empirical RUSLE technique increase the uncertainty of future soil erosion risk due to varying data sources applied without rigorous quantification of their uncertainties and propagation [1,154].

Overall, high potential soil erosion risk displayed within the vicinity of Jacha river and tributaries extending from the north and southeast parts of the district draining southwards continue to increase, as predicted by the RUSLE model widely in 2050. This is attributed to the increasing sealed surface area and the sloping topography contributing to increased overland flow and surface runoff [143,155]. Taking into account human activities, previous studies reiterated that sand poaching activities along riverbanks are associated with heavy trucks ferrying sand to construction sites, contributing to high soil compaction on unpaved roads [23,24,142], reducing the infiltration capacity, and hence increasing surface runoff processes. For Epworth district, activities such as sand poaching and extraction along the riverbanks will be inevitable due to the predicted built-up area expansion and due to the fact that for many locals, informal activities provide employment for the sustenance of their livelihoods. Therefore, there is a need to implement sound policies and sustainable environmental management approaches in order to curb environmental damage and the future extinction of water bodies and their ecosystem services. Uncertainties exist in this study about policy amendments regarding the functionality of the Local Boards and Authorities in regulating developmental plans. This, in turn, will affect LULC changes in the Epworth district of the Harare Metropolitan Province. However, this was held constant in the prediction of future LULC distribution patterns for Epworth district.

5. Conclusions

The study uses LULC distribution patterns between 1990 and 2008 to apply a Markov chain model which allows the development of a transition probability matrix and suitability maps, and later defines the complex dynamic spatial patterns of urban area by the flexible Cellular Automatons. The validation of the simulated 2018 LULC distribution patterns and the actual 2018 LULC map displayed strong spatial agreement, both quantitatively and through visual inspection. The strong agreement and consistency of the LULC spatial patterns from the cross validation displayed the reliability and usability of the CA–Markov model to predict 2034 and 2050 future LULC distribution patterns for Epworth district. The predicted findings show a continuous increase in urban built-up area over the years 2034 and 2050 at the expense of croplands and perturbed green spaces, predominantly with the expansion of high-density residential areas towards Epworth district peripheries.

Further, future potential soil erosion risk was predicted for the years 2034 and 2050 using the RUSLE model, which integrated R factors based on the average annual precipitation between 2019 and 2034 and 2035 and 2050, as provided by climate scenarios RCP4.5 and RCP8.5. The goodness of fit measures highlighted that the general circulation models (GCMs) are useful for the assessment of future soil erosion risk, following the evaluation of GCMs performance with gauged observations, which showed a good performance, ascertaining their feasibility. As such, ensemble average outcomes from multiple GCMs under both the RCP4.5 and RCP8.5 climate scenarios were incorporated in the regional statistical relations equation to derive the rainfall erosivity factor for use in the RUSLE model.

Future trends in climate variability reveal that the projected high rainfall for the RCP4.5 climate scenario between 2019 and 2050 compared to the RCP8.5 climate scenario will contribute to high localized soil erosion risk in vulnerable areas, including perturbed green spaces, agricultural land and stream banks. High soil loss rates were predicted in 2034 for both climate scenarios RCP4.5 and RCP8.5, in comparison with low soil loss rates in 2050 for both climate scenarios, and this is largely attributable to the predicted dynamic land use changes resulting in the reduction in surface area exposed to soil erosion processes over time. The predicted results also indicate that average annual soil loss rates will approximately halve in 2050 from an estimated 0–93 t ha^{-1} yr^{-1} in 2000, independent of whether the RCP4.5 or RCP8.5 climate scenario is applied. Nevertheless, for 2050, increasing soil erosion risks have been predicted along the flanks of the drainage networks.

Overall, this study highlights the application of the CA–Markov model in combination with the RUSLE model to derive useful simulations for predicting future LULC and soil erosion risk. In addition, based on the stipulated IPCC policy recommendations from the Fifth Assessment Report (AR5), governments and policy makers need to implement sound climate policies in order to curtail and curb environmental degradation and landscape fragmentation at the local scale.

Author Contributions: All authors: conception of the paper; A.K.M.: data acquisition, data analysis, writing of the original draft; B.S.: supervision, writing—review and editing, funding acquisition. All authors have read and agreed to the published version of the manuscript.

Funding: This research was funded by the Deutscher Akademischer Austauschdienst—DAAD, and "The APC was funded by Freie Universität Berlin".

Institutional Review Board Statement: Not Applicable.

Informed Consent Statement: Not applicable.

Data Availability Statement: SRTM DEM: (https://earthexplorer.usgs.gov/SRTM1Arc, accessed on 19 September 2020); GCMs: NASA Exchange Global Daily Downscaled Projections (NASA-GDDP) (Available online at https://earthobservatory.nasa.gov/images/86027/, accessed on 2 October 2020); In Situ datasets: obtained from Department of Meteorology in Harare on 01 July 2019; Global soil map and attributes retrieved from ISRIC (International Soil Reference Information Centre)—World Soil Information: (Available online at https://soilgrids.org, accessed: 17 July 2019). LULC maps: Marondedze and Schütt (2019) peer-reviewed article (doi:10.3390/land8100155).

Acknowledgments: The publication of this article was funded by Freie Universität Berlin. We thank our colleagues from Freie Universität Berlin for their valuable insights that greatly assisted this research work. We are grateful to the United States Geological Survey (USGS) for providing Landsat images and the terrain-corrected Shuttle Radar Topography Mission (SRTM) DEM. We extend our gratitude to the Department of Meteorology in Harare, Zimbabwe for providing rainfall data and NASA Earth Exchange Global Daily Downscaled Projections (NEX-GDDP) for providing statistically downscaled global climate models.

Conflicts of Interest: The authors declare no conflict of interest.

References

1. Borrelli, P.; Robinson, D.A.; Panagos, P.; Lugato, E.; Yang, J.E.; Alewell, C.; Wuepper, D.; Montanarella, L.; Ballabio, C. Land Use and Climate Change Impacts on Global Soil Erosion by Water (2015–2070). *Proc. Natl. Acad. Sci. USA* **2020**, *117*, 21994–22001. [CrossRef]
2. Mondal, A.; Khare, D.; Kundu, S.; Meena, P.K.; Mishra, P.K.; Shukla, R. Impact of Climate Change on Future Soil Erosion in Different Slope, Land Use, and Soil-Type Conditions in a Part of the Narmada River Basin, India. *J. Hydrol. Eng.* **2015**, *20*. [CrossRef]
3. Karydas, C.G.; Sekuloska, T.; Silleos, G.N. Quantification and Site-Specification of the Support Practice Factor When Mapping Soil Erosion Risk Associated with Olive Plantations in the Mediterranean Island of Crete. *Environ. Monit. Assess.* **2009**, *149*, 19–28. [CrossRef] [PubMed]
4. Islam, K.; Jashimuddin, M.; Nath, B.; Nath, T.K. Land Use Classification and Change Detection by Using Multi-Temporal Remotely Sensed Imagery: The Case of Chunati Wildlife Sanctuary, Bangladesh. *Egypt. J. Remote Sens. Space Sci.* **2018**, *21*, 37–47. [CrossRef]
5. Lambin, E.F. Modelling and Monitoring Land-Cover Change Processes in Tropical Regions. *Prog. Phys. Geogr. Earth Environ.* **1997**, *21*, 375–393. [CrossRef]
6. Brueckner, J.K.; Helsley, R.W. Sprawl and Blight. *J. Urban Econ.* **2011**, *69*, 205–213. [CrossRef]
7. Jat, M.K.; Choudhary, M.; Saxena, A. Application of Geo-Spatial Techniques and Cellular Automata for Modelling Urban Growth of a Heterogeneous Urban Fringe. *Egypt. J. Remote Sens. Space Sci.* **2017**, *20*, 223–241. [CrossRef]
8. Hegazy, I.R.; Kaloop, M.R. Monitoring Urban Growth and Land Use Change Detection with GIS and Remote Sensing Techniques in Daqahlia Governorate Egypt. *Int. J. Sustain. Built Environ.* **2015**, *4*, 117–124. [CrossRef]
9. Mushore, T.D.; Odindi, J.; Dube, T.; Mutanga, O. Prediction of Future Urban Surface Temperatures Using Medium Resolution Satellite Data in Harare Metropolitan City, Zimbabwe. *Build. Environ.* **2017**, *122*, 397–410. [CrossRef]
10. UN World Urbanization Prospects: The 2014 Revision, Highlights 2014. Available online: https://www.un.org/en/development/desa/publications/2014-revision-world-urbanization-prospects.html (accessed on 2 February 2021).
11. UN World Urbanization Prospects: The 2011 Revision United Nations. 2012. Available online: https://www.un.org/en/development/desa/population/publications/pdf/urbanization/WUP2011_Report.pdf (accessed on 4 February 2021).

12. World Economic Forum Global-Continent-Urban-Population-Urbanisation-Percent 2020. Available online: https://www.weforum.org/agenda/2020/11/global-continent-urban-population-urbanisation-percent/ (accessed on 24 March 2021).
13. UN. Population Division Then & Now: Urban Population Worldwide 2020. Available online: https://population.un.org/wup/ (accessed on 24 March 2021).
14. ZimStats (Zimbabwe National Statistics Agency) Census 2012: Preliminary Report. 2012. Available online: https://www.zimstat.co.zw/wp-content/uploads/publications/Population/population/census-2012-national-report.pdf (accessed on 14 October 2020).
15. Marondedze, A.K.; Schütt, B. Dynamics of Land Use and Land Cover Changes in Harare, Zimbabwe: A Case Study on the Linkage between Drivers and the Axis of Urban Expansion. *Land* **2019**, *8*, 155. [CrossRef]
16. Aburas, M.M.; Ho, Y.M.; Ramli, M.F.; Ash'aari, Z.H. The Simulation and Prediction of Spatio-Temporal Urban Growth Trends Using Cellular Automata Models: A Review. *Int. J. Appl. Earth Obs. Geoinf.* **2016**, *52*, 380–389. [CrossRef]
17. Myers, G. *African Cities: Alternative Visions of Urban Theory and Practice*; Zed Books Ltd.: London, UK, 2011; ISBN 1780321333.
18. Pontius, R.G. Quantification Error versus Location Error in Comparison of Categorical Maps. *Photogramm. Eng. Remote Sens.* **2000**, *66*, 1011–1016.
19. Sudhira, H.S.; Ramachandra, T.V.; Jagadish, K.S. Urban Sprawl: Metrics, Dynamics and Modelling Using GIS. *Int. J. Appl. Earth Obs. Geoinf.* **2004**, *5*, 29–39. [CrossRef]
20. Chalise, D.; Kumar, L.; Kristiansen, P. Land Degradation by Soil Erosion in Nepal: A Review. *Soil Syst.* **2019**, *3*, 12. [CrossRef]
21. Shikangalah, R.; Paton, E.; Jetlsch, F.; Blaum, N. Quantification of Areal Extent of Soil Erosion in Dryland Urban Areas: An Example from Windhoek, Namibia. *Cities Environ. (CATE)* **2017**, *10*, 22.
22. Kamusoko, C.; Gamba, J.; Murakami, H. Monitoring Urban Spatial Growth in Harare Metropolitan Province, Zimbabwe. *Adv. Remote Sens.* **2013**, *02*, 322–331. [CrossRef]
23. Marondedze, A.K.; Schütt, B. Assessment of Soil Erosion Using the RUSLE Model for the Epworth District of the Harare Metropolitan Province, Zimbabwe. *Sustainability* **2020**, *12*, 8531. [CrossRef]
24. USDA. NRCS Soil Quality—Urban Technical Note No. 1: Erosion and Sedimentation on Construction Sites 2000. Available online: http://www.aiswcd.org/wp-content/uploads/2013/04/u011.pdf (accessed on 18 May 2020).
25. Pruski, F.F.; Nearing, M.A. Climate-Induced Changes in Erosion during the 21st Century for Eight U.S. Locations: Climate-Induced Changes in Erosion. *Water Resour. Res.* **2002**, *38*, 34-1–34-11. [CrossRef]
26. IPCC, (Intergovernmental Panel on Climate Change). Climate Change 2014: Impacts, Adaptation, and Vulnerability. Contribution of Working Group II to the Fifth Assessment Report of the Intergovernmental Panel on Climate Change 2014. Available online: https://www.ipcc.ch/report/ar5/wg2/ (accessed on 15 February 2021).
27. Gupta, S.; Kumar, S. Simulating Climate Change Impact on Soil Erosion Using RUSLE Model—A Case Study in a Watershed of Mid-Himalayan Landscape. *J. Earth Syst. Sci.* **2017**, *126*, 43. [CrossRef]
28. Li, Z.; Fang, H. Impacts of Climate Change on Water Erosion: A Review. *Earth-Sci. Rev.* **2016**, *163*, 94–117. [CrossRef]
29. Segura, C.; Sun, G.; McNulty, S.; Zhang, Y. Potential Impacts of Climate Change on Soil Erosion Vulnerability across the Conterminous United States. *J. Soil Water Conserv.* **2014**, *69*, 171–181. [CrossRef]
30. Simonneaux, V.; Cheggour, A.; Deschamps, C.; Mouillot, F.; Cerdan, O.; Le Bissonnais, Y. Land Use and Climate Change Effects on Soil Erosion in a Semi-Arid Mountainous Watershed (High Atlas, Morocco). *J. Arid Environ.* **2015**, *122*, 64–75. [CrossRef]
31. Feng, X.; Wang, Y.; Chen, L.; Fu, B.; Bai, G. Modeling Soil Erosion and Its Response to Land-Use Change in Hilly Catchments of the Chinese Loess Plateau. *Geomorphology* **2010**, *118*, 239–248. [CrossRef]
32. Routschek, A.; Schmidt, J.; Kreienkamp, F. Impact of Climate Change on Soil Erosion—A High-Resolution Projection on Catchment Scale until 2100 in Saxony/Germany. *CATENA* **2014**, *121*, 99–109. [CrossRef]
33. Makwara, E.C.; Gamira, D. About to Lose All the Soil in Zaka's Ward 5, Zimbabwe: Rewards of Unsustainable Land Use. *Eur. J. Sustain. Dev.* **2012**, *1*, 457. [CrossRef]
34. Tundu, C.; Tumbare, M.J.; Kileshye Onema, J.-M. Sedimentation and Its Impacts/Effects on River System and Reservoir Water Quality: Case Study of Mazowe Catchment, Zimbabwe. *Proc. Int. Assoc. Hydrol. Sci.* **2018**, *377*, 57–66. [CrossRef]
35. Whitlow, R. Soil Erosion and Conservation Policy in Zimbabwe. *Land Use Policy* **1988**, *5*, 419–433. [CrossRef]
36. Mullan, D.; Favis-Mortlock, D.; Fealy, R. Addressing Key Limitations Associated with Modelling Soil Erosion under the Impacts of Future Climate Change. *Agric. For. Meteorol.* **2012**, *156*, 18–30. [CrossRef]
37. Herold, M.; Goldstein, N.C.; Clarke, K.C. The Spatiotemporal Form of Urban Growth: Measurement, Analysis and Modeling. *Remote Sens. Environ.* **2003**, *86*, 286–302. [CrossRef]
38. Tang, J.; Wang, L.; Yao, Z. Spatio-temporal Urban Landscape Change Analysis Using the Markov Chain Model and a Modified Genetic Algorithm. *Int. J. Remote Sens.* **2007**, *28*, 3255–3271. [CrossRef]
39. Tang, J.; Di, L. Past and Future Trajectories of Farmland Loss Due to Rapid Urbanization Using Landsat Imagery and the Markov-CA Model: A Case Study of Delhi, India. *Remote Sens.* **2019**, *11*, 180. [CrossRef]
40. Ahmed, B.; Ahmed, R. Modeling Urban Land Cover Growth Dynamics Using Multi-Temporal Satellite Images: A Case Study of Dhaka, Bangladesh. *ISPRS Int. J. Geo-Inf.* **2012**, *1*, 3–31. [CrossRef]
41. Halmy, M.W.A.; Gessler, P.E.; Hicke, J.A.; Salem, B.B. Land Use/Land Cover Change Detection and Prediction in the North-Western Coastal Desert of Egypt Using Markov-CA. *Appl. Geogr.* **2015**, *63*, 101–112. [CrossRef]

42. Hashem, N.; Balakrishnan, P. Change Analysis of Land Use/Land Cover and Modelling Urban Growth in Greater Doha, Qatar. *Ann. GIS* **2015**, *21*, 233–247. [CrossRef]
43. Sang, L.; Zhang, C.; Yang, J.; Zhu, D.; Yun, W. Simulation of Land Use Spatial Pattern of Towns and Villages Based on CA–Markov Model. *Math. Comput. Model.* **2011**, *54*, 938–943. [CrossRef]
44. Balzter, H. Markov Chain Models for Vegetation Dynamics. *Ecol. Model.* **2000**, *126*, 139–154. [CrossRef]
45. Yang, X.; Zheng, X.-Q.; Chen, R. A Land Use Change Model: Integrating Landscape Pattern Indexes and Markov-CA. *Ecol. Model.* **2014**, *283*, 1–7. [CrossRef]
46. Clarke, K.C.; Hoppen, S.; Gaydos, L. A Self-Modifying Cellular Automaton Model of Historical Urbanization in the San Francisco Bay Area. *Environ. Plan. B Plan. Des.* **1997**, *24*, 247–261. [CrossRef]
47. Fitawok, M.B.; Derudder, B.; Minale, A.S.; Van Passel, S.; Adgo, E.; Nyssen, J. Modeling the Impact of Urbanization on Land-Use Change in Bahir Dar City, Ethiopia: An Integrated Cellular Automata–Markov Chain Approach. *Land* **2020**, *9*, 115. [CrossRef]
48. White, R.; Engelen, G. Cellular Automata and Fractal Urban Form: A Cellular Modelling Approach to the Evolution of Urban Land-Use Patterns. *Environ. Plan. Econ. Space* **1993**, *25*, 1175–1199. [CrossRef]
49. Koomen, E.; Beurden, B.J. (Eds.) *Land-Use Modelling in Planning Practice*; GeoJournal Library; Springer: Dordrecht, The Netherlands, 2011; Volume 101, ISBN 978-94-007-1821-0.
50. Aburas, M.M.; Ho, Y.M.; Ramli, M.F.; Ash'aari, Z.H. Improving the Capability of an Integrated CA-Markov Model to Simulate Spatio-Temporal Urban Growth Trends Using an Analytical Hierarchy Process and Frequency Ratio. *Int. J. Appl. Earth Obs. Geoinf.* **2017**, *59*, 65–78. [CrossRef]
51. Alsharif, A.A.A.; Pradhan, B. Urban Sprawl Analysis of Tripoli Metropolitan City (Libya) Using Remote Sensing Data and Multivariate Logistic Regression Model. *J. Indian Soc. Remote Sens.* **2014**, *42*, 149–163. [CrossRef]
52. Sibanda, S.; Ahmed, F. Modelling Historic and Future Land Use/Land Cover Changes and Their Impact on Wetland Area in Shashe Sub-Catchment, Zimbabwe. *Model. Earth Syst. Environ.* **2020**. [CrossRef]
53. Wang, C.; Lei, S.; Elmore, A.J.; Jia, D.; Mu, S. Integrating Temporal Evolution with Cellular Automata for Simulating Land Cover Change. *Remote Sens.* **2019**, *11*, 301. [CrossRef]
54. Parsa, A.V.; Yavari, A.; Nejadi, A. Spatio-Temporal Analysis of Land Use/Land Cover Pattern Changes in Arasbaran Biosphere Reserve: Iran. *Model. Earth Syst. Environ.* **2016**, *2*, 1–13. [CrossRef]
55. Khawaldah, H.A.; Farhan, I.; Alzboun, N.M. Simulation and Prediction of Land Use and Land Cover Change Using GIS, Remote Sensing and CA-Markov Model. *Glob. J. Environ. Sci. Manag.* **2020**, *6*. [CrossRef]
56. Moss, R.H.; Edmonds, J.A.; Hibbard, K.A.; Manning, M.R.; Rose, S.K.; van Vuuren, D.P.; Carter, T.R.; Emori, S.; Kainuma, M.; Kram, T.; et al. The next Generation of Scenarios for Climate Change Research and Assessment. *Nature* **2010**, *463*, 747–756. [CrossRef] [PubMed]
57. van Vuuren, D.P.; Stehfest, E.; den Elzen, M.G.J.; Kram, T.; van Vliet, J.; Deetman, S.; Isaac, M.; Klein Goldewijk, K.; Hof, A.; Mendoza Beltran, A.; et al. RCP2.6: Exploring the Possibility to Keep Global Mean Temperature Increase below 2°C. *Clim. Change* **2011**, *109*, 95–116. [CrossRef]
58. Panagos, P.; Ballabio, C.; Meusburger, K.; Spinoni, J.; Alewell, C.; Borrelli, P. Towards Estimates of Future Rainfall Erosivity in Europe Based on REDES and WorldClim Datasets. *J. Hydrol.* **2017**, *548*, 251–262. [CrossRef]
59. Vrieling, A. Satellite Remote Sensing for Water Erosion Assessment: A Review. *CATENA* **2006**, *65*, 2–18. [CrossRef]
60. Nasrollahi, N.; AghaKouchak, A.; Cheng, L.; Damberg, L.; Phillips, T.J.; Miao, C.; Hsu, K.; Sorooshian, S. How Well Do CMIP5 Climate Simulations Replicate Historical Trends and Patterns of Meteorological Droughts? *Water Resour. Res.* **2015**, *51*, 2847–2864. [CrossRef]
61. Räisänen, J. How Reliable Are Climate Models? *Tellus Dyn. Meteorol. Oceanogr.* **2007**, *59*, 2–29. [CrossRef]
62. Ahmadalipour, A.; Moradkhani, H.; Svoboda, M. Centennial Drought Outlook over the CONUS Using NASA-NEX Downscaled Climate Ensemble: Drought Projection Using NASA-NEX Ensemble over the CONUS. *Int. J. Climatol.* **2017**, *37*, 2477–2491. [CrossRef]
63. Chemura, A.; Yalew, A.W.; Gornott, C. Quantifying Agroforestry Yield Buffering Potential Under Climate Change in the Smallholder Maize Farming Systems of Ethiopia. *Front. Agron.* **2021**, *3*, 609536. [CrossRef]
64. Demirel, M.C.; Moradkhani, H. Assessing the Impact of CMIP5 Climate Multi-Modeling on Estimating the Precipitation Seasonality and Timing. *Clim. Change* **2016**, *135*, 357–372. [CrossRef]
65. Murphy, J.M.; Sexton, D.M.H.; Barnett, D.N.; Jones, G.S.; Webb, M.J.; Collins, M.; Stainforth, D.A. Quantification of Modelling Uncertainties in a Large Ensemble of Climate Change Simulations. *Nature* **2004**, *430*, 768–772. [CrossRef]
66. Najafi, M.R.; Moradkhani, H. Ensemble Combination of Seasonal Streamflow Forecasts. *J. Hydrol. Eng.* **2016**, *21*, 04015043. [CrossRef]
67. Riahi, K.; Grübler, A.; Nakicenovic, N. Scenarios of Long-Term Socio-Economic and Environmental Development under Climate Stabilization. *Technol. Forecast. Soc. Change* **2007**, *74*, 887–935. [CrossRef]
68. Thomson, A.M.; Calvin, K.V.; Smith, S.J.; Kyle, G.P.; Volke, A.; Patel, P.; Delgado-Arias, S.; Bond-Lamberty, B.; Wise, M.A.; Clarke, L.E.; et al. RCP4.5: A Pathway for Stabilization of Radiative Forcing by 2100. *Clim. Change* **2011**, *18*. [CrossRef]
69. Luckow, P.; Wise, M.A.; Dooley, J.J.; Kim, S.H. Large-Scale Utilization of Biomass Energy and Carbon Dioxide Capture and Storage in the Transport and Electricity Sectors under Stringent CO_2 Concentration Limit Scenarios. *Int. J. Greenh. Gas Control* **2010**, *4*, 865–877. [CrossRef]

70. Chirisa, I.E.W.; Muhomba, K. Constraints to Managing Urban and Housing Land in the Context of Poverty: A Case of Epworth Settlement in Zimbabwe. *Local Environ.* **2013**, *18*, 950–964. [CrossRef]
71. Potts, D. "We Have A Tiger by the Tail": Continuities and Discontinuities in Zimbabwean City Planning and Politics. *Crit. Afr. Stud.* **2011**, *4*, 15–46. [CrossRef]
72. Tibaijuka, A.K. Report of the Fact-Finding Mission to Zimbabwe to Assess the Scope and Impact of Operation Murambatsvina 2005. Available online: https://www.environmentandurbanization.org/report-fact-finding-mission-zimbabwe-assess-scope-and-impact-operation-murambatsvina (accessed on 13 March 2021).
73. Butcher, C. Low Income Housing in Zimbabwe: A Case Study of the Epworth Squatter Upgrading Programme 1986. Available online: https://opendocs.ids.ac.uk/opendocs/handle/20.500.12413/10008 (accessed on 2 February 2021).
74. CSO Census 2002 Population Census: Provincial Profile Harare 2004. Available online: https://www.cso.ie/en/census/ (accessed on 14 October 2020).
75. Nyamapfene, K. *Soils of Zimbabwe*; Nehanda Publishers (Pvt) Ltd: Harare, Zimbabwe, 1991.
76. Thompson, J.G.; Purves, W.D. *A Guide to the Soils of Rhodesia. Technical Handbook No. 3*; Rhodesia Agricultural Journal; Ministry of Agriculture: Salisbury, Zimbabwe, 1978.
77. Eastman, J.R. *IDRISI Taiga Guide to GIS and Image Processing*; Clark Labs Clark University: Worcester, MA, USA, 2009.
78. Araya, Y.H.; Cabral, P. Analysis and Modeling of Urban Land Cover Change in Setúbal and Sesimbra, Portugal. *Remote Sens.* **2010**, *2*, 1549–1563. [CrossRef]
79. Eastman, J.R. *IDRISI 32 Guide to GIS and Image Processing*; Clark Labs Clark University: Worcester, MA, USA, 2000.
80. Kumar, S.; Radhakrishnan, N.; Mathew, S. Land Use Change Modelling Using a Markov Model and Remote Sensing. *Geomat. Nat. Hazards Risk* **2014**, *5*, 145–156. [CrossRef]
81. Subedi, P.; Subedi, K.; Thapa, B. Application of a Hybrid Cellular Automaton—Markov (CA-Markov) Model in Land-Use Change Prediction: A Case Study of Saddle Creek Drainage Basin, Florida. *Appl. Ecol. Environ. Sci.* **2013**, *1*, 126–132. [CrossRef]
82. Ye, B.; Bai, Z. Simulating Land Use/Cover Changes of Nenjiang County Based on CA-Markov Model. In *Computer and Computing Technologies in Agriculture, Volume I.*; Li, D., Ed.; The International Federation for Information Processing; Springer: Boston, MA, USA, 2008; Volume 258, pp. 321–329, ISBN 978 0-387-77250-9.
83. Guan, D.; Gao, W.; Watari, K.; Fukahori, H. Land Use Change of Kitakyushu Based on Landscape Ecology and Markov Model. *J. Geogr. Sci.* **2008**, *18*, 455–468. [CrossRef]
84. Hamad, R.; Balzter, H.; Kolo, K. Predicting Land Use/Land Cover Changes Using a CA-Markov Model under Two Different Scenarios. *Sustainability* **2018**, *10*, 3421. [CrossRef]
85. Santé, I.; García, A.M.; Miranda, D.; Crecente, R. Cellular Automata Models for the Simulation of Real-World Urban Processes: A Review and Analysis. *Landsc. Urban Plan.* **2010**, *96*, 108–122. [CrossRef]
86. White, R.; Engelen, G.; Uljee, I. The Use of Constrained Cellular Automata for High-Resolution Modelling of Urban Land-Use Dynamics. *Environ. Plan. B Plan. Des.* **1997**, *24*, 323–343. [CrossRef]
87. Jafari, M.; Majedi, H.; Monavari, S.; Alesheikh, A.; Kheirkhah Zarkesh, M. Dynamic Simulation of Urban Expansion Based on Cellular Automata and Logistic Regression Model: Case Study of the Hyrcanian Region of Iran. *Sustainability* **2016**, *8*, 810. [CrossRef]
88. Mondal, M.S.; Sharma, N.; Kappas, M.; Garg, P.K. Cellular Automata (CA) Contiguity Filters Impacts on CA Markov Modeling of Land Use Land Cover Change Predictions Results. *Int. Arch. Photogramm. Remote Sens. Spat. Inf. Sci.* **2020**, *XLIII-B3-2020*, 1585–1591. [CrossRef]
89. Mondal, M.S.; Sharma, N.; Kappas, M.; Garg, P.K. Modeling of Spatio-Temporal Dynamics of LULC–a Review and Assessment. *J. Geom.* **2012**, *6*, 29–39.
90. Zubair, A. *Change Detection in Land Use and Land Cover Using Remote Sensing Data and Gis (a Case Study of Ilorin and Its Environs in Kwara State)*; University of Ibadan: Ibadan, Nigeria, 2006.
91. Sayemuzzaman, M.; Jha, M.K. Modeling of Future Land Cover Land Use Change in North Carolina Using Markov Chain and Cellular Automata Model. *Am. J. Eng. Appl. Sci.* **2014**, *7*, 295–306. [CrossRef]
92. Ahmed, B.; Ahmed, R.; Zhu, X. Evaluation of Model Validation Techniques in Land Cover Dynamics. *ISPRS Int. J. Geo-Inf.* **2013**, *2*, 577–597. [CrossRef]
93. Merritt, W.S.; Letcher, R.A.; Jakeman, A.J. A Review of Erosion and Sediment Transport Models. *Environ. Model. Softw.* **2003**, *18*, 761–799. [CrossRef]
94. Ranzi, R.; Le, T.H.; Rulli, M.C. A RUSLE Approach to Model Suspended Sediment Load in the Lo River (Vietnam): Effects of Reservoirs and Land Use Changes. *J. Hydrol.* **2012**, *422–423*, 17–29. [CrossRef]
95. Karamage, F.; Zhang, C.; Liu, T.; Maganda, A.; Isabwe, A. Soil Erosion Risk Assessment in Uganda. *Forests* **2017**, *8*, 52. [CrossRef]
96. Prasannakumar, V.; Vijith, H.; Abinod, S.; Geetha, N. Estimation of Soil Erosion Risk within a Small Mountainous Sub-Watershed in Kerala, India, Using Revised Universal Soil Loss Equation (RUSLE) and Geo-Information Technology. *Geosci. Front.* **2012**, *3*, 209–215. [CrossRef]
97. Renard, K.G.; Foster, G.R.; Weesies, G.A.; McCool, D.K.; Yoder, D.C. *Predicting Soil Erosion by Water: A Guide to Conservation Planning with the Revised Universal Soil Loss Equation (RUSLE)*; USDA: Washington, DC, USA, 1997.
98. Smith, H.J. Application of Empirical Soil Loss Models in Southern Africa: A Review. *South Afr. J. Plant Soil* **1999**, *16*, 158–163. [CrossRef]

99. Meshesha, D.T.; Tsunekawa, A.; Tsubo, M.; Ali, S.A.; Haregeweyn, N. Land-Use Change and Its Socio-Environmental Impact in Eastern Ethiopia's Highland. *Reg. Environ. Change* **2014**, *14*, 757–768. [CrossRef]
100. Ai, L.; Fang, N.F.; Zhang, B.; Shi, Z.H. Broad Area Mapping of Monthly Soil Erosion Risk Using Fuzzy Decision Tree Approach: Integration of Multi-Source Data within GIS. *Int. J. Geogr. Inf. Sci.* **2013**, *27*, 1251–1267. [CrossRef]
101. Ferreira, V.; Samora-Arvela, A.; Panagopoulos, T. Soil Erosion Vulnerability under Scenarios of Climate Land-Use Changes after the Development of a Large Reservoir in a Semi-Arid Area. *J. Environ. Plan. Manag.* **2016**, *59*, 1238–1256. [CrossRef]
102. Wischmeier, W.H.; Smith, D.D. *Predicting Rainfall Erosion Losses: A Guide to Conservation Planning*; Department of Agriculture, Science and Education Administration: Corvallis, OR, USA, 1978.
103. Woldemariam, G.; Iguala, A.; Tekalign, S.; Reddy, R. Spatial Modeling of Soil Erosion Risk and Its Implication for Conservation Planning: The Case of the Gobele Watershed, East Hararghe Zone, Ethiopia. *Land* **2018**, *7*, 25. [CrossRef]
104. Hengl, T.; Mendes de Jesus, J.; Heuvelink, G.B.M.; Ruiperez Gonzalez, M.; Kilibarda, M.; Blagotić, A.; Shangguan, W.; Wright, M.N.; Geng, X.; Bauer-Marschallinger, B.; et al. SoilGrids250m: Global Gridded Soil Information Based on Machine Learning. *PLoS ONE* **2017**, *12*, e0169748. [CrossRef]
105. Sharpley, A.N.; Williams, J.R. *EPIC—Erosion/Productivity Imappct Calculator: 1. Model Documentation.*; Technical Bulletin; U.S. Department of Ariculture: Washington, DC, USA, 1990; Volume 1768.
106. Desmet, P.J.J.; Govers, A. A GIS Procedure for Automatically Calculating the USLE LS Factor on Topographically Complex Landscape Units. *J. Soil Water Conserv.* **1996**, *51*, 427–433.
107. Panagos, P.; Borrelli, P.; Meusburger, K. A New European Slope Length and Steepness Factor (LS-Factor) for Modeling Soil Erosion by Water. *Geosciences* **2015**, *5*, 117–126. [CrossRef]
108. Kefi, M.; Yoshino, K.; Setiawan, Y.; Zayani, K.; Boufaroua, M. Assessment of the Effects of Vegetation on Soil Erosion Risk by Water: A Case of Study of the Batta Watershed in Tunisia. *Environ. Earth Sci.* **2011**, *64*, 707–719. [CrossRef]
109. Panagos, P.; Borrelli, P.; Poesen, J.; Ballabio, C.; Lugato, E.; Meusburger, K.; Montanarella, L.; Alewell, C. The New Assessment of Soil Loss by Water Erosion in Europe. *Environ. Sci. Policy* **2015**, *54*, 438–447. [CrossRef]
110. Stocking, M.A.; Elwell, H.A. Rainfall Erosivity over Rhodesia. *Trans. Inst. Br. Geogr.* **1976**, *1*, 231. [CrossRef]
111. El-Swaify, S.A.; Gramier, C.L.; Lo, A. Recent Advances in Soil Conservation in Steepland in Humid Tropics. In Proceedings of the International Conference on Steepland Agriculture in the Humid Tropics, Kuala Lumpur, Malaysia, 17–21 August 1987.
112. Merritt, W.S.; Croke, B.F.W.; Jakeman, A.J.; Letcher, R.A.; Perez, P. A Biophysical Toolbox for Assessment and Management of Land and Water Resources in Rural Catchments in Northern Thailand. *Ecol. Model.* **2004**, *171*, 279–300. [CrossRef]
113. Thrasher, B.; Maurer, E.P.; McKellar, C.; Duffy, P.B. Technical Note: Bias Correcting Climate Model Simulated Daily Temperature Extremes with Quantile Mapping. *Hydrol. Earth Syst. Sci.* **2012**, *16*, 3309–3314. [CrossRef]
114. Sardari, M.R.A.; Bazrafshan, O.; Panagopoulos, T.; Sardooi, E.R. Modeling the Impact of Climate Change and Land Use Change Scenarios on Soil Erosion at the Minab Dam Watershed. *Sustainability* **2019**, *11*, 21. [CrossRef]
115. Bsaibes, A.; Courault, D.; Baret, F.; Weiss, M.; Olioso, A.; Jacob, F.; Hagolle, O.; Marloie, O.; Bertrand, N.; Desfond, V.; et al. Albedo and LAI Estimates from FORMOSAT-2 Data for Crop Monitoring. *Remote Sens. Environ.* **2009**, *113*, 716–729. [CrossRef]
116. Chen, J.-L.; Liu, H.-B.; Wu, W.; Xie, D.-T. Estimation of Monthly Solar Radiation from Measured Temperatures Using Support Vector Machines—A Case Study. *Renew. Energy* **2011**, *36*, 413–420. [CrossRef]
117. Chen, J.-L.; Li, G.-S.; Wu, S.-J. Assessing the Potential of Support Vector Machine for Estimating Daily Solar Radiation Using Sunshine Duration. *Energy Convers. Manag.* **2013**, *75*, 311–318. [CrossRef]
118. Araya, A.; Hoogenboom, G.; Luedeling, E.; Hadgu, K.M.; Kisekka, I.; Martorano, L.G. Assessment of Maize Growth and Yield Using Crop Models under Present and Future Climate in Southwestern Ethiopia. *Agric. For. Meteorol.* **2015**, *214–215*, 252–265. [CrossRef]
119. Willmott, C.J. Some Comments on the Evaluation of Model Perfomance. *Am. Meteorol. Soc.* **1982**, *63*. [CrossRef]
120. Nearing, M.A. Potential Changes in Rainfall Erosivity in the U.S with Climate Change during the 21st Century. *J. Soil Water Conserv.* **2001**, *56*, 229–232.
121. Zhang, X.-C. A Comparison of Explicit and Implicit Spatial Downscaling of GCM Output for Soil Erosion and Crop Production Assessments. *Clim. Change* **2007**, *84*, 337–363. [CrossRef]
122. Sperna Weiland, F.C.; van Beek, L.P.H.; Weerts, A.H.; Bierkens, M.F.P. Extracting Information from an Ensemble of GCMs to Reliably Assess Future Global Runoff Change. *J. Hydrol.* **2012**, *412–413*, 66–75. [CrossRef]
123. Vrochidou, A.-E.K.; Grillakis, M.G.; Tsanis, I.K. Drought Assessment Based on Multi-Model Precipitation Projections for the Island of Crete. *J. Earth Sci. Clim. Change* **2013**, *4*. [CrossRef]
124. Ferro, V.; Porto, P.; Yu, B. A Comparative Study of Rainfall Erosivity Estimation for Southern Italy and Southeastern Australia. *Hydrol. Sci. J.* **1999**, *44*, 3–24. [CrossRef]
125. Tawodzera, G. Vulnerability in Crisis: Urban Household Food Insecurity in Epworth, Harare, Zimbabwe. *Food Secur.* **2011**, *3*, 503–520. [CrossRef]
126. UNDP Urban Agriculture: Foods, Jobs and Sustainable Cities 1996. Available online: http://jacsmit.com/book/Chap01.pdf (accessed on 23 April 2021).
127. Wania, A.; Kemper, T.; Tiede, D.; Zeil, P. Mapping Recent Built-up Area Changes in the City of Harare with High Resolution Satellite Imagery. *Appl. Geogr.* **2014**, *46*, 35–44. [CrossRef]

128. Hagemann, S.; Göttel, H.; Jacob, D.; Lorenz, P.; Roeckner, E. Improved Regional Scale Processes Reflected in Projected Hydrological Changes over Large European Catchments. *Clim. Dyn.* **2009**, *32*, 767–781. [CrossRef]
129. Arumugam, P.; Chemura, A.; Schauberger, B.; Gornott, C. Near Real-Time Biophysical Rice (*Oryza Sativa* L.) Yield Estimation to Support Crop Insurance Implementation in India. *Agronomy* **2020**, *10*, 1674. [CrossRef]
130. Hudson, D.A.; Jones, R.G. Regional Climate Model Simulations of Present-Day and Future Climates of Southern Africa. *Tech Rep Hadley Cent. Tech. Note 39 Hadley Cent. Clim. Predict. Res. Met Off.* **2002**, *39*, 42.
131. Pinto, I.; Lennard, C.; Tadross, M.; Hewitson, B.; Dosio, A.; Nikulin, G.; Panitz, H.-J.; Shongwe, M.E. Evaluation and Projections of Extreme Precipitation over Southern Africa from Two CORDEX Models. *Clim. Change* **2016**, *135*, 655–668. [CrossRef]
132. Shongwe, M.E.; Oldenborgh, G.J.V. Projected Changes in Mean and Extreme Precipitation in Africa under Global Warming. Part II: East Africa. *J. Clim.* **2011**, *24*, 16. [CrossRef]
133. Abdulkareem, J.H.; Pradhan, B.; Sulaiman, W.N.A.; Jamil, N.R. Prediction of Spatial Soil Loss Impacted by Long-Term Land-Use/Land-Cover Change in a Tropical Watershed. *Geosci. Front.* **2019**, *10*, 389–403. [CrossRef]
134. Khosrokhani, M.; Pradhan, B. Spatio-Temporal Assessment of Soil Erosion at Kuala Lumpur Metropolitan City Using Remote Sensing Data and GIS. *Geomat. Nat. Hazards Risk* **2014**, *5*, 252–270. [CrossRef]
135. Kouli, M.; Soupios, P.; Vallianatos, F. Soil Erosion Prediction Using the Revised Universal Soil Loss Equation (RUSLE) in a GIS Framework, Chania, Northwestern Crete, Greece. *Environ. Geol.* **2009**, *57*, 483–497. [CrossRef]
136. Jones, R.J.A.; Le Bissonnais, Y.; Bazzoffi, P.; Sanchez Diaz, J.; Düwel, O.; Loj, G.; Øygarden, L.; Prasuhn, V.; Rydell, B.; Strauss, P. Interim Report Version 3.31, 28 October 2003. 2003, Volume 3, p. 28. Available online: https://esdac.jrc.ec.europa.eu/ESDB_Archive/pesera/pesera_cd/pdf/WP2ErosInterimRepV331_4CD.pdf (accessed on 11 April 2021).
137. Verheijen, F.G.A.; Jones, R.J.A.; Rickson, R.J.; Smith, C.J. Tolerable versus Actual Soil Erosion Rates in Europe. *Earth-Sci. Rev.* **2009**, *94*, 23–38. [CrossRef]
138. Morgan, R.P.C. *Soil Erosion and Conservation*, 3rd ed.; Blackwell Publishing Ltd: Oxford, UK, 2005.
139. Bamutaze, Y. Revisiting Socio-Ecological Resilience and Sustainability in the Coupled Mountain Landscapes in Eastern Africa. *Curr. Opin. Environ. Sustain.* **2015**, *14*, 257–265. [CrossRef]
140. Lufafa, A.; Tenywa, M.M.; Isabirye, M.; Majaliwa, M.J.G.; Woomer, P.L. Prediction of Soil Erosion in a Lake Victoria Basin Catchment Using a GIS-Based Universal Soil Loss Model. *Agric. Syst.* **2003**, *76*, 883–894. [CrossRef]
141. Alewell, C.; Egli, M.; Meusburger, K. An Attempt to Estimate Tolerable Soil Erosion Rates by Matching Soil Formation with Denudation in Alpine Grasslands. *J. Soils Sediments* **2015**, *15*, 1383–1399. [CrossRef]
142. Braud, I.; Breil, P.; Thollet, F.; Lagouy, M.; Branger, F.; Jacqueminet, C.; Kermadi, S.; Michel, K. Evidence of the Impact of Urbanization on the Hydrological Regime of a Medium-Sized Periurban Catchment in France. *J. Hydrol.* **2013**, *485*, 5–23. [CrossRef]
143. Dams, J.; Dujardin, J.; Reggers, R.; Bashir, I.; Canters, F.; Batelaan, O. Mapping Impervious Surface Change from Remote Sensing for Hydrological Modeling. *J. Hydrol.* **2013**, *485*, 84–95. [CrossRef]
144. Le Roux, J.J.; Sumner, P.D. Factors Controlling Gully Development: Comparing Continuous and Discontinuous Gullies. *Land Degrad. Dev.* **2012**, *23*, 440–449. [CrossRef]
145. Opeyemi, O.A.; Abidemi, F.H.; Victor, O.K. Assessing the Impact of Soil Erosion on Residential Areas of Efon-Alaaye Ekiti, Ekiti-State, Nigeria. *Int. J. Environ. Plan. Manag.* **2019**, *5*, 9.
146. Phil-Eze, P.O. Variability of Soil Properties Related to Vegetation Cover in a Tropical Rainforest Landscape. *J. Geogr. Plan.* **2010**, *3*, 174–188.
147. Renschler, C.S.; Mannaerts, C.; Diekkrüger, B. Evaluating Spatial and Temporal Variability in Soil Erosion Risk—Rainfall Erosivity and Soil Loss Ratios in Andalusia, Spain. *CATENA* **1999**, *34*, 209–225. [CrossRef]
148. Shongwe, M.E.; van Oldenborgh, G.J.; van den Hurk, B.J.J.M.; de Boer, B.; Coelho, C.A.S.; van Aalst, M.K. Projected Changes in Mean and Extreme Precipitation in Africa under Global Warming. Part I: Southern Africa. *J. Clim.* **2009**, *22*, 3819–3837. [CrossRef]
149. IPCC, (Intergovernmental Panel on Climate Change) Climate Change: The Physical Science Basis, Contribution of Working Group I to the Fourth Assessment Report of the Intergovernmental Panel on Climate Change 2007. Available online: https://www.ipcc.ch/report/ar4/wg1/ (accessed on 18 March 2021).
150. Boardman, J. Soil Erosion Science: Reflections on the Limitations of Current Approaches. *CATENA* **2006**, *68*, 73–86. [CrossRef]
151. Turnbull, L.; Parsons, A.J.; Wainwright, J.; Anderson, J.P. Runoff Responses to Long-Term Rainfall Variability in a Shrub-Dominated Catchment. *J. Arid Environ.* **2013**, *91*, 88–94. [CrossRef]
152. Shamshad, A.; Azhari, M.N.; Isa, M.H.; Hussin, W.M.A.W.; Parida, B.P. Development of an Appropriate Procedure for Estimation of RUSLE EI30 Index and Preparation of Erosivity Maps for Pulau Penang in Peninsular Malaysia. *CATENA* **2008**, *72*, 423–432. [CrossRef]
153. Phinzi, K.; Ngetar, N.S. The Assessment of Water-Borne Erosion at Catchment Level Using GIS-Based RUSLE and Remote Sensing: A Review. *Int. Soil Water Conserv. Res.* **2019**, *7*, 27–46. [CrossRef]
154. Falk, M.; Denham, R.J.; Mengersen, K.L. Estimating Un-Certainty in the Revised Universal Soil Loss Equation via Bayesian Melding. *J Agric Biol Env. Sta.* **2010**, *15*, 20–37. [CrossRef]
155. Cantón, Y.; Solé-Benet, A.; de Vente, J.; Boix-Fayos, C.; Calvo-Cases, A.; Asensio, C.; Puigdefábregas, J. A Review of Runoff Generation and Soil Erosion across Scales in Semiarid South-Eastern Spain. *J. Arid Environ.* **2011**, *75*, 1254–1261. [CrossRef]

Article

Correlation Analysis between Land-Use/Cover Change and Coastal Subsidence in the Yellow River Delta, China: Reviewing the Past and Prospecting the Future

Yi Zhang [1,*], Yilin Liu [2], Xinyuan Zhang [1], Haijun Huang [3,4], Keyu Qin [3], Zechao Bai [5] and Xinghua Zhou [1]

1. College of Ocean Science and Engineering, Shandong University of Science and Technology, Qingdao 266590, China; nancy2710@163.com (X.Z.); xhzhou@fio.org.cn (X.Z.)
2. College of Earth Science and Engineering, Shandong University of Science and Technology, Qingdao 266590, China; lyilin@msn.com
3. Key Laboratory of Marine Geology and Environment, Institute of Oceanology, Chinese Academy of Sciences, Qingdao 266071, China; hjhuang@qdio.ac.cn (H.H.); qdqky924@126.com (K.Q.)
4. College of Earth and Planetary Sciences, University of Chinese Academy of Sciences, Beijing 100049, China
5. School of Information Science and Technology, North China University of Technology, Beijing 100144, China; baizechao1991@163.com
* Correspondence: 7706465@163.com

Citation: Zhang, Y.; Liu, Y.; Zhang, X.; Huang, H.; Qin, K.; Bai, Z.; Zhou, X. Correlation Analysis between Land-Use/Cover Change and Coastal Subsidence in the Yellow River Delta, China: Reviewing the Past and Prospecting the Future. Remote Sens. 2021, 13, 4563. https://doi.org/10.3390/rs13224563

Academic Editors: Baojie He, Ayyoob Sharifi, Chi Feng and Jun Yang

Received: 8 October 2021
Accepted: 7 November 2021
Published: 13 November 2021

Publisher's Note: MDPI stays neutral with regard to jurisdictional claims in published maps and institutional affiliations.

Copyright: © 2021 by the authors. Licensee MDPI, Basel, Switzerland. This article is an open access article distributed under the terms and conditions of the Creative Commons Attribution (CC BY) license (https://creativecommons.org/licenses/by/4.0/).

Abstract: In recent years, noticeable subsidence depressions have occurred along the coastal zone of the Yellow River Delta. Consistent with these changes, dramatic human modifications within the coastal zone stand out, and the coastline is altered from an undisturbed natural area to an artificial coastline. However, very few studies have attempted to quantitatively analyze the relationship between subsidence depression and human activities. Here, the subsidence characteristics of the different land-use types in the Yellow River Delta are examined, and their spatiotemporal trends are quantified using a long-term satellite-observed time series of 30 years (1984–2017) regarding the land use map in combination with the InSAR-derived vertical ground deformations during three typical periods (P1: 1992–2000, P2: 2007–2010, and P3: 2016–2017). Noticeably, the highest subsidence rates were observed in areas where substantial human activities were observed, such as the subsidence in the salt fields ranging from 13 mm/year to 32 mm/year to 453 mm/year, respectively. Moreover, through the land-use prediction of Land Change Modeler (LCM), it is found that the salt field area will be further expanded in the future. The ecological vulnerability of the Yellow River Delta coastal zone should receive more attention in the future in terms of planning environmental protection strategies.

Keywords: land-use/cover change; coastal subsidence; underground brine exploitation; Sentinel-1A; Landsat; GIS

1. Introduction

There is no doubt that the amount of human interference in Earth systems has strongly increased during the last century and has now reached a new high level, with even greater effects than those of many natural processes on Earth [1]. In recent decades, most of the deltas in the world have undergone artificial transformations. In East Asia, the Yellow River Delta is highlighted as a hotspot due to its dramatic coastal land-use changes [2,3]. The delta's natural coastline has become dominated by artificial shorelines due to the boom of the nearshore salt and aquaculture industries, with natural areas decreasing from 70% in 1974 to 11% in 2015 [4].

On the other hand, deltaic sediments are naturally prone to sinking due to their high compressibility and low bearing capacity. It is clear that human activity has accelerated this natural process, primarily through the exploitation of groundwater and hydrocarbons [5–7]. In the Yellow River Delta, an increasing number of case studies have shown that groundwater (including underground brine water) pumping and hydrocarbon extraction are responsible for the large amount of sinking in coastal regions [8–14].

However, the relationship between land-use changes and subsidence has not been well analyzed, and it is still not clear whether there is a causal relationship between them [15,16]. This study demonstrates the evolution of the coastal land-use of the Yellow River Delta, which is represented by shrimp farms, oil fields, and salt pans, and these aspects have been subjected to tremendous alterations due to human activities over the past 30 years. Specifically, the study aims to evaluate the relationship between land-use evolution and land subsidence risk in this delta. Moreover, we attempt to predict the future trend of land use based on the land-use history and relevant terrain factors (see Section 3.1.3 for details), such as slope, aspect, and a digital elevation model (DEM). It is expected that this prediction can provide a scientific basis for land-use planning and the exploitation of groundwater.

2. Study Area

The Yellow River Delta is one of the most active areas of land use/cover change (LUCC) in China, and one of the fastest land-making deltas in the world (Figure 1). The delta is formed by the accumulation of delta sediments since the Yellow River diverted in 1855, so the thickness of sediments gradually thickens from land to sea, from 4 m to 16 m. The spatial distribution of sediment thickness is shown in the previous article [17].

Figure 1. The location of the study area is outlined by the yellow polygon within the standard false-color Landsat 8 OLI image of the Yellow River Delta in September 2017.

The Yellow River Delta is located between the Jiyang fault depression and the Chengning uplift, with rich reserves of oil, gas, brine, and water resources [11]. The groundwater in the Yellow River Delta is mainly composed of loose-rock pore water and mainly occurs in alluvial and marine sediments in the upper part of the Quaternary system [18]. Since the late Pleistocene, there have been three major transgression–regression events in the delta [17]. Consequently, a large amount of underground brine has been found in the Pleistocene aquifer. It has been shown that there is a large underground brine resource belt along the coast of Bohai Bay. According to its burial depth, the belt is generally divided into three categories: shallow brine (100 m to the surface), medium brine (100 to 400 m), and deep brine (400 m and deeper). Previous studies [18] have shown that the development of underground brine resources is limited to shallow underground brines. The salt pan area of the whole province is approximately 400 km², with approximately 5600 brine wells and an annual output of 6.53 million tons of raw salt.

3. Materials and Methods

3.1. Land-Use Maps

3.1.1. Remote Sensing Datasets

We built a novel constant time series of land-use maps by employing optical satellite remote sensing Landsat mission images. Landsat-series images were chosen due to the long period of available images, the appropriate ground resolution (30 m × 30 m), the broad range of spectral bands, and the free access to the images (Accessed date: 10 February 2019 http://earthexplorer.usgs.gov/). The Landsat tile (path 121, row 34) covering the entire delta was selected and outlined as the study area (Figure 1). To analyze the land-use evolution with the maximum access to the land-use history, quantify the synchronous subsidence rates obtained from InSAR-derived subsidence rates (see 3.3 for details), and reduce the classification errors caused by seasonal variation (especially vegetation cover), six Landsat satellite images (1984, 1992, 2000, 2007, 2010, and 2017) with low cloud coverage during autumn (August–October) were screened (Table 1). We use all satellite images passing through during the day, so it is in a descending mode.

Table 1. The detailed information of the six Landsat satellite images.

Sensor ID	Date Acquired	Path/Row	Resolution	Cloud Cover
Landsat5 TM	3 September 1984	121/34	30 m × 30 m	9.92%
Landsat5 TM	24 August 1992	121/34	30 m × 30 m	0.02%
Landsat5 TM	17 October 2000	121/34	30 m × 30 m	0.42%
Landsat7 ETM+	11 September 2007	121/34	30 m × 30 m	0.31%
Landsat7 ETM+	17 November 2010	121/34	30 m × 30 m	0
Landsat8 OLI	30 September 2017	121/34	30 m × 30 m	3.08%

For our research, we defined the three most representative manmade land-use classes to include shrimp ponds, salt pans, and oil wells in the deltaic coastal zone. To encompass the major human-induced zones, we produced a buffer polygon (~2500 km², with a 10 km radius) around the 2017 coastline (~350 km) as the study area in our analyses (Figure 1, yellow polygon).

Furthermore, the topographic map constructed in 1998 was employed to depict the locations of oil wells that are too small to be identified from the Landsat images.

3.1.2. Image Classification

The Landsat images were classified with a supervised method according to the following two primary steps: (1) the choice of training samples and (2) the use of an appropriate classification algorithm. We employed a "maximum likelihood classification" (MLC) algorithm to assign the land-use types to the patches in the Landsat images [19]. We define

$$X = \{x_i\}_{i=1}^{N},$$

as an original image pixel and

$$Y = \{y_i\}_{i=1}^{N},$$

as the classification result. The MLC algorithm can be described as follows:

$$\hat{Y} = a\{y_i\}_{i=1}^{N},$$

where N is the number of pixels in the original dataset and $\hat{Y}$ is the solution of the optimization problem. We should note that this algorithm is conducted under the presumption that each pixel to be classified is normally distributed in each class [20].

According to the results of the image classification, the transformation matrix can be realized to quantitatively examine the land-use changes. First, two different land-use maps with identical class names were dissolved by merging the matching records into a single map to improve the following processing. Then, these two dissolved maps were intersected based on the overlay analysis. The above two steps were accomplished in ArcGIS 10.2. Subsequently, these two maps were used to produce a transformation matrix in Excel.

3.1.3. Land-Use Prediction

The land change modeler (LCM) model is an integrated module in the IDRISI software. It has been developed by the Clark Laboratory and Conservation International for many years [21]. It is becoming one of the commonly used models to measure land-use changes [22]. The LCM model consists of a multilayer perceptron–artificial neural network (MLP-ANN), a Markov chain, cellular automata, and soft and hard prediction models. The model can predict the future land-use status through simulation of the existing land-use status and provides a good reference for decision-makers who plan and protect. Here, we randomly selected two-thirds of the samples as training samples, and the remaining one-third of the samples to verify the accuracy of the model. Moreover, the model can predict and analyze land-use changes in the environment provided by IDRISI and runs by following a set of rules in an orderly manner.

3.2. Land Subsidence Measurement

3.2.1. Rerunning Geodetic Leveling

The land subsidence in the study area in recent decades was monitored through repeated geodetic leveling by employing the benchmark network produced by the Yellow River Conservancy Commission from 1964 to 2007. The leveling dataset used for this study is available from Liu and Huang (2013) between 2000 and 2007 (Figure 1, red triangle) and from the Shandong Provincial Lubei Geo-engineering Exploration Institute from 2016 to 2017 (Figure 1, green triangle), with random errors of 3–5 mm/km and 1 mm/km, respectively. The original leveling measurements were digitized and interpolated to derive a homogeneous set of contour maps of the land subsidence rates (Figure 2b).

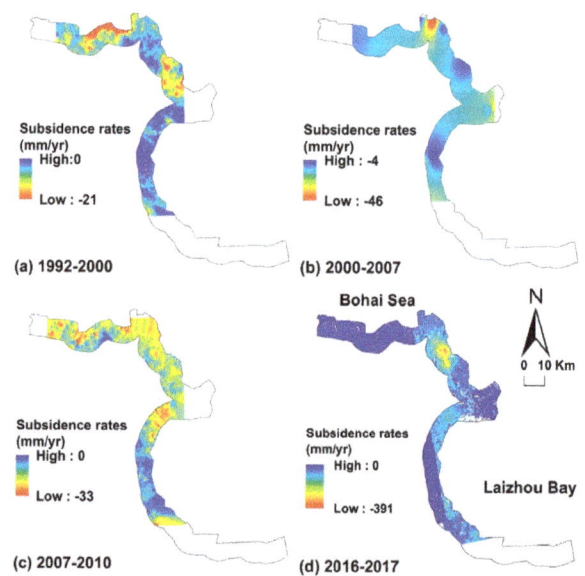

Figure 2. Land subsidence rate maps ((**a**,**c**,**d**,) derived from InSAR; (**b**) geodetic leveling) [11,13].

3.2.2. InSAR Observations

In comparison with the traditional investigation method of geodetic leveling, the InSAR technique has great advantages due to its broad coverage and high spatial–temporal resolution under all weather conditions. More recently, radar acquisitions by various satellites (ERS-1/2, ENVISAT-ASAR, and Sentinel-1A/B) were processed through synthetic aperture radar (SAR) interferometry over the delta [8–14]. In the ArcGIS software environment, the conversion from InSAR vector feature point data to grid data was realized, and then the land subsidence maps were generated by Kriging interpolation method. The land subsidence map was shown in Figure 2a (1992–2000) and 2c (2007–2010), with estimated errors caused by spatiotemporal variability in the surface scattering properties of 6.1 and 7.2 mm/year, respectively [11,13]. Except forSsentinel-1A, which was in ascending orbit mode, all satellites were in descending orbit mode.

As shown in Figure 2d, we further measured recent subsidence rates in P3, by analysis of interferometric synthetic aperture radar (InSAR) imagery, using 15 C-band ascending track Sentinel-1A images acquired over the period from Jan 2016 to Apr 2017. InSAR-derived subsidence rates are consistent with the leveling-based rates (mentioned in Section 3.2.1, Figure 1, green triangle). The comparative results are shown in Figure 3. The mean and standard deviation of the difference of the deformation rate between the two measurements are 3.52 mm/year and 6.87 mm/year, respectively. This small error indicates that the InSAR observations are in good agreement with the results from repeated geodetic leveling. The settlement extremes in Figure 3 are consistent with other references [8,14].

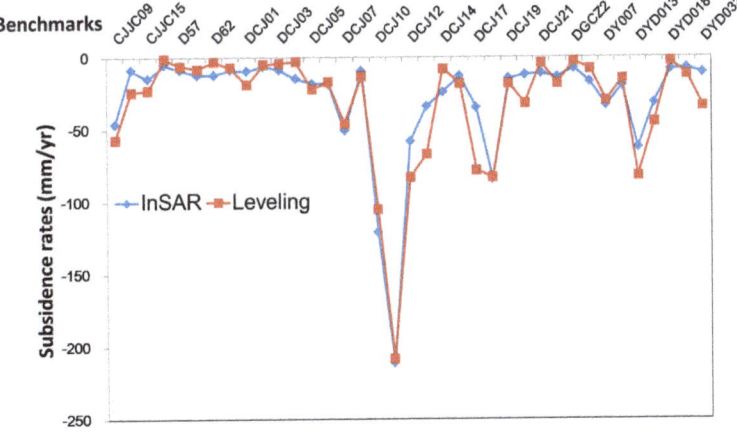

Figure 3. Comparison between InSAR and leveling measurements from 2016 to 2017.

3.3. Coupling Land-Use and Land-Cover Change with Subsidence

The land-use sequence maps were integrated with the corresponding land subsidence maps to quantify the subsidence rates for each land-use class during each time period. To improve the integration accuracy, the four corresponding Landsat images (1992, 2000, 2007, and 2010) were chosen for combination with the two available groups of InSAR-based subsidence measurements (Figure 2a,c).

First, the massive vector data (millions of InSAR feature points and image patches) were converted into a raster format by employing a conversion tool embedded in ArcGIS to improve the processing efficiency. Then, a relational equal-to operation was performed on two inputs (e.g., 1992 and 2000 classified raster maps) on a cell-by-cell basis using the map algebra functions in GIS. It was set to 1 in cells where the first raster (1992) was equal to the second (2000) and to 0 otherwise. Finally, the statistical values (such as the maximum, minimum, median, and average) of the input rasters (e.g., change or no change between

1992 and 2000) acquired within the zone of another dataset (e.g., land subsidence rates over 1992–2000) were retrieved through zonal statistical analysis with ArcGIS 10.2 software.

3.4. Analysis of Brine Exploitation Potential

The mining potential coefficient method is used to analyze and evaluate the potential of brine resources. The exploitation potential coefficient refers to the ratio between the allowable exploitation amount of regional brine and the current exploitation amount, and the calculation formula is as follows:

$$P = Q_z/Q_k$$

where P is the exploitation potential coefficient, Q_z is the exploitable quantity of brine (10^4 m^3/year), and Q_k is the current exploitation amount of brine (10^4 m^3/year).

Here, the mining potential coefficient is calculated according to the above formula. According to the zoning standards [23] shown in Table 2, the brine distribution area is divided into the potential area, compensation balance area, and overexploitation area.

Table 2. Evaluation criteria for brine resource potential.

Zoning Standards	P ≥ 1.2	0.8 < P < 1.2	P < 0.8
Evaluation	potential area	compensation balance area	overexploitation area

4. Results

4.1. Land-Use and Land-Cover Change

4.1.1. Land-Use Structure Change

Six maps of land-use/cover classifications were produced based on the Landsat satellite images, as shown in Figure 4. With the help of the topographic map dated 1998 (scale 1:50,000), the oil wells constructed since 2000 were digitized and added (black dots in Figure 4). The areal coverage of each land-use type is shown in Figure 5. The overall accuracies of the six land-use maps are mostly greater than 90% (Table 3).

Figure 4 shows that over approximately the past 30 years, the coastal land-use structure of the delta has undergone major changes. The most prominent change was the remarkable increase in the area of salt fields and shrimp ponds, which increased from less than 3% of the study area in 1984 to more than half of the area in 2017. Meanwhile, in response to the increased aquaculture and salt industries, residential areas have constantly grown. These developments directly reflect human activities and urbanization in the Yellow River Delta over the past few decades. Without artificial intervention, other land-use classes have remained relatively stable. Bare land mainly includes intertidal zones along the coast, and its size is greatly affected by the satellite image acquisition time. Water bodies include reservoirs, ponds, seasonal rivers, and tributaries of the Yellow River. The areas of farmland, forest, and grassland are primarily affected by seasonal alternations.

Table 3. Validation statistics of each land-use map.

Land-Use	1984	1992	2000	2007	2010	2017
Overall accuracy	94.8	92.6	91.6	92.0	80.5	90.1

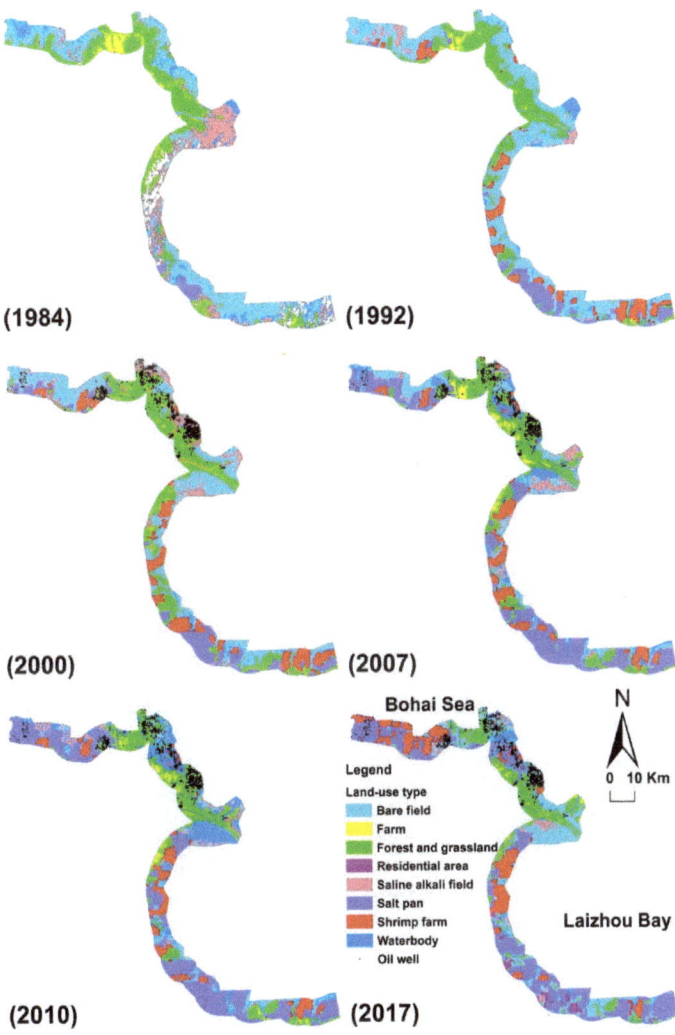

Figure 4. Land-use maps of the coastal area of the Yellow River Delta, China, derived from Landsat images from 1984, 1992, 2000, 2007, 2010, and 2017.

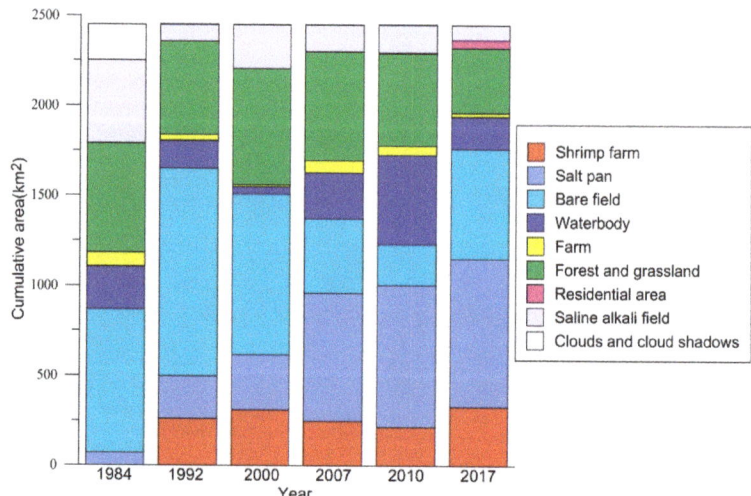

Figure 5. Changes in the area of each land-use type for each classified Landsat image.

4.1.2. Land-Use Dynamic Change

Using a transition matrix, we can quantitatively describe the land-use conversion during a certain period. Since the satellite images in 1984 suffered from heavy cloud cover (~10%), two transfer matrices were built for the periods of 1992–2007 (Table 4) and 2007–2017 (Table 5). The bins are colored from blue (small) to purple (large). The values on the diagonals are those without land-use changes. A zero value means that no change occurred. The most noticeable conversion that occurred was the growth of salt pans, mainly by reclaiming bare fields, with the area increasing almost threefold (Table 4, 10–29%) during P1 (1992–2007). In this period, farmland, forest, and saline alkali land increased slightly, while shrimp ponds remained basically unchanged. Subsequently, during the next 10 years (P2: 2007–2017, Table 5), the salt field area steadily continued to expand, and the shrimp pond area increased by a third. Due to persistent urbanization, the areas of farmland, forest, grassland, and saline alkali land have been greatly reduced.

Table 4. Land use/cover change transfer matrix of the study area in P1 (percentage).

	Bare field	Farm	Forest	Residential	Saline Land	Salt Pan	Shrimp Farm	Water	Total 1992
Bare field	13	0	8	0	4	13	2	6	47
Farm	0	1	1	0	0	0	0	0	1
Forest	1	2	14	0	0	2	0	2	21
Resident	0	0	0	0	0	0	0	0	0
Saline land	1	0	0	0	1	1	0	0	4
Salt pan	0	0	0	0	0	9	0	0	10
Shrimp farm	0	0	0	0	0	3	7	0	11
Water	1	0	1	0	1	1	0	2	6
Total in 2007	17	3	25	0	6	29	10	10	100

Table 5. Land use/cover change transfer matrix of the study area in P2 (percentage).

	Bare Field	Farm	Forest	Residential	Saline Land	Salt Pan	Shrimp Farm	Water	Total 2007
Bare field	8	0	0	0	0	2	4	2	17
Farm	1	0	1	0	0	0	0	0	3
Forest	8	0	11	0	0	4	1	1	25
Resident	0	0	0	0	0	0	0	0	0
Saline land	4	0	1	0	0	0	1	0	6
Salt pan	3	0	0	2	1	21	2	1	29
Shrimp farm	1	0	0	0	0	3	6	0	10
Water	2	0	1	0	1	3	1	3	10
Total in 2017	25	1	15	2	3	34	13	7	100

4.2. Subsidence Rates of Each Land-Use Type

The InSAR-derived subsidence rates measured from P1 and P2 to P3 for each land-use sequence are shown in Figure 6 (unchanged) and Figure 7 (changed). Since P3 is only one year (2016–2017), we assume that there is no significant change in land-use type during P3.

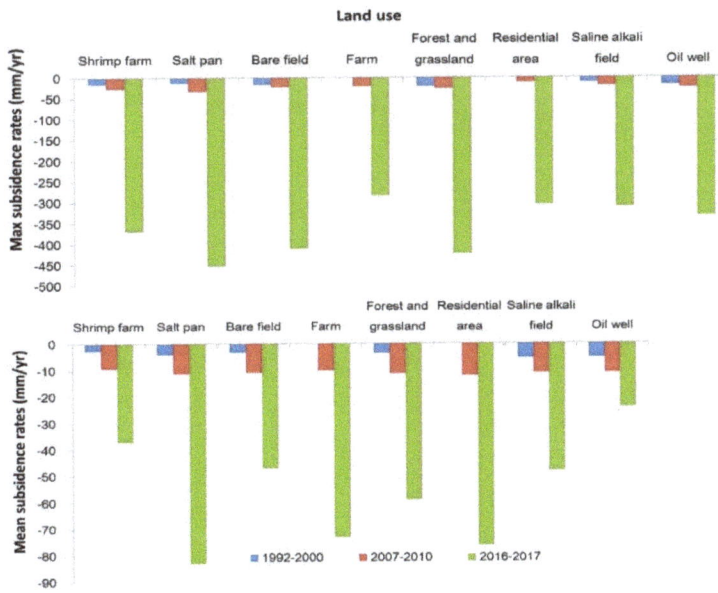

Figure 6. Statistics of the subsidence for unchanged land-use types from P1 (1992–2000) and P2 (2007–2010) to P3 (2016–2017).

The category shown in Figure 6 comprises the constant land-use sequences during both periods (P1, P2, and P3). After the slow growth of the first two periods (P1 and P2), the subsidence rate of the land-use class exhibited explosive growth in P3. The highest subsidence rates occurred for salt pans, which increased from 13 mm/year (P1) and 32 mm/year (P2) to 453 mm/year (P3). The lowest subsidence rates were observed for bare land, which increased from 16 to 22 mm/year in the two periods. The average land subsidence rate varied from 3–5 mm/year (P1) and 9–12 mm/year (P2) to 24–83 mm/year (P3). The standard deviation was less than 7 mm/year.

A total of 26 patterns of changes were identified in the other category. The maximum value occurred for saline-alkali fields (27 mm/year, P2) that were previously bare lands, closely followed by salt pans (25 mm/year, P2) that were previously bare fields. In particular, severe coastal subsidence appeared in both of the typical anthropogenic conversions to shrimp ponds and salt fields (Figure 7).

Significantly, due to the insufficient coverage of the land settlement dataset in this study, the settlement in P2 is too small. However, the existing literature [8] shows that in the P2 period, the subsidence rate in the coastal area of the Yellow River Delta reached 250 mm/year.

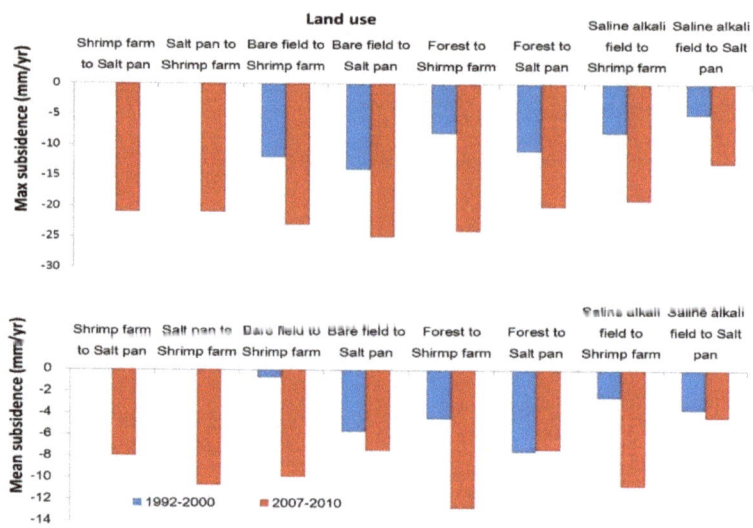

Figure 7. Statistics of the subsidence for changed land-use types in P1 (1992–2000) and P2 (2007–2010).

4.3. Evolution and Prediction of Land-Use/Cover and Subsidence

The land-use map of 2050 (Figure 8) predicted by the LCM model shows that the scale of salt fields will further expand, and this prediction agrees with the distribution characteristics of the brine resources in the coastal zone of the Yellow River Delta. It has been shown that there are two large underground brine reserves on the southwest bank of Laizhou Bay and the south bank of the Bohai Sea [24]. Currently, the distribution of mining intensity is uneven, and the Yangkou salt field on the southwest bank is in a state of overmining; however, the other areas were all deemed to be potential mining areas (see potential area in Figure 8). The predicted distribution area of the salt fields is consistent with the distribution of underground brine in this area. The prediction results show that in the next 30 years, the scale of salt fields will increase by 38% at the cost of a reduction in natural land (e.g., bare land, saline-alkali land, and forest grassland). With the development of urbanization, the scale of residential areas will increase by 7%. Compared with the actual land-use maps for 2010 and 2017, the prediction accuracy-based land-use maps from 1984–2007 reached 51% and 68%, respectively, due to insufficient relevance of the input maps.

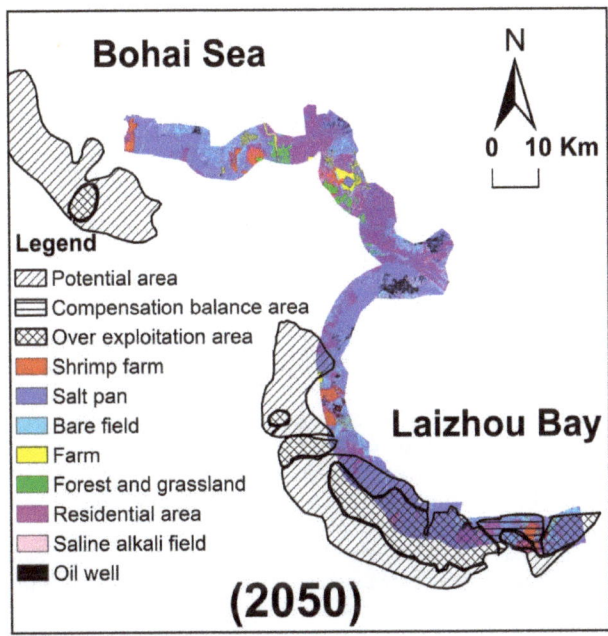

Figure 8. Distribution map of brine exploitation potential (see Section 3.4) and land-use distribution map in 2050 based on the LCM model.

5. Discussion

5.1. Expansion of Salt Pans and Exploitation of Underground Brine

Over the past 30 years, land-use changes in the coastal zone of the Yellow River Delta have featured the rapid expansion of salt fields and aquaculture. These features are in line with spatial–temporal patterns reported for the delta in other studies [2,25]. In particular, the expansion of salt pans is the most remarkable land-use transformation, with an area growth rate of 23 km^2/year, corresponding to the findings of Qiao [4]. The salt industry remained stable with slight variations from 1984 to 1992. Long-term unreasonable extraction has led to a decline in the underground brine concentration over time. For example, the concentration of brine in the Guangrao salt field decreased from 100–130 g/L at the beginning of 1959 to 40–70 g/L in 2007. Studies have shown that the underground brine in the salt field on the southeastern coast of Bohai Bay has declined by as much as 1 mm/year [17]. The appearance of land subsidence depressions in the salt fields denotes direct evidence of the decline in the underground brine levels [14].

5.2. Aquaculture and Oil Fields

In addition to salt fields, aquaculture (mainly shrimp ponds) and oil fields are two typical types of human activities in the delta, as well as two representative manmade subsidence factors supported by other studies [8–10]. Actually, all three of these land-use classes are closely related in space (Figure 4). In the Yellow River Delta, the terrain is relatively low-lying and flat, and groundwater resources are rich. Therefore, it is common to use underground salt water for shrimp culture. Recently, shrimp breeding ponds have been built in salt pan areas, increasing the yield of prawns and providing obvious economic benefits. Moreover, on the north bank of the Yellow River Delta, many oil wells have been interspersed among salt fields and shrimp ponds for decades. Long-term and high-intensity exploitation has contributed to declines in reservoir pressure. Furthermore, as the main method of oil exploration, water-driven exploitation accounts for 81.3% of the reservoir pressure effects. Therefore, artificial water injection pumped from shallow strata usually leads to more ground subsidence than does the exploitation of deep oil with a burial depth of 700–3500 m [13].

5.3. Land Subsidence per Land-Use Sequence

According to the InSAR measurement results of the three periods (P1, P2, and P3) and the spatial analysis of land-use classification data, the temporal and spatial characteristics of land subsidence under different land-use classes can be identified. In P1, in the early stage of the construction of the delta, the influence of human activities on the natural environment was not remarkable and was even less than the self-weight consolidation compaction effect of the sedimentary strata [26].

In P2, the average annual subsidence rates of all land-use classes changed markedly and became 2–3 times larger than those of the previous period. However, the differences in subsidence rates among different land-use types were still not significant, with values of less than 2 mm/year.

Although the average subsidence rates of various land-use classes were relatively similar (all less than 2 mm) during the period of either P1 or P2, the interperiod change of magnitude of subsidence increased by 2–3 times from P1 to P2. Remarkably, in P3, the subsidence rates were approximately an order of magnitude higher than those in P2, ranging from 24 mm/year (oil well) to 83 mm/year (salt field). This finding is consistent with the results of several recent studies [8,14]. In particular, due to the existence of subsidence depressions caused by groundwater exploitation (shown in Figure 2d), some land-use types around the subsidence center show a higher subsidence rate (Figure 6).

We believe that from P1 to P3, the salt field has further expanded in space, which is accompanied by excessive underground brine mining. Therefore, land subsidence disasters in the delta coastal zone are caused.

5.4. Prospect for the Future

To meet the needs of economic development, the scale of the salt industry is expected to expand further. According to Feng et al. [24], predatory exploitation of underground brine in 2005–2008 resulted in an average annual decrease in the brine level of 1.39 m. At present, shallow brine is the main resource in brine mining, while deep brine has not been developed. Therefore, the exploitation potential of brine is still great. Due to the unreasonable development of shallow brine resources, some environmental problems, such as the depletion of underground brine resources, waste of resources, ground fissures, ground subsidence funnels, and environmental pollution, have emerged.

Figure 9 shows the cumulative distribution function of the level-based (see locations in Figure 1) subsidence rates for different study areas during the same period (2016–2017). The 90% cumulative distribution of the coastal zone (red curve) is close to 100 mm/year, which is six times larger than that of the inland area (blue curve). Even during the early days of deltaic construction, the subsidence rates in the coastal zone were higher than those in the inland zone. For example, the 90% cumulative distribution of the subsidence rate from 2000 to 2007 (cyan curve) was 38 mm/year, which was more than twice that of 16 mm/year in 2016–2017 (blue curve).

According to the existing investigation, the average annual subsidence rate of the salt fields along the coastal zone varies from tens of millimeters to hundreds of millimeters [8,11,13,14,26,27]. Coastal zone subsidence combined with the absolute annual sea level rise in the Bohai Sea and extreme disaster events such as storm surges will likely have a great impact on the ecological environment and human life as well as on safety in the delta.

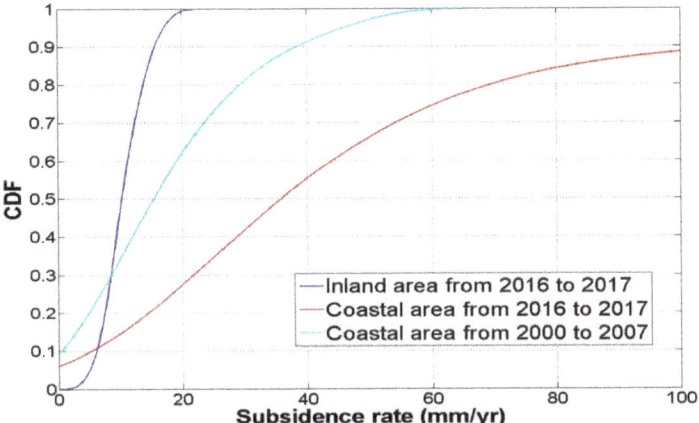

Figure 9. Cumulative distribution functions (CDF) of the subsidence rates in two periods.

6. Conclusions

In the past three decades, the coastal zone of the Yellow River Delta has changed from an undisturbed natural condition to a state dominated by salt fields and shrimp ponds characterized by artificial activity. At the expense of the development of the natural ecological environment, approximately 50% coverage of the coastal zone has been used for salt fields and shrimp ponds. Such a rapid expansion of salt fields has led to the excessive extraction of groundwater (i.e., underground brine), presumably leading to serious geophysical disasters such as land subsidence depressions and ground fissures.

Through the analysis of land-use changes and InSAR data, it was found that after the slow growth (increased by 2–3 times) of the first two periods (P1 and P2), the subsidence rate of the land-use class increased by an order of magnitude in P3. Moreover, extreme land subsidence includes different characteristics and often occurs in areas with strong human imprints, such as salt fields and shrimp ponds. With the development of underground brine production and the coastal brine industry, as well as the environmental impact of future sea level rise, the ecological vulnerability of the deltaic coastal area should deserve more attention from broad public and governmental managers. Addressing the relationship between economic development and environmental protection is a problem worthy of consideration. We should pay more attention to the development of underground brine.

Author Contributions: Conceptualization, Y.L. and X.Z. (Xinyuan Zhang); methodology, Y.Z.; software, K.Q.; validation, X.Z. (Xinghua Zhou); formal analysis, Y.L.; investigation, X.Z. (Xinyuan Zhang); resources, Z.B.; data curation, Y.Z.; writing—original draft preparation, Y.Z.; writing—review and editing, Y.Z.; visualization, Y.Z.; supervision, H.H.; project administration, H.H.; funding acquisition, Y.Z. All authors have read and agreed to the published version of the manuscript.

Funding: This research was funded by the National Natural Science Foundation of China (No. 42006148).

Institutional Review Board Statement: Not applicable.

Informed Consent Statement: Not applicable.

Data Availability Statement: The data presented in this study are available on request from the corresponding author.

Acknowledgments: The authors express their thanks to the people who helped with this work and acknowledge the valuable suggestions from the peer reviewers. In this study, the Landsat 5, 7, and 8 images were downloaded from the United States Geological Survey (USGS) Global Visualization Viewer (Accessed date: 10 February 2019 http://glovis.usgs.gov/), and we would like to thank the

European Space Agency for providing the Sentinel-1 data. We are grateful to them for allowing us to use their data.

Conflicts of Interest: The authors declare no conflict of interest.

References

1. Ehlers, E.; Krafft, T. *Earth System Science in the Anthropocene*; Springer: Berlin/Heidelberg, Germany, 2006; pp. 52–54.
2. Ye, Q.H.; Liu, G.; Tian, G.; Chen, S.; Huang, C.; Chen, S.; Liu, Q.; Chang, J.; Shi, Y. Geospatial-temporal analysis of land-use changes in the Yellow River Delta during the last 40 years. *Sci. China Ser. B* **2004**, *47*, 1008–1024. [CrossRef]
3. Zhao, Z.J.; Zhang, C.Y. Temporal and Spatial Change of Land Use/Cover in Yellow River Delta. *J. Basic Sci. Eng.* **2016**, *24*, 731–740.
4. Qiao, X.J.; Wang, Q.; Zhan, C.; Wang, X.; Wang, H.Y.; Du, G.Y.; Li, X.Y. Study on automatic extraction of coastline in the Yellow River Delta based on multispectral data. *Acta Oceanol. Sin.* **2016**, *38*, 59–71.
5. Galloway, D.L.; Burbey, T.J. Review: Regional land subsidence accompanying groundwater extraction. *Hydrogeol. J.* **2011**, *19*, 1459–1486. [CrossRef]
6. Tosi, L.; Teatini, P.; Strozzi, T. Natural versus anthropogenic subsidence of venice. *Sci. Rep.* **2013**, *3*, 2710. [CrossRef]
7. Gambolati, G.; Teatini, P. Geomechanics of subsurface water withdrawal and injection. *Water Resour. Res.* **2015**, *51*, 3922–3955. [CrossRef]
8. Higgins, S.; Overeem, I.; Tanaka, A.; Syvitski, J. Land subsidence at aquaculture facilities in the yellow river delta, China. *Geophys. Res. Lett.* **2013**, *40*, 3898–3902. [CrossRef]
9. Liu, Y.; Huang, H.J. Characterization and mechanism of regional land subsidence in the Yellow River Delta, China. *Nat. Hazards* **2013**, *68*, 687–709. [CrossRef]
10. Liu, P.; Li, Q.; Li, Z.; Hoey, T.; Liu, Y.; Wang, C. Land subsidence over oilfields in the Yellow River Delta. *Remote Sens.* **2015**, *7*, 1540–1564. [CrossRef]
11. Liu, Y.L.; Huang, H.J.; Dong, J.F. Large-area land subsidence monitoring and mechanism research using the small baseline subset interferometric synthetic aperture radar technique over the Yellow River Delta, China. *J. Appl. Remote Sens.* **2015**, *9*, 096019. [CrossRef]
12. Liu, Y.L.; Huang, H.J.; Liu, Y.X.; Bi, H.B. Linking land subsidence over the Yellow River delta, China, to hydrocarbon exploitation using multi-temporal InSAR. *Nat. Hazards* **2016**, *84*, 271–291. [CrossRef]
13. Zhang, J.Z.; Huang, H.J.; Bi, H.B. Land subsidence in the modern Yellow River Delta based on InSAR time series analysis. *Nat. Hazards* **2015**, *75*, 2385–2397. [CrossRef]
14. Zhang, B.W.; Wang, R.; Deng, Y.K.; Ma, P.F.; Lin, H.; Wang, J.L. Mapping the Yellow River Delta land subsidence with multitemporal SAR interferometry by exploiting both persistent and distributed scatterers. *ISPRS J. Photogramm. Remote Sens.* **2019**, *148*, 157–173. [CrossRef]
15. Corbau, C.; Simeoni, U.; Zoccarato, C.; Mantovani, G.; Teatini, P. Coupling land use evolution and subsidence in the Po Delta, Italy: Revising the past occurrence and prospecting the future management challenges. *Sci. Total Environ.* **2019**, *654*, 1196–1208. [CrossRef]
16. Minderhoud, P.S.J.; Coumou, L.; Erban, L.E.; Middelkoop, H.; Stouthamer, E.; Addink, E.A. The relation between land use and subsidence in the Vietnamese Mekong delta. *Sci. Total Environ.* **2018**, *634*, 715–726. [CrossRef] [PubMed]
17. Zhang, Y.; Huang, H.J.; Liu, Y.X.; Liu, Y.L. Self-weight consolidation and compaction of sediment in the Yellow River Delta, China. *Phys. Geogr.* **2017**, *1*, 1–15. [CrossRef]
18. Zou, Z.G.; Zhang, D.S.; Tan, Z.R. Ground brine resource and its exploitation in Shandong province. *Geol. Surv. Res.* **2008**, *31*, 214-21.
19. Wen, W.; Mendel, J.M. Maximum-likelihood classification for digital amplitude-phase modulations. *IEEE Trans. Commun.* **2000**, *48*, 189–193. [CrossRef]
20. Fisher, R.A. *Contributions to Mathematical Statistics*; John Wiley & Sons: Hoboken, NJ, USA, 1950; pp. 1–56.
21. Eastman, J.R. *Idrisi Guide to GIS and Image Processing*; Clark Labs: Worcester, MA, USA, 1999; pp. 1–144.
22. Anand, J.; Gosain, A.K.; Khosa, R. Prediction of land use changes based on Land Change Modeler and attribution of changes in the water balance of Ganga basin to land use change using the SWAT model. *Sci. Total Environ.* **2018**, *644*, 503–519. [CrossRef]
23. Yao, Y.Q.; Yuan, F. Analysis on development status and exploitation potential of shallow brine resources in Dongying City. *Shandong Land Resour.* **2013**, *1*, 41–44.
24. Feng, S.T.; Tan, X.F.; Liu, G. Analysis of exploitation potential of underground brine resources in Shandong Province. *Shandong Land Resour.* **2013**, *1*, 69–73.
25. Ma, T.; Li, X.; Bai, J.; Cui, B. Tracking three decades of land use and land cover transformation trajectories in China's large river deltas. *Land Degrad. Dev.* **2019**, *30*, 799–810. [CrossRef]
26. Zhang, Y.; Huang, H.J.; Liu, Y.X.; Liu, Y.L.; Bi, H.B. Spatial and temporal variations in subsidence due to the natural consolidation and compaction of sediment in the yellow river delta, china. *Mar. Georesour. Geotechnol.* **2019**, *37*, 152–163. [CrossRef]
27. Zhang, Y.; Huang, H.J.; Liu, Y.X.; Bi, H.B.; Zhang, Z.H.; Wang, K.F.; Yan, L.W. Impacts of soft soil compaction and groundwater extraction on subsidence in the Yellow River Delta. *Mar. Georesour. Geotechnol.* **2020**, *1*, 1–8. [CrossRef]

Article

Biophysical Effects of Temperate Forests in Regulating Regional Temperature and Precipitation Pattern across Northeast China

Yue Jiao [1,2], Kun Bu [2], Jiuchun Yang [2], Guangshuai Li [2,3], Lidu Shen [4], Tingxiang Liu [3], Lingxue Yu [2,5,*], Shuwen Zhang [2] and Hengqing Zhang [1]

1   School of Life Science, Liaoning Normal University, Dalian 116029, China; jiaoyue@iga.ac.cn (Y.J.); 100639@lnnu.edu.cn (H.Z.)
2   Remote Sensing and Geographic Information Research Center, Northeast Institute of Geography and Agroecology, Chinese Academy of Sciences, Changchun 130102, China; bukun@iga.ac.cn (K.B.); yangjiuchun@iga.ac.cn (J.Y.); liguangshuai@iga.ac.cn (G.L.); zhangshuwen@iga.ac.cn (S.Z.)
3   College of Geography Science, Changchun Normal University, Changchun 130031, China; liutingxiang@ccsfu.edu.cn
4   Institute of Applied Ecology, Chinese Academy of Sciences, Shenyang 110016, China; shenlidu@iae.ac.cn
5   Remote Sensing and Geographic Information Research Center, Changchun Jingyuetan Remote Sensing Observation Station, Chinese Academy of Sciences, Changchun 130102, China
*   Correspondence: yulingxue@iga.ac.cn

Abstract: The temperate forests in Northeast China are an important ecological barrier. However, the way in which temperate forests regulate the regional temperature and water cycling remains unclear. In this study, we quantitatively evaluated the role that temperate forests play in the regulation of the regional temperature and precipitation by combining remote sensing observations with a state-of-the-art regional climate model. Our results indicated that the forest ecosystem could slightly warm the annual air temperature by 0.04 ± 0.02 °C and bring more rainfall (17.49 ± 3.88 mm) over Northeast China. The temperature and precipitation modification function of forests varies across the seasons. If the trees were not there, our model suggests that the temperature across Northeast China would become much colder in the winter and spring, and much hotter in the summer than the observed climate. Interestingly, the temperature regulation from the forest ecosystem was detected in both forested regions and the adjacent agricultural areas, suggesting that the temperate forests in Northeast China cushion the air temperature by increasing the temperature in the winter and spring, and decreasing the temperature in the summer over the whole region. Our study also highlights the capacity of temperate forests to regulate regional water cycling in Northeast China. With high evapotranspiration, the forests could transfer sufficient moisture to the atmosphere. Combined with the associated moisture convergence, the temperate forests in Northeast China brought more rainfall in both forest and agricultural ecosystems. The increased rainfall was mainly concentrated in the spring and summer; these seasons accounted for 93.82% of the total increase in rainfall. These results imply that temperate forests make outstanding contributions to the maintainance of the sustainable development of agriculture in Northeast China.

Keywords: forest ecosystem; regional temperature and precipitation regulation; Northeast China; WRF regional climate model

1. Introduction

Forest ecosystems are some of the crucial components of the terrestrial ecosystem. They provide a variety of ecosystem services, including water and soil conservation, carbon concentrations, climate regulation, and biodiversity maintenance, etc. [1–3]. Reforestation or afforestation is regarded as one of the most effective natural climate solutions in terms of maintaining warming below 2 °C [4,5]. China and various other countries have launched a series of forest protection and restoration programs to make full use of the role that forests play in improving the vulnerable ecological environment [6,7]. The six reports from the Intergovernmental Panel on Climate Change (IPCC) documented unprecedented global warming and increasingly frequent extreme events over the past decade resulting from human activities [8], suggesting that forests' buffering impact on climate change has become more and more critical.

Generally, the forest ecosystem regulates the climate through biogeochemical processes and biogeophysical processes [9,10]. From the biochemical perspective, forest ecosystems are a crucial carbon sink, storing ~45% of terrestrial carbon [11]. Both tropical and temperate forests have high carbon storage capacities, and deforestation and forest fires convert the carbon sink into carbon dioxide sources, contributing to warming [12]. The biogeophysical processes are the energy, moisture, and momentum exchanges between the surface of the land and the atmosphere [13–15]. There is a general consensus regarding the cooling effects of tropical forests and the warming effects of boreal forests [16–19]. The competition between evapotranspirational cooling and albedo warming determines the final impact on temperature resulting from forests [20]. In tropical areas, deforestation significantly suppresses evapotranspiration (ET) and moderately increases albedo, exerting a warming influence on Earth's climate [21]. Contrarily, deforestation in boreal regions strongly increases albedo and causes a slight decrease in ET, resulting in a cooling effect [18,19,22]. Bonan et al. (2008) believe that the net climate impact of temperate forests is highly uncertain because of the competition between low albedo in winter and high ET in summer [12]. As a result, the climate effects caused by temperate forests are much more complex, with apparent regional and seasonal dependence [23–27]. For example, Peng et al. (2014) demonstrated that afforestation in China cools the surface temperature, except in dry regions [23]. He et al. (2015) studied the impact of temperate forests on the local surface temperature, and found that the conversion from forests to farmland may lead to warming in summer and cooling in winter [24].

The temperate forests in Northeast China are a natural barrier for the Northeast Plain and Hulunbuir grasslands of Inner Mongolia. They play an indispensable role in maintaining the regional ecological balance and ensuring ecological security and environmental quality at the local and national levels. Furthermore, the Northeast Plain, surrounded by temperate forests, is one of the main grain-producing areas in China. The forest ecosystem provides plenty of ecological services at the regional scale, including water conservation, wind prevention, sand fixation, and climate regulation, which have been widely documented. However, the way in which the forest activity alters the agricultural climate, which may further influence crop growth and grain yield in the Northeast China plain, still lacks comprehensive understanding.

Therefore, the primary objective of this study was to quantitatively evaluate the role of forest ecosystems in the regulation of the temperature and precipitation pattern across Northeast China. First, spatial-temporal continuous multisource remote sensing datasets were used to characterize the land use/land cover pattern and surface properties for all the ecosystems in Northeast China. Moreover, we incorporated these remote sensing-based parameters into the regional climate model in order to better simulate the regional temperature and precipitation pattern. Thereafter, we adopted scenario simulations based on a regional climate model to identify the impact of forests on the regional temperature and precipitation. Finally, the biogeophysical mechanisms through which forest ecosystems regulate the regional temperature and water cycling were further analyzed, i.e., the energy exchanges and water cycling processes between the land surface and the atmosphere. Our study will provide a scientific basis for temperate forest ecological protection and decision support for the sustainable development of agriculture in Northeast China.

2. Materials and Methods

2.1. Study Area

Northeast China (Figure 1) extends from a latitude of 38°40′ N to 53°34′ N and a longitude of 115°05′ E to 135°02′ E [28]. Its total area is 1.24 million square kilometers, accounting for 12.9% of the total land area of China. It is primarily made up of mountains and plains. The mountains are mainly located in the east, west, and north of Northeast China, while the plains are in the middle and south. The climate is characterized as a temperate monsoon climate, with long and cold winters, and mild and wet summers. The average minimum temperature in January is lower than −20 °C, while the average temperature in July ranges from 18 °C to 20 °C. Influenced by the summer monsoon from the Pacific, most of the precipitation is concentrated in the summer, which can reach 400–700 mm. The soil types are mainly black soil, chernozem, dark brown soil, and brown soil, which provide excellent conditions for both forests and crops. Forests and farmland are the main landscape types in Northeast China, accounting for 35% and 30% of the total area, respectively. The forests in the region are mainly distributed in the mountains, including the Great Khingan Range, the Lesser Khingan Mountains, and the Changbai Mountains. Moreover, the farmland is primarily located in the plain regions, including the Songnen plain, the Liaohe plain, and the Sanjiang plain.

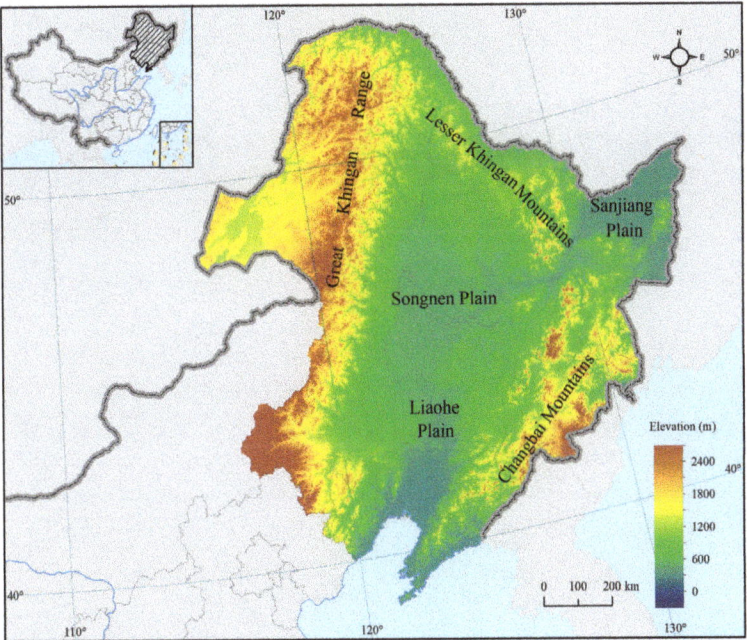

Figure 1. The geographic location of Northeast China.

The temperate forests in Northeast China are cold-temperate coniferous forests in the north and temperate summer green forests in the south [29]. As a result, the forest composition is mainly characterized as deciduous coniferous forest, deciduous broadleaf forest, and mixed forest (Figure 2). The forest in Northeast China comprises one belt of the "two screens and three belts" national ecological security pattern. It is an essential ecological barrier for the environment in Northeast China. Northeast China is also China's commodity grain production base, and thus, sustainable agricultural development is the primary objective in this region. Therefore, studying the way in which temperate forests modify the regional temperature and precipitation patterns is vital in order to fully understand the ecological functions of forests on both the individual forest and forest-agriculture ecosystem scale.

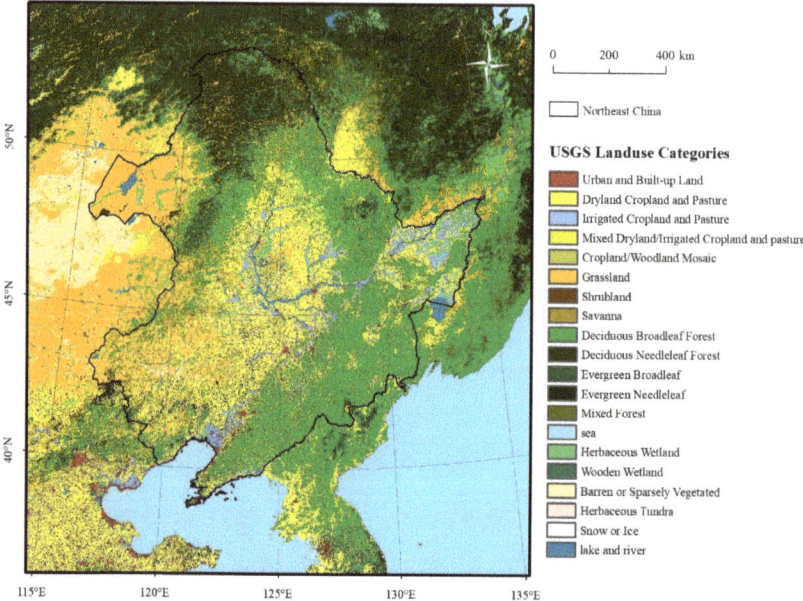

Figure 2. Spatial distribution of the land use/land cover in our simulation domain.

2.2. Data Sources

2.2.1. Land Use/Land Cover (LULC) Data

The land use/land cover data used in this study were divided into two parts: first, we selected the land use/land cover data in China from the resource and environmental science and data center of the Chinese Academy of Sciences (CAS). These LULC data were from Landsat Operational Land Imager, and had a spatial resolution of 1 km. Secondly, regarding the LULC pattern outside of China but within the simulation study area, we utilized the European Space Agency (ESA) climate change initiate (CCI) LULC data from 2015 [30]. Both LULC datasets were uniformly converted using the USGS LULC classification system, and then were projected onto the Lambert equal area projection system to match the model setup. Thereafter, the LULC percentage for each LULC category was calculated, and the primary LULC type in each 10 km grid was identified in order to update the original LULC-related datasets in the WRF model.

2.2.2. Other Surface Parameter Data

The spatially continuous surface parameter data used in this study mainly included leaf area index (LAI), the fraction of vegetation coverage (FVC), and albedo data. In the Noah land surface model, LAI and FVC are essential land surface parameters to express the energy, moisture, and carbon exchanges between the land surface and atmosphere. LAI reflects the physiological structure of plants and determines the water and carbon cycle for the vegetation canopy. For the water transfer process, LAI determines the canopy transpiration, respiration, and stomatal conductance, while for the carbon process, LAI is also an indispensable parameter for the calculation of vegetation photosynthesis and the soil carbon accumulation from falling leaves. In contrast, FVC is mainly involved in the energy and water cycle, which is used to calculate the physical processes, including radiation partitioning between vegetation canopy and soil, canopy evaporation interception, and precipitation interception and transmission to the surface. The LAI and FVC were obtained from the Global Land Surface Satellite Data Set (GLASS) [31], spanning from 1981 to 2018. The spatial resolution of the LAI and FVC data products is 0.05°, and the temporal resolution is eight days. In order to match the model requirements, the datasets were first integrated into monthly average data. Surface albedo data, which reflect the optical characteristics of the surface, were obtained from the fifth-generation reanalysis dataset of the European Center for medium range weather forecasting (ERA5-land) monthly dataset; they have a spatial resolution of 0.1°. In order to eliminate the impact of climate fluctuations on the surface parameters, we used the 3-year mean monthly averaged LAI, FVC, and albedo data from 2016 to 2018 to update the spatiotemporal data of the corresponding surface parameters in the WRF model.

2.2.3. Climate Forcing Data and Surface Meteorological Observational Data

We used the ERA5 data to provide the initial and boundary conditions for the model simulation. The ERA5 is a long-term, hourly climate reanalysis dataset, which has been widely used in climate change research [32]. The dataset provides near-real-time meteorological data concerning the surface and pressure levels from 1979, and has a spatial resolution of 0.25° × 0.25°. The meteorological observational data were obtained from the National Meteorological Science Data Center (http://data.cma.cn/, accessed on 8 May 2021). Our study area included 98 and 95 observational stations from 2016 to 2018 for temperature and precipitation, respectively. In this study, we used the annual average temperature and annual total precipitation to evaluate the accuracy of the model simulation results. We interpolated the meteorological observational data into the model resolution to eliminate the scale mismatch between the station and the model resolution using the Australian National University Spline (ANUSPLIN) method [33].

2.3. Experiment Design and Regional Climate Simulation

In this study, we used the weather research and forecasting (WRF) model to separate the biophysical impacts that forests have on the regional climate. The WRF model was developed by the National Center for Atmospheric Research (NCAR), the National Center for environmental prediction (NCEP), and other institutions, and has been widely used in numerical weather prediction and regional climate change research due to its flexibility in parametric scheme selection and simulation resolution setting [34]. The model has been registered and used by scientific researchers and relevant personnel in more than 160 countries. We chose the advanced research WRF (ARW) version 4.2 to conduct our experiments. As a result of the high spatial resolution of the ERA5 climate forcing data, we designed one domain covering Northeast China to carry out our simulation. The domain center is located at a latitude of 46° N and a longitude of 125° E. The horizontal resolution of the domain is 10 km, including 180 × 200 grids in the east-west and south-north directions (Figure 2).

We used two sets of simulation experiments, i.e., a forested scenario and an all-grassland scenario, to quantitatively evaluate the biophysical impact of the forest ecosystem on the local and regional climate. In the forested scenario, we used the LULC pattern from 2015, 3-year averaged monthly LAI, FVC, and albedo data from 2016 to 2018 to drive the regional climate model. Because there were no available LULC data from 2016 to 2018 in China from the Chinese Academy of Sciences, the LULC data from 2015 were used instead. In the all-grassland scenario, we assumed that all of the forests were degraded into grassland. All of the forests' types from the LULC data were replaced with grassland, and the surface parameters—including LAI, FVC, and albedo—for the degraded forest were updated with the statistical mean values of grassland (Table 1). The surface parameters of the three dominant forest types are also displayed in Table 1 in order to distinguish the differences in the surface properties between forest and grassland. It should be noted that the input albedo in the WRF model was the background albedo, suggesting that the actual albedo in the snow-covered months should be calculated based on both the background albedo and snow properties, including the snow depth. Therefore, the snow albedo (SNOWALB) of the original forest (0.44) was replaced with the regional statistical mean values of grassland (0.66) to update the snow albedo in the all-grassland scenario.

Table 1. The regional statistical monthly mean albedo, FVC (%), and LAI (m^2/m^2) for grassland and forests.

Vegetation Types	Parameters	January	February	March	April	May	June	July	August	September	October	November	December
Grass	Albedo	0.20	0.20	0.20	0.20	0.20	0.19	0.19	0.18	0.18	0.19	0.19	0.20
	FVC	1.54	1.26	1.07	1.70	10.79	22.80	32.82	33.44	18.14	4.51	2.29	1.86
	LAI	0.11	0.10	0.14	0.22	0.43	0.78	1.17	1.12	0.61	0.25	0.17	0.15
Mixed forest	Albedo	0.26	0.24	0.23	0.19	0.17	0.17	0.17	0.16	0.15	0.18	0.23	0.24
	FVC	4.76	4.42	5.04	7.43	26.82	53.92	67.48	67.43	40.06	12.40	7.36	5.88
	LAI	0.29	0.27	0.32	0.63	1.27	2.00	2.67	2.58	1.57	0.74	0.46	0.36
Deciduous coniferous forest	Albedo	0.33	0.32	0.28	0.21	0.13	0.13	0.14	0.13	0.12	0.18	0.27	0.32
	FVC	14.98	14.00	13.51	20.85	54.93	80.21	85.56	79.17	42.53	22.33	18.09	16.45
	LAI	0.67	0.61	0.58	0.75	2.05	3.84	4.36	3.95	2.11	0.81	0.75	0.69
Deciduous broadleaf forest	Albedo	0.34	0.33	0.29	0.16	0.15	0.16	0.16	0.15	0.13	0.14	0.25	0.29
	FVC	11.31	9.47	9.20	19.33	59.27	79.48	86.96	85.11	60.77	23.79	15.54	13.38
	LAI	0.49	0.50	0.50	0.79	2.52	4.02	4.44	4.16	2.64	0.92	0.58	0.49

In the two groups of scenario simulation experiments, we used the same initial and boundary conditions and the same surface physical parameterization schemes to isolate the contribution of forests to the regulation of the local and regional climate. Specifically, we used the ERA5 climate variables at 00:00 on 1 June 2015, to initialize the WRF model, and the data from 2015 to 2018 as the lateral boundary. The WRF model was started on 1 June 2015, and ran until 31 December 2018. The simulated model results from 1 June 2015 to 30 November 2015 were used for the model spin-up, and the results from the next three years and one month were used for further analysis. The differences between the forested and all-grassland scenarios were used to represent the biophysical impact of forests on the regulation of the regional climate. The scenario differences in air temperature at 2 m and the energy components—including incoming shortwave radiation, outgoing shortwave radiation, downward longwave radiation, upward longwave radiation, latent heat flux, sensible heat flux, and ground heat flux—were used to illustrate the temperature impact from forests, while the evapotranspiration, precipitation, and U-V wind were used to represent the impact of forests on water cycling.

3. Results and Discussion

3.1. Biophysical Impact of Temperate Forests on the Local and Regional Air Temperature

In this study, we estimated the biophysical impact of temperate forests on both the local and regional temperature by combining a land surface model and a high-resolution regional climate model. On the basis of the differences between the simulated results of the forested scenario and the all-grassland scenario, we separated the role that temperate forests play in the regulation of the local and regional temperature (Figure 3). From the perspective of biogeophysical processes, the temperature-regulating effect of the temperate forest in Northeast China was generally characterized as a slight warming effect (0.04 ± 0.02 °C), which was generally consistent with various previous studies [35–39]. Moreover, the temperature regulation effects of the temperate forests exhibited significant seasonal differences. Generally, the forest activity in Northeast China decreased the annual temperature range by cooling the air temperature in the summer and autumn, and warming the air temperature in the winter and spring. Specifically, the summer air temperature would rise by 0.88 ± 0.05 °C ($p < 0.01$) and the autumn temperature would increase by 0.05 ± 0.04 °C if the trees were not there. In contrast, the winter and spring would become much colder, with the air temperature decreasing by 0.65 ± 0.06 °C and 0.44 ± 0.11 °C, respectively, if there were no trees.

Figure 3 illustrates the spatial pattern of the air temperature changes across different seasons compared to the scenario in which trees are absent. Generally, the temperature benefits from forests are usually evaluated locally or regionally based on observational and model simulation results. Contrary to previous studies, our results highlight the role that temperate forests in Northeast China play in the regulation of the temperature in the surrounding agricultural ecosystems. We found that temperate forests could warm the cropland by 0.47 ± 0.11 °C and 0.38 ± 0.10 °C in the winter and spring, respectively, and that they could cool the cropland by 0.51 ± 0.05 °C in the summer. This indicates that the forests provide a cushioning effect that acts against the regional cold and high temperatures in the cropland regions.

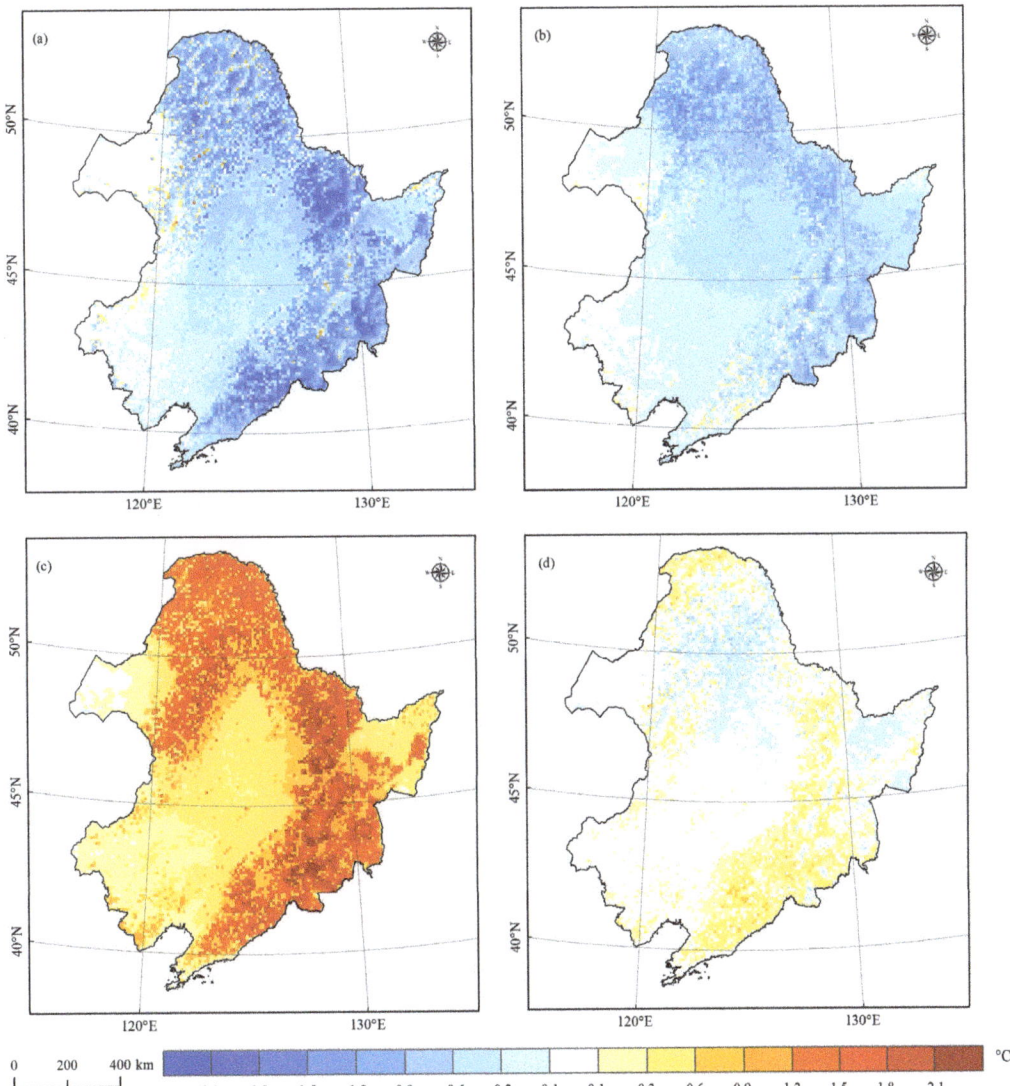

Figure 3. Differences (all-grassland scenario minus the forested scenario) in the air temperature (°C) at 2 m in December–January–February (**a**) March–April–May (**b**) June–July–August (**c**) and September–October–November (**d**) between the forested and all-grassland scenarios.

The surface energy budget and energy redistribution are often used to explain the mechanisms related to local temperature responses to land use and land cover changes. Given that the forest's air temperature regulation varied across different seasons, we quantitatively evaluated how the forest ecosystem modifies the local temperature by altering shortwave radiation, longwave radiation, sensible heat flux, latent heat flux, and ground heat flux (Figure 4).

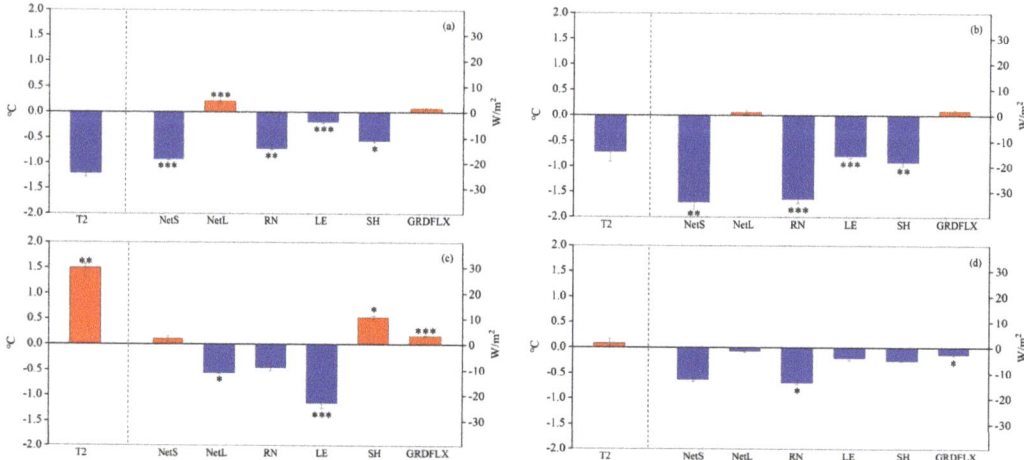

Figure 4. Differences (all-grassland scenario minus the forested scenario) in net shortwave radiation (NetS), net longwave radiation (NetL), net radiation (RN), latent heat flux (LE), sensible heat flux (SH), and ground heat flux (GRDFLX) for the forest ecosystem. (**a**) Winter; (**b**) spring; (**c**) summer; (**d**) autumn. *, **, and *** indicate that the differences are significant at $p < 0.05$, $p < 0.01$, and $p < 0.001$, respectively.

The simulated results from the regional climate model showed that forests significantly cooled the local summer air temperature, mainly due to their evapotranspiration being higher than grassland. Moreover, with higher surface roughness, the turbulent heat exchange between the forests and the atmosphere is more significant than when trees are absent. As a result, the latent heat flux would decline by 22.90 ± 2.05 W/m^2 ($p < 0.001$). In the winter, the higher albedo with the loss of forest cover due to snow cover could significantly increase solar radiation reflection and reduce the net shortwave radiation by 18.39 ± 0.50 W/m^2 ($p < 0.001$). Without the sheltering from forests, the decreased surface temperature would reduce the longwave radiation emitted by the surface and increase the net longwave radiation by 4.24 ± 0.35 W/m^2 ($p < 0.001$). As a result, the surface net radiation would significantly decrease by 14.16 ± 0.40 W/m^2 ($p < 0.01$) if the trees were not there. The ET differences between the forested and all-grassland scenarios are not evident (3.83 ± 0.48 W/m^2, $p < 0.001$) compared with the net solar radiation, implying that the albedo-climate feedback plays a dominant role in determining the winter temperature changes. These results are in general accordance with the study of He et al. (2015) [23]. Using remote sensing observations and a space-for-time approach, He et al. (2015) concluded that the annual net climate effect, which signifies the land surface temperature change, is not evident because of the contrary effects of the energy budget change in summer and winter [23].

Unlike He et al.'s study, we also observed spring warming effects from forests. With lower snow albedos, forests can absorb more shortwave radiation (33.96 ± 2.78 W/m^2, $p < 0.01$) in spring, the change magnitude of which is even greater than that in winter. This can be explained by the limited incoming solar radiation in winter, although with higher albedo differences. The growing season starts significantly earlier in the forests of Northeast China compared with the grassland. As a result, the latent heat flux in the forested scenario is higher by 15.76 ± 0.82 W/m^2 ($p < 0.001$) due to stronger evapotranspiration. Due to forest loss, the sensible heat flux would reduce by 18.22 ± 1.65 W/m^2 ($p < 0.01$) to balance the reduced net radiation. This means that the forest cooling effects due to evapotranspiration can be fully counteracted by the albedo warming effects in spring.

3.2. The Influence of Temperate Forests on the Regional Water Cycle

Evapotranspiration is one of the most important factors for the maintainance of the land surface water cycle. The simulated results indicated that temperate forests support higher rates of evapotranspiration compared with grassland. Without the forests, the total annual ET would decrease by 59.91 ± 0.44 mm ($p < 0.001$) in Northeast China. The decline in evapotranspiration was mainly concentrated in the summer, which registered decreases of 30.64 ± 2.68 mm ($p < 0.01$), followed by the spring (19.63 ± 1.00 mm, $p < 0.05$), autumn (5.60 ± 1.86 mm), and winter (4.10 ± 0.60 mm, $p < 0.001$). The spatial pattern of ET changes between the forested and all-grassland scenarios, as illustrated in Figure 5, showed that the significant ET changes were mainly concentrated in the forested areas.

Moreover, the ET responses showed apparent regional heterogeneity. The ET capacity of forests is highly dependent on latitude, forest type, season, and elevation [11,40–42]. Through all four seasons, the reduction in ET in the temperate forest areas at low latitudes was much more significant than that in the Great Khingan Range forests, which are located at a higher latitude. Because the deciduous broad-leaved forests at low latitudes have a larger leaf area index than the deciduous coniferous forests at a higher latitude, the evapotranspiration capacity of their vegetation canopy is higher. Furthermore, warmer temperatures have the potential to increase ET [43]. Compared with the all-grassland scenario, the temperate forests cool the summer air temperature, inhibiting the local ET in some agricultural areas without land surface changes. The summer ET in regions, including the west side of the Lesser Khingan and Changbai Mountains and the Sanjiang plain, would increase by approximately 10–30 mm if the forests were not there.

By comparing and analyzing the differences in precipitation in the two scenarios (Figure 6), we found that the forest ecosystem in Northeast China plays a critical role in the regulation of the regional precipitation patterns. Our climate model simulations showed that the forests increased the annual precipitation in Northeast China by 17.49 ± 3.88 mm compared with the all-grassland scenario. The precipitation improvement was most evident in the summer and spring seasons, which were 9.94 ± 5.08 mm and 6.46 ± 2.07 mm, respectively. The spatial pattern of the precipitation differences between the two scenarios, as illustrated in Figure 6, demonstrated that the precipitation regulation from forests was not only focused on the forested areas but also the surrounding regions. For example, the forest activity could increase the rainfall by 25.81 ± 7.86 mm in the forest ecosystem, and 55.68% and 29.19%, in the spring and summer, respectively. Regarding the agricultural ecosystem, our simulated results demonstrated that the annual precipitation would decrease by 21.31 ± 5.76 mm if the trees were not there, of which 15.60 ± 6.23 mm and 4.65 ± 0.88 mm occurred in the summer and spring. In some specific regions of the Songnen and Sanjiang plains, the loss of the forests would decrease the summer precipitation by more than 30 mm.

Sufficient water vapor and water vapor convergence are the conditions which are necessary for precipitation formation [42,44]. With higher surface roughness and surface drag coefficient, forests also modify the wind field and atmospheric circulation (Figure 7). The winter season in our study area is significantly affected by the extreme cold in Siberia. Without the forest barrier, the northwesterly wind on the northeast plain would intensify, bringing more cold air and reducing the air temperature in this region (Figure 3a). In the spring, although cyclonic conditions form in the border area between Northeast China and Russia, the evapotranspiration would decrease significantly if the forests were converted to grassland, making precipitation formation difficult due to a lack of water vapor.

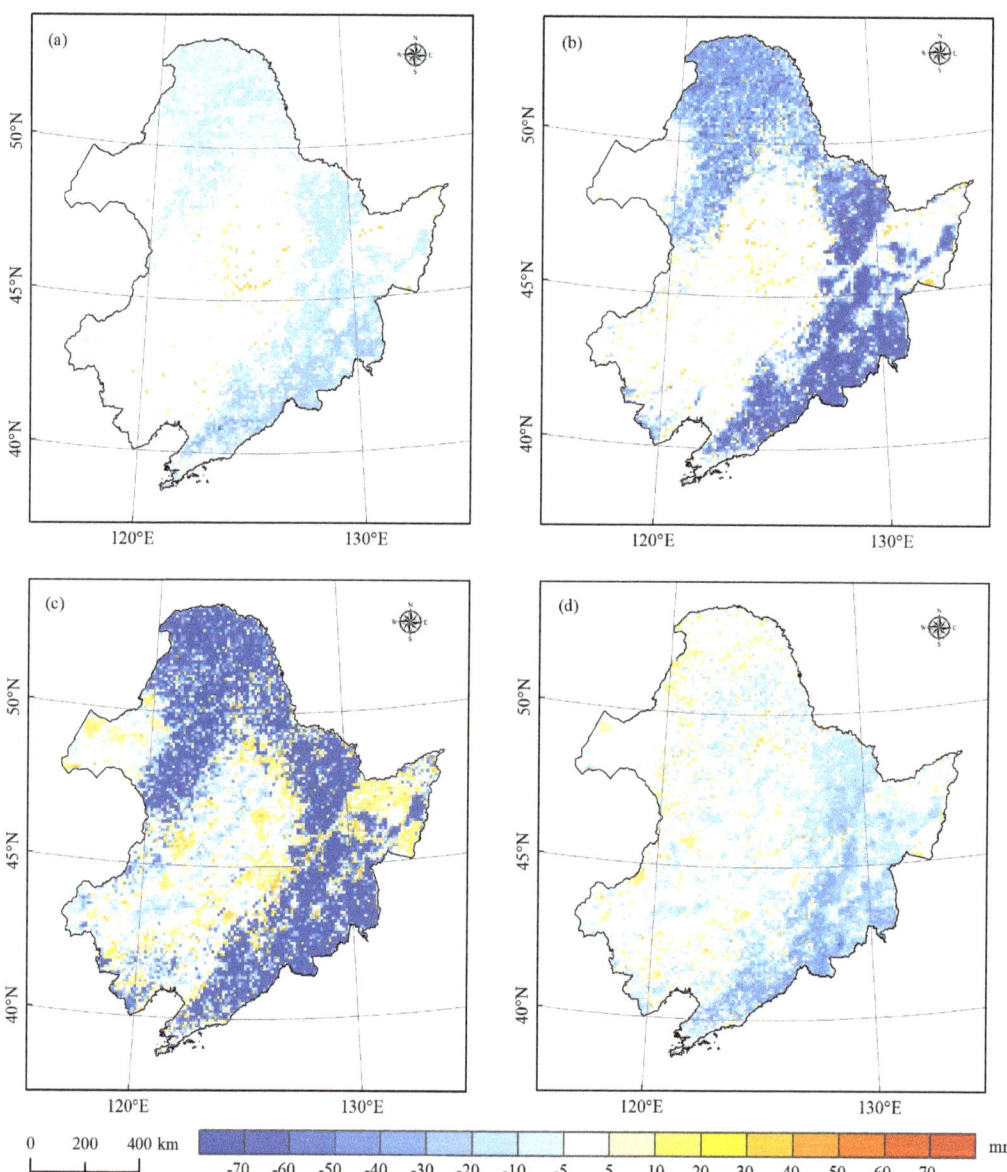

Figure 5. Differences (all-grassland scenario minus the forested scenario) in evapotranspiration (mm) in December–January–February (**a**) March–April–May (**b**) June–July–August (**c**) and September–October–November (**d**) between the forested and all-grassland scenarios.

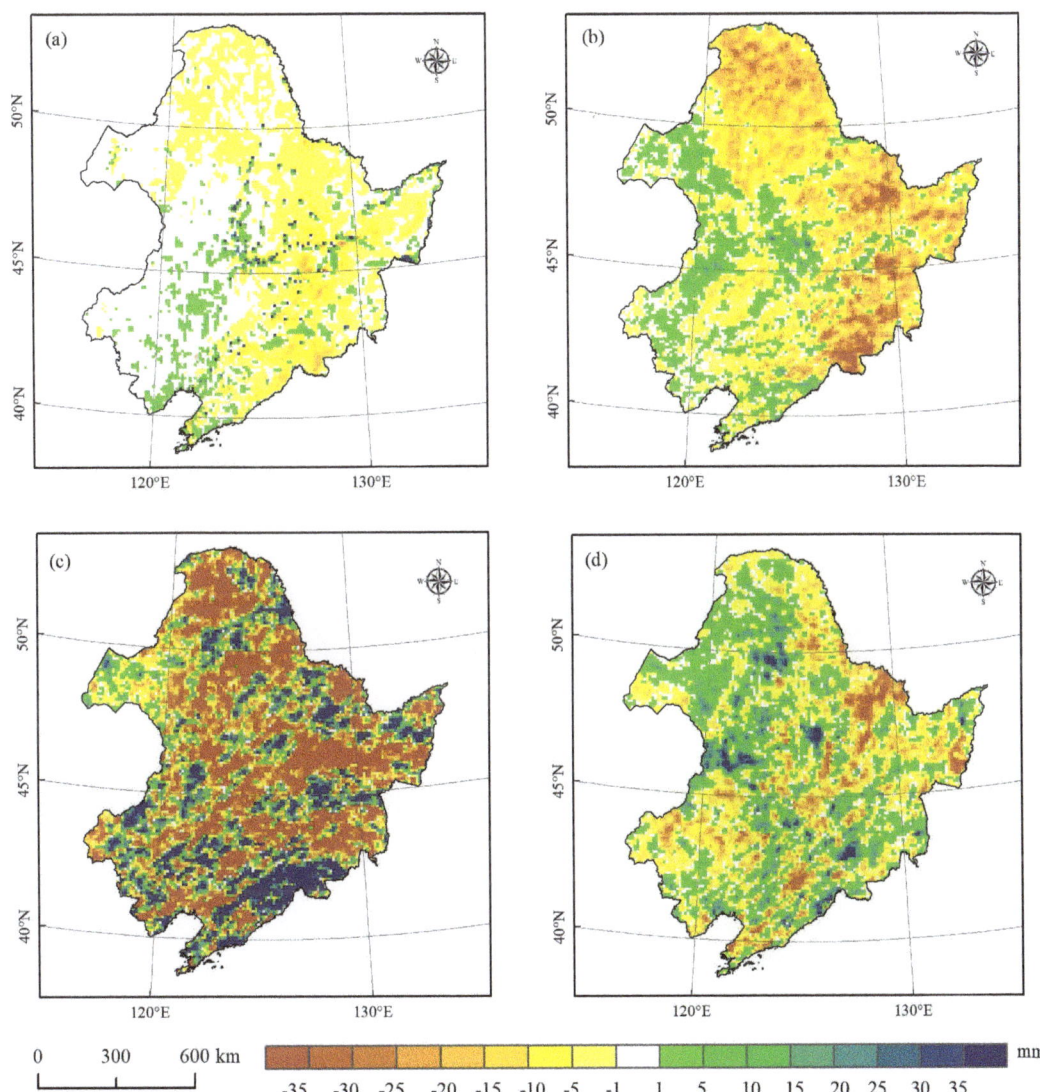

Figure 6. Differences (All-grassland scenario minus forested scenario) in precipitation (mm) in December–January–February (**a**) March–April–May (**b**) June–July–August (**c**) September–October–November (**d**) between the forested and all-grassland scenarios.

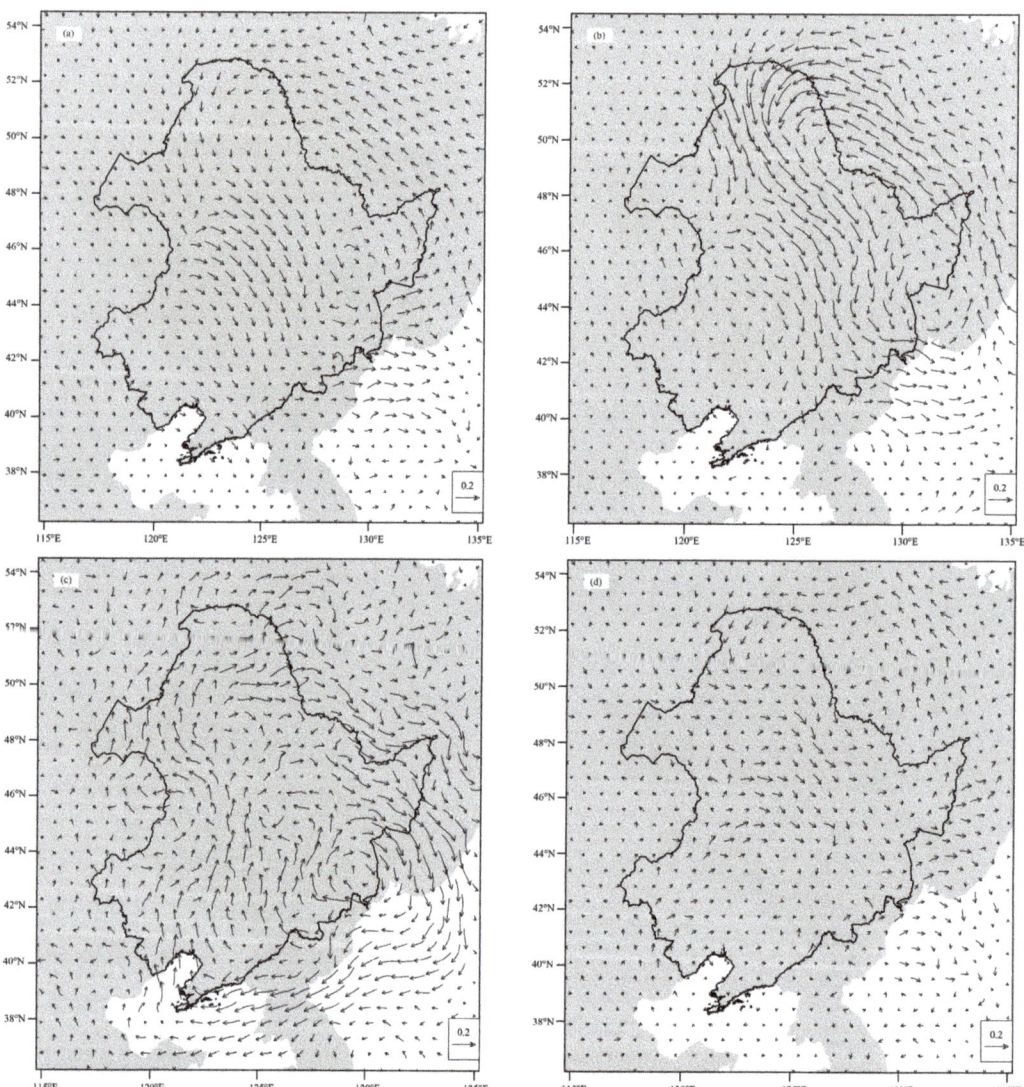

Figure 7. Differences (all-grassland scenario minus the forested scenario) in 700-hPa U-V wind (m/s) in December–January–February (**a**) March–April–May (**b**) June–July–August (**c**) and September–October–November (**d**) between the forested and all-grassland scenarios.

In the summer, the forests can provide sufficient water vapor with higher evapotranspiration capacity, combining with the moisture convergence, generating more rainfall (Figure 6c). This is consistent with a recent study from O'Connor et al. (2021). They reported that the forest land cover in the upwind precipitationshed can reduce the monthly precipitation variability downwind [45]. The forest cover in the Lesser Khingan and Changbai mountains promoted atmospheric moisture recycling and caused increased precipitation in the plain regions compared to the scenario in which there is an absence of trees.

3.3. Model Validation and Uncertainty Analysis

By comparing the simulated model annual average temperature and precipitation data with the meteorological observational data, we found that the physical parameterization schemes and near-real-time surface parameters adopted in this study can accurately simulate the temperature and precipitation characteristics in Northeast China (Figure 8). Specifically, we found that the temperature simulated by the model had a cold bias of 0.07 °C, and the correlation coefficient with the observed temperature was 0.97. Regarding precipitation, the model overestimated the precipitation by 0.42 mm/d, and the correlation coefficient with the observed precipitation was 0.85. Considering the capability of the climate model to simulate precipitation, we assumed that the precipitation differences between scenarios are able to describe the precipitation benefits from forest ecosystems.

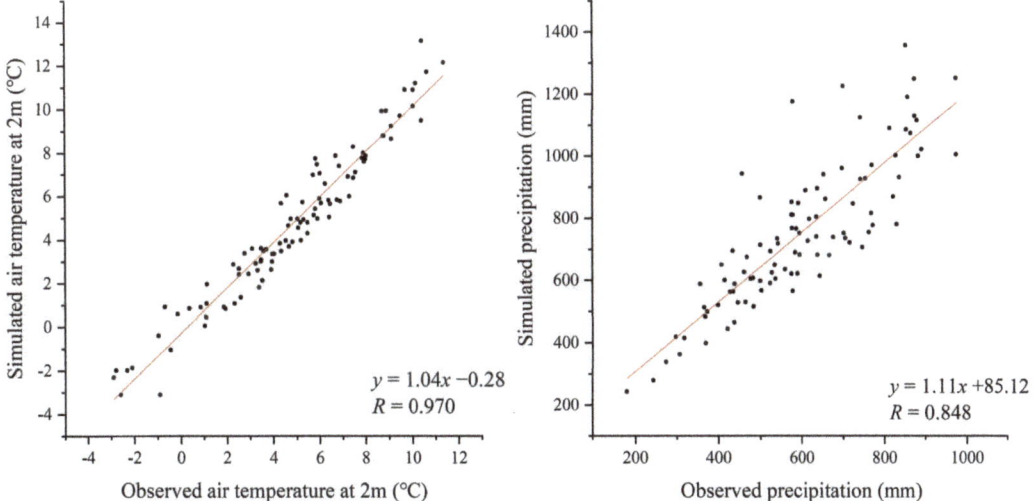

Figure 8. Simulated and observed air temperature at 2 m, and precipitation.

4. Conclusions

The temperate forest in Northeast China is an essential ecological barrier from the perspective of regional climate regulation. In this study, we quantitatively evaluated the role that temperate forests play in the regulation of the regional temperature and precipitation pattern by combining remote sensing observations and a state-of-the-art regional climate model (WRF). Our results indicated that the forest ecosystem slightly warms the annual air temperature by 0.04 ± 0.02 °C and brings more rainfall (17.49 ± 3.88 mm) over Northeast China. The temperature and precipitation modification function of forests varies across the seasons. If the trees were not there, our model suggests that the temperature across Northeast China would become much colder in the winter and spring, and much hotter in the summer than the observed climate. Interestingly, the temperature regulation from forest ecosystems was detected in both forested regions and the adjacent agricultural areas, suggesting that the temperate forests in Northeast China cushion the air temperature by increasing the temperature in the winter and spring, and decreasing the temperature in the summer over the whole region. Our study also highlights the capacity of temperate forests to regulate regional water cycling in Northeast China. With high evapotranspiration, the forests could transfer sufficient moisture to the atmosphere. Combined with the associated moisture convergence, the temperate forests in Northeast China brought more rainfall in both forest and agricultural ecosystems. The increased rainfall mainly occurred in the summer and spring; these seasons accounted for 93.82% of the total increase in rainfall.

It should be pointed out that there may exist some limitations in this study. First, our research assumed that the surface properties reflected by remote sensing products are reliable. In other words, the quality of the remote sensing products may affect the simulated results presented here. Second, this study used idealized experimental scenarios (forested and all-grassland) to evaluate the role that temperate forests play in the regulation of the regional temperature, precipitation, and atmospheric circulation in Northeast China. Future studies should assess the impact that the forest structure and changes in forest quality have on regional climates by using long-term observations and simulations to support regional ecological development and sustainable agricultural development.

Author Contributions: Conceptualization, L.Y.; investigation, Y.J.; methodology, T.L.; software, K.B.; visualization, K.B.; validation, Y.J. and G.L.; writing—original draft preparation, Y.J.; writing—review and editing, L.Y., T.L., L.S. and J.Y.; resources, S.Z. and H.Z.; project administration, L.Y. All authors have read and agreed to the published version of the manuscript.

Funding: This study was supported by the National Natural Science Foundation of China (41601093, 42071025), the Science and Technology Basic Resources Investigation Program of China (2017FY101301), a Strategic Priority Research Program (A) of the Chinese Academy of Sciences (XDA2003020301), and the 14th Five-year Network Security and Informatization Plan of the Chinese Academy of Sciences (WX145XQ06-07).

Institutional Review Board Statement: The study did not involve humans or animals.

Informed Consent Statement: The study did not involve humans.

Data Availability Statement: The land use/land cover data in China were from the resource and environmental science and data center of the Chinese Academy of Sciences (http://www.resdc.cn/ (accessed on 8 May 2021)). The CCI land cover data in 2015 are available from ESA (http://www.esa-landcover-cci.org/ (accessed on 8 May 2021)). Monthly LAI and FVC data were obtained from the product GLASS, which is available from the University of Maryland (http://www.glass.umd.edu/ (accessed on 8 May 2021)). The ERA5 and ERA5-Land reanalysis data were obtained from ECMWF (https://cds.climate.copernicus.eu/cdsapp#!/search?type=dataset&text=ERA5 (accessed on 8 May 2021)). The meteorological observation data were obtained from the China Meteorological Data Service Center (CMDC; http://data.cma.cn/en (accessed on 8 May 2021)) and the National Climate Data Center of the United States (https://gis.ncdc.noaa.gov/ (accessed on 8 May 2021)).

Acknowledgments: We sincerely appreciate the anonymous reviewers for their constructive comments and suggestions.

Conflicts of Interest: The authors declare no conflict of interest.

References

1. Evaristo, J.; McDonnell, J.J. Global analysis of streamflow response to forest management. *Nature* **2019**, *570*, 455, reprinted in *Nature* **2020**, *578*, 326. [CrossRef]
2. Farley, K.A.; Jobbagy, E.G.; Jackson, R.B. Effects of afforestation on water yield: A global synthesis with implications for policy. *Glob. Chang. Biol.* **2005**, *11*, 1565–1576. [CrossRef]
3. Jackson, R.B.; Randerson, J.T.; Canadell, J.G.; Anderson, R.G.; Avissar, R.; Baldocchi, D.D.; Bonan, G.B.; Caldeira, K.; Diffenbaugh, N.S.; Field, C.B.; et al. Protecting climate with forests. *Environ. Res. Lett.* **2008**, *3*, 044006. [CrossRef]
4. Griscom, B.W.; Adams, J.; Ellis, P.W.; Houghton, R.A.; Lomax, G.; Miteva, D.A.; Schlesinger, W.H.; Shoch, D.; Siikamaki, J.V.; Smith, P.; et al. Natural climate solutions. *Proc. Natl. Acad. Sci. USA* **2017**, *114*, 11645–11650. [CrossRef]
5. Bastin, J.F.; Finegold, Y.; Garcia, C.; Mollicone, D.; Rezende, M.; Routh, D.; Zohner, C.M.; Crowther, T.W. The global tree restoration potential. *Science* **2019**, *365*, 76–79. [CrossRef] [PubMed]
6. Liu, J.G.; Li, S.X.; Ouyang, Z.Y.; Tam, C.; Chen, X.D. Ecological and socioeconomic effects of China's policies for ecosystem services. *Proc. Natl. Acad. Sci. USA* **2008**, *105*, 9477–9482. [CrossRef] [PubMed]
7. Cai, D.; Ge, Q.; Wang, X.; Liu, B.; Goudie, A.S.; Hu, S. Contributions of ecological programs to vegetation restoration in arid and semiarid China. *Environ. Res. Lett.* **2020**, *15*, 114046. [CrossRef]
8. IPCC. *Climate Change 2021: The Physical Science Basis. Contribution of Working Group I to the Sixth Assessment Report of the Intergovernmental Panel on Climate Change*; Cambridge University Press: Cambridge, UK, 2021.
9. Pongratz, J.; Reick, C.H.; Raddatz, T.; Claussen, M. Biogeophysical versus biogeochemical climate response to historical anthropogenic land cover change. *Geophys. Res. Lett.* **2010**, *37*, L08702. [CrossRef]

10. Devaraju, N.; Bala, G.; Nemani, R. Modelling the influence of land-use changes on biophysical and biochemical interactions at regional and global scales. *Plant Cell Environ.* **2015**, *38*, 1931–1946. [CrossRef] [PubMed]
11. Bonan, G.B. Forests and climate change: Forcings, feedbacks, and the climate benefits of forests. *Science* **2008**, *320*, 1444–1449. [CrossRef]
12. Bala, G.; Caldeira, K.; Wickett, M.; Phillips, T.J.; Lobell, D.B.; Delire, C.; Mirin, A. Combined climate and carbon-cycle effects of large-scale deforestation. *Proc. Natl. Acad. Sci. USA* **2007**, *104*, 9911. [CrossRef]
13. Alkama, R.; Cescatti, A. Biophysical climate impacts of recent changes in global forest cover. *Science* **2016**, *351*, 600–604. [CrossRef]
14. Perugini, L.; Caporaso, L.; Marconi, S.; Cescatti, A.; Quesada, B.; de Noblet-Ducoudre, N.; House, J.I.; Arneth, A. Biophysical effects on temperature and precipitation due to land cover change. *Environ. Res. Lett.* **2017**, *12*, 053002. [CrossRef]
15. Zeng, Z.Z.; Piao, S.L.; Li, L.Z.X.; Zhou, L.M.; Ciais, P.; Wang, T.; Li, Y.; Lian, X.; Wood, E.F.; Friedlingstein, P.; et al. Climate mitigation from vegetation biophysical feedbacks during the past three decades. *Nat. Clim. Chang.* **2017**, *7*, 432–436. [CrossRef]
16. Comarazamy, D.E.; Gonzalez, J.E.; Luvall, J.C.; Rickman, D.L.; Bornstein, R.D. Climate Impacts of Land-Cover and Land-Use Changes in Tropical Islands under Conditions of Global Climate Change. *J. Clim.* **2013**, *26*, 1535–1550. [CrossRef]
17. Gullison, R.E.; Frumhoff, P.C.; Canadell, J.G.; Field, C.B.; Nepstad, D.C.; Hayhoe, K.; Avissar, R.; Curran, L.M.; Friedlingstein, P.; Jones, C.D.; et al. Tropical forests and climate policy. *Science* **2007**, *316*, 985–986. [CrossRef]
18. Betts, R.A. Offset of the potential carbon sink from boreal forestation by decreases in surface albedo. *Nature* **2000**, *408*, 187–190. [CrossRef]
19. Lee, X.; Goulden, M.L.; Hollinger, D.Y.; Barr, A.; Black, T.A.; Bohrer, G.; Bracho, R.; Drake, B.; Goldstein, A.; Gu, L.H.; et al. Observed increase in local cooling effect of deforestation at higher latitudes. *Nature* **2011**, *479*, 384–387. [CrossRef] [PubMed]
20. Davin, E.L.; de Noblet-Ducoudré, N. Climatic Impact of Global-Scale Deforestation: Radiative versus Nonradiative Processes. *J. Clim.* **2010**, *23*, 97–112. [CrossRef]
21. Feddema, J.J.; Oleson, K.W.; Bonan, G.B.; Mearns, L.O.; Buja, L.E.; Meehl, G.A.; Washington, W.M. The importance of land-cover change in simulating future climates. *Science* **2005**, *310*, 1674–1678. [CrossRef]
22. Potter, S.; Solvik, K.; Erb, A.; Goetz, S.J.; Johnstone, J.F.; Mack, M.C.; Randerson, J.T.; Román, M.O.; Schaaf, C.L.; Turetsky, M.R.; et al. Climate change decreases the cooling effect from postfire albedo in boreal North America. *Glob. Chang. Biol.* **2020**, *26*, 1592–1607. [CrossRef]
23. He, T.; Shao, Q.Q.; Cao, W.; Huang, L.; Liu, L.L. Satellite-Observed Energy Budget Change of Deforestation in Northeastern China and its Climate Implications. *Remote Sens.* **2015**, *7*, 11586–11601. [CrossRef]
24. Peng, S.S.; Piao, S.L.; Zeng, Z.Z.; Ciais, P.; Zhou, L.M.; Li, L.Z.X.; Myneni, R.B.; Yin, Y.; Zeng, H. Afforestation in China cools local land surface temperature. *Proc. Natl. Acad. Sci. USA* **2014**, *111*, 2915–2919. [CrossRef]
25. Arora, V.K.; Montenegro, A. Small temperature benefits provided by realistic afforestation efforts. *Nat. Geosci.* **2011**, *4*, 514–518. [CrossRef]
26. Cao, Q.; Wu, J.G.; Yu, D.Y.; Wang, W. The biophysical effects of the vegetation restoration program on regional climate metrics in the Loess Plateau, China. *Agric. Forest Meteorol.* **2019**, *268*, 169–180. [CrossRef]
27. Yu, L.; Liu, Y.; Liu, T.; Yan, F. Impact of recent vegetation greening on temperature and precipitation over China. *Agric. Forest Meteorol.* **2020**, *295*, 108197. [CrossRef]
28. Yu, L.X.; Zhang, S.W.; Tang, J.M.; Liu, T.X.; Bu, K.; Yan, F.Q.; Yang, C.B.; Yang, J.C. The effect of deforestation on the regional temperature in Northeastern China. *Theor. Appl. Climatol.* **2015**, *120*, 761–771. [CrossRef]
29. Zhang, S.W.; Zhang, Y.Z.; Li, Y.; Chang, L.P. *Temporal and Spatial Characteristics of Land Use/Cover in Northeast China*; Science Press: Beijing, China, 2006.
30. ESA. Land Cover CCI Product User Guide Version 2. Tech. Rep. 2017. Available online: maps.elie.ucl.ac.be/CCI/viewer/download/ESACCI-LC-Ph2-PUGv2_2.0.pdf (accessed on 8 May 2021).
31. Liang, S.L.; Zhao, X.; Liu, S.H.; Yuan, W.P.; Cheng, X.L.; Xiao, Z.Q.; Zhang, X.T.; Liu, Q.; Cheng, J.; Tang, H.; et al. A long-term Global LAnd Surface Satellite (GLASS) data-set for environmental studies. *Int. J. Digit. Earth* **2013**, *6*, 5–33. [CrossRef]
32. NCAR. ERA5 Reanalysis (0.25 Degree Latitude-Longitude Grid). 2019. Available online: https://rda.ucar.edu/datasets/ds633.0/ (accessed on 8 May 2021).
33. Hutchinson, M.F. Interpolation of Rainfall Data with Thin Plate Smoothing Splines—Part I: Two Dimensional Smoothing of Data with Short Range Correlation. *J. Geogr. Inf. Decis. Anal.* **1998**, *2*, 139–151.
34. Skamarock, W.C.; Klemp, J.B.; Dudhia, J.; Gill, D.; Barker, D.M.; Duda, M.G.; Huang, X.Y.; Wang, W.; Powers, J.G. *A Description of the Advanced Research WRF Version 3*; NCAR Technical Note, NCAR/TN-475+STR; National Center for Atmospheric Research: Boulder, CO, USA, 2008.
35. Betts, R.A. Afforestation cools more or less. *Nat. Geosci.* **2011**, *4*, 504–505. [CrossRef]
36. Bonan, G.B. Effects of land use on the climate of the United States. *Clim. Chang.* **1997**, *37*, 449–486. [CrossRef]
37. Betts, R.A. Biogeophysical impacts of land use on present-day climate: Near-surface temperature change and radiative forcing. *Atmos. Sci. Lett.* **2001**, *2*, 39–51. [CrossRef]
38. Feddema, J.; Oleson, K.; Bonan, G.; Mearns, L.; Washington, W.; Meehl, G.; Nychka, D. A comparison of a GCM response to historical anthropogenic land cover change and model sensitivity to uncertainty in present-day land cover representations. *Clim. Dyn.* **2005**, *25*, 581–609. [CrossRef]

39. Brovkin, V.; Claussen, M.; Driesschaert, E.; Fichefet, T.; Kicklighter, D.; Loutre, M.F.; Matthews, H.D.; Ramankutty, N.; Schaeffer, M.; Sokolov, A. Biogeophysical effects of historical land cover changes simulated by six Earth system models of intermediate complexity. *Clim. Dyn.* **2006**, *26*, 587–600. [CrossRef]
40. Cerasoli, S.; Yin, J.; Porporato, A. Cloud cooling effects of afforestation and reforestation at midlatitudes. *Proc. Natl. Acad. Sci. USA* **2021**, *118*, e2026241118. [CrossRef]
41. Zeng, Z.Z.; Wang, D.S.; Yang, L.; Wu, J.; Ziegler, A.D.; Liu, M.F.; Ciais, P.; Searchinger, T.D.; Yang, Z.L.; Chen, D.L.; et al. Deforestation-induced warming over tropical mountain regions regulated by elevation. *Nat. Geosci.* **2021**, *14*, 23–29. [CrossRef]
42. Yu, L.; Xue, Y.; Diallo, I. Vegetation greening in China and its effect on summer regional climate. *Sci. Bull.* **2021**, *66*, 13–17. [CrossRef]
43. Pascolini-Campbell, M.; Reager, J.T.; Chandanpurkar, H.A.; Rodell, M. A 10 per cent increase in global land evapotranspiration from 2003 to 2019. *Nature* **2021**, *593*, 543–547. [CrossRef]
44. Ge, J.; Pitman, A.J.; Guo, W.D.; Zan, B.L.; Fu, C.B. Impact of revegetation of the Loess Plateau of China on the regional growing season water balance. *Hydrol. Earth Syst. Sci.* **2020**, *24*, 515–533. [CrossRef]
45. O'Connor, J.C.; Dekker, S.C.; Staal, A.; Tuinenburg, O.A.; Rebel, K.T.; Santos, M.J. Forests buffer against variations in precipitation. *Glob. Chang. Biol.* **2021**, *27*, 4686–4696. [CrossRef]

MDPI
St. Alban-Anlage 66
4052 Basel
Switzerland
Tel. +41 61 683 77 34
Fax +41 61 302 89 18
www.mdpi.com

Sustainability Editorial Office
E-mail: sustainability@mdpi.com
www.mdpi.com/journal/sustainability

www.ingramcontent.com/pod-product-compliance
Lightning Source LLC
LaVergne TN
LVHW070202100526
838202LV00015B/1985